Modelling, Robustness and Sensitivity Reduction in Control Systems

NATO ASI Series

Advanced Science Institutes Series

A series presenting the results of activities sponsored by the NATO Science Committee, which aims at the dissemination of advanced scientific and technological knowledge, with a view to strengthening links between scientific communities.

The Series is published by an international board of publishers in conjunction with the NATO Scientific Affairs Division

A Life Sciences B Physics	Plenum Publishing Corporation London and New York
C Mathematical and Physical Sciences	D. Reidel Publishing Company Dordrecht, Boston, Lancaster and Tokyo
D Behavioural and Social Sciences E Applied Sciences	Martinus Nijhoff Publishers Boston, The Hague, Dordrecht and Lancaster
F Computer and Systems Sciences G Ecological Sciences H Cell Biology	Springer-Verlag Berlin Heidelberg New York London Paris Tokyo

Series F: Computer and Systems Sciences Vol. 34

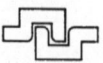

Modelling, Robustness and Sensitivity Reduction in Control Systems

Edited by

Ruth F. Curtain

Mathematics Institute, University of Groningen
9700 AV Groningen, The Netherlands

Springer-Verlag Berlin Heidelberg GmbH

Published in cooperation with NATO Scientific Affairs Divison

Proceedings of the NATO Advanced Research Workshop on Modelling, Robustness and Sensitivity Reduction in Control Systems held in Groningen, The Netherlands, December 1–5, 1986

ISBN 978-3-642-87518-2 ISBN 978-3-642-87516-8 (eBook)
DOI 10.1007/978-3-642-87516-8

Library of Congress Cataloging in Publication Data. NATO Advanced Research Workshop on Modelling, Robustness, and Sensitivity Reduction in Control Systems (1986 : Groningen, Netherlands) Modelling, robustness, and sensitivity reduction in control systems. (NATO ASI series. Series F, Computer and systems sciences ; vol. 34) "Published in cooperation with NATO Scientific Affairs Division." "Proceedings of the NATO Advanced Research Workshop on Modelling, Robustness and Sensitivity Reduction in Control Systems held in Groningen, The Netherlands, December 1–5, 1986."—T.p. verso. 1. Automatic control—Congresses. 2. Control theory—Congresses. I. Curtain, Ruth F. II. North Atlantic Treaty Organization. Scientific Affairs Division. III. Title. IV. Series: NATO ASI series. Series F, Computer and systems sciences ; vol. 34. TJ212.2.N36 1986 629.8 87-12817
ISBN 0-387-17845-7 (U.S.)

Originally published by Springer-Verlag Berlin Heidelberg in 1987.

2145/3140-543210

PREFACE

This volume contains the proceedings of the NATO Advanced Research Workshop on "Modelling, Robustness and Sensitivity Reduction in Control Sytems" which was held at the University of Groningen, the Netherlands, during the first week of December, 1986.

Modelling is a fundamental and difficult problem in all the sciences; to design a controller one needs a model. While for some applications one has a good physical model, often one only has measurements of the inputs and outputs of the system available. **Modelling from measurement data** was one important theme of the workshop. To control theorists, this has traditionally meant the stochastic approach of "System Identification" but here the newer deterministic approaches shared the spotlight.

Of course all models are approximate, but one sometimes requires a lower order, simpler model which still retains the main features of the original model with respect to the problem of control design. **Approximation** in this sense is often called **model reduction** and this theme was discussed during the workshop. Given that we only have approximate models available, the concept of robustness has always played an important role in controller design. Robust controllers are those which can control not only the given nominal model, but also neighbouring perturbations while at the same time guaranteeing an acceptable performance; they are robust with respect to model uncertainties. Typical performance requirements are tracking ability, stability and the suppression of disturbances, usually with respect to certain frequency bands, and so another very desirable property of a controller is that its performance has a low sensitivity to external disturbances. **Robustness and sensitivity reduction** of controllers were two related themes of the workshop.

During the last decade major advances have been made in the theory of approximation (model reduction) and robustness and sensitivity reduction of controllers by exploiting known results in two areas of mathematics: in classical mathematical analysis such as the work in interpolation theory by Nevanlinna, Pick, Fejér and Carathéodory, and in more recent developments in Operator Theory, such as the work of Adamjan, Arov and Krein in the seventies. This synthesis has resulted in a new research area in Systems and Control Theory known as **H^∞- Control** which was the main theme of this workshop and is closely related to the other themes of approximation, robustness and sensitivity reduction. These proceedings contain new contributions in these areas which range from abstract mathematical papers to some very concrete and challenging applications; an interesting interplay between mathematics and engineering.

As is well known, NATO workshops are primarily supported by the NATO Scientific Affairs Division and we are grateful to them for their sponsorship and generous financial support. This workshop was designated as belonging to the "Double Jump" programme, which means that the sectors: university, industry and government research institutions should all be involved in the workshop. A glance at the list of participants will verify that this was the case as far as participation in the scientific part of the workshop is concerned. With respect to the financial support, we have the pleasure of thanking the following long list of government agencies and companies: the Dutch Academy of Sciences, the Dutch Organization for the Advancement of Pure Scientific Research, the British Science and Engineering Research Council, the University and the Province of Groningen, de Nederlandse Aardolie Maatschappij (the Dutch Oil Company), de N.V. Nederlandse Gasunie (the Dutch Gas Company), Hollandse Signaalapparaten B.V. and Hoogovens Groep B.V.

Finally we would like to thank the Mathematics Institute of the University of Groningen for the support in organizing the workshop in particular the assistance of the workshop secretary, Janieta Schlukebir.

Ruth F. Curtain

Ruth F. Curtain

for the International Committee:

Prof. R.F. Curtain, Mathematics Institute, University of Groningen
Dr. K. Glover, Dept. Engineering, University of Cambridge, U.K.
Prof. B. Francis, Dept. Engineering, University of Toronto, Canada
Prof. J.C. Doyle, Honeywell SRC, MN 17-2367, 2600 Ridgeway Parkway, Minneapolis, U.S.A./ Dept. Electrical Engineering, California Institute of Tecnnology, Pasadena, CA 91125, USA.

TABLE OF CONTENTS

3. Modelling and Signal Processing

4. Modelling and System Identification

5. H^∞-techniques

A Guide To H^∞-Control Theory

Bruce A. Francis
Department of Electrical Engineering
University of Toronto
Toronto, Canada M5S 1A4
USA

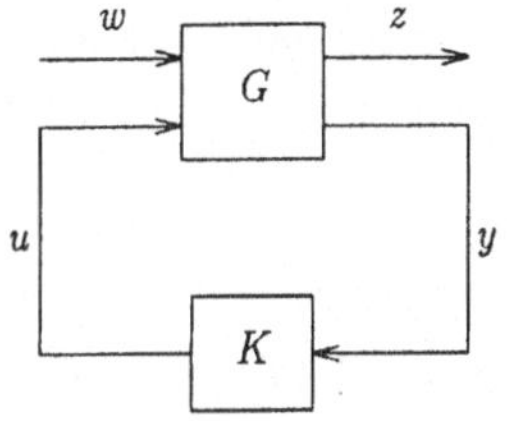

Figure 1. The standard block diagram

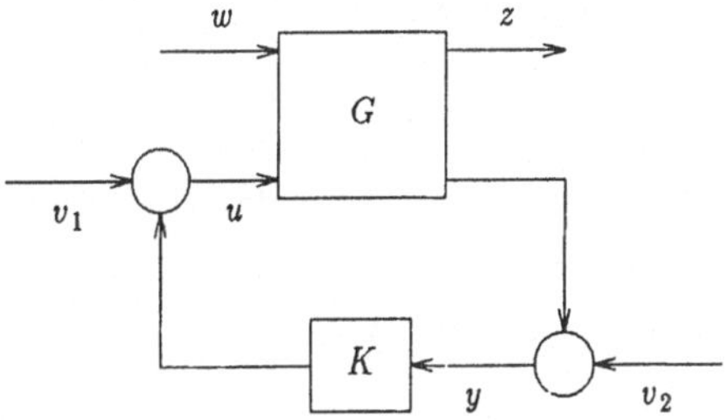

Figure 2. Diagram for stability definition

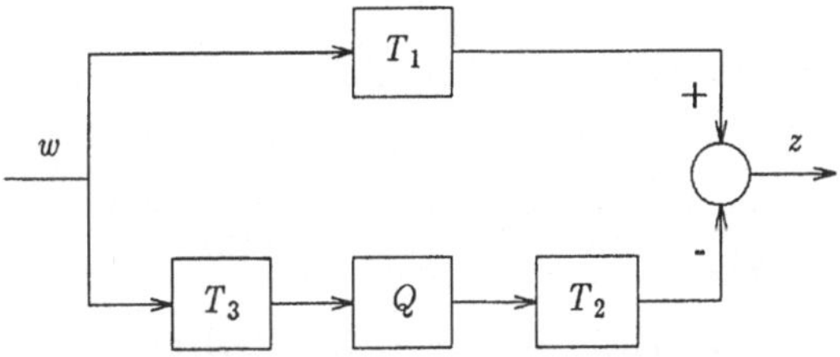

Figure 3. Model-matching

NATO ASI Series, Vol. F34
Modelling, Robustness and Sensitivity Reduction
in Control Systems. Edited by R. F. Curtain

1. Introduction

This paper is intended as a tutorial on the most basic $\mathbf{H}_\infty$- control problem. The set-up is linear, time-invariant, finite-dimensional, continuous-time. The main theme is that the theory is most simply and elegantly developed in the framework of operators, while computations are most easily performed using state-space methods. (Thus state-space methods serve merely as slaves in an input-output setting.) The results are summarized in the form of algorithms, primarily to demonstrate that the computations can be done using off-the-shelf software.

Pioneered by Zames (1981), $\mathbf{H}_\infty$- optimization in control theory has been developed by many researchers and from several viewpoints. The state-space approach to computations was initiated primarily by Silverman and Bettayeb (1980) and Doyle (1984). The reader may consult Francis and Doyle (1987) and Dorato (1987) for reference lists and historical accounts.

The main text consists of five parts. In Section 2 the standard problem is posed and the model-matching problem (MMP)

$$\underset{Q}{\text{minimize}} \; \|T_1 - T_2 Q T_3\|_\infty$$

is offered as an example. Here T_i and Q are real-rational $\mathbf{H}_\infty$-matrices. The reader is then reminded that the standard problem can be reduced to MMP using the familiar parametrization of Youla, Jabr, and Bongiorno (1976).

The rest of the paper deals with MMP. The classification scheme of Limebeer and Hung (1986) is introduced, yielding three model-matching problems, MMP(i) (i=1-3), of increasing difficulty. In general one solves these problems by first computing the minimal model-matching error (the minimum norm above) and then computing an optimal Q.

Section 3 begins with a discussion of when an optimal Q exists. Mild sufficient conditions are given and then a three-step, high-level algorithm is developed for solving MMP(1). The most difficult step is the Nehari problem of approximating an $\mathbf{L}_\infty$-matrix by an $\mathbf{H}_\infty$-matrix.

In Section 4 the Nehari problem in the scalar-valued case (T_i and Q are scalar-valued functions) is solved completely using the theory of Sarason (1967) and

Adamjan, Arov, and Krein (1971), with state-space formulas by Silverman and Bettayeb (1980).

Section 5 deals with the factorization of a rational matrix. The canonical factorization theorem of Bart, Gohberg, Kaashoek, and van Dooren (1980) is presented and used to obtain spectral factorization, inner-outer factorization, and J-spectral factorization.

Finally, in Section 6 the Nehari problem in the matrix-valued case is solved using the theory of Ball and Helton (1983), with state-space formulas by Ball and Ran (1986).

The notation is fairly standard: $\mathbf{L}_\infty$ is the space of essentially-bounded matrix functions on the imaginary axis; $\mathbf{H}_2$ and $\mathbf{H}_\infty$ are the Hardy spaces for the right half-plane; and prefix $\mathbf{R}$ denotes real-rational. For a state-space realization, $[A, B, C, D]$ stands for the transfer matrix $D + C(sI - A)^{-1}B$.

2. The standard problem and the model-matching problem

The standard set-up is shown in Figure 1. In this figure w, u, z, and y are vector-valued signals: w is the exogenous input, typically consisting of command signals, disturbances, and sensor noises; u is the control signal; z is the output to be controlled, its components typically being tracking errors, filtered actuator signals, etc.; and y is the measured output. The transfer matrices G and K are, by assumption, real-rational and proper: G represents a generalized plant, the fixed part of the system, and K represents a controller. Partition G as

$$G = \begin{bmatrix} G_{11} & G_{12} \\ G_{21} & G_{22} \end{bmatrix}.$$

Then Figure 1 stands for the algebraic equations

$$z = G_{11} w + G_{12} u$$

$$y = G_{21} w + G_{22} u$$

$$u = Ky.$$

To define what it means for K to stabilize G, introduce two additional inputs, v_1 and v_2, as in Figure 2. It simplifies the theory to guarantee that the nine transfer matrices from w, v_1, v_2 to z, u, y exist and are proper for every proper real-rational K. A simple sufficient condition for this is that G_{22} be strictly proper. Accordingly, this will be *assumed* hereafter. If these nine transfer matrices are stable, i.e. they belong to $\mathbf{RH}_\infty$, then we say that *K stabilizes G*. (This is the usual notion of internal stability.)

The *standard problem* is this: find a real-rational proper K to minimize the $\mathbf{H}_\infty$-norm of the transfer matrix from w to z under the constraint that K stabilize G. Observe that the transfer matrix from w to z is a linear-fractional, hence non-linear, transformation of K:

$$z = \left[G_{11} + G_{12}K(I-G_{22}K)^{-1}G_{21}\right]w.$$

There are several well-studied special cases of the standard problem, for example the weighted sensitivity, the mixed sensitivity, and the robust stability problems, but perhaps the simplest special case is the model-matching problem, abbreviated MMP. In Figure 3 the transfer matrix T_1 represents a "model" which is to be matched by the cascade $T_2\ Q\ T_3$ of three transfer matrices T_2, T_3, and Q. Here, T_i (i=1–3) are given and the "controller" Q is to be designed. It is assumed that $T_i \in \mathbf{RH}_\infty$ (i=1–3) and it is required that $Q \in \mathbf{RH}_\infty$. Thus the four blocks in Figure 3 represent stable linear systems.

For our purposes the *model-matching criterion* is

$$\sup \left\{\|z\|_2 \colon w \in \mathbf{H}_2,\ \|w\|_2 \le 1\right\} = \text{minimum}.$$

Since the $\mathbf{H}_2$-induced norm equals the $\mathbf{H}_\infty$-norm of the transfer matrix, this is equivalent to

$$\|T_1 - T_2\ Q\ T_3\|_\infty = \text{minimum}.$$

This model-matching problem can be recast as a standard problem by defining

$$G := \begin{bmatrix} T_1 & T_2 \\ T_3 & 0 \end{bmatrix}$$

$$K := -Q,$$

so that Figure 3 becomes equivalent to Figure 1. The constraint that K stabilize G

is then simply that $Q \in \mathbf{RH}_\infty$.

This version of the model-matching problem is not so important *per se*; its significance for us arises from the fact that the standard problem can in fact be transformed into the model-matching problem, which is considerably simpler. How to do this is by now standard: one parametrizes all K 's stabilizing G as a linear-fractional transformation of a free parameter matrix Q in $\mathbf{RH}_\infty$; then the transfer matrix from w to z is an affine function of Q, i.e. it's of the form $T_1 - T_2 Q T_3$. The theory behind this conversion is omitted; however we summarize the procedure in the form of a state-space algorithm, due primarily to Doyle (1984), to compute T_i (i=1--3) from G.

The algorithm starts with a minimal realization of G:

$$G(s) = [A,\ B,\ C,\ D].$$

Since the input and output of G are partitioned as

$$\begin{bmatrix} w \\ u \end{bmatrix},\ \begin{bmatrix} z \\ y \end{bmatrix},$$

the matrices B, C, and D have corresponding partitions:

$$B = [B_1\ \ B_2]$$

$$C = \begin{bmatrix} C_1 \\ C_2 \end{bmatrix}$$

$$D = \begin{bmatrix} D_{11} & D_{12} \\ D_{21} & D_{22} \end{bmatrix}.$$

Then

$$G_{ij}(s) = [A,\ B_j,\ C_i,\ D_{ij}],\quad i,j = 1,2.$$

Note that D_{22}=0 because G_{22} is strictly proper.

Procedure 1

Step 1. Choose F and H so that

$$A_F := A + B_2 F,\ A_H := A + H C_2$$

are stable.

Step 2. Set

$$\underline{A} = \begin{bmatrix} A_F & -B_2F \\ 0 & A_H \end{bmatrix}$$

$$\underline{B} = \begin{bmatrix} B_1 \\ B_1+HD_{21} \end{bmatrix}$$

$$\underline{C} = [C_1+D_{12}F \quad -D_{12}F]$$

$$T_1(s) = [\underline{A}, \underline{B}, \underline{C}, D_{11}]$$

$$T_2(s) = [A_F, B_2, C_1+D_{12}F, D_{12}]$$

$$T_3(s) = [A_H, B_1+HD_{21}, C_2, D_{21}] .$$

Limebeer and Hung (1986) introduced a useful classification scheme for MMP. It involves the relative dimensions of the matrices T_1 and T_2. Let's say a matrix is wide if the number of its rows is $\leq$ the number of its columns, and strictly wide if the inequality is $<$. Similarly for tall and strictly tall. The classification scheme is this:

MMP(1): T_1 is wide and T_2 is tall

MMP(2): T_1 is strictly tall or T_2 is strictly wide (exclusive or)

MMP(3): T_1 is strictly tall and T_2 is strictly wide.

It turns out that MMP(2) is harder than MMP(1), and MMP(3) harder than MMP(2). Difficulty here refers to the complexity of computing optimal Q's.

3. Existence and a high-level algorithm for MMP(1)

To each Q in $\mathbf{RH}_\infty$ there corresponds a *model-matching error*, $\|T_1 - T_2QT_3\|_\infty$. Let α denote the *infimal model-matching error*:

$$\alpha := \inf\{\|T_1 - T_2QT_3\|_\infty : Q \in \mathbf{RH}_\infty\} . \tag{1}$$

A matrix Q in $\mathbf{RH}_\infty$ satisfying

$$\alpha = \|T_1 - T_2 Q T_3\|_\infty$$

will be called *optimal.*

This section is first concerned with the question of when an optimal Q exists. This theorem provides a sufficient condition:

An optimal Q exists if the ranks of the two matrices $T_2(j\omega)$ and $T_3(j\omega)$ are constant for all $0 \leq \omega \leq \infty$.

These rank conditions are not necessary for existence, but from now on we assume they hold. To see the underlying idea in this theorem, first note from (1) that

$$\alpha = \text{dist}\,(T_1,\, T_2 \mathbf{RH}_\infty T_3) \tag{2}$$

where

$$T_2 \mathbf{RH}_\infty T_3 := \{T_2 Q T_3 : Q \in \mathbf{RH}_\infty\}.$$

Since $\mathbf{RH}_\infty \subset \mathbf{H}_\infty$ we have

$$\alpha \geq \text{dist}\,(T_1,\, T_2 \mathbf{H}_\infty T_3). \tag{3}$$

But since T_1 is itself real-rational, the two distances in (2) and (3) are equal, so

$$\alpha = \text{dist}\,(T_1,\, T_2 \mathbf{H}_\infty T_3). \tag{4}$$

The rank conditions above guarantee that $T_2 \mathbf{H}_\infty T_3$ is weak-star closed in $\mathbf{H}_\infty$ (see for instance Theorem II.7.5 in Garnett (1981)), and this in turn implies that the distance in (4) is achieved, i.e.

$$\alpha = \|T_1 - T_2 Q T_3\|_\infty$$

for some Q in $\mathbf{H}_\infty$. Proving that the distance is achieved by a real-rational Q is harder, and in fact hasn't been completely worked out. In the general case, the constructive procedure for MMP yields a Q in $\mathbf{RH}_\infty$ so that $T_2 Q T_3$ is arbitrarily close to T_1.

Deep results on uniqueness have recently been obtained by Foias and Tannenbaum (1986).

There's a simple formula duc to Young (1086) for α as the norm of a certain operator. Define two subspaces $\mathbf{X}$ and $\mathbf{Y}$ of $\mathbf{L}_2$,

$$\mathbf{X} := T_3^{-1} \mathbf{H}_2$$

$$:= \{f \in \mathbf{L}_2 : T_3 f \in \mathbf{H}_2\}$$

$$\mathbf{Y} := \text{orthogonal complement of } T_2\mathbf{H}_2 \text{ in } \mathbf{L}_2 .$$

Now define the operator Ξ from $\mathbf{X}$ to $\mathbf{Y}$ as follows:

$$\Xi f := \text{orthogonal projection of } T_1 f \text{ onto } \mathbf{Y},\ \ f \in \mathbf{X}.$$

Then Young's formula is $\alpha = \|\Xi\|$. An alternative formula is that of Feintuch and Francis (1985).

There's a more concrete formula in the simplest case, MMP(1), when T_2 has full row rank (i.e. its normal rank equals its number of rows) and T_3 has full column rank. For a matrix function $G(s)$, let $G^{\sim}(s)$ denote $G(-s)'$, the transpose of $G(-s)$. Recall that G in $\mathbf{RH}_\infty$ is *inner* if $G^{\sim}G = I$ and *outer* if it has a right-inverse which is analytic in Re $s>0$. Pre-multiplication of an $\mathbf{L}_\infty$-matrix by an inner matrix preserves $\mathbf{L}_\infty$-norms:

$$G \text{ inner },\ F \in \mathbf{L}_\infty \Rightarrow \|GF\|_\infty = \|F\|_\infty .$$

The following fact also is useful:

Every matrix G in $\mathbf{RH}_\infty$ has a factorization $G = G_i G_o$ with G_i inner and G_o outer. If rank $G(j\omega)$ is constant for all $0 \leq \omega \leq \infty$, then G_o is right-invertible in $\mathbf{RH}_\infty$.

How to compute inner-outer factorizations by state-space methods is postponed till Section 5. With the above assumptions on T_2 we can write

$$T_2 = U_i U_o ,\ \ U_i \text{ inner},\ \ U_o \text{ outer}.$$

Similarly, inner-outer factorization of T_3' yields

$$T_3 = V_o V_i ,\ \ V_i' \text{ inner},\ \ V_o' \text{ outer}.$$

In fact, U_i and V_i are square and invertible (in $\mathbf{RL}_\infty$, not $\mathbf{RH}_\infty$). Thus for Q in $\mathbf{RH}_\infty$

$$\begin{aligned}\|T_1 - T_2 Q T_3\|_\infty &= \|T_1 - U_i U_o Q V_o V_i\|_\infty \\ &= \|U_i^{\sim} T_1 V_i^{\sim} - U_o Q V_o\|_\infty \\ &= \|R - X\|_\infty\end{aligned}$$

where

$$R := U_i^{\sim} T_1 V_i^{\sim} \,, \; X := U_o Q V_o \,.$$

We conclude that

$$\alpha = \text{dist}\,(R, \mathbf{H}_\infty)\,.$$

Moreover, we have a three-step, high-level algorithm solving MMP(1):

(i) find the distance from the $\mathbf{L}_\infty$-matrix R to $\mathbf{H}_\infty$

(ii) find a matrix X in $\mathbf{RH}_\infty$ attaining this distance

(iii) solve $X = U_o Q V_o$ for Q in $\mathbf{RH}_\infty$.

Nehari's theorem (Nehari, 1957) is an elegant solution to the distance problem (i). This theorem says that the distance equals the norm of the Hankel operator with symbol R, denoted Γ_R and defined as

$$\Gamma_R : \mathbf{H}_2 \rightarrow \mathbf{H}_2^\perp \,, \;\; \Gamma_R f = \Pi_{\mathbf{H}_2^\perp} R f \,.$$

Our second state-space procedure, due to Silverman and Bettayeb (1980), computes this norm, and hence α.

Procedure 2

Step 1. Find a minimal realization of the antistable part of R:

$$R(s) = [A,\, B,\, C,\, 0] + (\text{a matrix in } \mathbf{RH}_\infty)\,.$$

Step 2. Solve the Lyapunov equations

$$A L_c + L_c A' = B B'$$

$$A' L_o + L_o A = C' C$$

for the controllability and observability gramians L_c and L_o.

Step 3. $\alpha = \|\Gamma_R\| =$ the square root of the largest eigenvalue of $L_c L_o$.

Problem (ii) above is called the Nehari problem and is treated in Sections 4 and 6.

Problem (iii) is easy enough to be dispatched now. We are given X, U_o, and V_o in $\mathbf{RH}_\infty$ with U_o right-invertible in $\mathbf{RH}_\infty$ and V_o left-invertible in $\mathbf{RH}_\infty$, and we want to solve

$$X = U_o Q V_o$$

for Q in $\mathbf{RH}_\infty$. An obvious way to proceed is to find a right-inverse $\hat{U}_o$ of U_o and a left-inverse $\hat{V}_o$ of V_o, and then to set $Q = \hat{U}_o X \hat{V}_o$. So the basic problem is to do right-inversion in $\mathbf{RH}_\infty$ (left-inversion is then accomplished by transposing matrices). Minto (1985) derived the following procedure for this problem. The input is a minimal realization $[A, B, C, D]$ of a matrix $U(s)$ right-invertible in $\mathbf{RH}_\infty$.

Procedure 3

Step 1. Compute a matrix F such that

$$A + BF \text{ is stable}$$

$$C + DF = 0 .$$

Step 2. Set

$$\hat{U}(s) = [A + BF,\ B,\ F, I] D'(DD')^{-1} .$$

4. The Nehari problem in the scalar-valued case

We are given a scalar-valued function R in $\mathbf{RL}_\infty$ and we have computed its distance α to $\mathbf{RH}_\infty$. We now want to compute an X in $\mathbf{RH}_\infty$ attaining this distance. In this scalar-valued case, X is in fact unique.

The following are results of Sarason (1967) and Adamjan, Arov, and Krein (1971). The solution for X is in terms of an eigenvector of $\Gamma_R^* \Gamma_R$ corresponding to the largest eigenvalue; the latter equals α^2, by Nehari's theorem. If g is such an eigenvector, then the optimal X satisfies the equation

$$(R - X)g = \Gamma_R g ,$$

and hence

$$X = R - \Gamma_R g / g .$$

Silverman and Bettayeb (1980) developed a state-space procedure based on the previous formula. It's a continuation of Procedure 2.

Procedure 2 (Cont'd)

Step 4. Find an eigenvector w of $L_c L_o$ corresponding to the largest eigenvalue.

Step 5. Set

$$X = R - \alpha[A,\ w,\ C,\ 0]/[-A',\ \alpha^{-1} L_o w,\ B',\ 0] .$$

Other approaches to the problem of this section are Nevanlinna-Pick interpolation (e.g. Garnett, 1981) and the polynomial approach of Kwakernaak (1985).

5. Factorization theory

The purposes of this section are to develop a state-space procedure for inner-outer factorizations (as promised in Section 3) and to develop a special factorization, called J-spectral factorization, required for the Nehari problem in the matrix-valued case. Both of these are based on canonical factorization.

5.1 Canonical factorization

Consider a scalar-valued real-rational function $G(s)$ which is proper and has no poles on the imaginary axis; thus $G \in \mathbf{RL}_\infty$. Suppose in addition that G has no zeros on the imaginary axis nor at infinity. Then $G^{-1} \in \mathbf{RL}_\infty$ too. Now consider the problem of factoring G as $G = G_+ G_-$ where G_+ has all its poles and zeros in Re $s > 0$ and G_- has all its poles and zeros in Re $s < 0$. Furthermore, we require that G_+ and G_- be proper and have proper inverses. When does such a factorization exist? The necessary and sufficient condition is

$$\{\text{no. poles in Re } s < 0\} = \{\text{no. zeros in Re } s < 0\},$$

or equivalently

$$\{\text{no. poles}\} = \{\text{no. zeros in Re } s < 0\} + \{\text{no. poles in Re } s > 0\}.$$

The purpose of this subsection is to derive the analogous condition in the matrix case and give a procedure for doing such a factorization.

Let $G(s)$ be a square matrix such that G, $G^{-1} \in \mathbf{RL}_\infty$. Thus G and its inverse are proper and have no poles on the imaginary axis. Our goal is to factor G as $G = G_+ G_-$, where the factors G_+ and G_- are square and have the properties

$$G_-, G_-^{-1} \in \mathbf{RH}_\infty$$

$$G_+^\sim, (G_+^{-1})^\sim \in \mathbf{RH}_\infty .$$

The latter condition means that G_+ and its inverse are proper and analytic in Re $s < 0$. For ease of reference let's call a factorization as just described a *canonical factorization* of G.

We begin with a minimal realization,

$$G(s) = [A, B, C, D] .$$

Since $G(\infty) = D$ and $G^{-1} \in \mathbf{RL}_\infty$, D is invertible. Define

$$A^\times := A - BD^{-1}C$$

and write the state-space equations for G:

$$\dot{x} = Ax + Bu$$

$$y = Cx + Du .$$

Re-arrange to get y as input and u as output:

$$\dot{x} = A^\times x + BD^{-1}y$$

$$u = -D^{-1}Cx + D^{-1}y .$$

Thus

$$G(s)^{-1} = [A^\times, BD^{-1}, -D^{-1}C, D^{-1}] .$$

Next we recall the notions of modal subspaces. Suppose A is of dimension $n \times n$. Let $\alpha(s)$ denote the characteristic polynomial of A and factor it as

$\alpha(s)=\alpha_-(s)\alpha_+(s)$, where α_- has all its zeros in $\operatorname{Re} s<0$ and α_+ has all its zeros in $\operatorname{Re} s>0$. (There are no zeros on the imaginary axis.) Then the *modal subspaces* of $\mathbf{R}^n$ relative to A are

$$\mathbf{X}_-(A):=\operatorname{Ker}\alpha_-(A)$$

$$\mathbf{X}_+(A):=\operatorname{Ker}\alpha_+(A),$$

where Ker denotes kernel (null space). These two modal subspaces are *complementary*, i.e. they're independent and their sum is all of $\mathbf{R}^n$.

Now consider the two modal subspaces $\mathbf{X}_-(A^\times)$ and $\mathbf{X}_+(A)$. The former is associated with left half-plane zeros of G and the latter with right half-plane poles of G. The factorization theorem of Bart, Gohberg, Kaashoek, and van Dooren (1980) is this:

G has a canonical factorization iff $\mathbf{X}_-(A^\times)$ and $\mathbf{X}_+(A)$ are complementary.

Their proof is constructive:

Procedure 4

Step 1. Obtain real matrices T_1 and T_2, each with full column rank, such that

$$\mathbf{X}_-(A^\times)=\operatorname{Im} T_1$$

$$\mathbf{X}_+(A)=\operatorname{Im} T_2\,.$$

Here Im denotes image (range space). Define $T:=[T_1\ T_2]$ and note that T is square and nonsingular by the complementarity condition.

Step 2. Introduce the partitions

$$T^{-1}AT=\begin{bmatrix}A_1 & A_2\\ A_3 & A_4\end{bmatrix}$$

$$T^{-1}B=\begin{bmatrix}B_1\\ B_2\end{bmatrix}$$

$$CT=[C_1\ \ C_2]$$

corresponding to the above partition of T, e.g. A_1 is square and its dimension equals that of $\mathbf{X}_-(A^\times)$.

Step 3. Set

$$G_+(s) := [A_4, B_2, C_2, D]$$

$$G_-(s) := [A_1, B_1, D^{-1}C_1, I].$$

5.2 Inner-outer factorization

Suppose $G \in \mathbf{RH}_\infty$. We want to factor G as $G = G_i G_o$ with G_i inner and G_o outer. It's true that every matrix in $\mathbf{RH}_\infty$ has such a factorization. We shall give a state-space procedure for doing this, but only when G is tall and $G(j\omega)$ has full column rank for all $0 \leq \omega \leq \infty$. The case where G is wide is harder. The formulas to be developed are due to Doyle (1984) by a different method.

The idea is this: do a canonical factorization $G^\sim G = G_+ G_-$; modify G_+ and G_- so that $G_+ = G_-^\sim$; set $G_o := G_-$ and then $G_i := G G_o^{-1}$. The matrix G_i is inner because

$$G^\sim G = G_+ G_-$$

$$= G_-^\sim G_-$$

$$= G_o^\sim G_o$$

$$\therefore (G_o^\sim)^{-1} G^\sim G G_o^{-1} = I$$

$$\therefore G_i^\sim G_i = I.$$

Start with a minimal realization,

$$G(s) = [A, B, C, D].$$

In view of the above rank assumption on G, $G^\sim G$ and its inverse belong to $\mathbf{RL}_\infty$. Also, D is injective, so $E := (D'D)^{-1}$ exists. A realization of $G^\sim G$ is

$$G^\sim G(s) = [\underline{A}, \underline{B}, \underline{C}, \underline{D}]$$

$$\underline{A} = \begin{bmatrix} A & 0 \\ -C'C & -A' \end{bmatrix}$$

$$\underline{B} = \begin{bmatrix} B \\ -C'D \end{bmatrix}$$

$$\underline{C} = [D'C \;\; B']$$

$$\underline{D} = D'D\,.$$

The A-matrix of the inverse of $G^{\sim}G$ is

$$\begin{aligned}\underline{A}^{\times} &:= \underline{A} - \underline{B}\underline{D}^{-1}\underline{C} \\ &= \begin{bmatrix} A & 0 \\ -C'C & -A' \end{bmatrix} - \begin{bmatrix} B \\ -C'D \end{bmatrix} E[D'C \;\; B'] \\ &= \begin{bmatrix} A-BED'C & -BEB' \\ -C'(I-DED')C & -(A-BED'C)' \end{bmatrix}.\end{aligned}$$

According to the previous section, $G^{\sim}G$ has a canonical factorization iff $\mathbf{X}_-(\underline{A}^{\times})$ and $\mathbf{X}_+(\underline{A})$ are complementary. This is the case. In fact, these subspaces have the form

$$\mathbf{X}_+(\underline{A}) = \{0\} \oplus \mathbf{R}^n = \mathrm{Im}\begin{bmatrix} 0 \\ I \end{bmatrix}$$

$$\mathbf{X}_-(\underline{A}^{\times}) = \mathrm{Im}\begin{bmatrix} I \\ X \end{bmatrix}.$$

To construct a canonical factorization of $G^{\sim}G$ we follow Procedure 4. Define

$$T := \begin{bmatrix} I & 0 \\ X & I \end{bmatrix}.$$

Then

$$T^{-1}\underline{A}T = \begin{bmatrix} A & 0 \\ ? & -A' \end{bmatrix}$$

(? denotes an irrelevant block)

$$T^{-1}\underline{B} = \begin{bmatrix} B \\ -(D'C+B'X)' \end{bmatrix}$$

$$\underline{C}T = [D'C+B'X \;\; B']\,.$$

Thus $G^{\sim}G = G_+G_-$ where

$$G_+(s) := [-A',\, -(D'C+B'X)',\, B',\, D'D]$$

$$G_-(s) := [A,\, B,\, E(D'C+B'X),\, I]\,.$$

Observe that $G_+ = (D'DG_-)^{\sim}$, so

$$G^{\sim} G = G_-^{\sim} D'DG_- .$$

Define $G_o := E^{-1/2} G_-$ to get $G^{\sim} G = G_o^{\sim} G_o$ and

$$G_o(s) = E^{-1/2} [A, B, E(D'C+B'X), I] .$$

The following procedure summarizes what's above. The input data is a minimal realization of G.

Procedure 5

Step 1. Set

$$E := (D'D)^{-1}$$

$$\underline{A}^{\times} := \begin{bmatrix} A - BED'C & -BEB' \\ -C'(I-DED')C & -(A-BED'C)' \end{bmatrix} .$$

Step 2. Compute V_1 and V_2 (both square) such that

$$\operatorname{Im} \begin{bmatrix} V_1 \\ V_2 \end{bmatrix} = \mathbf{X}_-(\underline{A}^{\times})$$

and set $X = V_2 V_1^{-1}$.

Step 3. Set

$$F := -E(D'C + B'X)$$

$$G_i(s) := [A + BF,\ BE^{1/2},\ C + DF,\ DE^{1/2}]$$

$$G_o(s) := [A,\ B,\ -E^{-1/2} F,\ E^{-1/2}] .$$

5.3 J-spectral factorization

The material in this subsection is due to Ball and Ran (1986). Start with a real-rational matrix $R(s)$ having the properties

R is strictly proper

$$R^{\sim} \in \mathbf{RH}_{\infty}$$

$$\|\Gamma_R\| < 1 .$$

Define the matrices

$$G := \begin{bmatrix} I & R \\ 0 & I \end{bmatrix}$$

$$J := \begin{bmatrix} I & 0 \\ 0 & -I \end{bmatrix} .$$

Our goal is to achieve the following *J-spectral factorization* of G:

$$G^{\sim} J G = G_{-}^{\sim} J G_{-}$$

$$G_{-} , G_{-}^{-1} \in \mathbf{RH}_{\infty} .$$

As in the previous section, the idea is to do a canonical factorization of $G^{\sim} JG$.

Starting with a minimal realization

$$R(s) = [A, B, C, 0] ,$$

we get the realizations

$$G(s) = [A, [0 \ B], \begin{bmatrix} C \\ 0 \end{bmatrix}, I] \tag{1}$$

$$(G^{\sim} JG)(s) = [\underline{A}, \underline{B}, \underline{C}, J]$$

where

$$\underline{A} := \begin{bmatrix} -A' & C'C \\ 0 & A \end{bmatrix}$$

$$\underline{B} := \begin{bmatrix} C' & 0 \\ 0 & B \end{bmatrix}$$

$$\underline{C} := \begin{bmatrix} 0 & C \\ -B' & 0 \end{bmatrix} .$$

Then

$$\underline{A}^{\times} := \underline{A} - \underline{B} J^{-1} \underline{C}$$

$$= \begin{bmatrix} -A' & 0 \\ -BB' & A \end{bmatrix} .$$

It can be shown that the modal subspaces are

$$X_-(\underline{A}^\times)=\text{Im}\begin{bmatrix} I \\ L_c \end{bmatrix},\ X_+(\underline{A})=\text{Im}\begin{bmatrix} L_o \\ I \end{bmatrix}$$

where L_c and L_o are the controllability and observability gramians for the triple (A, B, C). (These modal subspaces can be proved to be complementary because of our assumption $\|\Gamma_R\| < 1$.)

A straightforward application of Procedure 4 leads to the factorization $G^\sim JG = G_+ G_-$ where

$$G_+(s)=[A, [-L_cNC' \ \ N'B], \begin{bmatrix} C \\ -B'L_o \end{bmatrix}, J]$$

$$G_-(s)=[-NA'N^{-1}, [NC' \ \ -NL_oB], \begin{bmatrix} CL_c \\ B' \end{bmatrix}, I] \tag{2}$$

$$N:=(I-L_oL_c)^{-1}.$$

Since $G_+ = G_-^\sim J$, we're done.

6. The Nehari problem in the matrix-valued case

The material in this section is based on (Ball and Helton, 1983) and (Ball and Ran, 1986). Since low-level proofs are not readily available elsewhere, they're included here.

The full Nehari problem, find a matrix X in $\mathbf{RH}_\infty$ closest to a given matrix R in $\mathbf{RL}_\infty$, is relatively hard, so we'll content ourselves with a simpler problem. First note that by scaling we may arrange that the distance from R to $\mathbf{RH}_\infty$ is slightly less than 1. Our simpler problem is this: given R in $\mathbf{RL}_\infty$ with dist$(R, \mathbf{RH}_\infty)<1$, find all X's in $\mathbf{RH}_\infty$ such that $\|R-X\|_\infty \le 1$. Only some of these X's are actually closest to R, i.e. satisfy

$$\|R-X\|_\infty = \text{dist}(R, \mathbf{RH}_\infty).$$

We've seen that the distance equals $\|\Gamma_R\|$, so the standing assumption in this section is that $\|\Gamma_R\|<1$. We'll also assume for now that R is strictly proper and analytic in $\text{Re}\, s \le 0$, i.e. $R^\sim \in \mathbf{RH}_\infty$, and then remove this assumption for the final algorithm.

We need some preliminary concepts. Let **X** and **Y** be two Hilbert spaces. There is a natural way to add them together to get a third Hilbert space, their *external direct sum* $\mathbf{X}\oplus\mathbf{Y}$. We shall represent vectors in $\mathbf{X}\oplus\mathbf{Y}$ like this: $\begin{pmatrix} x \\ y \end{pmatrix}$. As a set, $\mathbf{X}\oplus\mathbf{Y}$ consists of all such vectors as x ranges over **X** and y over **Y**. Vector addition and scalar multiplication are defined componentwise, and the inner product is defined as follows:

$$\left\langle \begin{pmatrix} x_1 \\ y_1 \end{pmatrix}, \begin{pmatrix} x_2 \\ y_2 \end{pmatrix} \right\rangle := <x_1, x_2> + <y_1, y_2>.$$

Now introduce in addition an *indefinite inner-product* on $\mathbf{X}\oplus\mathbf{Y}$:

$$\left[\begin{pmatrix} x_1 \\ y_1 \end{pmatrix}, \begin{pmatrix} x_2 \\ y_2 \end{pmatrix} \right] := <x_1, x_2> - <y_1, y_2>.$$

A more compact way of defining $[\,,\,]$ is to introduce the operator J on $\mathbf{X}\oplus\mathbf{Y}$ (related in an obvious way to the matrix J of Subsection 5.3):

$$J\begin{pmatrix} x \\ y \end{pmatrix} := \begin{pmatrix} x \\ -y \end{pmatrix}.$$

Then $[z_1, z_2] := <z_1, Jz_2>$. The external direct sum $\mathbf{X}\oplus\mathbf{Y}$ together with the above indefinite inner-product is called a *Krein space*.

A vector z in $\mathbf{X}\oplus\mathbf{Y}$ is *negative* if $[z,z]\leq 0$, and a subspace of $\mathbf{X}\oplus\mathbf{Y}$ is *negative* if all its vectors are negative.

Consider an operator Φ from **Y** to **X**. Its *graph* is a subspace of $\mathbf{X}\oplus\mathbf{Y}$, namely,

$$\left\{ \begin{pmatrix} \Phi y \\ y \end{pmatrix} : y \in \mathbf{Y} \right\}.$$

As an example, let F be a matrix in $\mathbf{RH}_\infty$ and consider the operator on $\mathbf{H}_2$ of multiplication by F. Its graph is

$$\left\{ \begin{pmatrix} Fg \\ g \end{pmatrix} : g \in \mathbf{H}_2 \right\},$$

or equivalently

$$\begin{bmatrix} F \\ I \end{bmatrix} \mathbf{H}_2 ,$$

where we use the notation

$$M\mathbf{H}_2 := \{Mg : g \in \mathbf{H}_2\} .$$

This graph is a subspace of $\mathbf{H}_2 \oplus \mathbf{H}_2$. It's convenient to denote it by $\mathbf{G}_F$. If $F \in \mathbf{RL}_\infty$, then the corresponding graph

$$\mathbf{G}_F := \begin{bmatrix} F \\ I \end{bmatrix} \mathbf{H}_2$$

lives in $\mathbf{L}_2 \oplus \mathbf{H}_2$.

Now we need an elementary fact.

Lemma 1. Let $F \in \mathbf{RL}_\infty$. Then $F\mathbf{H}_2 \subset \mathbf{H}_2$ iff $F \in \mathbf{RH}_\infty$. If F is square and $F\mathbf{H}_2 = \mathbf{H}_2$, then $F^{-1} \in \mathbf{RH}_\infty$.

Proof. The implication

$$F \in \mathbf{RH}_\infty => F\mathbf{H}_2 \subset \mathbf{H}_2$$

is easy. For the other way, suppose $F\mathbf{H}_2 \subset \mathbf{H}_2$. Since each column of the matrix $(s+1)^{-1} I$ belongs to $\mathbf{H}_2$, the same is true of $(s+1)^{-1} F(s)$. Therefore, this latter matrix is strictly proper and analytic in $\mathrm{Re}\, s \geq 0$. Hence F is proper and analytic in $\mathrm{Re}\, s \geq 0$, i.e. $F \in \mathbf{RH}_\infty$.

Finally, suppose F is square and $F\mathbf{H}_2 = \mathbf{H}_2$. Then there exists a matrix G, each of whose columns belongs to $\mathbf{H}_2$, such that

$$F(s)G(s) = (s+1)^{-1} I .$$

This implies that F has an inverse in $\mathbf{RH}_\infty$, namely, $(s+1)G(s)$. □

In terms of $S := R - X$, a problem equivalent to our simplified Nehari problem is this: find all S's in $\mathbf{RL}_\infty$ such that $\|S\|_\infty \leq 1$ and $R - S \in \mathbf{RH}_\infty$. The next lemma gives geometric characterizations of these two conditions. Define the $\mathbf{RL}_\infty$-matrix

$$G := \begin{bmatrix} I & R \\ 0 & I \end{bmatrix} . \tag{1}$$

Lemma 2. Let $S \in \mathbf{RL}_\infty$. Then $\|S\|_\infty \leq 1$ iff $\mathbf{G}_S$ is negative, and $R - S \in \mathbf{RH}_\infty$ iff

$$\mathbf{G}_S \subset G(\mathbf{H}_2 \oplus \mathbf{H}_2).$$

Proof. Suppose $\|S\|_\infty \leq 1$. A vector in $\mathbf{G}_S$ has the form $\begin{pmatrix} Sf \\ f \end{pmatrix}$ for some f in $\mathbf{H}_2$.

This vector is negative:

$$\left[\begin{pmatrix} Sf \\ f \end{pmatrix}, \begin{pmatrix} Sf \\ f \end{pmatrix}\right] = \|Sf\|_2^2 - \|f\|_2^2$$

$$\leq (\|S\|_\infty^2 - 1)\|f\|_2^2$$

$$\leq 0.$$

The converse is equally easy.

Now suppose $R - S \in \mathbf{RH}_\infty$. Then by Lemma 1

$$(S - R)\mathbf{H}_2 \subset \mathbf{H}_2,$$

so

$$\mathbf{G}_S = \begin{bmatrix} S \\ I \end{bmatrix} \mathbf{H}_2$$

$$= \begin{bmatrix} I & R \\ 0 & I \end{bmatrix} \begin{bmatrix} S - R \\ I \end{bmatrix} \mathbf{H}_2$$

$$\subset \begin{bmatrix} I & R \\ 0 & I \end{bmatrix} (\mathbf{H}_2 \oplus \mathbf{H}_2)$$

$$= G(\mathbf{H}_2 \oplus \mathbf{H}_2).$$

Conversely, if

$$\mathbf{G}_S \subset G(\mathbf{H}_2 \oplus \mathbf{H}_2),$$

then

$$\begin{bmatrix} S \\ I \end{bmatrix} \mathbf{H}_2 \subset \begin{bmatrix} I & R \\ 0 & I \end{bmatrix} (\mathbf{H}_2 \oplus \mathbf{H}_2).$$

Pre-multiply by $[-I \;\; R]$ to get

$$(R-S)\mathbf{H}_2 \subset \mathbf{H}_2 ,$$

which implies by Lemma 1 that $R-S \in \mathbf{RH}_\infty$. □

In view of Lemma 2 we would like to be able to characterize negative graphs contained in $G(\mathbf{H}_2 \oplus \mathbf{H}_2)$. The J-spectral factorization of Subsection 5.3 was introduced for this very purpose. Following that subsection we have

$$G^\sim JG = G_-^\sim JG_- \tag{2}$$

$$G_-, G_-^{-1} \in \mathbf{RH}_\infty$$

$$J := \begin{bmatrix} I & 0 \\ 0 & -I \end{bmatrix}.$$

Define

$$L := GG_-^{-1} . \tag{3}$$

Then from (2)

$$L^\sim JL = J \tag{4}$$

and from (3)

$$L(\mathbf{H}_2 \oplus \mathbf{H}_2) = G(\mathbf{H}_2 \oplus \mathbf{H}_2) . \tag{5}$$

A square matrix M in $\mathbf{RL}_\infty$ having the property $M^\sim JM=J$ is said to be *J—unitary*. Such a matrix is invertible in $\mathbf{RL}_\infty$; in fact the inverse of M is $JM^\sim J$. The usefulness of J-unitary matrices derives from the fact that (under mild conditions) they map negative graphs into negative graphs. The precise statement (the key lemma) is as follows.

Lemma 3. Let X be an $\mathbf{RL}_\infty$-matrix with $\|X\|_\infty \leq 1$. Suppose M is a J-unitary matrix having the properties

$$M\mathbf{G}_X \subset \mathbf{L}_2 \oplus \mathbf{H}_2 \tag{6}$$

$$\{0\} \oplus \mathbf{H}_2 \subset M(\mathbf{L}_2 \oplus \mathbf{H}_2) . \tag{7}$$

Then there exists Y in $\mathbf{RL}_\infty$ such that $\|Y\|_\infty \leq 1$ and $\mathbf{G}_Y = M\mathbf{G}_X$.

Proof. Define

$$\begin{bmatrix} Y_1 \\ Y_2 \end{bmatrix} := M \begin{bmatrix} X \\ I \end{bmatrix}. \tag{8}$$

Then from (6)

$$\begin{bmatrix} Y_1 \\ Y_2 \end{bmatrix} \mathbf{H}_2 = M\mathbf{G}_X \subset \mathbf{L}_2 \oplus \mathbf{H}_2 ,$$

so pre-multiplying by $[0\ I]$ we get $Y_2\mathbf{H}_2 \subset \mathbf{H}_2$. Then by Lemma 1, $Y_2 \in \mathbf{RH}_\infty$. We shall show that $Y_2^{-1} \in \mathbf{RH}_\infty$. This takes three steps.

Claim #1: $M\mathbf{G}_X$ is a negative subspace of $\mathbf{L}_2 \oplus \mathbf{H}_2$.

A vector in $M\mathbf{G}_X$ has the form

$$M \begin{pmatrix} Xf \\ f \end{pmatrix}$$

for some f in $\mathbf{H}_2$. Then

$$\begin{aligned} \left[M \begin{pmatrix} Xf \\ f \end{pmatrix}, M \begin{pmatrix} Xf \\ f \end{pmatrix} \right] &= \left\langle M \begin{pmatrix} Xf \\ f \end{pmatrix}, JM \begin{pmatrix} Xf \\ f \end{pmatrix} \right\rangle \\ &= \left\langle \begin{pmatrix} Xf \\ f \end{pmatrix}, M^\sim JM \begin{pmatrix} Xf \\ f \end{pmatrix} \right\rangle \\ &= \left\langle \begin{pmatrix} Xf \\ f \end{pmatrix}, J \begin{pmatrix} Xf \\ f \end{pmatrix} \right\rangle \\ &= \left[\begin{pmatrix} Xf \\ f \end{pmatrix}, \begin{pmatrix} Xf \\ f \end{pmatrix} \right]. \end{aligned}$$

The last quantity is ≤ 0 because $\mathbf{G}_X$ is negative.

Claim #2: $Y_2\mathbf{H}_2$ is a closed subspace of $\mathbf{H}_2$.

Suppose $\{f_k\}$ is a sequence in $Y_2\mathbf{H}_2$ which converges to some f in $\mathbf{H}_2$. Then

$$\begin{pmatrix} h_k \\ f_k \end{pmatrix} \in \begin{bmatrix} Y_1 \\ Y_2 \end{bmatrix} \mathbf{H}_2 = M\mathbf{G}_X$$

for certain vectors h_k in $\mathbf{L}_2$. Since $\{f_k\}$ is Cauchy and $M\mathbf{G}_X$ is negative, it follows that $\{h_k\}$ is Cauchy too, so it converges to some h in $\mathbf{L}_2$. Since $\mathbf{G}_X$ is closed and $M, M^{-1} \in \mathbf{RL}_\infty$, it follows that $M\mathbf{G}_X$ is closed. Thus

$$\begin{pmatrix} h \\ f \end{pmatrix} \in \begin{bmatrix} Y_1 \\ Y_2 \end{bmatrix} \mathbf{H}_2 ,$$

so $f \in Y_2 \mathbf{H}_2$.

Claim #3: $Y_2 \mathbf{H}_2 = \mathbf{H}_2$.

Suppose otherwise. Since $Y_2 \mathbf{H}_2$ is closed, we have

$$\mathbf{H}_2 = (Y_2 \mathbf{H}_2)^{\perp} \oplus (Y_2 \mathbf{H}_2).$$

Let g be a nonzero vector in $(Y_2 \mathbf{H}_2)^{\perp}$ and define

$$\begin{pmatrix} f_1 \\ f_2 \end{pmatrix} := M^{-1} \begin{pmatrix} 0 \\ g \end{pmatrix}. \tag{9}$$

Then (7) implies that $f_2 \in \mathbf{H}_2$, so g and $Y_2 f_2$ are orthogonal. Now we get

$$\begin{aligned} 0 &= -< g, Y_2 f_2 > \\ &= \left\langle \begin{pmatrix} 0 \\ g \end{pmatrix}, J \begin{pmatrix} Y_1 f_2 \\ Y_2 f_2 \end{pmatrix} \right\rangle \\ &= \left\langle M \begin{pmatrix} f_1 \\ f_2 \end{pmatrix}, JM \begin{pmatrix} X f_2 \\ f_2 \end{pmatrix} \right\rangle \quad \text{from (8) and (9)} \\ &= \left\langle \begin{pmatrix} f_1 \\ f_2 \end{pmatrix}, M^{\sim} JM \begin{pmatrix} X f_2 \\ f_2 \end{pmatrix} \right\rangle \\ &= \left\langle \begin{pmatrix} f_1 \\ f_2 \end{pmatrix}, J \begin{pmatrix} X f_2 \\ f_2 \end{pmatrix} \right\rangle \\ &= < f_1, X f_2 > - \| f_2 \|_2^2 . \end{aligned}$$

Thus

$$\begin{aligned} \| f_2 \|_2^2 &= < f_1, X f_2 > \\ &\leq \| f_1 \|_2 \| X f_2 \|_2 \\ &\leq \| f_1 \|_2 \| f_2 \|_2 , \end{aligned}$$

or

$$\| f_2 \|_2 \leq \| f_1 \|_2 . \tag{10}$$

But $\begin{pmatrix}0\\g\end{pmatrix}$ is strictly negative and M^{-1} is J-unitary. This implies that $\begin{pmatrix}f_1\\f_2\end{pmatrix}$ is strictly negative too, i.e.

$$\|f_1\|_2 < \|f_2\|_2 . \tag{11}$$

Inequalities (10) and (11) are contradictory.

It follows from the third claim and Lemma 1 that $Y_2^{-1} \in \mathbf{RH}_\infty$. Defining $Y := Y_1 Y_2^{-1}$, we have $Y \in \mathbf{RL}_\infty$ and

$$\begin{bmatrix}Y_1\\Y_2\end{bmatrix}\mathbf{H}_2 = \mathbf{G}_Y . \tag{12}$$

From the first claim $\mathbf{G}_Y$ is negative, so from Lemma 2 $\|Y\|_\infty \leq 1$. Finally, (8) and (12) imply that $\mathbf{G}_Y = M\mathbf{G}_X$. □

Now we have the solution to our problem as posed in terms of S.

Theorem 1. The set of all matrices S in $\mathbf{RL}_\infty$ such that $\|S\|_\infty \leq 1$ and $R - S \in \mathbf{RH}_\infty$ is given by the formulas

$$S = X_1 X_2^{-1}$$

$$\begin{bmatrix}X_1\\X_2\end{bmatrix} = L\begin{bmatrix}Y\\I\end{bmatrix}$$

$$Y \in \mathbf{RH}_\infty , \ \|Y\|_\infty \leq 1 .$$

Proof. First suppose

$$S \in \mathbf{RL}_\infty , \ \|S\|_\infty \leq 1 , \ R - S \in \mathbf{RH}_\infty .$$

By Lemma 2 and (5), $\mathbf{G}_S$ is negative and

$$\mathbf{G}_S \subset L(\mathbf{H}_2 \oplus \mathbf{H}_2) . \tag{13}$$

Define $M := L^{-1}$. From (13)

$$M\mathbf{G}_S \subset \mathbf{H}_2 \oplus \mathbf{H}_2 , \tag{14}$$

so (6) holds with S substituted for X. Also

$$
\begin{aligned}
L(\{0\}\oplus\mathbf{H}_2) &\subset L(\mathbf{H}_2\oplus\mathbf{H}_2)\\
&= G(\mathbf{H}_2\oplus\mathbf{H}_2) \text{ from (5)}\\
&= \begin{bmatrix} I & R\\ 0 & I\end{bmatrix}(\mathbf{H}_2\oplus\mathbf{H}_2)\\
&\subset \mathbf{L}_2\oplus\mathbf{H}_2 ,
\end{aligned}
$$

so (7) holds. Noting that M is J-unitary, invoke Lemma 3 to get the existence of Y in $\mathbf{RL}_\infty$ such that

$$\|Y\|_\infty \le 1, \ \ \mathbf{G}_Y = M\mathbf{G}_S .$$

Since $\mathbf{G}_Y \subset \mathbf{H}_2\oplus\mathbf{H}_2$ from (14), we have by Lemma 1 that actually $Y\in\mathbf{RH}_\infty$. Define

$$\begin{bmatrix} X_1\\ X_2\end{bmatrix} := L\begin{bmatrix} Y\\ I\end{bmatrix},$$

so that

$$
\begin{aligned}
\begin{bmatrix} X_1\\ X_2\end{bmatrix}\mathbf{H}_2 &= L\mathbf{G}_Y\\
&= \mathbf{G}_S\\
&= \begin{bmatrix} S\\ I\end{bmatrix}\mathbf{H}_2 . \qquad (15)
\end{aligned}
$$

Pre-multiply (15) by $[0 \ \ I]$ to get $X_2\mathbf{H}_2=\mathbf{H}_2$. Thus $X_2^{-1}\in\mathbf{RH}_\infty$ by Lemma 1. Now pre-multiply (15) by $[I \ \ -S]$ to get $S=X_1X_2^{-1}$.

Conversely, suppose $Y\in\mathbf{RH}_\infty$ and $\|Y\|_\infty\le 1$. Then

$$
\begin{aligned}
L\mathbf{G}_Y &\subset L(\mathbf{H}_2\oplus\mathbf{H}_2)\\
&= G(\mathbf{H}_2\oplus\mathbf{H}_2)\\
&\subset \mathbf{L}_2\oplus\mathbf{H}_2 ,
\end{aligned}
$$

and

$$
\begin{aligned}
\{0\}\oplus\mathbf{H}_2 &= G(\{0\}\oplus\mathbf{H}_2)\\
&\subset G(\mathbf{H}_2\oplus\mathbf{H}_2)
\end{aligned}
$$

$$= L(\mathbf{H}_2 \oplus \mathbf{H}_2)$$

$$\subset L(\mathbf{L}_2 \oplus \mathbf{H}_2).$$

Invoke Lemma 3 again: there exists S in $\mathbf{RL}_\infty$ such that $\|S\|_\infty \leq 1$ and $\mathbf{G}_S = L\mathbf{G}_Y$. Define

$$\begin{bmatrix} X_1 \\ X_2 \end{bmatrix} := L \begin{bmatrix} Y \\ I \end{bmatrix}.$$

Then

$$\begin{bmatrix} X_1 \\ X_2 \end{bmatrix} \mathbf{H}_2 = \mathbf{G}_S,$$

so $S = X_1 X_2^{-1}$ as before. □

The formulas in Theorem 1 yield S as a linear fractional transformation of Y. To see this, partition L as

$$L = \begin{bmatrix} L_1 & L_2 \\ L_3 & L_4 \end{bmatrix}.$$

Then

$$S = (L_1 Y + L_2)(L_3 Y + L_4)^{-1}.$$

One possible candidate for Y is $Y = 0$, in which case S is simply $L_2 L_4^{-1}$.

Let's summarize in the form of an algorithm. The input is a matrix R in $\mathbf{RL}_\infty$ such that $\|\Gamma_R\| < 1$, and the output is a matrix X in $\mathbf{RH}_\infty$ such that $\|R - X\|_\infty \leq 1$.

Procedure 6

Step 1. Find a minimal realization of the antistable part of R:

$$R(s) = [A,\ B,\ C,\ 0] + (\text{a matrix in } \mathbf{RH}_\infty).$$

Step 2. Solve the Lyapunov equations

$$AL_c + L_c A' = BB'$$

$$A'L_o + L_o A = C'C$$

and set $N := (I - L_o L_c)^{-1}$.

Step 3. Set

$$L_1(s) = [A, -L_c N C', C, I]$$

$$L_2(s) = [A, N'B, C, 0]$$

$$L_3(s) = [-A', NC', -B', 0]$$

$$L_4(s) = [-A', NL_o B, B', I].$$

Step 4. Select Y in $\mathbf{RH}_\infty$ with $\|Y\|_\infty \leq 1$ (for example $Y=0$) and set

$$X = R - (L_1 Y + L_2)(L_3 Y + L_4)^{-1}.$$

(The realizations in Step 3 are derived directly from equations (5.1), (5.2), and (3).)

7. Conclusion

How to use the above theory in control design is a continuing area of research. For example, the theory can be used to develop curves showing the trade-off between performance and robustness, e.g. (O'Young and Francis, 1986). Elementary design problems, with perhaps competing performance objectives, can often be approximated by standard $\mathbf{H}_\infty$- problems, e.g. (Helton, 1985). More difficult design problems cannot be so approximated and for these there is Doyle's μ-synthesis, in which the standard $\mathbf{H}_\infty$-problem is embedded.

REFERENCES

Adamjan, V.M., D.Z. Arov, and M.G. Krein (1971). "Analytic properties of Schmidt pairs for a Hankel operator and the generalized Schur-Takagi problem," *Math. USSR Sbornik*, vol. 15, pp. 31-73.

Ball, J.A. and J.W. Helton (1983). "A Beurling-Lax theorem for the Lie group U(m,n) which contains most classical interpolation theory," *J. Op. Theory*, vol. 9, pp. 107-142.

Ball, J.A. and A.C.M. Ran (1986). "Optimal Hankel norm model reductions and Wiener-Hopf factorizations I: the canonical case," *SIAM J. Control and Opt.* To appear.

Bart, H., I. Gohberg, M.A. Kaashoek, and P. van Dooren (1980). "Factorizations of transfer matrices," *SIAM J. Cont. Opt.*, vol. 18, pp. 675-696.

Dorato, P. (1987). *Robust Control,* IEEE Press. To appear.

Doyle, J.C. (1984). "Lecture Notes in Advances in Multivariable Control," *ONR/Honeywell Workshop*, Minneapolis, MN.

Feintuch, A. and B.A. Francis (1985). "Uniformly optimal control of linear feedback systems," *Automatica*, vol. 21, pp. 563-574.

Foias, C. and A. Tannenbaum (1986). "On the uniqueness of a minimal norm representative of an operator in the commutant of the compressed shift," Tech. Rept., Dept. Elect. Eng., McGill Univ., Montreal.

Francis, B.A. and J.C. Doyle (1987). "Linear control theory with an H_∞- optimality criterion," *SIAM J. Control Opt.* To appear.

Garnett, J.B. (1981). *Bounded Analytic Functions,* Academic Press, New York.

Glover, K. (1984). "All optimal Hankel-norm approximations of linear multivariable systems and their L_∞-error bounds," *Int. J. Cont.*, vol. 39, pp. 1115-1193.

Helton, J.W. (1985). "Worst case analysis in the frequency-domain: an H_∞- approach to control," *IEEE Trans. Auto. Cont.*, vol. AC-30, pp. 1154-1170.

Kwakernaak, H. (1985). "Minimax frequency domain performance and robustness optimization of linear feedback systems," *IEEE Trans. Auto. Cont.*, vol. AC-30, pp. 994-1004.

Limebeer, D.J.N. and Y.S. Hung (1986). "An analysis of the pole-zero cancellations in H_∞ optimal control problems of the first kind," Tech. Rept., Dept. Elect. Eng., Imperial College, London.

Minto, D. (1985). "Design of reliable control systems: theory and computations," Ph.D. Thesis, Dept. Elect. Eng., Univ. Waterloo, Waterloo.

Nehari, Z. (1957). "On bounded bilinear forms," *Ann. of Math.*, vol. 65, pp. 153-162.

O'Young, S. and B.A. Francis (1986). "Optimal performance and robust stabilization," *Automatica*, vol. 22, pp. 171-183.

Sarason, D. (1967). "Generalized interpolation in H_∞," *Trans. AMS*, vol. 127, pp. 179-203.

Silverman, L. and M. Bettayeb (1980). "Optimal approximation of linear systems," *Proc. JACC.*

Youla, D.C., H.A. Jabr, and J.J. Bongiorno Jr. (1976). "Modern Wiener-Hopf design of optimal controllers: part II," *IEEE Trans. Auto. Cont.*, vol. AC-21, pp. 319-338.

Young, N. (1986). "An algorithm for the super-optimal sensitivity-minimising controller," *Proc. Workshop on New Perspectives in Industrial Control System Design using H_∞- Methods*, Oxford.

Zames, G. (1981). "Feedback and optimal sensitivity: model reference transformations, multiplicative seminorms, and approximate inverses," *IEEE Trans. Auto. Cont.*, vol. AC-23, pp. 301-320.

NONLINEAR INTERPOLATION THEORY IN H^{∞}

Joseph A. Ball
Department of Mathematics
Virginia Polytechnic Institute
Blacksburg, Virginia 24061-4097

Ciprian Foias
Department of Mathematics
Indiana University
Bloomington, Indiana 47405

J. William Helton
Department of Mathematics
University of California at San Diego
La Jolla, California 92093

Allen Tannenbaum
Department of Electrical Engineering
University of Minnesota
Minneapolis, Minnesota 55455

and

Department of Mathematics
Ben-Gurion University of the Negev
Beer Sheva, Israel 84105

Abstract

Recently there has been an interesting cross-fertilisation between the areas of mathematical interpolation theory, operator theory and control theory which has led to new results in all areas. In particular the Sarason-Sz. Nagy Foias commutant lifting theorem provides a very broad framework for generalized Nevalinna-Pick interpolation which in turn gives a solution to the H^{∞}-optimal weighted sensitivity problem in control theory. In this paper we give a generalization of this "commutant lifting theorem" for certain classes of nonlinear analytic operators and discuss some possible applications.

NATO ASI Series, Vol. F34
Modelling, Robustness and Sensitivity Reduction in Control Systems. Edited by R. F. Curtain

1. Introduction

Recently there has been a surge of renewed interest in complex interpolation theory. From the strictly mathematical point of view, it has been found that operator theory can provide far-reaching and powerful generalizations of classical Nevanlinna-Pick interpolation (see e.g. the work of Adamjan-Arov-Krein [1], [2], Ball-Helton [4], Sarason [20], and Sz Nagy-Foias [21], [22]).

However a strong motivating factor in this new research in (generalized) interpolation theory is its widespread and manifold applications, especially in engineering. Indeed these methods have appeared in control and systems theory to solve the H^∞-optimal weighted sensitivity problem [25], [26], [9], in the area of model reduction [5], [11], and for robust design and stabilization [13], [15], [23]. (See [10] for a much more extensive list of references on the uses of interpolation theory in control.) In circuit theory, there have been applications of Nevanlinna-Pick interpolation to the broadband matching problem [24], [12], and these methods have been even used in studying certain problems in geophysics [8] and biomathematics [7].

Now Sarason [20] (in a special case) and Sz Nagy-Foias [21] (in complete generality) proved an extremely broad interpolation theorem which has come to be known as the "commutant lifting theorem". More precisely, let $T: H \to H$, and $T': H' \to H'$ be contractions on the Hilbert spaces H, H'. Let $U: K \to K$, $U': K' \to K'$ be the minimal isometric dilations of T and T', respectively. (This means that U is such that $P_H U = T P_H$ where $P_H: K \to H$ is projection, $K = \bigvee_{n=0}^{\infty} U^n H$, and similarly for U'.) Then the commutant lifting theorem [21] asserts that there exists a contraction $\hat{A}: K \to K'$ such that $U'\hat{A} = \hat{A}U$, and $P_H'\hat{A} = AP_H$. Moreover $\hat{A}$ may be chosen to be such that $\|\hat{A}\| = \|A\|$. (For the relation of this result with Nevanlinna-Pick interpolation, see [8], [20].)

In this paper we shall give a local nonlinear analogue of the commutant lifting theorem. That is, we will consider nonlinear analytic maps defined in a ball around the origin in a Hilbert space which commute with certain contractions, and show that these admit nonlinear intertwing dilations as in the linear case (see Theorem (3.8) below). Unlike the linear case however, we cannot guarantee that the norm of the dilation equals the norm of the original operator, but we can give upper bounds (see (3.8), (4.5)).

Now Sarason in [20] proves a version of the (linear) commutant lifting theorem in the special case in which $H = H' = H^2 \ominus mH^2$ ($m \in H^\infty$ a nonconstant inner function), $K = K' = H^2$, $U = U'$ = the unilateral right shift on H^2, and $T = T'$ = the compressed shift. When we specialize our nonlinear commutant lifting theorem to this special case, we will get a nonlinear analogue of the Hankel operator, and just as in [16] and [20] for the linear Hankel, we will show that our Hankel operator admits an L^∞-symbol (in a certain precise sense, see (4.5)).

Finally, we will discuss a possible control-theoretic application of our results in which we use our nonlinear commutant lifting theorem to study a certain nonlinear weighted optimization problem of the kind that appears in robust control. Hopefully, the results of this paper will prove useful for a number of engineering applications similar to those that we mentioned above.

For complete details of the results discussed in this paper see [27].

2. Preliminaries

In this section, we would like to collect a few basic and elementary facts about analytic mappings on Hilbert spaces. We are basically following the treatment of the book Hille-Phillips [14] to which the reader may refer for all of the details.

Let G and H denote complex Hilbert spaces. Set

$$B_{r_0}(G) := \{g \in G: \|g\| < r_0\}$$

(the open ball of radius r_0 in G about the origin). Then we say that a mapping

$$\phi: B_{r_0}(G) \to H$$

is holomorphic if the complex function $(z_1,\ldots,z_n) \to \langle \phi(z_1g_1 + \ldots + z_ng_n), h \rangle$ is holomorphic in a neighborhood of $(1,1,\ldots,1) \in \mathbf{C}^n$ as a function of the complex variables $z_1,\ldots,z_n$ for all $g_1,\ldots,g_n \in G$ such that $\|g_1 + \ldots + g_n\| < r_0$, and for all $h \in H$. (Note that we denote the Hilbert space norms in G and H by $\| \ \|$ and the innter products by $\langle , \rangle$.)

We will now assume that $\varphi(0) = 0$. It is then easy to see that if $\varphi: B_r(G) \to H$ is holomorphic, then φ admits a convergent Taylor series expansion, i.e.

$$\varphi(g) = \varphi_1(g) + \varphi_2(g,g) + \dots + \varphi_n(g,\dots,g) + \dots$$

where $\varphi_n : G \times \dots \times G \to H$ is an n-linear map. Clearly, without loss of generality we may assume that the n-linear map $(g_1,\dots,g_n) \to \varphi_n(g_1,\dots,g_n)$ is symmetric in the arguments $g_1,\dots,g_n$.

Now set

$$\tilde{\varphi}_n(g_1 \otimes \dots \otimes g_n) := \varphi_n(g_1,\dots,g_n).$$

Then $\tilde{\varphi}_n$ extends in a unique manner to a dense subset of $G^{\otimes n} := G \otimes \dots \otimes G$ (tensor product taken n times). Notice by $G^{\otimes n}$ we mean the Hilbert space completion of the algebraic tensor product of the G's. Clearly if $\tilde{\varphi}_n$ has finite norm on this dense subset, then $\tilde{\varphi}_n$ extends by continuity to a bounded linear operator $\tilde{\varphi}_n: G^{\otimes n} \to H$. By abuse of notation we will set $\varphi_n := \tilde{\varphi}_n$.

Finally, we will conclude this section with a well-known result on the radius of convergence of a Taylor series. Indeed, recall that by a <u>majorizing sequence</u> for the holomorphic map φ, we mean a positive sequence of numbers $\{\alpha_n\}$ n=1,2,... such that $\|\varphi_n\| \le \alpha_n$ for $n \ge 1$. Suppose that $\rho := \lim\sup_{n\to\infty}(\alpha_n)^{1/n} < \infty$. Then it is completely standard ([14]) that the Taylor series expansion of φ converges at least on the ball $B_r(G)$ of radius $r = 1/\rho$.

3. Nonlinear Commutant Lifting Theorem

In this section we formulate and prove a local nonlinear analogue of the commutant lifting theorem as discussed in the Introduction. We use the notation of Sections 1 and 2 here. In particular, $\varphi: B_{r_0}(G) \to H$ denotes a holomorphic map with $\varphi(0) = 0$, $T: H \to H$ denotes a (linear) contraction with minimal isometric dilation $U: K \to K$, and $P_H: K \to H$ will be orthogonal projection. Finally $S: G \to G$ will denote an isometry on G.

We are now ready to state our first result:

<u>Propositon</u> (3.1) Notation as above. Suppose $\varphi: B_{r_0}(G) \to H$ intertwines $T: H \to H$ and $S: G \to G$, i.e. $T\varphi(g) = \varphi(Sg)$ for $g \in B_{r_0}(G)$. Suppose moreover

that $\{\alpha_n\}$ n=1,2,... is a majorizing sequence for φ, and that $\rho := \lim_{n\to\infty} \sup(\alpha_n)^{1/n} < \infty$. Set $r = 1/\rho$. (Without loss of generality, we may assume that $r < r_0$.) Then there exists a holomorphic Ψ: $B_r(G) \to K$ such that $\Psi(0) = 0$, $U\Psi(g) = \Psi(Sg)$ and $\varphi(g) = P_H\Psi(g)$ for $g \in B_r(G)$, and such that $\{\alpha_n\}$ n=1,2,... is a majorizing sequence for Ψ.

Proof. Consider the Taylor expansion

$$\varphi(g) = \varphi_1(g) + \varphi_2(g,g) + \ldots + \varphi_n(g,\ldots,g) + \ldots .$$

Recall that we are setting (by abuse of notation)

$$\varphi_n(g_1 \otimes \ldots \otimes g_n) = \varphi_n(g_1, \ldots, g_n).$$

Then since in the expansion of $\varphi(z_1g_1 + .. + z_ng_n)$ with $\|g_1\| = \ldots = \|g_n\| \le \frac{1}{n} r_0$, the coefficient of $z_1z_2\ldots z_n$ is $n!\varphi_n(g_1 \otimes \ldots \otimes g_n)$, we can recover $\varphi_n(g_1 \otimes \ldots \otimes g_n)$ from φ via the Cauchy formula

$$(1)\quad \varphi_n(g_1 \otimes \ldots \otimes g_n) = \frac{1}{n!}\left(\frac{1}{2\pi}\right)^n \int_0^{2\pi} \ldots \int_0^{2\pi} \varphi(e^{i\vartheta_1}g_1 + \ldots + e^{i\vartheta_n}g_n) \cdot e^{-i(\vartheta_1 + \ldots + \vartheta_n)}\, d\vartheta_1 \ldots d\vartheta_n.$$

Now the intertwining relation

$$T\varphi = \varphi \circ S$$

combined with (1) leads immediately to

$$T\varphi_n(g_1 \otimes \ldots \otimes g_n) = \varphi_n(Sg_1 \times \ldots \times Sg_n)$$

and hence by linearity we get

$$(2)\qquad T\varphi_n = \varphi_n S^{\otimes n}$$

on all of $G^{\otimes n}$. (By definition, $S^{\otimes n}$: $G^{\otimes n} \to G^{\otimes n}$ is the unique isometry with the property that $S^{\otimes n}(g_1 \otimes \ldots \otimes g_n) = Sg_1 \otimes \ldots \otimes Sg_n$.)

From (2), using the (linear) commutant lifting theorem [21], we see that there exists a linear map

$$\Psi_n:\ G^{\otimes n} \to K$$

such that $U\Psi_n = \Psi_n S^{\otimes n}$, $\|\Psi_n\| = \|\varphi_n\| \le \alpha_n$, and $P_H\Psi_n = \varphi_n$ for every $n \ge 1$. Defining Ψ: $B_r(G) \to K$ by

$$\Psi(g) := \sum_{n=1}^{\infty} \Psi_n(g \times \ldots \times g),$$

it is trivial to check that Ψ has the required properties.

□

Remarks (3.2) (i) The hypothesis in (3.1) that U is a minimal isometric dilation of T is unnecessary. It suffices to assume that U is any isometric dilation of T since the usual (linear) commutant lifting theorem is true in this case as well.

(ii) The problem with (3.1) is in finding a majorizing sequence for a given holomorphic φ: $B_{r_0}(G) \to H$, $\varphi(0) = 0$. We will need to make some additional assumptions on φ in order to get better bounds on the dilation Ψ. Since we are interested in engineering applications, we would like our assumptions to fit a "natural" class of physical systems. Fortunately, for such a class of nonlinear operators, we can get a nice majorizing sequence and non-trivial bounds on Ψ. This leads us to the following definition:

Definition (3.3) φ: $B_{r_0}(G) \to H$, $\varphi(0) = 0$ has fading memory if its nonlinear part $\varphi - \varphi'(0)$ admits a factorization

$$\varphi - \varphi'(0) = \hat{\varphi} \circ W$$

where $\hat{\varphi}$ is a holomorphic mapping defined in some neighborhood of $0 \in G$, and W is a linear Hilbert-Schmidt operator.

Remark (3.4) System-theoretically the property of "fading memory" implies that two given input signals which are close in the recent past but not necessarily close in the remote past will yield present outputs which are close. For more details about this important class of operators see [6]. For operators with fading memory, it is completely elementary to construct a majorizing sequence (for the proof see [27]):

Lemma (3.5) Let φ: $B_{r_0}(G) \to H$, $\varphi(0) = 0$, have fading memory. Suppose moreover that if we write

$$\varphi - \varphi'(0) = \hat{\varphi} \circ W$$

as in (3.3), then $\hat{\varphi}$: $B_{r_1}(G) \to B_{r_2}(H)$. Then the sequence

$$\alpha_1 := \|\varphi'(0)\|$$

$$\alpha_n := r_2 \frac{e^n}{r_1{}^n} \|W\|_2^n \quad \text{for } n \geq 2$$

($\|W\|_2$ = Hilbert-Schmidt norm of W) is a majorizing sequence for φ.

We can now prove a version of (3.1) for operators with fading memory:

Proposition (3.6) Notation and hypotheses as in (3.1). Suppose moreover that φ has fading memory, and that if $\hat{\varphi}\circ W$ is the nonlinear part of φ (as in (3.5)), then $\hat{\varphi}: B_{r_1}(G) \to B_{r_2}(H)$. Then there exists a holomorphic $\Psi: B_r(G) \to K$ where

$$r = \frac{r_1}{e\|W\|_2} \leq r_0,$$

with all the properties of the dilation Ψ given in (3.1), and moreover such that $\Psi : B_u(G) \to B_v(K)$ for $u < r$, where

$$v := \|\varphi'(0)\|u + \frac{r_2 e^2 \|W\|_2^2 u^2}{r_1^2 - r_1 e\|W\|_2 u}.$$

Proof. From (3.1) Φ has a dilation Ψ such that a majorizing sequence for Φ is a majorizing sequence for Ψ. Since φ has fading memory, it has the majorizing sequence $\{\alpha_n\}$ n=1,2,... given in (3.5), and clearly

$$\rho := \limsup_{n\to\infty}(\alpha_n)^{1/n} = \frac{e}{r_1}\|W\|_2$$

so that the radius of convergence of Ψ is at least $r = r_1/(e\|W\|_2)$ $(\leq r_0)$.

Finally if $\|g\| \leq u < r$, then by (3.1) and (3.2) we have

$$\|\Psi(g)\| \leq \|\Psi_1(g)\| + \sum_{n=2}^{\infty} \|\Psi_n(g\otimes\ldots\otimes g)\| \leq \|\varphi'(0)\|u + \sum_{n=2}^{\infty} r_2 \frac{e^n}{r_1^n}\|W\|_2^n u^n$$

$$= \|\varphi'(0)\|u + \frac{r_2 e^2\|W\|_2^2 u^2}{r_1^2 - r_1 e\|W\|_2 u}.$$

(Note that by definition $\varphi'(0)g = \varphi_1(g)$ where φ_1 is the first term of the Taylor series expansion of φ.)

□

We can now easily formulate and prove the following local nonlinear generalization of the commutant lifting theorem:

Theorem (3.7) (i) Let $T: H \to H$, $T': H' \to H'$ be (linear) contractions with minimal isometric dilations $U: K \to K$, $U': K' \to K'$, respectively. Denote the corresponding projections by $P_H: K \to H$, $P_H': K' \to H'$. Let $\varphi: B_{r_0}(H) \to H'$ be holomorphic, $\Phi(0) = 0$, and suppose that $\{\alpha_n\}$ n=1,2,... is a majorizing sequence. Finally assume that $T'\varphi = \varphi\circ T$ on $B_{r_0}(H)$. Then there exists a holomorphic $\Psi : B_r(K) \to K'$, $\Psi(0) = 0$ where $r = 1/\rho$ and

$$\rho = \limsup_{n\to\infty} (\alpha_n)^{1/n}$$

such that

(a) $U'\Psi = \Psi\circ U$ on $B_r(K)$,

(b) $P_H{}'\Psi = \varphi\circ P_H$ on $B_r(K)$,

(c) $\{\alpha_n\}$ $n=1,2,\ldots$ is a majorizing sequence for ψ.

(ii) Hypotheses and notation as in (i). Suppose moreover that φ has fading memory, and that if

$$\varphi - \varphi'(0) = \hat{\varphi}\circ W$$

where W is Hilbert-Schmidt, then $\hat{\varphi}: B_{r_1}(H) \to B_{r_2}(H')$. Then φ has a dilation $\Psi: B_r(K) \to K'$ ($\Psi(0) = 0$), where

$$r = \frac{r_1}{e\|W\|_2} \quad (\leq r_0),$$

with the properties (a), (b), (c) given in (i) above, and such that for $u < r$, $\Psi: B_u(K) \to B_v(K')$ where

$$v = \|\varphi'(0)\|u + \frac{r_2 e^2 \|W\|_2^2 u^2}{r_1^2 - r_1 e\|W\|_2 u}.$$

Proof. We prove (i) and (ii) together. Indeed just apply (3.1) and (3.6) with $G = K$, $S = U$, and

$$\varphi_{new}(k) := \varphi(P_H k)$$

for $k \in B_{r_0}(K)$.

□

4. Nonlinear Generalized Interpolation in H^∞

For many applications in control engineering a certain special case due to Sarason [20] of the commutant lifting theorem (see also Section 1) has been found to be very useful. Since his theorem generalized certain results in classical interpolation theory with bounded analytic functions, Sarason referred to his technique as "generalized interpolation in H^∞".

Therefore in this sense the results of this section may be considered to be as a form of "nonlinear generalized interpolation". We will also define a "nonlinear Hankel operator" here giving us an opportunity to make contact with some results from [1].

We begin with the following proposition:

Proposition (4.1) For G and K complex Hilbert spaces let $S: G \to G$ be a unilateral shift, and $U: K \to K$ be an isometry. Let $\Psi: B_r(G) \to K$ denote a holomorphic map such that $\Psi(0) = 0$ and $U\Psi(g) = \Psi(Sg)$ for $g \in B_r(G)$. Then there exists a holomorphic mapping

$$\mathcal{M}: B_r(G) \to L((I-SS^*)G,K)$$

such that

$$\Psi(g) = \sum_{n=0}^{\infty} U^n \mathcal{M}(S^{*n}g)(I-SS^*)S^{*n}g \tag{4}$$

for $g \in B_r(G)$. The series given in (4) is boundedly weakly convergent in K for all $g \in B_r(G)$. Conversely, let $\Psi: B_r(G) \to K$ be defined as in (4) for some $\mathcal{M}$ as above (such that the series boundedly weakly converges in K for each $g \in B_r(G)$) Then Ψ is holomorphic in $B_r(G)$, $\Psi(0) = 0$, and $U\Psi(g) = \Psi(Sg)$ for $g \in B_r(G)$.

Proof. Let $\Psi: B_r(G) \to K$ be holomorphic, $\Psi(0) = 0$, and $U\Psi(g) = \Psi(Sg)$ for $g \in B_r(G)$. Define a mapping $T: B_r(G) \to K$ by

$$\Gamma(g) := \Psi(g) - U\Psi(S^*g). \tag{5}$$

Then we set

$$\mathcal{M}(g) := \int_0^1 \Gamma'((1-t)SS^*g + tg)dt$$

$$= \int_0^1 \Gamma'(SS^*g + t(1-SS^*)g)dt \; .$$

It is easy to check that $\mathcal{M}: B_r(G) \to L((I-SS^*)G,K)$ and is holomorphic (see Hille-Phillips [14]).

Moreover (see Rudin [19]), we have that

$$\Gamma(g) := \mathcal{M}(g)(I-SS^*)g. \tag{6}$$

Now substituting (6) into (5), we get

$$\Psi(g) = \mathcal{M}(g)(I-SS^*)g + U\Psi(S^*g). \tag{7}$$

Then iterating this procedure, we see that

$$(8) \qquad \Psi(g) = \sum_{n=0}^{N} U^n \Omega(S^*g)(I-SS^*)S^{*n}g + U^{N+1}\Psi(S^{*N+1}g).$$

But since U is an isometry and S is a shift,

$$\|U^{N+1}\Psi(S^{*N+1})g\| = \|\Psi(S^{*N+1}g)\| \to \|\Psi(0)\| = 0.$$

This proves that Ψ admits the representation (4).

Conversely if Ψ admits the representation (4) it is immediate that $\Psi(0) = 0$, and using the facts that $S^*S = I$ and $(I-S^*S)S = 0$, we get

$$\Psi(Sg) = \sum_{n=0}^{\infty} U^n\Omega(S^{*n}Sg)(1-SS^*)S^{*n}Sg = \sum_{n=1}^{\infty} U^n\Omega(S^{*n-1}g)(1-SS^*)S^{*n-1}g = U\Psi(g).$$

□

Remarks (4.2) (i) If Ψ is linear, say $\Psi(g) = \theta g$ where $\theta \in L(G,K)$, then $\Omega(g) \equiv \theta|(1-SS^*)G$. Conversely, if $\Omega(g) \equiv \Omega(0)$ is independent of $g \in B_r(G)$, then Ψ is linear.

(ii) If the shift S has multiplicity one, then Ω is uniquely determined by Ψ. Indeed suppose Ψ admits the representation (4) via the mappings Ω_1 and Ω_2. Set $\Omega = \Omega_1 - \Omega_2$. Then via (4), Ω will represent the 0 mapping. That means $\Omega(g)(1-SS^*)g = 0$ for $g \in B_r(G)$, and thus $\Omega(g) \equiv 0$ since the rank of $1-SS^*$ is one.

Proposition (4.1) leads us to the following definition:

Definition (4.3) Let Ψ and Ω be as in (4.1). Then we call Ω a symbol of Ψ. As we have seen if S has multiplicity 1, then the symbol of Ψ is unique.

Remarks (4.4) Our terminology is completely consistent (and based on) the classical Hankel operator terminology ([17]). Indeed, for $m \in H^\infty$ a nonconstant inner function, set $H(m) := H^2 \ominus mH^2$, and let $S: H^2 \to H^2$ denote the unilateral right shift, $S(m): H(m) \to H(m)$ the compressed shift, and $P: H^2 \to H(m)$ the projection. Then a (linear) Hankel operator $A: H^2 \to H(m)$ is bounded operator such that $S(m)A = AS$. (Via the isometry $H^2 \ominus mH^2 \to \bar{m} H^2 \ominus H^2 (\subset (H^2)^\perp)$, it is easy to see that our definition of "Hankel operator" is equivalent to the usual definition as given e.g. in [17].)

Now a basic result (see [16], [20]) is that the Hankel A admits a symbol, i.e. there exists $w \in H^\infty$ such that

$$PM_w = A$$

where $M_w: H^2 \to H^2$ denotes multiplication by w. Moreover the symbol w may be chosen such that $\|w\|_\infty = \|A\|$ ([20]). We should note that since $mH^2 \subset \ker A$, we could have equivalently defined A as a bounded linear operator on $H(m)$ such that $S(m)A = AS(m)$. This is the set-up of [20] and [22].

Now if in Theorem (3.8), we choose $H = H' = H(m)$, $U = U' = S$ on $K = K' = H^2$, then φ represents a local nonlinear analogue of the Hankel, and Ψ: $B_r(H^2) \to H^2$ its dilation. By Proposition (4.1) Ψ admits a symbol Ω (we are taking $G = H^2$), and hence as in the linear case we may define ω to be a <u>symbol</u> of the nonlinear Hankel φ. In order to complete our analogy, we will need to say in what sense ω is in H^∞, and to discuss its norm.

In order to do this, let's specialize (4.1) to the case at hand. Let $\omega : B_r(H^2) \to H^2$ be such that

(9) $$\Psi(h)(z) := \sum_{n=0}^{\infty} z^n h_n \omega(z, S^n h)$$

is boundedly weakly convergent in H^2 for all $h(z) \in B_r(H^2)$,

(10) $$h(z) = h_0 + zh_1 + \dots + z^n h_n + \dots$$

Now if

(11) $$\sum_{n=0}^{\infty} h_n S(m)^n \, P[\omega(z, S^{*n}h) - \omega(z, S^{*n}Ph)] = 0$$

for all $h \in B_r(H^2)$ and $z \in \partial D$ (= unit circle), then

(12) $$\varphi(h) := P\Psi(h), \quad h \in B_r(H(m))$$

defines a holomorphic map φ: $B_r(H(m)) \to H(m)$ such that $\varphi(0) = 0$ and $\varphi \circ S(m) = S(m)\varphi$ on $B_r(H(m))$. Indeed by (11) we have that

(13) $$P\Psi(h) = \varphi(Ph), \quad h \in B_r(H^2),$$

and by (4.1) we see that

$$\begin{aligned} S(m)\varphi(Ph) &= S(m)P\Psi(h) \\ &= PS\Psi(h) \\ &= P\Psi(Sh) \\ &= \varphi(PSh) \\ &= \varphi(S(m)Ph) \text{ for } h \in B_r(H^2) . \end{aligned}$$

The converse of this is the following nonlinear version of Sarason's theorem [20]:

<u>Theorem</u> (4.5) Let φ: $B_{r_0}(H(m)) \to H(m)$ be holomorphic, $\varphi(0) = 0$, $S(m)\varphi = \varphi \circ S(m)$. Then there exists a dilation of φ, Ψ: $B_r(H^2) \to H^2$ with the

properties described in (3.8) which admits the functional representation (9) and which satisfies (11). Moreover

(14) $\mathcal{M}(\cdot;\, 0) \in H^{\infty}$, and $\|\mathcal{M}(\cdot;\, 0)\|_{\infty} = \|\Psi'(0)\|$.

Proof. Everything follows from (3.7) and (4.1) except (14). So for (14) we note that

$$\mathcal{M}(\cdot;\, 0) = \frac{\partial \Psi}{\partial h}\Big|_{h=0} ,$$

and since $S\Psi = \Psi S$ we have that $\Psi'(0)S = S\Psi'(0)$ and $\Psi'(0)1 = \mathcal{M}(\cdot,\, 0)$, i.e. $\Psi'(0)$ is the multiplication operator in H^2 given by $h \to \mathcal{M}(\cdot,\, 0)h$. Thus $\|\mathcal{M}(\cdot,\, 0)\|_{\infty} = \|\Psi'(0)\|$, concluding the proof of the theorem.

5. Control-Theoretic Application

As we have mentioned, a major reason for much of the recent research in generalized interpolation theory is because of its applicability to certain problems in engineering. We would like to conclude this paper with a possible application of the nonlinear commutant lifting theorem (and in particular the nonlinear Sarason theorem) to a certain question in robust nonlinear control. Specifically, we will consider a nonlinear version of the problem of weighted sensitivity H^{∞}-minimization as discussed in [25], [26].

Suppose that we are given a fixed discrete-time, linear invariant, single input/single output system which via the discrete Fourier transform is modelled by a complex function $m(z)$. For simplicity we will assume that $m(z)$ is stable and all-pass. This means that $m(z) \in H^{\infty}$ and is an inner function. In control, $m(z)$ is called the plant and should be considered as the transfer function of a given fixed physical system.

We now make the following definition:

Definition (5.1) Let $S: H^2 \to H^2$ denote the unilateral right shift. Then we say that an input/output operator Ψ is weakly causal if there exists an $r > 0$ such that $\Psi: B_r(H^2) \to H^2$ is holomorphic, $\Psi(0) = 0$, and $S\Psi = \Psi \circ S$ on $B_r(H^2)$. We will set

C_w := {space of weakly causal operators}.

Next referring to Figure 1, we want to construct an input/output operator C (called a feedback compensator) such that the operators $(I+mC)^{-1}$ and $C(I+mC)^{-1}$ are weakly causal (and of course well-defined). We will say that such a C weakly stabilizes the closed loop given in Figure 1. Moreover we want to find such a weakly stabilizing C which minimizes the effect of the disturbances (denoted by d in Figure 1) on the output y. This is a typical kind of disturbance attenuation problem in which we are requiring a weak form of stability for the closed loop.

In order to formulate a precise mathematical problem now, we will first have to define a disturbance model, and moreover we will have to give a more explicit form for our feedback compensators.

Following the ideas in [25], we take the disturbances to be in the following set:

$$M := \{d : d = Wv,\ v \in B_r(H^2)\}$$

where $W: B_r(H^2) \to H^2$ is weakly causal, and admits a weakly causal inverse (locally).

As for the feedback compensator C, we will always assume that C has the form

$$C = q\circ(I-mq)^{-1} \qquad q \in \mathbf{C}_w. \tag{15}$$

It should be noted that in point of fact, if we had more rigorously defined the notion of feedback compensator as in [3], then assumption (15) actually becomes a proposition (i.e. one can prove that C is a weakly stabilizing compensator if and only if it admits the form (15)). See [3] for details.

We are now at long last ready to formulate an explicit control (and mathematical) disturbance attenuation problem. Indeed once more referring to Figure 1 and using (15), the transfer operator from v to y, $S_q v = y$, is

$$(I+mC)^{-1}\circ W = (I-mq)\circ W =: S_q.$$

S_q is called the weighted sensitivity operator. Note that by our invertibility assumptions on W, we can "absorb" the W into the q in $S_q = (1-mq)\circ W = W - m(q\circ W)$, and define an equivalent operator

$$\hat{S}_q := W - mq \qquad q \in \mathbf{C}_w.$$

Now for $u < r$, define $v_q > 0$ to be the smallest number such that

$$\hat{S}_q : B_u(H^2) \to B_{v_q}(H^2).$$

Then what we are interested in computing (or more realistically non-trivially bounding) is

$$v := \inf_q v_q.$$

Clearly for input signals whose energy is bounded by u, v is the minimal weighted sensitivity over the given set of feedback compensators. This is a nonlinear version of the type of disturbance attenuation problem considered in [25].

But mathematically this is precisely the kind of problem we have been considering. Indeed let $P: H^2 \to H^2 \ominus mH^2 =: H(m)$ denote orthogonal projection, and set

$$\Phi := PW|H(m).$$

Then clearly a dilation of Φ (in the sense of (4.5)) is of the form $\Psi = W - mq$, $q \in C_W$. But via (3.8) and (4.5) we can now bound v. Explicitly if we make the very natural assumption that Φ is of fading memory we can get the bounds derived above on v.

We should also note that we could have posed the above problem as a nonlinear version of the Nehari problem in which we want to compute the distance of a operator to the space of weakly causal operators. See [10] for this kind of approach to sensitivity minimization in the linear case.

In short, we believe that problems in control engineering such as sensitivity minimization should be quite amenable to the techniques discussed in this paper and should make an interesting and important area for future research.

References

[1] V. M. Adamjan, D. Z. Arov, and M. G. Krein, "Analytic properties of Schmidt pairs for a Hankel operator and the generalized Schur-Takagi problem", Math. USSR Sbornik, 15 (1971), pp. 31-73.
[2] V. M. Adamjan, D. Z. Arov, and M. G. Krein, "Infinite block Hankel matrices and related extension problems", AMS Translations 111 (1978), pp. 133-156.
[3] V. Anantharam and C. A. Desoer, "On the stabilization of nonlinear systems", IEEE Trans. Auto. Cont. AC-29 (1984), pp. 569-573.
[4] J. A. Ball and J. W. Helton, "A Beurling-Lax theorem for the Lie group U(m,n) which contains most classical interpolation theory", J. Operator Theory 9 (1983), pp. 107-142.

[5] J. A. Ball and A. C. M. Ran, "Optimal Hankel norm model reductions and Wiener-Hopf factorizations I: the canonical case", SIAM J. Control and Opt., to appear.

[6] S. Boyd and L. Chua, "Fading memory and the problem of approximating nonlinear operators with Volterra series", IEEE Trans. Circuits and Systems CAS-32 (1985), pp. 1150-1161.

[7] J. W. Evans and J. W. Helton, "Applications of Pick-Nevanlinna theory to retention-solubility studies in the lungs", Math. Biosci. 63 (1983), pp. 215-240.

[8] C. Foias, "Contractive intertwining dilations and waves in layered media", Proceedings of International Congress of Mathematicians, Helsinki (1978), vol. 2, pp. 605-613.

[9] C. Foias and A. Tannenbaum, "On the Nehari problem for a certain class of L^∞-functions appearing in control theory", J. Functional Analysis, to appear.

[10] B. A. Francis, A Course in H^∞ Control Theory, Lecture Notes in Control and Inf. Sci., Springer-Verlag, to appear, 1986.

[11] K. Glover, "All optimal Hankel-norm approximations of linear multivariable systems and their L^∞-error bounds", Int. J. Control 39 (1984), pp. 1115-1193.

[12] J. W. Helton, "Broadbanding: gain equalization directly from data", IEEE Trans. on Circuits and Systems CAS-28 (1981), pp. 1125-1137.

[13] J. W. Helton, "Worst case analysis in the frequency domain: an H^∞-approach to control", IEEE Trans. Auto. Control AC-30 (1985), pp. 1154-1170.

[14] E. Hille and R. S. Phillips, Functional Analysis and Semigroups, AMS Colloquium Publications, vol. XXXI, Providence, Rhode Island, 1957.

[15] H. Kimura, "Robust stabilization for a class of transfer functions", IEEE Trans. Auto. Cont. AC-29 (1984), pp. 788-793.

[16] Z. Nehari, "On bounded bilinear forms", Annals of Mathematics 65 (1957), pp. 153-162.

[17] S. C. Power, Hankel Operators on Hilbert Space, Pitman Advanced Publishing Program, Boston, 1982.

[18] M. Rosenblum and J. Rovnyak, Hardy Classes and Operator Theory, Oxford University Press, New York, 1985.

[19] W. Rudin, Functional Analysis, McGraw-Hill, New York, 1973.

[20] D. Sarason, "Generalized interpolation in H^∞ ", Trans. AMS 127 (1967), pp. 179-203.

[21] B. Sz.-Nagy and C. Foias, "Dilation des commutants d'opérateurs", C.R. Acad. Sci. Paris, série A, 265 (1968), pp. 493-495.

[22] B. Sz.-Nagy and C. Foias, Harmonic Analysis of Operators on Hilbert Space, North-Holland Publishing Company, Amsterdam, 1970.

[23] A. Tannenbaum, Invariance and System Theory: Algebraic and Geometric Aspects, Lecture Notes in Mathematics, vol. 845, Springer-Verlag, New York, 1981.

[24] D. C. Youla and M. Saito, "Interpolation with positive real functions", J. of Franklin Institute 284 (1967), pp. 77-108.

[25] G. Zames, "Feedback and optimal sensitivity: model reference transformations, multiplicative seminorms, and approximate inverses", IEEE Trans. Auto. Cont. AC-26 (1981), pp. 301-320.

[26] G. Zames and B. A. Francis, "Feedback, minimax sensitivity, and optimal robustness", IEEE Trans. Auto. Cont. AC-28 (1983), pp. 585-601.

[27] J. A. Ball, C. Foias, J. W. Helton, and A. Tannenbaum, "On a local nonlinear commutant lifting theorem", Technical Report, Department of Electrical Engineering, University of Minnesota, October 1986, submitted for publication.

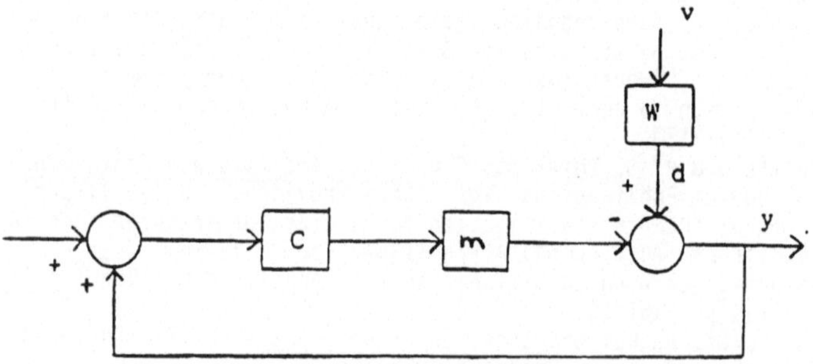

Figure 1

Super-optimal Hankel norm approximations

N.J. Young
Department of Mathematics
University Gardens
University of Glasgow
GLASGOW G12 8QW
Scotland

For any $m \times n$ rational transfer function matrix F and any non-negative integer k there exists a unique rational function $\hat{F}$ having at most k unstable poles and minimising $s^{\infty}(F - \hat{F})$: that is, minimising the sequence $(s_0^{\infty}(F - \hat{F}), s_1^{\infty}(F - \hat{F}), \ldots)$ with respect to the lexicographic ordering, where

$$s_j^{\infty}(G) := \sup_{|z|=1} s_j(G(z)),$$

and $s_j(.)$ denotes the jth singular value of a matrix.

Recent developments allow the design of controllers which have good robustness properties with respect to modelling uncertainty, but which are often of high order. It therefore becomes important to have methods of model reduction - that is, of approximating a given plant by a lower order plant without an unacceptable loss of performance. The celebrated paper of Glover [6] gives a complete solution of this problem in the case that the measure of the error $E(z)$ between two transfer function matrices is taken to be the Hankel norm of E. One can motivate this and other H^{∞} type criteria (cf. Francis and Doyle [5]) by saying that if the largest singular value $s_0(E(z))$ of $E(z)$ is uniformly small on the imaginary axis or unit circle then, for an input signal u of unit energy, Eu will have small energy. Now for a matrix A with many rows and columns, $s_0(A)$ is quite a crude measure of the "size" of A: much more information is given by the sequence $s_0(A) \geq s_1(A) \geq \ldots$ of singular values of A. Correspondingly we may introduce a finer measure of the size of the transfer function matrix $E(z)$: $s^{\infty}(E)$ is defined to be the sequence

$$s^{\infty}(E) = (s_0^{\infty}(E), s_1^{\infty}(E), s_2^{\infty}(E), \ldots)$$

where

$$s_i^{\infty}(E) = \sup_{|z|=1} s_i(E(z))$$

The physical interpretation of s^{∞} is given by Foo and Postlethwaite [2]. Whereas H^{∞}-minimisation criteria usually allow a large degree of non-uniqueness (typically, the more rows and columns in E, the more degrees

NATO ASI Series, Vol. F34
Modelling, Robustness and Sensitivity Reduction in Control Systems. Edited by R. F. Curtain

of freedom in the minimal norm approximant), s^∞-criteria often yield unique optimal solutions. This is the case, for example, for the Nevanlinna-Pick or model matching problem for rational analytic functions in the closed unit disc: if A, B, C are such functions of types $m \times n$, $m \times m$, $n \times n$ respectively then the minimum of

$$s^\infty(A + BQC)$$

with respect to the lexicographic ordering is attained at a unique rational $m \times n$ matrix Q analytic in the closed unit disc (see [4], [5]). The expression "super-optimal" for this Q has gained acceptance. An equivalent statement is that for any rational function G bounded on the imaginary axis there is a unique anti-causal function Q for which $s^\infty(G - Q)$ is minimised with respect to the lexicographic ordering.

The Nevanlinna-Pick problem can be regarded as a special case of the model reduction problem with Hankel norm criterion. To formulate this mathematically we introduce the space $L^\infty_{m\times n}$ of essentially bounded Lebesgue measurable $\mathbb{C}^{m\times n}$-valued functions on the unit circle ∂D with essential supremum norm. Thus, for $F \in L^\infty_{m\times n}$,

$$\|F\|_\infty = \sup_{|z|=1} s_0(F(z)).$$

We also require the space L^2_m of square integrable Lebesgue measurable $\mathbb{C}^m$-valued functions on ∂D with its natural inner product, and the subspace H^2_m of L^2_m comprising the functions whose negative Fourier coefficients vanish. If $F \in L^\infty_{m\times n}$ then the Hankel operator

$$H_F : H^2_n \to L^2_m \ominus H^2_m$$

is defined by

$$H_F\, x = P_-(Fx)$$

where P_- is the orthogonal projection operator from L^2_m to H^2_m. We shall say that F has k poles in D if H_F has finite rank k. We denote by $H^\infty_{m\times n,k}$ the set of functions in $L^\infty_{m\times n}$ having k or fewer poles in D. $H^\infty_{m\times n,0}$ is abbreviated to $H^\infty_{m\times n}$. We consider the following problem.

(MR): *Given a rational function $F \in L^\infty_{m\times n}$ and a non-negative integer k, find a function $\hat{F} \in H^\infty_{m\times n,k}$ such that $\|F - \hat{F}\|_\infty$ is minimised.*

The main motivation for this problem is from engineering, so it is appropriate to re-cast the problem in terminology which reflects the idea of model reduction. F is to be thought of as the transfer function of a linear system of undesirably high McMillan degree, and a rational approximation $\tilde{F}$ of McMillan degree no greater than k is sought. Ideally we should like to choose $\tilde{F}$ to minimise $\|F - \tilde{F}\|_\infty$, but no one knows how to do this satisfactorily. We can, however, find an $\hat{F} \in H^\infty_{m\times n,k}$ which minimises $\|F - \hat{F}\|_\infty$. On taking $\tilde{F}$ to be the causal part of $\hat{F}$ (i.e. $\tilde{F} = \underline{P}\,\hat{F}$, in a self-explanatory notation) we obtain a rational approximant of McMillan degree at most k, and the minimality of $\|F - \hat{F}\|_\infty$ can be re-expressed by saying that $\tilde{F}$ minimises the Hankel norm $\|F - \tilde{F}\|_H$, where

$$\|G\|_H := \inf_{Q\in H^\infty_{m\times n}} \|G + Q\|_\infty .$$

$\|.\|_H$ is the quotient norm on L^∞/H^∞. Thus our problem can be re-stated:

Given a rational $m \times n$ transfer function matrix F and a non-negative integer k, find a rational function $\tilde{F}$ of McMillan degree at most k such that the Hankel norm of the error $\|F - \tilde{F}\|_H$ is minimised.

Typically there are many $\tilde{F}$'s (or $\hat{F}$'s) which solve these two equivalent problems. Glover [6] gives a state space algorithm which generates all solutions, and also proves estimates for the L^∞-norm of the error incurred by minimising the wrong norm. In this lecture we present an existence and uniqueness result for the model reduction problem obtained by replacing the L^∞-norm minimisation objective in the formulation above (MR) by s^∞- minimisation with respect to the lexicographic ordering. The statement can be illustrated with a very simple example.
Consider the function $F \in L^\infty_{2\times 2}$,

$$F(z) = \begin{bmatrix} 2z^{-1} & 0 \\ 0 & z^{-1} \end{bmatrix}$$

which has McMillan degree 2. The problem (MR) is meaningful with k = 0 or 1. For $k = 0$ it is easy to see that

$$\inf_{\hat{F}\in H^\infty_{2\times 2}} \|F - \hat{F}\|_\infty = 2,$$

attained when

$$\hat{F} = \begin{bmatrix} 0 & 0 \\ 0 & g \end{bmatrix}$$

for any $g \in H^\infty$ such that $\|g - z^{-1}\| \leq 2$ (this follows from the fact that the best approximation in H^∞ to $2z^{-1}$ is the zero function). For such an $\hat{F}$ we have

$$s^\infty(F - \hat{F}) = (2,\ \|g - z^{-1}\|_\infty,\ 0,\ 0,\ \ldots).$$

The minimum of $s^\infty(F - \hat{F})$ is thus attained at the unique function $\hat{F} = 0$.

The existence and uniqueness of s^∞-minimisers in the case $k = 0$ was proved in [11]. Now we are interested in the analogous result for positive k, so let us take $k = 1$ in the example. The approximating function $\hat{F}$ is now allowed to have one pole, so we can use this to approximate the worst entry of F, i.e. $2z^{-1}$. We find that

$$\inf_{\hat{F} \in H^\infty_{2\times 2,1}} \|F - \hat{F}\|_\infty = 1,$$

attained when

$$\hat{F} = \begin{bmatrix} g/z & 0 \\ 0 & 0 \end{bmatrix}$$

for any $g \in H^\infty$ such that $\|g - 2\|_\infty \leq 1$. For such $\hat{F}$

$$s^\infty(F - \hat{F}) = (1,\ \|g - 2\|_\infty,\ 0,\ 0,\ \ldots),$$

minimised by the unique function

$$\hat{F}(z) = \begin{bmatrix} 2z^{-1} & 0 \\ 0 & 0 \end{bmatrix}.$$

Such simple diagonal examples seem to suggest that s^∞-minimisation picks out a natural best approximant among L^∞-norm minimisers.

THEOREM *Let $F \in L^\infty_{m\times n}$ be rational and let k be a non-negative integer. There exists a unique $\hat{F} \in H^\infty_{m\times n,k}$ which minimises $s^\infty(F - \hat{F})$ with respect to the lexicographic ordering. The singular values of $F - \hat{F}$ are constant a.e. on the unit circle.*

The full proof involves a good deal of technical detail and will be published elsewhere. Here I sketch the main ideas. Suppose $m, n > 1$. The starting point for all work on this problem is the following remarkable fact.

LEMMA 1 *Let $F \in L^\infty_{m\times n}$ be rational and let k be a non-negative integer. Then*

$$\inf_{\hat{F} \in H^\infty_{m\times n,k}} \|F - \hat{F}\|_\infty = s_k(H_F),$$

and the infimum is attained. The operator $H_F^* H_F$ has an eigenvector v corresponding to the eigenvalue $s_k(H_F)^2$, and every $\hat{F} \in H^\infty_{m\times n,k}$ for which the infimum is attained satisfies

$$(F - \hat{F})v = H_F v.$$

Furthermore,

$$s_k(H_F)\|v(z)\| = \|(H_F v)(z)\|$$

for almost all $z \in \partial D$.

This was proved by Adamyan, Arov and Krein for scalar F and general k [1] and for matrix F and $k = 0$ [2]. The case of matrix F and general k was first treated by Kung and Lin [8]. Several other approaches have since been found (Ball and Helton [3], Treil [9], Glover [6]).

A rational $F \in L^\infty_{m\times n}$ can be written

$$F = B^* A C^*$$

where $A \in H^\infty_{m\times n}$ and B, C are inner functions of types $m \times m$, $n \times n$ respectively (cf. [5] for terminology). Then

$$B(F - \hat{F})C = A - B\hat{F}C,$$

and so minimising $s^\infty(F - \hat{F})$ as $\hat{F}$ varies over $H^\infty_{m\times n,k}$ is equivalent to minimising $s^\infty(A + BQC)$ over $Q \in H^\infty_{m\times n,k}$, since multiplying by inner functions does not change the singular values of $(F - \hat{F})(z)$, $z \in \partial D$. Let $t = s_k(H_F)$, so that, by Lemma 1,

$$\inf_{Q\in H^\infty_{m\times n,k}} \|A + BQC\|_\infty = t.$$

A normal families argument (or any of the cited references) shows that the infimum is attained, so that the set

$$\mathcal{G}_0 = \{G \in A + BH^\infty_{m\times n,k}\, C : \|G\|_\infty = t\}$$

is non-empty. Lemma 1 further tells us that there exist unit vectors $v \in L^2_n$, $w \in L^2_m$ such that

$$Gv = tw$$

for *every* $G \in \mathcal{G}_0$, and moreover,

$$\|v(z)\| = \|w(z)\| \quad \text{a.e. on } \partial D. \tag{1}$$

Let us consider the special case that v, w are of the special forms

$$v = \begin{bmatrix} v_1 \\ 0 \end{bmatrix}, \qquad w = \begin{bmatrix} w_1 \\ 0 \end{bmatrix} \tag{2}$$

where v_1, w_1 are scalar functions and the zeros are columns of length $n - 1$, $m - 1$ respectively. Then for $G \in \mathcal{G}_0$,

$$G\begin{bmatrix} v_1 \\ 0 \end{bmatrix} = t\begin{bmatrix} w_1 \\ 0 \end{bmatrix}$$

and hence

$$G = \begin{bmatrix} tw_1/v_1 & * \\ 0 & * \end{bmatrix}$$

By (1), the (1, 1) entry of G has modulus t a.e. on ∂D, while $\|G\|_\infty = t$. The (1, j) entry of G must therefore be identically zero for $j > 1$, and so the general $G \in \mathcal{G}_0$ has block diagonal form:

$$G = A + BQC = \begin{bmatrix} g_0 & 0 \\ 0 & * \end{bmatrix},$$

and we can arrange that $g_0 \in H^\infty$. The general case can be reduced to the special case (2) by multiplication on either side by suitable inner functions. This diagonalization procedure serves to make s^∞ more accessible to the intuition. The details are much as in the case $k = 0$, as described in [11].

Fix some $G_1 \in \mathcal{G}_0$, say

$$G_1 = A + BQ_1C = \begin{bmatrix} g_0 & 0 \\ 0 & h_1 \end{bmatrix},$$

$Q_1 \in H^\infty_{m\times n,k}$. Then for any $G \in \mathcal{G}_0$ we have

$$G - G_1 = B(QC - Q_1C) = \begin{bmatrix} 0 & 0 \\ 0 & * \end{bmatrix}.$$

Thus QC and Q_1C have the same first column - say

$$QC = [F_0 \quad F_1], \qquad Q_1C = [F_0 \quad H_1],$$

with $F_0 \in L^\infty_{m\times 1}$. Then

$$\begin{aligned} G &= G_1 + B(QC - Q_1C) \\ &= \begin{bmatrix} g_0 & 0 \\ 0 & h_1 \end{bmatrix} + B[0 \quad F_1 - H_1]. \\ &= \begin{bmatrix} g_0 & 0 \\ 0 & g_1 \end{bmatrix}, \quad \text{say.} \end{aligned}$$

Clearly

$$s^\infty(G) = (t, \; s^\infty(g_1)). \tag{3}$$

This time $s = n - 1$, and we take Ω to be the $(n-1)$-square inner function $Y^T C_1$. It follows from the choice of Φ that $\Phi F_0 = KC_0$ for some $K \in H^\infty_{m\times n}$. We take $R = KC_1$.

We have now parametrized the F_1 satisfying (i). Write

$$F_1 = \Phi^*(R + F\Omega),$$
$$H_1 = \Phi^*(R + H\Omega)$$

where $F, H \in H^\infty_{m\times(n-1),k-\ell}$. To satisfy (ii) we require

$$B\Phi^*(F - H)\Omega \in \begin{bmatrix} 0 \\ L^\infty_{(m-1)\times(n-1)} \end{bmatrix}. \qquad \text{(ii}'\text{)}$$

Pick $m \times m$ inner functions $\tilde{B}$, $\tilde{\Phi}$, with degree $\tilde{\Phi}$ = degree Φ such that

$$B\Phi^* = \tilde{\Phi}^*\tilde{B}.$$

LEMMA 3 *There exist inner functions Φ, Ψ of types $(m-1)\times(m-1)$, $(n-1)\times(n-1)$ respectively and $h \in H^\infty_{(m-1)\times(n-1),k-\ell-r}$ such that as F varies in $H^\infty_{m\times(n-1),k-\ell}$ subject to (ii'), $B\Phi^*(F - H)\Omega$ describes the set*

$$\begin{bmatrix} 0 \\ \tilde{\Phi}^* \,\Theta(h + H^\infty_{(m-1)\times(n-1),k-\ell-r})\Psi^*\,\Omega \end{bmatrix}$$

where

$$r = \text{degree } \Psi \leqslant k - \ell.$$

The construction is similar to that of Lemma 2.

The lemmas show that the set of (2,2)-block entries of all

$$G \in A + BH^\infty_{m\times n,k}C$$

which are of the form

$$G = \begin{bmatrix} g_0 & 0 \\ 0 & * \end{bmatrix}$$

is precisely

$$h_1 + \tilde{\Phi}^* \,\Theta(h + H^\infty_{(m-1)\times(n-1),k-\ell-r})\Psi^*\,\Omega. \qquad (4)$$

In view of (3), to minimise $s^\infty(G)$ as G varies over $\mathcal{G}_0$ we need to minimise s^∞ over (4). Since $\tilde{\Phi}$, Θ, Ψ and Ω are inner functions this is equivalent to minimising s^∞ over

$$(\Theta^* \,\tilde{\Phi}\, h_1\, \Omega^*\, \Psi + h) \;+\; H^\infty_{(m-1)\times(n-1),k-\ell-r},$$

and this is a model reduction problem of type $(m-1)\times(n-1)$. The

This time $s = n - 1$, and we take Ω to be the $(n-1)$-square inner function $Y^T C_1$. It follows from the choice of Φ that $\Phi F_0 = KC_0$ for some $K \in H^\infty_{m\times n}$. We take $R = KC_1$.

We have now parametrized the F_1 satisfying (i). Write

$$F_1 = \Phi^*(R + F\Omega),$$
$$H_1 = \Phi^*(R + H\Omega)$$

where $F, H \in H^\infty_{m\times(n-1),k-\ell}$. To satisfy (ii) we require

$$B\Phi^*(F - H)\Omega \ \in \begin{bmatrix} 0 \\ L^\infty_{(m-1)\times(n-1)} \end{bmatrix}. \qquad (ii')$$

Pick $m \times m$ inner functions $\tilde{B}$, $\tilde{\Phi}$, with degree $\tilde{\Phi}$ = degree Φ such that

$$B\Phi^* = \tilde{\Phi}^*\tilde{B}.$$

LEMMA 3 *There exist inner functions Φ, Ψ of types $(m-1)\times(m-1)$, $(n-1) \times (n-1)$ respectively and $h \in H^\infty_{(m-1)\times(n-1),k-\ell-r}$ such that as F varies in $H^\infty_{m\times(n-1),k-\ell}$ subject to (ii'), $B\Phi^*(F - H)\Omega$ describes the set*

$$\begin{bmatrix} 0 \\ \tilde{\Phi}^* \ \Theta(h + H^\infty_{(m-1)\times(n-1),k-\ell-r})\Psi^* \ \Omega \end{bmatrix}$$

where

$$r = \text{degree } \Psi \leqslant k - \ell.$$

The construction is similar to that of Lemma 2.

The lemmas show that the set of (2,2)-block entries of all

$$G \in A + BH^\infty_{m\times n,k}C$$

which are of the form

$$G = \begin{bmatrix} g_0 & 0 \\ 0 & * \end{bmatrix}$$

is precisely

$$h_1 + \tilde{\Phi}^* \ \Theta(h + H^\infty_{(m-1)\times(n-1),k-\ell-r})\Psi^* \ \Omega. \qquad (4)$$

In view of (3), to minimise $s^\infty(G)$ as G varies over $\mathcal{G}_0$ we need to minimise s^∞ over (4). Since $\tilde{\Phi}$, Θ, Ψ and Ω are inner functions this is equivalent to minimising s^∞ over

$$(\Theta^* \ \tilde{\Phi} \ h_1 \ \Omega^* \ \Psi + h) \ + \ H^\infty_{(m-1)\times(n-1),k-\ell-r},$$

and this is a model reduction problem of type $(m-1) \times (n-1)$. The operations performed preserve rationality at every step, and the theorem is easily shown to be true if m or n is 1. The desired conclusion thus follows by induction on m.

The method of proof yields a high level algorithm for the construction of s^{∞}-minimisers: I intend to pursue this in a future paper. Observe also that the continuous - and discrete - time versions of the model reduction problem (with L^{∞}-norm type criteria) are equivalent, so that one can immediately deduce the obvious continuous-time analogue of the theorem using the Cayley transform (see [6]).

REFERENCES

1. V. M. Adamyan, D. Z. Arov and M. G. Krein, Analytic properties of Schmidt pairs for a Hankel operator and the generalized Schur-Takagi problem, *Mat. Sb.* 86 (128) (1971), 34-75; *Math. USSR Sb.* 15 (1971), 31-73.
2. V. M. Adamyan, D. Z. Arov and M. G. Krein, Infinite Hankel block matrices and related extension problems, *Izv. Akad. Nauk Armyan. SSR Ser. Mat.* 6 (1971), 87-112; *Amer. Math. Soc. Transl. (2)* 111 (1978) 133-156.
3. J. A. Ball and J. W. Helton, A Beurling-Lax theorem for the Lie group $U(m,n)$ which contains most classical interpolation theory, *J. Operator Theory* 9 (1983), 107-142.
4. Y. K. Foo and I. Postlethwaite, An H^{∞} minimax approach to the design of robust control systems, *Systems and Control Letters*, 5 (1984), 81-88.
5. B. A. Francis and J. C. Doyle, Linear control theory with an H^{∞} optimality criterion, *SIAM J. Control and Optimisation*, to appear.
6. K. Glover, All optimal Hankel-norm approximations of linear multi-variable systems and their L^{∞}-error bounds, *Int. J. Control* 39 (1984), 1115-1193.
7. K. Hoffman, "Banach Spaces of Analytic Functions", Prentice Hall, New Jersey, 1962.
8. S. Kung and D. W. Lin, Optimal Hankel norm model reductions: multivariable systems, *I.E.E.E. Trans. Automatic Control* 26 (1981), 832-852.
9. S. R. Treil, Vektornyi variant teoremy Adamyana-Arova-Kreina

Funkc. Analiz i Prilozheniya, 1985.

10. N. J. Young, Interpolation by analytic matrix fuctions, in "Operators and Function Theory", Proceedings of NATO ASI, edited by S. C. Power, D Reidel Publishing Co. 1985, 351-383.
11. N. J. Young, The Nevanlinna-Pick problem for matrix-valued functions, *J. Operator Theory* 15 (1986), 239-265.

Hankel Norm Approximation for Infinite Dimensional Systems and Wiener-Hopf Factorization

A.C.M. Ran
Subfaculteit Wiskunde en Informatica
Vrije Universiteit
Postbus 7161, 1007 MC Amsterdam
The Netherlands

Abstract

Recently Hankel norm approximation for finite dimensional systems has been studied extensively, both for continuous time systems and discrete time systems. One of the approaches is to combine the work of J.A. Ball and J.W. Helton with results on Wiener-Hopf factorization. This approach will be extended in the present paper to a certain class of infinite dimensional systems. The results we obtain here have close connections with recent results of R.F. Curtain and K. Glover.

1. Introduction

Before discussing the problem we consider in this paper, let us introduce some function spaces which we shall use in the sequel.

Let $H^{\infty+}_{p\times q}$ denote the class of $p\times q$ matrix functions, analytic in the right half plane and uniformly bounded there, $H^{\infty-}_{p\times q}$ denotes the analogous class for the left half plane. For ℓ a non-negative integer we let $H^{\infty-}_{p\times q}(\ell)$ denote the class of functions $\hat{G}+F$ where $\hat{G}$ has McMillan degree ℓ and $F\in H^{\infty-}_{p\times q}$. We let $L^{\infty}_{p\times q}$ denote the class of measurable uniformly bounded functions K on the imaginary axis with norm

$$\|K\|_{L^\infty}=\sup_{-\infty<w<\infty}\|K(jw)\|.$$

We shall also need the Hilbert spaces $L^2_q[0,\infty)$ (resp. L^2_q), of $\mathbb{C}^q$-valued functions which are norm square-integrable on $[0,\infty)$ (resp. on the imaginary axis) and $L^2_{k\times m}[0,\infty)$ the Hilbert space of $k\times m$ matrices all of whose entries are in $L^2[0,\infty)$. Further we also need the Hardy subspaces H^{2-}_q and H^{2+}_q of L^2_q, which are defined as the spaces of $\mathbb{C}^q$-valued functions analytic on the left (resp. right) half plane such that the L^2_q-norm of its restrictions to vertical line s in the left (resp. right) half plane are uniformly bounded (see e.g. [Hof]).

In this paper we consider a fixed system given by the input-output map

$$(1.1)\qquad y(t)=\int_0^t h(t-s)u(s)ds.$$

Here the inputs u are elements of $L^2_q[0,\infty)$, the outputs y are in $L^2_p[0,\infty)$ and we assume that for some $\omega>0$

$$(1.2)\qquad h\in e^{-\omega t}L^2_{p\times q}[0,\infty),$$

NATO ASI Series, Vol. F34
Modelling, Robustness and Sensitivity Reduction in Control Systems. Edited by R. F. Curtain

and, moreover, h is of the form

$$(1.3)\qquad h(t)=\int_t^{\infty} g(s)ds$$

for some $g \in L^2_{p\times q}[0,\infty)$.

In Section 2 we shall describe a realization $h(t)=Ce^{At}B$ for h, where $B\colon \mathbb{C}^q \to L^2_p[0,\infty)$ is bounded, A generates an exponentially decaying semigroup on $L^2_p[0,\infty)$ and $C\colon L^2_p[0,\infty)\to \mathbb{C}^p$ is bounded. This realization is taken from [BGK2]. Introduce $G(s)=C(sI-A)^{-1}B$. Then the problem we consider in this paper is that of approximating $G(s)$ by functions $K(s) \in H^{\infty -}_{p\times q}(\ell)$ in L^∞-norm. It is well-known (see e.g. [AAK1], [KL] and [BH]) that

$$\inf\{\|G-K\|_{L^\infty} \mid K \in H^{\infty-}_{p\times q}(\ell)\}=\sigma_{\ell+1}(G),$$

where $\sigma_{\ell+1}(G)$ is the $(\ell+1)$-st Hankel singular value of G. The Hankel singular values $\sigma_1(G)\geqq\sigma_2(G)\geqq\ldots\geqq 0$ are defined as the singular values of the Hankel operator $\Gamma\colon L^2_q[0,\infty)\to L^2_p[0,\infty)$ given by

$$(1.4)\qquad (\Gamma v)(t)=\int_0^{\infty} h(t+s)v(s)ds.$$

Thus, given $\sigma>0$, the smallest integer ℓ for which there is a $K \in H^{\infty-}_{p\times q}(\ell)$ such that $\|G-K\|_{L^\infty}\leqq\sigma$ is the first ℓ for which $\sigma\geqq\sigma_{\ell+1}(G)$. In case $\sigma_{\ell+1}(G)\leqq\sigma<\sigma_\ell(G)$ there is a linear fractional map parametrization for the set of all such functions K, obtained in [BH]. The case where $G(s)$ is a rational matrix function has been studied extensively recently. There, the linear fractional map can be expressed explicitly in terms of matrices A, B, C appearing in a realization of $G(s)$ (see e.g. [Gl], [BR 1,2]). In this paper we shall assume $\sigma_{\ell+1}(G)<\sigma<\sigma_\ell(G)$ and follow the line of argument in [BR1] to obtain explicitly the linear fractional map parametrizing all $K \in H^{\infty-}_{p\times q}(\ell)$ with $\|G-K\|_{L^\infty}\leqq\sigma$. This is done in the first part of Section 3. As the argument follows [BR1] closely, only the new parts of the proof are presented in detail. The formulas thus obtained can be made even more explicit by plugging in the specific form of the operators A, B, C in our realization. We obtain in this way formulas for the coefficients in the linear fractional map in terms of the function h we started with and the corresponding Hankel operator Γ of (1.4). For the case $\sigma>\|\Gamma\|$ (the Nehari problem) the formulas in Theorem 3.5 are related to those obtained in [AAK2], [DG] and [B]. There also formulas for the solution of the Nehari problem are given in terms of the function h and the Hankel operator Γ. Note that in Theorem 3.5 we also allow $\sigma<\|\Gamma\|$.

The present paper has connections with [CG2]. Before discussing that connection let us summarize part of the results of [CG2]. In that paper for a function $h \in L_1\cap L_2$ a realization $h(t)=\tilde{C}e^{\tilde{A}t}\tilde{B}$ (different from ours) is given, under the assumption that the corresponding Hankel operator Γ given by (1.4) is Hilbert-Schmidt. Assuming further that Γ is even nuclear they show that finite dimensional truncations A_n, B_n, C_n of $\tilde{A}$, $\tilde{B}$, $\tilde{C}$ converge in the sense that (among other things) $\|G(s)-G_n(s)\|_{L^\infty}\to 0$, where $G_n(s)=C_n(s-A_n)^{-1}B_n$. Also error estimates are provided. Then the results of [Gl] are applied to $G_n(s)$: all optimal Hankel norm approximants of $G_n(s)$ are obtained via a linear fractional map given by a rational matrix function θ_n. It is then shown that these matrix functions θ_n converge in L^∞-norm to some function θ as $n \to \infty$, and again error bounds are provided. This function θ solves the optimal Hankel norm approximation problem for $G(s)$.

The aim of this paper is to show that a function θ as desired can be obtained explicitly and directly from a realization of h, by the method employed in [BR1]. This goal, which is mainly of theoretical interest, is realized here for the class of functions h given by (1.2) and (1.3). Further, the formulas for θ in terms of h and Γ^{cc} presented in Theorem 3.5 below, and the derivation of these formulas directly from Theorem 3.4 provide some additional insight in the problem.

Acknowledgements

It is a pleasure to thank professor R.F. Curtain for presenting a lecture on the material in [CG2] at the Vrije Universiteit. This paper had its origins in that lecture. Also I would like to thank professor M.A. Kaashoek for many discussions on the contents of this paper, and professor J.A. Ball for some useful suggestions.

2. The realization for h and some of its properties

Let h be a function satisfying (1.2) and (1.3). According to [BGK2] h has a realization $h(t) = Ce^{At}B$ which is given by the following. The operator $C: L_p^2[0,\infty) \to \mathbb{C}^p$ is defined by

$$(2.1) \qquad Cf = -\int_0^\infty e^{-\omega t} f(t)dt.$$

Next, introduce a subspace

$$(2.2) \qquad D = \{f \in L_p^2 \mid f(t) = \int_t^\infty g(s)ds,\ g \in L_p^2\}$$

and

$$(2.3) \qquad A: D \to L_p^2[0,\infty), \quad Af = -\omega f + f'.$$

Then A generates an exponentially decaying semigroup on $L_p^2[0,\infty)$ given by

$$(2.4) \qquad e^{At}f = e^{-\omega t} f(\cdot + t) \qquad t \geqq 0.$$

Finally, the operator $B: \mathbb{C}^q \to L_p^2[0,\infty)$ is given by

$$(2.5) \qquad (By)(t) = Ae^{\omega t}h(t)y = e^{\omega t}h'(t)y.$$

(Note that indeed $e^{\omega t}h(t) \in D$ because of (1.3)). A little calculation shows that indeed

$$Ce^{At}By = -\int_0^\infty h'(s+t)yds = h(t)y.$$

We shall call $\theta = (A, B, C)$ the *realization triple for* h. (Indeed θ is a realization triple in the sense of [BGK4] with separating projection 0.) Introduce the following operators: the *reachability operator*

$$\Gamma_\theta : L_q^2[0,\infty) \to L_p^2[0,\infty),$$

$$(2.6) \quad (\Gamma_\theta u)(t) = \int_0^\infty (e^{As} B u(s))(t) ds = e^{\omega t} \int_0^\infty h'(t+s) u(s) ds$$

and the *observability operator*

$$\Lambda_\theta : L_p^2[0,\infty) \to L_p^2[0,\infty)$$

$$(2.7) \quad (\Lambda_\theta f(t) = C e^{At} f = -\int_t^\infty e^{-\omega s} f(s) ds.$$

These maps are bounded (see [CG1], Section 3 and the references given there) and in fact Γ_θ is compact (see [BGK3]).

A straightforward computation shows that the Hankel operator $\Gamma : L_q^2 \to L_p^2$ defined by

$$(2.8) \quad (\Gamma u)(t) = \int_0^\infty h(t+s) u(s) ds$$

is given by

$$\Gamma = \Lambda_\theta \Gamma_\theta .$$

Next, introduce the controllability and observability gramians $\hat{P}$ and $\hat{Q}$ by

$$(2.9) \quad \hat{P} = \int_0^\infty e^{At} B B^* (e^{At})^* dt = \Gamma_\theta \Gamma_\theta^*$$

and

$$(2.10) \quad \hat{Q} = \int_0^\infty (e^{At})^* C^* C e^{At} dt = \Lambda_\theta^* \Lambda_\theta .$$

Clearly $\hat{P}$ and $\hat{Q}$ are bounded, in fact $\hat{P}$ is compact. Further we have the following formulas for $\hat{P}$ and $\hat{Q}$.

LEMMA 2.1. *The following formulas hold*

$$(2.11) \quad (\hat{Q} f)(t) = e^{-\omega t} \int_0^t \int_s^\infty e^{-\omega \alpha} f(\alpha) d\alpha ds ,$$

$$(2.12) \quad (\hat{P} f)(t) = e^{\omega t} \int_0^\infty \int_0^\infty h'(t+s) h'(s+\alpha)^* e^{\omega \alpha} f(\alpha) d\alpha ds ,$$

$$(2.13) \quad \hat{Q} = A^{*-1} M_{e^{-2\omega t}} A^{-1},$$

(2.14) $$A^{-1}\hat{P}A^{*-1} = M_{e^{\omega t}}\Gamma\Gamma^{*}M_{e^{\omega t}}.$$

Here M_f is the operator of multiplication by f.

PROOF. From (2.7) one computes

$$(\Lambda_{\theta}^{*}f)(t) = -e^{-\omega t}\int_0^t f(s)ds.$$

Together with (2.10) and (2.7) this gives (2.11). Also, from (2.6) one has

$$(\Gamma_{\theta}^{*}f)(t) = \int_0^\infty h'(t+s)^{*}e^{\omega s}f(s)ds.$$

Then (2.12) becomes clear.

Next, we prove (2.13). A simple computation gives that

(2.15) $$(A^{-1}f)(t) = -e^{\omega t}\int_t^\infty f(s)e^{-\omega s}ds = -(M_{e^{\omega t}}\Lambda_{\theta}f)(t).$$

Since $\hat{Q} = \Lambda_{\theta}^{*}\Lambda_{\theta}$ this gives (2.13).

Finally, to prove (2.14) we compute A^{*-1} first. From (2.15) we have

(2.16) $$(A^{*-1}f)(t) = -e^{-\omega t}\int_0^t f(s)e^{\omega s}ds.$$

Using (2.12) we have

(2.17) $$(M_{e^{-\omega t}}A^{-1}\hat{P}A^{*-1}M_{e^{-\omega t}}f)(t) =$$

$$= M_{e^{-\omega t}}A^{-1}e^{\omega s}\int_0^\infty\int_0^\infty h'(\tau+s)h'(\tau+\alpha)^{*}e^{\omega\alpha}((A^{*-1}M_{e^{-\omega t}})f)(\alpha)d\alpha d\tau$$

$$= -M_{e^{-\omega t}}A^{-1}e^{\omega s}\int_0^\infty\int_0^\infty h'(\tau+s)h'(\tau+\alpha)^{*}\int_0^\alpha f(\beta)d\beta d\alpha d\tau.$$

Now by partial integration we have

(2.18) $$\int_0^\infty h'(\tau+\alpha)^{*}\int_0^\alpha f(\beta)d\beta d\alpha = -\int_0^\infty h(\tau+\alpha)^{*}f(\alpha)d\alpha.$$

So (2.17) equals

$$M_{e^{-\omega t}}A^{-1}e^{\omega t}\int_0^\infty\int_0^\infty h'(\tau+s)h(\tau+\alpha)^{*}f(\alpha)d\alpha d\tau =$$

$$= -\int_t^\infty\int_0^\infty\int_0^\infty h'(\tau+s)h(\tau+\alpha)^{*}f(\alpha)d\alpha d\tau ds =$$

$$= -\int_0^\infty\int_0^\infty\int_t^\infty h'(\tau+s)h(\tau+\alpha)^{*}f(\alpha)ds d\alpha d\tau =$$

$$= \int_0^\infty \int_0^\infty h(\tau+t)h(\tau+\alpha)^* f(\alpha)\,d\alpha\, d\tau =$$

$$= (\Gamma\Gamma^* f)(t). \quad \square$$

The next proposition is the main result of this section.

PROPOSITION 2.2. *The operator $\hat{P}$ maps $D(A^*)$ into D, the operator $\hat{Q}$ maps D into $D(A^*)$. Further $\hat{P}$ and $\hat{Q}$ satisfy the following Lyapunov equations:*

(2.19) $\quad (A\hat{P}+\hat{P}A^*)x = -BB^*x, \quad x \in D(A^*),$

(2.20) $\quad (A^*\hat{Q}+\hat{Q}A)x = -C^*Cx, \quad x \in D.$

PROOF. Since $D(A^*) = \mathrm{Im}(A^{*-1})$ and $D = \mathrm{Im}\, A^{-1}$ the first part of the proposition easily follows from Lemma 2.1.

To prove (2.19) it now suffices to show

$$\hat{P}A^{*-1}+A^{-1}\hat{P} = -A^{-1}BB^*A^{*-1}.$$

Now $(A^{-1}By)(t) = e^{\omega t}h(t)y$, which implies

$$B^*A^{*-1}g = \int_0^\infty e^{\omega t}h(t)^* g(t)\,dt.$$

Hence

$$(A^{-1}BB^*A^{*-1}g)(t) = e^{\omega t}h(t)\int_0^\infty e^{\omega s}h(s)^* g(s)\,ds.$$

By (2.12) and (2.16) we have

$$(\hat{P}A^{*-1}g)(t) = -e^{\omega t}\int_0^\infty\int_0^\infty h'(t+s)h'(s+\alpha)^* \int_0^\alpha g(\beta)e^{\omega\beta}\,d\beta\, d\alpha\, ds =$$

$$= e^{\omega t}\int_0^\infty\int_0^\infty h'(t+s)h(s+\alpha)^* g(\alpha)e^{\omega\alpha}\,d\alpha\, ds,$$

where the last equality is by (2.18). Also,

$$(A^{-1}\hat{P}g)(t) = -e^{\omega t}\int_t^\infty\int_0^\infty\int_0^\infty h'(\beta+s)h'(s+\alpha)^* e^{\omega\alpha}g(\alpha)\,d\alpha\, ds\, d\beta =$$

$$= e^{\omega t}\int_0^\infty\int_0^\infty h(t+s)h'(s+\alpha)^* g(\alpha)e^{\omega\alpha}\,d\alpha\, ds.$$

So

$$((A^{-1}\hat{P}+\hat{P}A^{*-1})g)(t) = e^{\omega t}\int_0^\infty\int_0^\infty \frac{d}{ds}h(t+s)h(s+\alpha)^* e^{\omega\alpha}g(\alpha)\,d\alpha\, ds$$

$$= -e^{\omega t}h(t)\int_0^\infty h(\alpha)^* e^{\omega\alpha} g(\alpha)d\alpha =$$

$$= -(A^{-1}BB^*A^{*-1}g)(t).$$

Hence (2.19) holds.

To prove (2.20) use (2.13), (2.15) and (2.16) to see

$$(A^*\hat{Q}f + \hat{Q}Af)(t) = (M_{e^{-2\omega t}}A^{-1}f + A^{*-1}M_{e^{-2\omega t}}f)(t)$$

$$= -e^{-\omega t}\int_t^\infty f(s)e^{-\omega s}ds - e^{-\omega t}\int_0^t f(s)e^{-\omega s}ds =$$

$$= -e^{-\omega t}\int_0^\infty f(s)e^{-\omega s}ds.$$

Now $(C^*y)(t) = -e^{-\omega t}y$ as one easily checks, so $(C^*Cf)(t) = e^{-\omega t}\int_0^\infty f(s)e^{-\omega s}ds$. Hence (2.20) holds. □

In the sequel the operator

$$Z = (I - \sigma^{-2}\hat{Q}\hat{P})^{-1}$$

plays an important role. By (2.13) and (2.14) we have

$$Z = A^{*-1}M_{e^{-\omega t}}(I - \sigma^{-2}\Gamma\Gamma^*)^{-1}M_{e^{\omega t}}A^*. \tag{2.21}$$

3. The Hankel norm approximation problem

We start by summarizing from [BH] and [BR1] two results we shall need.

THEOREM 3.1. [BH] *Let h be a function satisfying (1.2) and (1.3), and let $h(t) = Ce^{At}B$ be the realization given by (2.1)-(2.5). Let Γ be the Hankel operator defined by h as in (1.4), let $G(s) = C(s-A)^{-1}B$ and, finally, let $\sigma_1(G) \geqq \sigma_2(G) \geqq \ldots \geqq 0$ be the Hankel singular values of G, i.e. the eigenvalues of $(\Gamma^*\Gamma)^{1/2}$. Then*

$$\inf\{\|G-K\| \mid K \in H^{\infty-}_{p\times q}(\ell)\} = \sigma_{\ell+1}(G).$$

Moreover, if σ and ℓ are chosen such that $\sigma_\ell(G) > \sigma > \sigma_{\ell+1}(G)$ then there is a $(p+q)\times(p+q)$ matrix function

$$\theta(s) = \begin{bmatrix} \theta_{11}(s) & \theta_{12}(s) \\ \theta_{21}(s) & \theta_{22}(s) \end{bmatrix}$$

such that any function $\hat{F}$ of the form $\hat{F} = G-K$ where $K \in H^{\infty-}_{p\times q}(\ell)$ with $\|\hat{F}\|_{L^\infty} \leqq \sigma$ has the form

$$(3.1)\quad \hat{F}(s) = (\theta_{11}(s)H(s)+\theta_{12}(s))(\theta_{21}(s)H(s)+\theta_{22}(s))^{-1}$$

for a unique $H \in H^{\infty,-}_{p\times q}$ *with* $\|H\|_{L^\infty} \leqq 1$. *Conversely, if* $H \in H^{\infty-}_{p\times q}$ *with* $\|H\|_{L^\infty} \leqq 1$ *then* (3.1) *defines a function* $\hat{F}$ *of the form* $\hat{F} = G-K$ *for a* $K \in H^{\infty-}_{p\times q}(\ell)$ *with* $\|\hat{F}\|_{L^\infty} \leqq \sigma$.

The matrix function $\theta(s)$ *is any matrix function satisfying the two conditions*

$$(3.2)\quad \theta(-\bar{s})^* \begin{bmatrix} I_p & 0 \\ 0 & -\sigma^2 I_q \end{bmatrix} \theta(s) = \begin{bmatrix} I_p & 0 \\ 0 & -I_q \end{bmatrix},$$

$$(3.3)\quad \theta H^{2-}_{p+q} = \begin{bmatrix} I_p & G \\ 0 & I_q \end{bmatrix} H^{2-}_{p+q}.$$

The following lemma reduces the problem of constructing θ to a symmetric Wiener-Hopf factorization problem. Here and in the sequel we shall denote for a matrix function $W(s)$ the function $W(-\bar{s})^*$ by $W^*(s)$.

LEMMA 3.2. (cf. Lemma 2.2 in [BR1]) *Suppose* $G(s)$ *is as above. Then a matrix function* $\theta(s)$ *exists satisfying* (3.2) *and* (3.3) *if and only if the function*

$$(3.4)\quad W(s) := \begin{bmatrix} I_p & 0 \\ G^*(s) & I_q \end{bmatrix} \begin{bmatrix} I_p & 0 \\ 0 & -\sigma^2 I_q \end{bmatrix} \begin{bmatrix} I_p & G(s) \\ 0 & I_q \end{bmatrix}$$

has a symmetric right canonical factorization

$$(3.5)\quad W(s) = X^*(s) \begin{bmatrix} I_p & 0 \\ 0 & -I_q \end{bmatrix} X(s),$$

where $X(s)$ *is analytic and invertible in the closed left half plane. If* X *is the factor in such a factorization of* W, *then the function*

$$(3.6)\quad \theta(s) = \begin{bmatrix} I_p & G(s) \\ 0 & I_q \end{bmatrix} X(s)^{-1}$$

satisfies (3.2) *and* (3.3).

The next step is to factorize $W(s)$ as in (3.5) and provide a formula for $X(s)^{-1}$. For the case where $G(s)$ is a rational matrix function this was done in [BR1] using the factorization result in [BGK1]. Here we shall use [BGK4], Theorem II.4.1 (see also [BGK2], Theorem 8.1) to obtain the formula for $X(s)^{-1}$. This formula turns out to be the same as in the rational case (see [BR1], Lemma 2.3).

LEMMA 3.3. *Suppose* $G(s) = C(sI-A)^{-1}B$ *is as above. Then the function* $X(s)^{-1}$, *where* X *is the factor in the factorization* (3.5) *is given by*

$$(3.7)\quad X(s)^{-1} = \begin{bmatrix} I_p & 0 \\ 0 & \sigma^{-1} I_q \end{bmatrix} - \sigma^{-2} \begin{bmatrix} -C\hat{P} \\ B^* \end{bmatrix} (sI+A^*)^{-1}(I-\sigma^{-2}\hat{Q}\hat{P})^{-1}[C^*, \sigma^{-1}\hat{Q}B]$$

where $\hat{P}$ *and* $\hat{Q}$ *are the controllability and observability gramians, respectively, given by* (2.9) *and* (2.10).

PROOF. First we construct a realization for $W(s)$ using (3.4). Clearly

$$(3.8)\qquad \begin{bmatrix} I_p & G(s) \\ 0 & I_q \end{bmatrix} = \begin{bmatrix} I_p & 0 \\ 0 & I_q \end{bmatrix} + \begin{bmatrix} C \\ 0 \end{bmatrix} (s-A)^{-1}[0 \quad B]$$

Writing out the corresponding realization for $\begin{bmatrix} I & 0 \\ G^* & I \end{bmatrix}$ and multiplying realizations in the usual way we arrive at the following realization for $W(s)$:

$$W(s) = \tilde{D} + \tilde{C}(s-\tilde{A})^{-1}\tilde{B},$$

where

$$\tilde{A} = \begin{bmatrix} -A^* & C^*C \\ 0 & A \end{bmatrix}, \quad \tilde{B} = \begin{bmatrix} C^* & 0 \\ 0 & B \end{bmatrix},$$

$$\tilde{C} = \begin{bmatrix} 0 & C \\ -B^* & 0 \end{bmatrix}, \quad \tilde{D} = \begin{bmatrix} I_p & 0 \\ 0 & \sigma^2 I_q \end{bmatrix}.$$

As usual, let $\tilde{A}^\times = \tilde{A} - \tilde{B}\tilde{D}^{-1}\tilde{C}$, then

$$\tilde{A}^\times = \begin{bmatrix} -A^* & 0 \\ -\sigma^{-2}BB^* & A \end{bmatrix}.$$

We have to show that both $(\tilde{A},\tilde{B},\tilde{C})$ and $(\tilde{A}^\times,\tilde{B},\tilde{C})$ are realization triples in the sense of [BGK4]. Note that $\tilde{B}$ and $\tilde{C}$ are bounded. Hence it is sufficient to show that $\tilde{A}$ and $\tilde{A}^\times$ are exponentially dichotomous of exponential type ω (see the remarks preceding Proposition I.2.1 in [BGK4]). To see this note that

$$\begin{bmatrix} I & \hat{Q} \\ 0 & I \end{bmatrix} \tilde{A} \begin{bmatrix} I & \hat{Q} \\ 0 & I \end{bmatrix} = \begin{bmatrix} -A^* & 0 \\ 0 & A \end{bmatrix}$$

because of (2.20), and that

$$\begin{bmatrix} I & 0 \\ \sigma^{-2}\hat{P} & I \end{bmatrix} \tilde{A}^\times \begin{bmatrix} I & 0 \\ \sigma^{-2}\hat{P} & I \end{bmatrix} = \begin{bmatrix} -A^* & 0 \\ 0 & A \end{bmatrix},$$

because of (2.19). Note that $D(A^*) = \mathrm{Im}(A^{*-1})$ is dense. Hence $\begin{bmatrix} -A^* & 0 \\ 0 & A \end{bmatrix}$ is exponentially dichotomous of exponential type ω, and since $\hat{P}$ and $\hat{Q}$ are bounded it follows that $\tilde{A}$ and $\tilde{A}^\times$ are exponentially dichotomous of exponential type ω too.

From here on the proof follows the lines of [BR1], the proof of Lemma 2.3; applying

Theorem II.4.1 in [BGK4] or Theorem 8.1 in [BGK2] instead of the factorization results for rational matrix functions presented in [BGK1]. □

Now we are in a position to state our main result.

THEOREM 3.4. *Suppose the conditions of Theorem 3.1 are satisfied, and let* σ *and* ℓ *be chosen such that* $\sigma_\ell(G)>\sigma>\sigma_{\ell+1}(G)$. *Then a* $(p+q)\times(p+q)$*-matrix function* $\theta(s)$ *satisfying* (3.2) *and* (3.3) *is given by the formula*

$$\theta(s) = \begin{bmatrix} \theta_{11}(s) & \theta_{12}(s) \\ \theta_{21}(s) & \theta_{22}(s) \end{bmatrix},$$

where

(3.9) $\quad \theta_{11}(s) = I_p+\sigma^{-2}C(s-A)^{-1}\hat{P}ZC^*$

(3.10) $\quad \theta_{12}(s) = \sigma^{-1}C(s-A)^{-1}Z^*B$

(3.11) $\quad \theta_{21}(s) = -\sigma^{-2}B^*(s+A^*)^{-1}ZC^*$

(3.12) $\quad \theta_{22}(s) = \sigma^{-1}I_q-\sigma^{-3}B^*(s+A^*)^{-1}Z\hat{Q}B.$

Here $\hat{P}$ *and* $\hat{Q}$ *are given by* (2.7) *and* (2.8), *and* $Z = (I-\sigma^{-2}\hat{Q}\hat{P})^{-1}$.

The proof of this theorem is the same as the proof for the rational case presented in [BR1]. One uses (3.6)-(3.8) and multiplies realizations to find a realization for $\theta(s)$. Partitioning $\theta(s)$ one obtains the block entries $\theta_{ij}(s)$ from this realization. Then one simplifies the formulas by using the Lyapunov equations (2.19) and (2.20) for $\hat{P}$ and $\hat{Q}$.

Finally we shall show that the formulas in Theorem 3.4 can be stated even more explicitly in terms of the function h.

THEOREM 3.5. *Under the conditions of Theorem 3.4 and with notation as above we have*

(3.13) $$\theta_{11}(s)y = y+\sigma^{-2}\int_0^\infty e^{-s\alpha}\{(I-\sigma^{-2}\Gamma\Gamma^*)^{-1}\int_0^\infty h(\cdot+\tau)h(\tau)^*y\,d\tau\}(\alpha)\,d\alpha,$$

(3.14) $$\theta_{12}(s)y = +\sigma^{-1}\int_0^\infty e^{-s\alpha}\{(I-\sigma^{-2}\Gamma\Gamma^*)^{-1}h(\cdot)y\}(\alpha)\,d\alpha,$$

(3.15) $$\theta_{21}(s)y = +\sigma^{-2}\int_0^\infty e^{s\alpha}h(\alpha)^*\left[y+\sigma^{-2}\int_0^\alpha e^{-st}\{(I-\sigma^{-2}\Gamma\Gamma^*)^{-1}\int_0^\infty h(\cdot+\tau)h(\tau)^*y\,d\tau\}(t)\,dt\right]d\alpha$$

(3.16) $$\theta_{22}(s)y = \sigma^{-1}y+\sigma^{-3}\int_0^\infty h(\alpha)^*e^{s\alpha}\int_0^\alpha e^{-st}\{(I-\sigma^{-2}\Gamma\Gamma^*)^{-1}h(\cdot)y\}(t)\,dt\,d\alpha.$$

PROOF. Using (3.9) and (2.21) we have

$$\begin{aligned}\theta_{11}(s) &= I_p + \sigma^{-2} C(s-A)^{-1} \hat{P} Z C^* \\ &= I_p + \sigma^{-2} C(s-A)^{-1} Z^* \hat{P} C^* \\ &= I_p + \sigma^{-2} CA(s-A)^{-1} M_{e^{\omega t}} (I - \sigma^{-2} \Gamma\Gamma^*)^{-1} M_{e^{-\omega t}} A^{-1} \hat{P} C^* .\end{aligned}$$

Computing $(s-A)^{-1}$ and CA gives

$$((s-A)^{-1} g)(t) = + e^{(s+\omega)t} \int_t^\infty e^{-(s+\omega)\alpha} g(\alpha) d\alpha,$$

$$CAf = f(0), \qquad f \in D.$$

So for any g we have

$$CA(s-A)^{-1} M_{e^{\omega t}} g = + \int_0^\infty e^{-s\alpha} g(\alpha) d\alpha.$$

Further, by (2.12) and (2.15) we obtain

$$(M_{e^{-\omega t}} A^{-1} \hat{P} C^* y)(t) = - \int_t^\infty \int_0^\infty \int_0^\infty h'(s+\tau) h'(\tau+\alpha)^* e^{\omega\alpha} (C^* y)(\alpha) d\alpha d\tau ds.$$

Using $(C^* y)(t) = -e^{-\omega t} y$ and interchanging the order of integration gives

$$(M_{e^{-\omega t}} A^{-1} \hat{P} C^* y)(t) = \int_0^\infty h(t+\tau) h(\tau)^* y d\tau.$$

Plugging all this into the formula for $\theta_{11}(s)$ obtained above yields (3.13).

Computation of $\theta_{12}(s)$ basically follows the same pattern. We have

$$\begin{aligned}\theta_{12}(s) &= \sigma^{-1} C(s-A)^{-1} Z^* B \\ &= \sigma^{-1} CA(s-A)^{-1} M_{e^{\omega t}} (I - \sigma^{-2} \Gamma\Gamma^*)^{-1} M_{e^{-\omega t}} A^{-1} B.\end{aligned}$$

Noting that $(M_{e^{-\omega t}} A^{-1} B y)(t) = h(t) y$ then gives (3.14).

To compute θ_{21} and θ_{22} first note that

$$(M_{e^{-\omega t}} (s-A)^{-1} A^{-1} B y)(t) = + e^{st} \int_t^\infty e^{-s\alpha} h(\alpha) d\alpha;$$

and a direct computation then gives for $s \in i\mathbb{R}$:

$$B^* A^{*-1} (s+A^*)^{-1} M_{e^{-\omega t}} f = - \int_0^\infty h(\alpha)^* e^{s\alpha} \int_0^\alpha e^{-st} f(t) dt d\alpha.$$

Thus from (3.12) and (2.21) we obtain

$$\begin{aligned}\theta_{22}(s) y &= \sigma^{-1} y - \sigma^{-3} B^* (s+A^*)^{-1} Z \hat{Q} B y = \\ &= \sigma^{-1} y - \sigma^{-3} B^* A^{*-1} (s+A^*)^{-1} M_{e^{-\omega t}} (I - \sigma^{-2} \Gamma\Gamma^*)^{-1} M_{e^{\omega t}} A^* \hat{Q} B y.\end{aligned}$$

From (2.13) we have $M_{e^{\omega t}}A^*\hat{Q}By = M_{e^{-\omega t}}A^{-1}By = h(\cdot)y$. Plugging all this into the formula for $\theta_{22}(s)$ gives (3.16).

Finally, we compute $\theta_{21}(s)$. By (3.11)

$$\theta_{21}(s)y = -\sigma^{-2}B^*(s+A^*)^{-1}ZC^*y =$$

$$= -\sigma^{-2}B^*(s+A^*)^{-1}C^*y - \sigma^{-4}B^*(s+A^*)^{-1}\hat{Q}\hat{P}ZC^*y =$$

$$= -\sigma^{-2}B^*(s+A^*)^{-1}C^*y - \sigma^{-4}B^*(s+A^*)^{-1}\hat{Q}Z^*\hat{P}C^*y =$$

$$= -\sigma^{-2}B^*(s+A^*)^{-1}C^*y - \sigma^{-4}B^*A^{*-1}(s+A^*)^{-1}M_{e^{-\omega t}}(I-\sigma^{-2}\Gamma\Gamma^*)^{-1}M_{e^{-\omega t}}A^{-1}\hat{P}C^*y\ ,$$

where the last equality usese (2.13) and (2.21).
The first term in this expression equals

$$-\sigma^{-2}B^*(s+A^*)^{-1}C^*y = +\sigma^{-2}\int_0^\infty e^{s\alpha}h(\alpha)^*y\,d\alpha$$

as one easily checkes. Using the formulas for $M_{e^{-\omega t}}A^{-1}\hat{P}C^*$ and $B^*A^{*-1}(s+A^*)^{-1}M_{e^{-\omega t}}$ obtained earlier one easily derives (3.15). □

References

[AAK1] Adamjan, V.M.; Arov, D.Z. and Krein, M.G., Analytic properties of Schmidt pairs for a Hankel operator and the generalized Schur-Takagi problem, Math. USSR Sb. 15 (1971), 31-73.

[AAK2] Adamjan, V.M.; Arov, D.Z. and Krein, M.G., Infinite Hankel block matrices and related extension problems, Izv. Akad. Nauk Armjan. SSR Ser. Mat. 6 (1971), 87-112. English translation: Amer. Math. Soc. Transl. (2) Vol 111, 133-156, 1978.

[B] Ball, J.A., Nevanlinna-Pick interpolation: Generalizations and Applications, Indiana University Lecture Notes, March 1986.

[BH] Ball, J.A. and Helton, J.W., A Beurling-Lax theorem for the Lie group $U(m,n)$ which contains most classical interpolation theory, J. Operator Theory 9 (1983), 107-142.

[BR1] Ball, J.A. and Ran, A.C.M., Hankel norm approximation of a rational matrix function in terms of its realization, in Modelling, Identification and Robust Control (eds. C.I. Byrnes, A. Lindquist), North Holland, Amsterdam etc., 1986, pag. 285-296.

[BR2] Ball, J.A. and Ran, A.C.M., Optimal Hankel norm model reductions and Wiener-Hopf factorization I: The canonical case, to appear, SIAM J. Control and Opt.

[BGK1] Bart, H.; Gohberg, I. and Kaashoek, M.A., Minimal Factorization of Matrix and Operator Functions, OT1 Birhäuser, Basel, 1979.

[BGK2] Bart, H.; Gohberg, I. and Kaashoek, M.A., Wiener-Hopf factorization, inverse Fourier transform and exponentially dichotomous operators, J. Funct. Anal. 68 (1986), 1-42.

[BGK3] Bart, H.; Gohberg, I. and Kaashoek, M.A., Fredholm theory of Wiener-Hopf equations in terms of realization of their symbols, Integral Equations and Operator Theory 8 (1985), 590-613.

[BGK4] Bart, H.; Gohberg, I. and Kaashoek, M.A., Wiener-Hopf equations with symbols analytic in a strip, in OT21, Birkhäuser, Basel, pag. 39-74.

[CG1] Curtain. R.F. and Glover, K., Balanced realizations for infinite dimensional systems, in Operator Theory and Systems, (eds. H. Bart, I. Gohberg and M.A. Kaashoek), OT 19, Birkhäuser, Basel, 1986, pag. 87-104.

[CG2] Curtain, R.F. and Glover, K., Realization and Approximation of linear infinite dimensional systems with error bounds, Cambridge University report CUED/F-CAMS/TR.258 (1986).

[DG] Dym, H. and Gohberg, I., Unitary interpolants, factorization indices and infinite Hankel block matrices, J. Funct. Anal. 54 (1983), 229-289.
Hankel integral operators and isometric interpolants on the line, J. Funct. Anal. 54 (1983), 290-307.

[Gl] Glover, K., All optimal Hankel-norm approximations of linear multivariable systems and their L^∞-error bounds, Int. J. Control 39 (1984), 1115-1193.

[Hof] Hoffman, K., Banach Spaces of Analytic Functions, Prentice-Hall, Englewoods Cliffs, 1962.

[KL] Kung, S.-Y. and Lin, D.W., Optimal Hankel-norm model reductions: multivariable systems, I.E.E.E. Trans. Autom. Control 26 (1981), 832-852.

Imaginary-Axis Zeros in Multivariable H^∞-Optimal Control*

Michael G. Safonov
Department of Electrical Engineering
University of Southern California
Los Angeles, California 90089-0781
U.S.A.

Abstract

When a plant has $j\omega$-axis zeros or $j\omega$-axis poles, algorithms for computing H^∞-optimal control laws fail. Closely related problems arise with strictly proper plants; these plants may be interpreted as having $j\omega$-axis zeros at $\omega = \infty$. These intrinsic problems with H^∞ arise because the optimal control system has an irrational transfer function with point discontinuities on the $j\omega$-axis at the offending $j\omega$-axis zeros and poles of the plant. The difficulties with $j\omega$-axis poles and zeros are discussed and the methods for perturbing the H^∞-problem to produce near-optimal rational control laws are proposed.

1. Introduction

Consider the problem of finding $Q \in (H^\infty)^{pxq}$ such that

$$\| T_{11} + T_{12}\, Q\, T_{21} \|_\infty \leq 1 \qquad (1)$$

where T_{11}, T_{12}, T_{21} are compatibly dimensioned, but not necessarily square, matrices over H^∞. The available algorithms for solving this problem [1-4] fail if either T_{12} or T_{21} has any zeros of the $j\omega$-axis, including zeros at $\omega = \infty$. Earlier H^∞-optimization algorithms [5-11] dealing with special cases of the problem (1) also all fail for T_{12} and/or T_{21} having $j\omega$-axis zeros, though a sequence of nearly optimal solutions is developed for the SISO case by Francis and Zames [16].

The reason for the failure of H^∞-algorithms for the case of $j\omega$-axis zeros has nothing to do with the specific algorithm being employed. The failure arises from an intrinsic ill-conditioning of the H^∞ problem when there are $j\omega$-axis zeros. The source of this ill-conditioning may be seen by considering the H^∞-solution in the simple case $T_{11} = T_{21} = 1$ and $T_{12} = (s-a)/(s+1)$. For Re(a) < 0 the minimum norm

*Research supported in part by National Science Foundation Grant ECS-8500961.

NATO ASI Series, Vol. F34
Modelling, Robustness and Sensitivity Reduction in Control Systems. Edited by R. F. Curtain

solution is

$$T_{11} + T_{12}\, Q\, T_{21} \;=\; 0 \tag{2}$$

$$\Rightarrow \qquad Q \;=\; -\,\frac{s+1}{s-a}\;. \tag{3}$$

However, when Re(a) > 0 then minimum norm solution is

$$T_{11} + T_{12}\, Q\, T_{21} \;=\; 1 \tag{4}$$

$$\Rightarrow \quad Q \;=\; 0\;. \tag{5}$$

Clearly, there is a discontinuity when Re(a) = 0.

Further insight into the effects of $j\omega$-axis zeros on H^∞—problems may be seen from the interpolation perspective: Each RHP zero z_i of T_{12} and T_{21} becomes an interpolation constraint on the values that

$$Y \;:=\; T_{11} + T_{12}\, Q\, T_{21} \tag{6}$$

can assume as Q ranges over H^∞. Indeed when the sets of RHP zeros z_i of $T_{12}, T_{21} \in H^\infty$ are simple and disjoint, the interpolation conditions

$$Y(z_i) \;=\; T_{11}(z_i) \tag{7}$$

are the only constraints on the realizable values of Y. The condition (7) is equivalent to the condition

$$\frac{1}{2\pi}\int_{-\infty}^{\infty} \frac{1}{j\omega + \bar{z}_i}\, Y(j\omega)\,d\omega \;=\; T_{11}(z_i)\;, \tag{8}$$

which obviously restricts the values that Y assumes.

In the case where $\mathrm{Re}(z_i) = 0$ so that z_i is on the $j\omega$-axis, the interpolation constraint affects $Y(j\omega)$ only at the isolated point $j\omega = z_i$; that is, it only mandates a point discontinuity in $Y(j\omega)$ at frequency $j\omega = z_i$ but is otherwise unconstraining. To see that point discontinuities in H^∞ functions of arbitrary magnitude k and phase θ are possible, one need only consider the sequence of functions

$$f^{(n)}_{k,\theta}(j\omega) \;:=\; \left(\frac{s + j\omega_o + \dfrac{k^{1/2}}{n}\, e^{j\theta/4}}{s + j\omega_o + \dfrac{1}{n}\, e^{-j\theta/4}} \right)^2 \in H^\infty \tag{9}$$

for which

$$f_{k,\theta}(s) \;:=\; \lim_{n\to\infty} f_{k,\theta}^{(n)}(s) \;=\; \begin{cases} ke^{j\theta} & , \text{ if } s = j\omega_0 \\ 1 & , \text{ otherwise} \end{cases} \quad . \tag{10}$$

It follows that in computing H^∞ optimal functions $Y(s)$ one could simply ignore the $j\omega$-axis interpolation points in computing the optimal $Y(j\omega)$ and then simply change $Y(j\omega)$ afterward at the isolated $j\omega$-axis points z_i to $Y(j\omega) = T_{11}(z_i)$.

In H^∞ optimal control problems the issue of $j\omega$-axis zeros of T_{12} and T_{21} is quite common. Consider, for example, the "mixed sensitivity" H^∞ problem (e.g., [10,12])

$$\min \left\| \begin{bmatrix} W_1 S \\ W_2 T \end{bmatrix} \right\|_\infty \tag{11}$$

where $S := (I+PC)^{-1}$, $T := I-S$, and C is a stabilizing feedback for the plant P having coprime stable-rational matrix fraction descriptions

$$P \;=\; N_r D_r^{-1} \;=\; D_\ell^{-1} N_\ell \tag{12}$$

with associated Bezout equations

$$\begin{bmatrix} V_r & U_r \\ -N_\ell & D_\ell \end{bmatrix} \begin{bmatrix} D_r & -U_\ell \\ N_r & V_\ell \end{bmatrix} = \begin{bmatrix} I & 0 \\ 0 & I \end{bmatrix} \tag{13}$$

where V_r, U_r, N_r, D_r, U_ℓ, V_ℓ, N_ℓ, D_ℓ, W_1, W_2, W_1^{-1}, and W_2^{-1} are proper rational matrices over H^∞. In this case the Youla lemma (Theorem 3 of [17]) yields

$$\begin{bmatrix} W_1 S \\ W_2 T \end{bmatrix} = T_{11} + T_{12}\, Q\, T_{21} \tag{14}$$

with

$$\left[\begin{array}{c|c} T_{11} & T_{12} \\ \hline T_{21} & 0 \end{array}\right] = \left[\begin{array}{c|c} W_1 V_\ell D_\ell & -W_1 N_r \\ W_2 N_r U_r & W_2 N_r \\ \hline D_\ell & 0 \end{array}\right] \tag{15}$$

Thus zeros of D_ℓ are zeros of T_{21} and the zeros of N_r are zeros of T_{12}. In particular, every closed RHP pole of P is a zero of T_{21} and every closed RHP zero of P (including zeros at ∞) is a zero of T_{12}.

It follows that the issue of $j\omega$-axis zeros in H^∞-control problems is far from exceptional. Every strictly proper plant will have zeros of T_{12} at $j\omega = \infty$ and every feedback system with integral action has a pole at $\omega = 0$ and hence a zero of the T_{21} of (15) at $j\omega = 0$.

In this paper two approaches to perturbing H^∞-problems to compensate for $j\omega$-axis zeros are suggested. In Section 2, we propose a simple conformal mapping of the s-plane which perturbs the $j\omega$-axis (including $\omega = \infty$), moving the offending points slightly into the interior of the RHP. In Section 3, a less general but in some ways more flexible approach is suggested involving the use of weightings with $j\omega$-axis poles to "cancel" the offending $j\omega$-axis and infinite zeros.

2. Bilinear Conformal Mapping of the s-Plane

Consider the following "bilinear map" on the Laplace transform variable s

$$\tilde{s} = \frac{s+a}{1+bs} \tag{16}$$

where a,b are nonnegative real numbers. The function $\tilde{s}(s)$ maps the circle, denoted $\mathcal{C}$, in the complex s-plane having diameter $[-b^{-1}, -a]$ onto the $j\omega$-axis in the $\tilde{s}$-plane (see Fig. 1) and it maps the $j\omega$-axis into a similar circle in the RHP. This function is a conformal map of the exterior of this s-plane circle onto the right half $\tilde{s}$-plane. The inverse conformal map is

$$s = \frac{\tilde{s}-a}{1-b\tilde{s}} . \tag{17}$$

Furthermore, if $\tilde{G}(\tilde{s})$ is any matrix-valued rational function which is

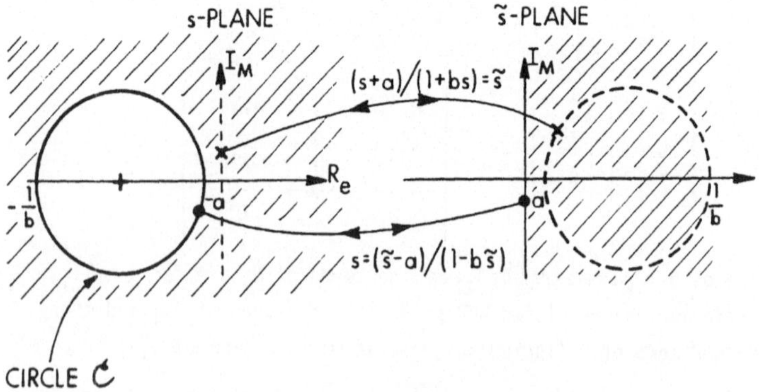

Figure 1

analytic in the right half $\tilde{s}$ plane then

$$G(s) := \tilde{G}\left(\frac{s+a}{1+bs}\right) \tag{18}$$

is analytic outisde the circle $\mathcal{C}$ and hence in the right half s-plane. Furthermore

$$\|\tilde{G}\|_\infty := \sup_\omega \bar{\sigma}(\tilde{G}(j\omega)) \equiv \sup_{s\in\mathcal{C}} \bar{\sigma}(G(s)) \leq \sup_\omega \bar{\sigma}(G(j\omega)) = \|G\|_\infty \tag{19}$$

where the inequality is a consequence of the maximum modulus theorem. On the other hand, since

$$s = \lim_{\substack{a\to 0\\ b\to 0}} \tilde{s} \tag{20}$$

it follows that

$$\lim_{\substack{a\to 0\\ b\to 0}} \|\tilde{G}\|_\infty = \|G\|_\infty . \tag{21}$$

The implication is clear: If one makes the change of variables $\tilde{s} = (s+a)/(1+bs)$ then for sufficiently small $a,b > 0$ any solution $\tilde{Q}$ to the transformed H^∞ problem of finding $\tilde{Q} \in (H^\infty)^{nxq}$ such that

$$\|\tilde{T}_{11} + \tilde{T}_{12}\, Q\, \tilde{T}_{21}\|_\infty \leq P \tag{22}$$

will be nearly the same as the solution Q to the original H^∞- problem (1). Here

$$\tilde{T}_{ij}(\tilde{s}) = T_{ij}((\tilde{s}-a)/(1-b\tilde{s})); \quad i,j = 1,2 . \tag{23}$$

The big difference is that for any sufficiently small $a,b > 0$ there are no imaginary-axis zeros of $\tilde{T}_{12}(\tilde{s})$ or $\tilde{T}_{21}(\tilde{s})$. Rather, each imaginary axis zero $z_i = j\omega_i$ is shifted into the right-half $\tilde{s}$-plane to the new location

$$\tilde{z}_i = \frac{j\omega_i + a}{1 + j\omega_i b} \tag{24}$$

and zeros at infinity are shifted to $\tilde{z}_i = 1/b$ thereby making $\tilde{T}_{12}$ and $\tilde{T}_{21}$ biproper transfer functions. This works for any $a,b > 0$ provided only that (i) T_{12} and T_{21} are not rank deficient, and (ii) that a and b are sufficiently small so that no pole of T_{11}, T_{12} or T_{21} falls outside the circle $\mathcal{C}$ and no zero of T_{12} or T_{21} falls on $\mathcal{C}$.

The effect of the change of variables on a system is easy to compute, even in a state-space framework. For example, if P(s) has state-space realization

$$P(s) = C(Is-A)^{-1}B + D \tag{25}$$

then the conformal map $\tilde{s} = (s+a)/(1+bs)$ yields

$$\tilde{P}(\tilde{s}) = \tilde{C}(I\tilde{s}-\tilde{A})\tilde{B} + \tilde{D} \tag{26}$$

where

$$\left[\begin{array}{c|c} \tilde{A} & \tilde{B} \\ \hline \tilde{C} & \tilde{D} \end{array}\right] = \left[\begin{array}{c|c} (A+aI)(I+bA)^{-1} & (1-ba)(I+bA)^{-1}B \\ \hline C(I+bA)^{-1} & D-bC(I+bA)^{-1}B \end{array}\right]. \tag{27}$$

Similarly, the inverse transformation $s = (\tilde{s}-a)(1-b\tilde{s})$ produces the same effect on the state-space matrices, except that a and b are replaced by -a and -b, respectively.

The upshot is that a simple way to deal effectively with H^∞-problems having $j\omega$-axis zeros is to perform the transformation (27) on the state-space representation of the weighted plant under consideration at the outset, then solve the H^∞-problem (1) for the transformed problem, and finally perform the inverse transformation on the state-space representation of the control law. The resultant H^∞-optimal design is assured of stability because its poles are inside the circle $\mathcal{C}$ of Fig. 1 and it is assured of satisfying the design requirement (1) by virtue of the inequality (19).

In choosing the parameters $a,b > 0$ the least conservative design will be achieved when they are the small, with conservativeness vanishing all together in the limit as $a,b \to 0$. In practice the numerical precision of the computer performing the calculations are the only limit on just how small a and b may be.

3. Cancelling Zeros with Mixed Sensitivity Weighting Functions

Another approach to the problem of $j\omega$-axis zeros in T_{12} and T_{21} that can be effective in many instances is to simply select a suitable "cost-weighting" which cancels the offending zeros. For example, when the cost is the weighted sensitivity function

$$\| SW \|_\infty$$

where $S := (I + PC)^{-1} = (V_\ell - N_r Q)D_\ell$, any $j\omega$-axis zeros of D_ℓ may be elminated by choosing a weighting function W which cancels these zeros.

Similarly, in mixed sensitivity problems such as (14) - (15)

$$\left\| \begin{bmatrix} W_1 S \\ W_2 T \end{bmatrix} \right\|_\infty \equiv \left\| \begin{bmatrix} W_1 (V_\ell - N_r Q) D_\ell \\ W_2 N_r (U_r + Q D_\ell) \end{bmatrix} \right\|_\infty .$$

In this case, any $j\omega$-axis and any infinite zeros of $T_{12} = \begin{bmatrix} -W_1 N_r \\ +W_2 N_r \end{bmatrix}$ may be cancelled by a suitably selected weighting function W_2 having poles at the $j\omega$-axis and infinite zeros of N_r. This works because $T_{11} = \begin{bmatrix} W_1 V_\ell D_\ell \\ W_2 N_r \end{bmatrix}$ so that the $j\omega$-axis and infinite poles of W_2 cancel in this term too.

For example, if $N_r \in H^\infty$ has multiplicity m_i zeros at $s = \pm j\omega_i$ $(i=1,\ldots,m)$, then for any $\varepsilon_i > 0$ the weighting function

$$W_\varepsilon(s) := \prod_i \left(1 + \frac{\varepsilon_i s}{s^2 + \omega_i^2}\right)^{m_i}$$

has the properties that

i) $W_\varepsilon(s) N_r(s) \in H^\infty$

ii) $\lim_{\varepsilon \to 0} W_\varepsilon(j\omega)\, N_r(j\omega) = N_r(j\omega) \quad . \quad \forall\, \omega \neq \omega_i \; ; \quad i = 1,\ldots,m.$

Thus, the transformed mixed sensitivity problem

$$\tilde{W}_2(s) \leftarrow W_\varepsilon(s) W_2(s)$$

has the same solution as the original problem in the limit as $\varepsilon \to 0$, and "nearly" the same solution (except in small neighborhood of each point ω_i $(i=1,\ldots,m)$) for any sufficiently small ε_i's. In the case of m_0 zeros at $z_i = 0$, a multiplier of the form

$$W_\varepsilon(s) = \left(1 + \frac{\varepsilon_0}{s}\right)^{m_0}$$

would be appropriate, whereas in the case of m_∞ zeros at ∞ an appropriate multiplier would be

$$W_\varepsilon(s) = (1 + \varepsilon_\infty s)^{m_\infty}$$

Thus, in general, for single-input-single output plants the weight $W_\varepsilon(s)$ would take the form

$$W_\varepsilon(s) = \left(1 + \frac{\varepsilon_0}{s}\right)^{m_0} (1 + \varepsilon_\infty s)^{m_\infty} \prod_i \left(1 + \frac{\varepsilon s}{s^2 + \omega_i^2}\right)^{m_i} .$$

In the multivariable case, the generation of suitable multipliers to cancel $j\omega$-axis zeros becomes complicated slightly by the fact that associated with each zero is zero vector. However, the difficulties are not insurmountable. Indeed, an H^{∞}-aircraft control design example has been treated via this approach by Safonov and Chiang [13]. The example there involves a 2-input 2-output plant with 2 zeros at ∞, one of multiplicity 1 and the other of multiplicity 2. However, a general methodology for dealing with $j\omega$-axis zeros and zeros at ∞ is not developed in [13].

Our objective here is, given a state-space description of a transfer matrix $P(s)$, to describe a general methodology for finding a class of rational weighting matrices $W_{\varepsilon}(s)$ parameterized by ε such that $W_{\varepsilon}(s)P(s)$ is in H^{∞} and has no $j\omega$-axis or infinite zeros. Further, it is desired that $\lim_{\varepsilon\to 0} W_{\varepsilon}(j\omega) = I$, except at those values of ω at which $P(s)$ has $j\omega$-axis or infinite zeros. The problem of finding W_{ε} such that $P(s)W_{\varepsilon}(s)$ has no $j\omega$-axis or infinite zeros is similar and so will not be explicitly treated.

Following is an algorithm for generating suitable matrices $W_{\varepsilon}(s)$. We assume that we have a state space realization for $P^T(s)$:

$$P^T(s) = C(Is-A)^{-1}B + D$$

where $(\ )^T$ denotes transpose.

As is well known (e.g., [14]) the zeros z_i and zero vectors u_i of $P(s)$ are solutions of

$$\begin{bmatrix} Iz_i-A & B \\ C & D \end{bmatrix} \begin{bmatrix} x_i \\ u_i \end{bmatrix} = 0 ;$$

The solutions z_i, x_i, u_i may be found, for example, by applying the QZ algorithm to find orthogonal constant matrices Q and Z such that

$$Q\begin{bmatrix} Is-A & B \\ C & D \end{bmatrix}Z = (Es + F)$$

where E and F are upper triangular matrices of the forms

$$E = \begin{bmatrix} \alpha_1 & x & x & x & \cdots & & \cdots x \\ 0 & \alpha_2 & x & x & \cdots & & \cdots x \\ & \cdots & & & & \cdots & \\ & \cdots & & & & \cdots & \\ 0 & \cdots & & & 0 & \alpha_m & x \cdots x \end{bmatrix}$$

$$F = \begin{bmatrix} \beta_1 & x & x.. & ... & x \\ 0 & \beta_2 & x.. & ... & x \\ \vdots & & & \vdots & \\ 0... & 0 & \beta_m & x... & x \end{bmatrix} .$$

The zeros z_i of P are

$$z_i = \begin{cases} \infty & ; \text{ if } \alpha_i = 0 \text{ and } \beta_i \neq 0 \\ \beta_i/\alpha_i & , \text{ if } \alpha_i \neq 0 \end{cases} .$$

and the first column of the orthogonal matrix Z is the vector (x_1^T, u_1^T), i.e.,

$$\begin{bmatrix} x_1 \\ u_1 \end{bmatrix} = \begin{bmatrix} z_{11} \\ \vdots \\ z_{1m} \end{bmatrix} .$$

Givens rotations [15] can be used to re-order the (α_i, β_i) pairs so that all of the $(z_i\ \ , x_i, u_i)$ solutions may be computed. (Note: If for some i, $\alpha_i = \beta_i = 0$ then $T_{21}(s)u_i \equiv 0 \quad \forall\ s$, i.e., $T_{21}(s)$ is rank deficient.)

Having thus obtained the zero/zero-vector pairs (z_i, u_i) an initial guess for the weight matrix $W_\varepsilon(s)$ may now be constructed as follows

$$W(s) = I + \frac{1}{2} \sum_{j\omega_i \in Z_I} \frac{\varepsilon_i}{s - j\omega_i} u_i u_i^* + \left(\frac{\varepsilon_i}{s + j\omega_i} \right) \bar{u}_i \bar{u}_i^* + \sum_{z_i \in Z_\infty} \frac{\varepsilon_i s}{2} (u_i u_i^* + \bar{u}_i \bar{u}_i^*)$$

where

Z_I := the set of finite imaginary-axis zeros z_i of P(s)

Z_∞ := the set of infinite zeros z_0 of P(s).

Here $\bar{u}_i$ denotes the complex-conjugate of u_i and u_i^* denotes the transpose of $\bar{u}_i$.

At this point, provided that all the $j\omega$-axis and infinite zeros of P(s) have multiplicity one, the matrix $W_\varepsilon P$ has all its $j\omega$-axis and infinite zeros "cancelled," as desired. However, if one or more of these zeros of P have multiplicity greater than one, then one may iterate the procedure as follows

i) Let $W_{OLD} \leftarrow W_\varepsilon$

$P \leftarrow W_{OLD}P$

ii) Compute W_ε for P, as before

iii) If $W_\varepsilon P$ has no jω-axis or infinite zeros, stop; otherwise go to step (i).

On completion of this iteration $W_\varepsilon P$ is analytic on the jω-axis and at infinity, as desired.

4. Conclusions

In H^∞–optimal control problems, jω-axis plant poles and zeros and strictly proper plants lead to certain intrinsic ill-conditioning problems. Two methods have been presented for overcoming these problems. The first method involves a simple conformal map of the s-plane which for small a,b > 0 shifts the jω-axis and infinite zeros and poles slightly, so that they lie into interior of the right-half s-plane. This approach is completely general and works in all situations.

The second method proposed involving weighting functions to cancel poles is more specialized, but works well for mixed sensitivity problems involving strictly proper plants and plants with jω-axis zeros. It offers the advantage of a separate ε_i for each zero, providing much greater flexibility than the s-plane conformal map procedure. The second method has been demonstrated in a multivariable aircraft design example [13].

References

[1] R. Y. Chiang and M. G. Safonov, "The LINF Computer Program for L^∞-Controller Design," USC E.E. Report EECG-0785-1, July 1986.
[2] M. G. Safonov, E. Jonckheere, M. Verma, and D.J.N. Limebeer, "Synthesis of Positive Real Multivariable Feedback Systems," to appear Int. J. Control.
[3] J. C. Doyle, "Advances in Multivariable Control," ONR/Honeywell Workshop, Minneapolis, MN, October 1984.
[4] J. C. Doyle, "Synthesis of Robust Controllers and Filters," Proc. IEEE Conf. on Decision and Control, San Antonio, TX, December 1985.
[5] G. Zames, "Feedback and Optimal Sensitivity: Model Reference Transformations, Multiplicative Seminorms, and Approximate Inverses," IEEE Trans. on Automatic Control, Vol. AC-26, No. 2, pp. 301-320, April 1981.

[6] M. G. Safonov and M. Verma, "Multivariable L^∞- Sensitivity Optimization and Hankel Approximation," Proc. American Control Conf., San Francisco, CA, June 22-24, 1983; also, IEEE Trans. on Autom. Contr., AC-30, pp. 279-280, 1985.
[7] B.-C. Chang and J. B. Pearson, "Optimal Disturbance Reduction in Linear Multivariable Systems," IEEE Trans. on Automatic Control, AC-29, pp. 863-887, 1984.
[8] G. Zames and B. A. Francis, "Feedback, Minimax Sensitivity and Optimal Robustness," IEEE Trans. on Automatic Control, Vol. AC-28, pp. 585-600, May 1983.
[9] B. Francis, J. W. Helton, and G. Zames, "H^∞-Optimal Feedback Controllers for Linear Multivariable Systems," IEEE Trans. on Automatic Control, Vol. AC-29, No. 10, pp. 888-900, Oct. 1984.
[10] H. Kwakernaak, "Robustness Optimization of Linear Feedback Systems," IEEE Trans. on Automatic Control, Vol. 30, pp. 994, 1985.
[11] M. G. Safonov and M. Verma, "L^∞—Optimization and Hankel Approximation," IEEE Trans. on Automatic Control, Vol. AC-30, No. 3, pp. 279-280, March 1985.
[12] M. S. Verma and E. A. Jonckheere, "L^∞—Compensation with Mixed Sensitivity as a Broadband Matching Problem," Systems and Control Letters, 4, pp. 125-129, 1984.
[13] M. G. Safonov and R. Y. Chiang, "CACSD Using the State-Space L^∞-Theory - A Design Example," Proc. IEEE Conf. in CACSD, Washington, D.C., September 24-26, 1986.
[14] T. Kailath, Linear Systems, Englewood Cliffs, NJ: Prentice-Hall, 1980, pp. 448-450.
[15] G. H. Golub and C. F. Van Loan, Matrix Computations, Baltimore, MD: Johns Hopkins University Press, 1983.
[16] B. A. Francis and G. Zames, "On H^∞-Optimal Sensitivity Theory for SISO Feedback Systems," IEEE Trans. on Automatic Control, AC-29, pp. 9-16, 1984.

A Polynomial Approach to H_∞- Optimization of Control Systems

Huibert Kwakernaak

University of Twente
P.O. Box 217, 7500 AE Enschede
The Netherlands

Abstract. Following Francis and Doyle (1986), the "standard" H_∞ control system optimization problem is converted to a generalized Nehari problem, whose solution is subsequently obtained by solving a set of polynomial matrix equations. By way of example a two-degree-of-freedom control problem is worked out.

1. Introduction

We consider the "standard" H_∞ control system optimization problem as formulated by Francis and Doyle (1986). In the block diagram of Fig. 1, the real rational plant transfer function G is partitioned as

$$G = \begin{bmatrix} G_{11} & G_{21} \\ G_{21} & G_{22} \end{bmatrix}. \tag{1}$$

Then the transfer function H from the external input w to the control error z is

$$H = G_{11} + G_{12}K(I - G_{22}K)^{-1}G_{21}. \tag{2}$$

The standard problem is to minimize $\|H\|_\infty$ with respect to the compensator transfer matrix K, subject to stability of the closed-loop system.

Our approach in this paper is first to follow Francis and Doyle in converting the standard problem to a generalized Nehari problem, and then to reduce the Nehari problem to the solution of a set of polynomial matrix equations.

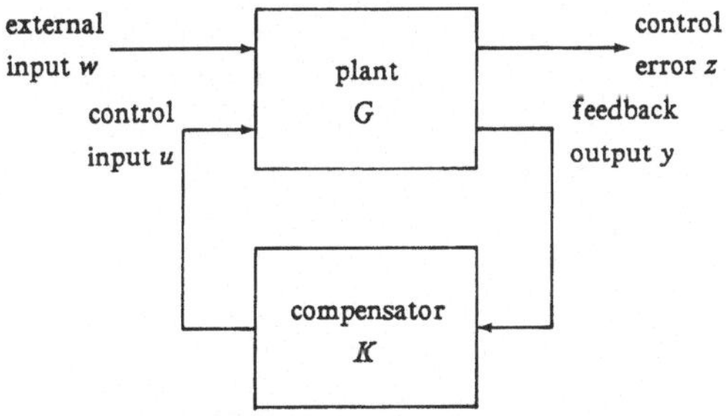

Fig. 1. Configuration of the standard control problem.

NATO ASI Series, Vol. F34
Modelling, Robustness and Sensitivity Reduction
in Control Systems. Edited by R. F. Curtain

2. Parametrization and Reformulation of the Problem

It is assumed that the plant G is stabilizable, i.e., that there exists a real rational compensator transfer matrix K that stabilizes the feedback system. Then (Francis and Doyle, 1986) K stabilizes G if and only if K stabilizes G_{22}.

First suppose that G_{22} is stable. Then K stabilizes G_{22} if and only if $K(I-G_{22}K)^{-1}$ is stable (Zames, 1981). Thus we may as well consider the problem of minimizing $\|H\|_\infty$ with respect to all stable $\tilde{K}$, where

$$H = T_0 + T_1\tilde{K}\,T_2, \tag{3}$$

with $T_0 := G_{11}$, $T_2 := G_{12}$, $T_2 := G_{21}$, and $\tilde{K} := K(I-G_{22}K)^{-1}$.

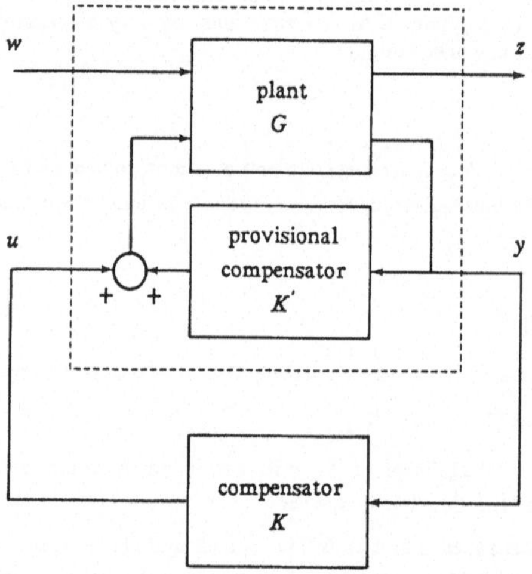

Fig. 2. Standard configuration with a provisional stabilizing compensator K'

When G_{22} is not stable, one can first determine a provisional stabilizing compensator K' as in Fig. 2. The subblocks of the transfer matrix G' of the extended plant can easily found to be

$$G'_{11} = G_{11} + G_{12}K'(I-G_{22}K')^{-1}G_{21}, \tag{4}$$

$$G'_{12} = G_{12}(I-K'G_{22})^{-1}, \tag{5}$$

$$G'_{21} = (I-G_{22}K')^{-1}G_{21}, \tag{6}$$

$$G'_{22} = (I-G_{22}K')^{-1}G_{22}. \tag{7}$$

We again consider the problem of minimizing $\|H\|_\infty$ as given by (3) with respect to all stable $\tilde{K}$, where now $T_0 := G'_{11}$, $T_1 := G'_{12}$, $T_2 := G'_{21}$, and $\tilde{K} := K(I-G'_{22}K)^{-1}$.

We thus study the problem of minimizing $\|H\|_\infty$ as given by (3) with respect to all stable $\tilde{K}$. Following Francis and Doyle (1986), write $T_1 = T_1^i T_1^o$ and $T_2 = T_2^{co} T_2^{ci}$, with T_1^i inner, T_1^o outer, T_2^{co} co-outer, and T_2^{ci} co-inner. Defining $\hat{K} := T_1^o \tilde{K} T_2^{co}$, we then consider the equivalent problem of minimizing the ∞-norm of $H = T_o + T_1^i \hat{K} T_2^{ci}$ with respect to all real rational stable $\hat{K}$. Still following Francis and Doyle, let

$$E_1 = \begin{bmatrix} T_1^{i*} \\ I - T_1^i T_1^{i*} \end{bmatrix}, \quad E_2 = \begin{bmatrix} T_2^{ci*} & I - T_2^{ci*} T_2^{ci} \end{bmatrix}, \tag{8}$$

where if A is a real rational matrix then $A^*(s) := A^T(-s)$, with the superscript ^{T} denoting the transpose. E_1 is inner, E_2 co-inner, and $\|H\|_\infty = \|\tilde{H}\|_\infty$, where $\tilde{H} := E_1 H E_2$. Furthermore,

$$E_1 T_1^i = \begin{bmatrix} I \\ 0 \end{bmatrix}, \quad T_2^{ci} E_2 = \begin{bmatrix} I & 0 \end{bmatrix}, \tag{9}$$

so that

$$\tilde{H} = R + \begin{bmatrix} I \\ 0 \end{bmatrix} \hat{K} \begin{bmatrix} I & 0 \end{bmatrix}, \tag{10}$$

where $R := E_1 T_0 E_2$. Partitioning

$$R = \begin{bmatrix} R_{11} & R_{12} \\ R_{21} & R_{22} \end{bmatrix}, \tag{11}$$

we thus consider the problem of minimizing

$$\|H\|_\infty = \left\| \begin{bmatrix} R_{11} + \hat{K} & R_{12} \\ R_{21} & R_{22} \end{bmatrix} \right\|_\infty \tag{12}$$

with respect to all real rational stable $\hat{K}$. This is what is known as the *generalized Nehari problem.*

In the following, R_{11} is assumed to be square. If it is not, it can be augmented to a square matrix by adding a number of zero rows to R_{11} and R_{12} or a number of zero columns to R_{11} and R_{21}. It can easily be shown that by deleting the augmented rows or columns from any (square) $\hat{K}$ that solves the augmented problem a (nonsquare) $\hat{K}$ results that solves the original problem.

3. Sufficient Condition for Optimality

We consider the problem of minimizing (12), where from now on the circumflex on K is omitted. Defining the quadratic expression

$$V_K = \begin{bmatrix} R_{11} + \hat{K} & R_{12} \\ R_{21} & R_{22} \end{bmatrix}^* \begin{bmatrix} R_{11} + \hat{K} & R_{12} \\ R_{21} & R_{22} \end{bmatrix}, \tag{13}$$

we note that $\|H\|_\infty^2 = \|V_K\|_\infty$, so that henceforth we consider the minimization of $\|V_K\|_\infty$. We define another quadratic expression

$$Z_{K,L} := V_K + L^* L, \tag{14}$$

where L is real rational, not necessarily stable, and has less than full rank. Let Φ be a real rational spectral density matrix (i.e., $\Phi^* = \Phi$, Φ is strictly proper and $\Phi(i\omega) \geqslant 0$ for all real ω), and consider the *auxiliary problem*, which is the problem of minimizing

$$\operatorname{tr} \int_{-\infty}^{\infty} \Phi(i\omega) Z_{K,L}(i\omega)\, d\omega \tag{15}$$

with respect to all real rational stable K and all real rational L of less than full rank.

Lemma 1. Suppose that there exists a Φ such that the auxiliary problem has an *equalizing* solution K_o, L_o, i.e., there exists a real constant λ_o such that $Z_{K_o,L_o} = \lambda_o^2 I$. Then K_o minimizes $\| V_K \|_\infty$. ■

Proof. We have $\| V_{K_o} \|_\infty = \| Z_{K_o,L_o} - L_o^* L_o \|_\infty = \| \lambda_o^2 I - L_o^* L_o \|_\infty = \lambda_o^2$. Suppose that there exists a K_* such that $\| V_{K_*} \|_\infty < \| V_{K_o} \|_\infty = \lambda_o^2$. Then $V_{K_*}(i\omega) < \lambda_o^2 I$ for all real ω and hence

$$\operatorname{tr} \int_{-\infty}^{\infty} Z_{K_*,0}(i\omega)\Phi(i\omega)\, d\omega = \operatorname{tr} \int_{-\infty}^{\infty} V_{K_*}(i\omega)\Phi(i\omega)\, d\omega$$

$$< \operatorname{tr} \int_{-\infty}^{\infty} \lambda_o^2 \Phi(i\omega)\, d\omega = \operatorname{tr} \int_{-\infty}^{\infty} Z_{K_o,L_o}(i\omega)\Phi(i\omega)\, d\omega, \tag{16}$$

which contradicts the assumption that K_o, L_o solves the auxiliary problem. Hence the hypothesis that there exists a K_* such that $\| V_{K_*} \|_\infty < \| V_{K_o} \|_\infty$ is false. ■

4. Equalizing Solutions

The lemma of the preceding section suggests considering equalizing solutions, i.e., K, L for which there exists a real constant λ such that $Z_{K,L} = \lambda^2 I$. Partitioning $L = [L_1 \;\; L_2]$, where L_2 is square and has the same number of columns as R_{12} and R_{22}, it is easily found that $Z_{K,L} = \lambda^2 I$ is equivalent to

$$(R_{11} + K)^*(R_{11} + K) + R_{21}^* R_{21} + L_1^* L_1 = \lambda^2 I, \tag{17}$$

$$(R_{11} + K)^* R_{12} + R_{21}^* R_{22} + L_1^* L_2 = 0, \tag{18}$$

$$R_{12}^* R_{12} + R_{22}^* R_{22} + L_2^* L_2 = \lambda^2 I. \tag{19}$$

From (19) it follows that for λ sufficiently large L_2 is invertible so that from (18) we have $L_1^* = -[(R_{11} + K)^* R_{12} + R_{21}^* R_{22}] L_2^{-1}$. Substitution of this into (17), completion of the square (in $R_{11}+K$) and some algebraic manipulation yield with the use of (19) an expression of the form

$$(K + S_\lambda)^* \Gamma_\lambda^{-*} \Gamma_\lambda^{-1} (K + S_\lambda) = \lambda^2 \Delta_\lambda^* \Delta_\lambda. \tag{20}$$

Here Γ_λ and Δ_λ are left respectively right spectral factors defined by

$$\Gamma_\lambda \Gamma_\lambda^* = I - \frac{1}{\lambda^2} R_{12} (I - \frac{1}{\lambda^2} R_{22}^* R_{22})^{-1} R_{12}^*, \tag{21}$$

$$\Delta_\lambda^* \Delta_\lambda = I - \frac{1}{\lambda^2} R_{21}^* (I - \frac{1}{\lambda^2} R_{22} R_{22}^*)^{-1} R_{21}. \tag{22}$$

and where

$$S_\lambda = R_{11} + \frac{1}{\lambda^2} R_{12}(I - \frac{1}{\lambda^2} R_{22}^* R_{22})^{-1} R_{22}^* R_{21}. \tag{23}$$

If (20) is satisfied there exists an inner matrix U such that $\Gamma_\lambda^{-1}(K+S_\lambda) = \lambda U \Delta_\lambda$, or

$$K = \lambda \Gamma_\lambda U \Delta_\lambda - S_\lambda. \tag{24}$$

The problem now is to determine an inner matrix U such that K is stable. To this end, define the polynomial matrices D_λ, N_λ, $\hat{D}_\lambda$, $D_{+,\lambda}$ and $D_{-,\lambda}$ by letting

$$\Gamma_\lambda^{-1} S_\lambda \Delta_\lambda^{-1} = D_\lambda^{-1} N_\lambda, \tag{25}$$

with D_λ and N_λ not necessarily left coprime,

$$D_\lambda D_\lambda^* = \hat{D}_\lambda \hat{D}_\lambda^*, \tag{26}$$

with $\hat{D}_\lambda$ stable, and

$$D_\lambda = D_{+,\lambda} D_{-,\lambda}, \tag{27}$$

with $D_{+,\lambda}$ antistable and $D_{-,\lambda}$ stable.

Lemma 2. Suppose that there exist polynomial matrices Q_λ, $\hat{P}_\lambda$ and P_λ, with P_λ stable (i.e., $\det(P_\lambda)$ has all its roots in the left-half complex plane), satisfying the polynomial matrix equations

$$\frac{1}{\lambda} N_\lambda P_\lambda = \hat{D}_\lambda \hat{P}_\lambda - D_{+,\lambda} Q_\lambda, \quad P_\lambda^* P_\lambda = \hat{P}_\lambda^* \hat{P}_\lambda. \tag{28}$$

Then

$$K_\lambda = \lambda \Gamma_\lambda D_{-,\lambda}^{-1} Q_\lambda P_\lambda^{-1} \Delta_\lambda, \quad U_\lambda = D_\lambda^{-1} \hat{D}_\lambda \hat{P}_\lambda P_\lambda^{-1} \tag{29}$$

satisfy (24), and hence K_λ is a stable equalizing solution. ■

The proof of the lemma is by substitution.

5. Reduced Solutions

Equalizing solutions may be obtained by solving the polynomial matrix equations (28). Inspection of (29) shows that the equalizing solution K_λ is stable if and only if P_λ is stable. Consider (28) for $\lambda = \infty$. Then the first polynomial equation reduces to $0 = \hat{D}_\infty \hat{P}_\infty - D_{+,\infty} Q_\infty$, which may be solved for $\hat{P}_\infty$ and Q_∞. By substitution of $\hat{P}_\infty$ into the second equation of (28) and spectral factorization a stable P_∞ may be found. As in Kwakernaak (1986), the implicit function theorem may be invoked to show that if the matrix fraction (25) is suitably arranged, namely such that $\hat{P}_\infty$ can be chosen to be column reduced with column degrees that mutually differ by at most one, (28) has a solution such that P_λ is stable for all sufficiently large λ.

As λ decreases, the roots of P_λ (i.e., the roots of $\det(P_\lambda)$) move away from their initial locations. Eventually one of these roots may cross over from the left into the right half of the complex plane. Let λ_{opt} be the value of λ for which the first root of P_λ crosses over from the left- to the right-half plane, either via the imaginary axis or via infinity. Then since K_λ as given by (29) is equalizing, it remains finite on the imaginary axis also when $\lambda = \lambda_{opt}$, which means that the root of $P_{\lambda_{opt}}$ on the imaginary axis must be canceled by a right factor of $Q_{\lambda_{opt}}$ that has that same root. Since also $\hat{P}_{\lambda_{opt}}$ must have this right factor, it can be canceled from the solution triple $Q_{\lambda_{opt}}$, $\hat{P}_{\lambda_{opt}}$, $P_{\lambda_{opt}}$. We call the result a *reduced solution* of (29). In the following section it is shown that such a reduced solution solves the H_∞-optimization problem.

In some cases there is no reduced solution. Let λ_1 be the value of $|\lambda|$ below which the spectral factorization (21) does not exists, and λ_2 that below which the factorization (22) does not exist. It can be shown that

$$\lambda_1^2 = \left\| \begin{bmatrix} R_{12} \\ R_{22} \end{bmatrix} \right\|_\infty, \qquad \lambda_2^2 = \left\| \begin{bmatrix} R_{21} & R_{22} \end{bmatrix} \right\|_\infty, \tag{30}$$

which both are lower bounds for $\|H\|_\infty$ (Francis and Doyle, 1986). When no root of P_λ crosses over into the right-half complex plane until λ reaches the value $\max(\lambda_1, \lambda_2)$, the solution for $\lambda = \max(\lambda_1, \lambda_2)$ is optimal but not reduced.

6. Optimality of Reduced Solutions

Equalizing solutions only solve the H_∞ problem when they minimize the auxiliary problem for some Φ. Using standard variational arguments it is found that sufficient conditions for the pair K, L to minimize (15) are that

$$\Phi L^* = 0, \qquad \Phi_{11}(R_{11} + K)^* + \Phi_{12}R_{12}^* \;\text{ is stable,} \tag{31}$$

where

$$\Phi = \begin{bmatrix} \Phi_{11} & \Phi_{12} \\ \Phi_{21} & \Phi_{22} \end{bmatrix}. \tag{32}$$

The first of the conditions (31) is satisfied when $\Phi_{12} = \Phi_{21}^* = -\Phi_{11}L_{11}^*L_2^{-*}$, $\Phi_{22} = L_2^{-1}L_1\Phi_{11}L_1^*L_2^{-*}$. Assuming this to hold, the remaining sufficient condition for optimality takes the form that $\Phi_{11}[(R_{11}+K)^* - L_1^*L_2^{-*}R_{12}^*]$ be stable. Using (17, 18, 19) to eliminate L_1 and L_2 it is easily found that for equalizing solutions the optimality condition is that $\Phi_{11}(K+S_\lambda)^*\Gamma_\lambda^{-*}\Gamma_\lambda^{-1}$ be stable. With (24) we finally obtain that the equalizing solution K_λ minimizes the auxiliary criterion if

$$\lambda\Phi_{11}\Delta_\lambda^* U_\lambda^* \Gamma_\lambda^{-1} \tag{33}$$

is stable.

Define $\tilde{D}_{-,\lambda}$ stable and $\tilde{D}_{+,\lambda}$ antistable such that $D_\lambda = \tilde{D}_{-,\lambda}\tilde{D}_{+,\lambda}$. Then it may easily be verified that $\hat{D}_\lambda = \tilde{D}_{-,\lambda}D_{\sim,\lambda}$, where $D_{\sim,\lambda}$ is a left spectral factor defined by $D_{\sim,\lambda}D_{\sim,\lambda}^* = \tilde{D}_{+,\lambda}\tilde{D}_{+,\lambda}^*$. It follows that $D_\lambda^{-1}\hat{D}_\lambda = \tilde{D}_{+,\lambda}^{-1}D_{\sim,\lambda}$, and substitution of U_λ as given by (29) into (33) yields

$$\lambda\Phi_{11}\Delta_\lambda^* P_\lambda^{-*}\hat{P}_\lambda^* D_{\sim,\lambda}^* \tilde{D}_{+,\lambda}^{-*}. \tag{34}$$

Lastly, let $E_{+,\lambda}$ be antistable and $E_{-,\lambda}$ stable such that $E_{+,\lambda}E_{-,\lambda} = D_{\sim,\lambda}\hat{P}_\lambda$. Then it is easily checked that (34) is stable if

$$\Phi_{11} = \Delta_\lambda^{-1} P_\lambda E_{-,\lambda}^{-1} \, C \, E_{-,\lambda}^{-*} \Delta_\lambda^{-*}, \tag{35}$$

with C a constant nonnegative-definite symmetric matrix to be determined.

Thus it looks as if *any* equalizing solution minimizes the auxiliary criterion, provided Φ_{11} is chosen as in (35). The hitch is that Φ_{11} is not necessarily a spectral density matrix. It is para-Hermitian (i.e., $\Phi_{11}^* = \Phi_{11}$) and nonnegative-definite on the imaginary axis, but it may not be strictly proper. As Δ_λ is biproper, the properness of Φ_{11} is determined by that of $P_\lambda E_{-,\lambda}^{-1}$. It may be verified by substitution that $P_\infty = \tilde{D}_{+,\infty}^*$, $\hat{P}_\infty = D_{\sim,\infty}^*$ and $Q_\infty = D_{-,\infty}\tilde{D}_{+,\infty}^*$ satisfy (28) for $\lambda = \infty$, and that $E_{+,\infty} = \tilde{D}_{+,\infty}$ and $E_{-,\infty} = \tilde{D}_{+,\infty}^*$ form a correct factorization of $\tilde{D}_{\sim,\infty}\hat{P}_\infty$. Thus, for $\lambda = \infty$ we have $P_\lambda E_{-,\lambda}^{-1} = I$, which is proper but not strictly proper. If the various factorizations are correctly attended to with due regard for column and row reducedness, $P_\lambda E_{-,\lambda}^{-1}$ can be kept proper but not strictly proper as λ decreases, until λ_{opt} is reached. At this point the cancellation in $P_{\lambda_{\text{opt}}}$ makes it possible to construct a constant symmetric nonnegative-definite matrix C such that $P_{\lambda_{\text{opt}}} E_{-,\lambda_{\text{opt}}}^{-1} \, C \, E_{-,\lambda_{\text{opt}}}^{-*} P_{\lambda_{\text{opt}}}^*$ and hence also Φ_{11} is strictly proper (see Kwakernaak, 1986). It follows that the reduced solution is optimal.

The solution of the H_∞ optimization problem is highly nonunique. Let K be any (not necessarily equalizing) optimal solution, and K_o, L_o an equalizing optimal solution pair. Then clearly $V_K(i\omega) \leq \lambda_{\text{opt}}^2 I = Z_{K_o,L_o}(i\omega)$ for all real ω. It follows that the pair K, 0 also minimizes the auxiliary criterion, and hence

$$\text{tr} \int_{-\infty}^{\infty} V_K(i\omega)\,\Phi(i\omega)\,d\omega = \text{tr} \int_{-\infty}^{\infty} V_{K_o}(i\omega)\,\Phi(i\omega)\,d\omega. \tag{36}$$

In a straightforward fashion this leads to the conclusion that $(K-K_o)\Phi_{11} = 0$. If Φ_{11} is singular, which is normally the case, nontrivial $K-K_o$ exist. This necessary condition is the starting point for determining nonequalizing solutions and other equalizing solutions from the given equalizing solution K_o.

11. Numerical Solution

The H_∞-optimization problem can be numerically solved by first solving the equations (28) at $\lambda = \infty$ as described in the previous section, and then following the solution as a function of λ, while decreasing λ, until one of the roots of P_λ crosses over into the right-half plane. Newton's algorithm has been successfully used for solving equations similar to (28) (Boekhoudt, 1984). Near the optimal solution the Jacobian of the equations becomes singular, which impedes convergence. This difficulty may be overcome by identifying the factor that is about to cancel as the optimal solution is approached, removing this factor approximately, and solving for the reduced solution directly by including λ as a variable. The latter problem is well-conditioned and an exact solution of the H_∞ problem is obtained, within the usual limitations of numerical accuracy.

Alternatively, a search method similar to that proposed by Doyle (Francis and Doyle, 1986) may be envisaged. To this end, the equations (28) are modified to

$$\frac{1}{\lambda} N_\mu P_\lambda = \hat{D}_\mu \hat{P}_\lambda - D_{+,\mu} Q_\lambda, \quad P_\lambda^* P_\lambda = \hat{P}_\lambda^* \hat{P}_\lambda. \tag{37}$$

First a value of μ is picked and the coefficient matrices N_μ, $\hat{D}_\mu$ and $D_{+,\mu}$ are computed. Then a reduced solution of (37) is obtained by the method described above, while varying λ alone. Keeping the algorithm in the reduced solution mode, μ and correspondingly the coefficient matrices are varied until λ and μ coincide. In this approach, the Newton equations that need be solved are always well-conditioned and the algorithm is simpler than the first. The line search can be made efficient and accurate.

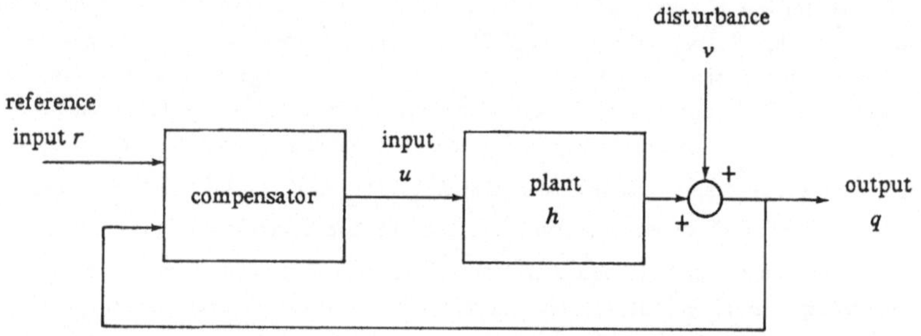

Fig. 3. Two-degree-of-freedom SISO control system.

12. Application: Two-Degree-of-Freedom Optimization of SISO Systems

By way of application we consider in this section the two-degree-of-freedom SISO system of Fig. 3. We wish to optimize the closed-loop system response to the reference input r and the disturbance v. To bring the problem into standard form, let

$$v = V_1 w_1, \qquad r = V_2 w_2, \tag{38}$$

where w_1 and w_2 are the two components of the external input w, and V_1 and V_2 are suitable shaping filters. Furthermore, we choose the components of the control error z as

$$z_1 = W_1(r - q), \qquad z_2 = W_2 u, \tag{39}$$

where W_1 and W_2 are frequency dependent weighting functions, q is the plant output and u the plant input. It is seen that z_1 is the weighted tracking error and z_2 the weighted plant input. As $r = V_2 w_2$ and $q = v + hu = V_1 w_1 + hu$, we may rewrite (39) in the form

$$z_1 = W_1(V_2 w_2 - V_1 w_1 - hu), \qquad z_2 = W_2 u. \tag{40}$$

The two components y_1 and y_2 of the observed output y are

$$y_1 = q = V_1 w_1 + hu, \qquad y_2 = r = V_2 w_2. \tag{41}$$

Finally, the controlled input is of course the plant input u. Joining the various relations it follows

$$\begin{bmatrix} z_1 \\ z_2 \\ y_1 \\ y_2 \end{bmatrix} = \begin{bmatrix} -W_1V_1 & W_1V_2 & -W_1h \\ 0 & 0 & W_2 \\ V_1 & 0 & h \\ 0 & V_2 & 0 \end{bmatrix} \begin{bmatrix} w_1 \\ w_2 \\ u \end{bmatrix} = G \begin{bmatrix} w_1 \\ w_2 \\ u \end{bmatrix}. \tag{42}$$

When the plant is unstable, we need a stabilizing provisional compensator, whose transfer matrix we take of the form

$$K' = [k \;\; 0], \tag{43}$$

where k stabilizes the plant h. We find for the extended system

$$G'_{11} = \begin{bmatrix} -W_1 V_1 \dfrac{1}{1-hk} & W_1 V_2 \\ W_2 V_1 \dfrac{k}{1-hk} & 0 \end{bmatrix}, \qquad G'_{12} = \begin{bmatrix} -W_1 \dfrac{h}{1-hk} \\ W_2 \dfrac{1}{1-hk} \end{bmatrix}, \tag{44}$$

$$G'_{21} = \begin{bmatrix} V_1 \dfrac{1}{1-hk} & 0 \\ 0 & V_2 \end{bmatrix}, \qquad G'_{22} = \begin{bmatrix} \dfrac{h}{1-hk} \\ 0 \end{bmatrix}. \tag{45}$$

In polynomial form we now assume

$$V_1 = \frac{m_1}{d}, \quad V_2 = \frac{m_2}{d_2}, \quad W_1 = \frac{p_1}{q_1}, \quad W_2 = \frac{p_2}{q_2}, \quad h = \frac{n}{d}, \quad k = \frac{y}{x}, \tag{46}$$

with n and d coprime, and m_1, m_2, d_2, p_1, q_1, p_2 and q_2 stable polynomials. We furthermore denote

$$dx - ny =: \chi, \tag{47}$$

where the polynomial χ is stable, because by assumption k stabilizes h. With all this we obtain

$$T_0 = G'_{11} = \begin{bmatrix} -\dfrac{p_1}{q_1}\dfrac{m_1 x}{\chi} & \dfrac{p_1}{q_1}\dfrac{m_2}{d_2} \\ \dfrac{p_2}{q_2}\dfrac{m_1 y}{\chi} & 0 \end{bmatrix}, \tag{48}$$

$$T_1 = G'_{12} = \begin{bmatrix} -\dfrac{p_1}{q_1}\dfrac{nx}{\chi} \\ \dfrac{p_2}{q_2}\dfrac{dx}{\chi} \end{bmatrix}, \qquad T_2 = G'_{21} = \begin{bmatrix} \dfrac{m_1 x}{\chi} & 0 \\ 0 & \dfrac{m_2}{d_2} \end{bmatrix}. \tag{49}$$

It is not difficult to do the inner-outer factorizations of T_1 and T_2. They result in

$$T_1^{i} = \begin{bmatrix} -p_1 n \, q_2 \\ p_2 d \, q_1 \end{bmatrix} \frac{x}{\hat{x}\gamma}, \qquad T_1^{o} = \frac{\hat{x}\gamma}{q_1 q_2 \chi}, \tag{50}$$

$$T_2^{co} = \begin{bmatrix} \dfrac{\hat{x}\, m_1}{\chi} & 0 \\ 0 & \dfrac{m_2}{d_2} \end{bmatrix}, \qquad T_2^{ci} = \begin{bmatrix} \dfrac{x}{\hat{x}} & 0 \\ 0 & 1 \end{bmatrix}, \tag{51}$$

where the polynomials $\hat{x}$ and γ are stable spectral factors, respectively defined by

$$\hat{x}^{*}\hat{x} = x^{*}x, \qquad \gamma^{*}\gamma = p_1^{*}p_1 q_2^{*}q_2 n^{*}n + p_2^{*}p_2 q_1^{*}q_1 d^{*}d. \tag{52}$$

After some algebra we obtain the matrix R in the form

$$R = \begin{bmatrix} R_{11} \\ R_{21} \end{bmatrix}, \tag{53}$$

where

$$R_{11} = \left[\left(\frac{x^*}{\hat{x}^*}\right)^2 \frac{m_1 e}{\gamma^* \chi q_1 q_2} \quad -\frac{x^*}{\hat{x}^*} \frac{m_2 p_1^* p_1 q_2^* n^*}{\gamma^* q_1 d_2} \right], \tag{54}$$

$$R_{21} = \frac{p_1 p_2}{\gamma^* \gamma} \begin{bmatrix} q_1^* p_2^* d^* \\ p_1^* q_2^* n^* \end{bmatrix} \left[-m_1 \frac{x^*}{\hat{x}^*} \quad m_2 \frac{d}{d_2} \right]. \tag{55}$$

Here $e := p_1^* p_1 q_2^* q_2 n^* x + p_2^* p_2 q_1^* q_1 d^* y$. We supplement R_{11} with a row of zeros to a 2×2 matrix and observe that $R_{12} = R_{22} = 0$ so that $\Gamma_\lambda = I$. By first computing $\Delta_\lambda^* \Delta_\lambda$ according to (22) and then inverting it, is may be found that

$$\Delta_\lambda^{-1} = \frac{1}{\pi_\lambda} \begin{bmatrix} \hat{x} & 0 \\ 0 & 1 \end{bmatrix}^{-1} \begin{bmatrix} 1 & 0 \\ 0 & d_2 \end{bmatrix} A_\lambda, \tag{56}$$

where the stable polynomial π_λ and the stable polynomial matrix A_λ are defined by

$$\pi_\lambda^* \pi_\lambda = \gamma^* \gamma d_2^* d_2 - \frac{1}{\lambda^2} p_1^* p_1 p_2^* p_2 (m_1^* m_1 d_2^* d_2 + m_2^* m_2 d^* d), \tag{57}$$

$$A_\lambda A_\lambda^* = \begin{bmatrix} (\gamma^* \gamma d_2^* d_2 - \frac{1}{\lambda^2} p_1^* p_1 p_2^* p_2 m_2^* m_2 d^* d) x^* x & -\frac{1}{\lambda^2} p_1^* p_1 p_2^* p_2 m_1^* m_1 d\, x \\ -\frac{1}{\lambda^2} p_1^* p_1 p_2^* p_2 m_1^* m_1 d^* x^* & \gamma^* \gamma - \frac{1}{\lambda_2} p_1^* p_1 p_2^* p_2 m_1^* m_1 \end{bmatrix}. \tag{58}$$

The polynomial matrices N_λ, $\hat{D}_\lambda$ and $D_{+,\lambda}$ that occur in the polynomial equations (28) may now easily be determined. They can be chosen as

$$N_\lambda = \begin{bmatrix} x_+^* m_1 e & -p_1^* p_1 q_2^* q_2 m_2 n^* \chi x_+^* x \\ 0 & 0 \end{bmatrix} A_\lambda, \tag{59}$$

$$\hat{D}_\lambda = \begin{bmatrix} \hat{x} x_+^* \gamma \chi q_1 q_2 \pi_\lambda & 0 \\ 0 & \psi \end{bmatrix}, \quad D_{+,\lambda} = \begin{bmatrix} \gamma^* x_+^2 & 0 \\ 0 & \psi^* \end{bmatrix}, \tag{60}$$

where $x = x_+ x_-$, with x_+ antistable and x_- stable, and where ψ is an arbitrary stable polynomial whose degree is chosen so as to ensure that the column degrees of $\hat{P}_\infty$ mutually differ by at most 1. The matrix $D_{-,\lambda}$ that is needed to determine K_λ is given by

$$D_{-,\lambda} = \begin{bmatrix} \chi q_1 q_2 x_- \pi_\lambda & 0 \\ 0 & 1 \end{bmatrix}. \tag{61}$$

All that remains is to find a reduced solution of the polynomial equations (28).

As an example, consider the following plant transfer function and weighting functions:

$$h(s) = \frac{1}{s-2}, \quad V_1 = \frac{s+1}{s+2}, \quad V_2 = \frac{.25s+1}{2s+1}, \quad W_1 = 1, \quad W_2 = .1. \tag{62}$$

It is not difficult to find that the optimal compensator that solves the one-degree-of-freedom problem (i.e., the two-degree-of-freedom problem with $V_2 = 0$) is given by $K' = [-3, 0]$. We choose this as the provisional stabilizing compensator. It is straighforward but tedious to determine the coefficient matrices N_λ, $\hat{D}_\lambda$ and $D_{+\lambda}$ as given by (59) and (60). The matrix polynomial equations (37) were solved numerically using the search method indicated in Section 11, and without difficulty the solution $\lambda = \mu = 1.0522$ was obtained, with the polynomial matrices P_λ, $\hat{P}_\lambda$ and Q_λ given by

$$P_\lambda = \begin{bmatrix} 3.5509 & -.17260 \\ -.44910 & 1.8274 \end{bmatrix} \quad \hat{P}_\lambda = \begin{bmatrix} 3.5792 & -.40053 \\ 0 & 1.7913 \end{bmatrix},$$

$$Q_\lambda = \begin{bmatrix} -.36107-.43230s-.074740s^2-.00351256s^3 & 1.4684+1.6568s+.19277s^2+.0043569s^3 \\ 0 & 1.7913 \end{bmatrix}. \tag{63}$$

After back substitution, which is again straightforward and tedious, the optimal compensator is found to be given by $K = [k_1, k_2]$, where

$$k_1 = -3.0034 \frac{(s+.48968)(s+.5)(s+1)^2(s+4)(s+10.198)(s+10.195)(s+35.232)}{(s+.48924)(s+.5)(s+1)^2(s+4)(s+10.198)(s+10.197)(s+35.234)},$$

$$k_2 = 3.0914 \frac{(s+.49192)(s+.5)(s+1)^3(s+10.198)(s+10.196)(s+35.358)}{(s+.48924)(s+.5)(s+1)^2(s+4)(s+10.198)(s+10.197)(s+35.234)}. \tag{64}$$

After performing several obvious cancellations it looks as if the optimal compensator is given by

$$k_1 = -3.0034, \quad k_2 = 3.0914 \frac{s+1}{s+4}. \tag{65}$$

13. Conclusions

The paper demonstrates that the "standard" H_∞- optimal control problem, after converting it to a generalized Nehari problem, can be solved by the polynomial approach developed in earlier papers (Kwakernaak 1985, 1986). The numerical example that was treated shows that the method has an inherent difficulty, namely that by the introduction of the provisional stabilizing compensator and the subsequent reduction to Nehari form by pre- and post-multiplication of the closed-loop transfer matrix H by suitable inner matrices, a number of spurious factors are introduced. The result is a Nehari problem of a much higher degree than the original problem. In the final solution the spurious factors cancel. The details of this have been investigated by Limebeer and Hung (1986). To cancel with confidence, it is necessary to compute the final solution with quite good accuracy.

It seems much more attractive to find methods for solving the standard problem that avoid the introduction of spurious factors and do not unnecessarily increase the complexity of the problem. One such method, based on the polynomial approach, is currently being developed.

Acknowledgement. The author gratefully acknowledges Piet Boekhoudt's invaluable help in working out the numerical example.

References

Boekhoudt, P. (1984).

Minimax frekwentiedomeinoptimalisatie van multivariabele lineaire teruggekoppelde systemen. Student project report, Department of Applied Mathematics, Twente University of Technology.

Francis, B. and J. C. Doyle (1986).

Linear control theory with an H_∞ optimality criterion. To appear, *SIAM J. Optimization.*

Kwakernaak, H. (1985)

Minimax frequency domain performance and robustness optimization of linear feedback systems. *IEEE Trans. Aut. Control*, **AC-30**, pp. 994-1004.

Kwakernaak, H. (1986).

A polynomial approach to minimax frequency domain optimization of multivariable feedback systems. *Int. J. Control*, **44**, pp. 117-156.

Limebeer, D.J.N. and Y.S. Hung (1986).

An analysis of the pole-zero cancellations in H^∞- optimal control problems of the first kind. Imperial College, Department of Electrical Engineering.

Zames, G.A. (1981).

Feedback and optimal sensitivity: Model reference transformation, seminorms, and approximate inverses. *IEEE Trans. Aut. Control*, **AC-26**, pp. 301-320 (previously presented at the 1979 Allerton Conf.).

BOUNDS ON THE ACHIEVABLE ACCURACY IN MODEL REDUCTION

Keith Glover and Jonathan R. Partington
University Engineering Department
Trumpington Street
Cambridge CB2 1PZ
United Kingdom

Abstract

The problem of approximating a given, possibly infinite dimensional, transfer function, by one of prescribed McMillan degree is considered. Firstly lower bounds on the achievable error are given for a variety of norms (L^∞ and frequency weighted L^∞ for the transfer function, L^1 for the impulse response, largest singular value, Hilbert-Schmidt and nuclear norms for the Hankel operator). Upper bounds are derived for the optimal Hankel norm method, truncated balanced realizations, modal expansions and some frequency weighted methods. These all involve singular values of Hankel operators and the asymptotic behaviour of these singular values is analysed for infinte dimensional systems. Finally improved estimates of the achievable L^∞-error are studied.

1. Introduction

In this paper we consider the problem of approximating a linear time invariant dynamical system by one of McMillan degree k. That is, consider an input/output relation defined by the convolution integral

$$y(t) = \int_{-\infty}^{t} h(t-\tau)\ u(\tau)\ d\tau \tag{1.1}$$

where the input $u(t) \in \mathbb{C}^m$, the output $y(t) \in \mathbb{C}^p$, $h(t) \in \mathbb{C}^{p\times m}$ is the impulse response, and $G(s)$ = Laplace transform of $h(t)$ is the transfer function. The particular function spaces that u and y belong to will vary according to application although

NATO ASI Series, Vol. F34
Modelling, Robustness and Sensitivity Reduction in Control Systems. Edited by R. F. Curtain

stability of G(s) will be assumed. The model reduction problem is then to approximate G(s) by $\hat{G}(s)$ of McMillan degree k such that an appropriate norm is either minimized or made suitably small. In particular we will be concerned with obtaining lower bounds on the achievable approximation error (given G(s) and k), and further deriving upper bounds on the errors when particular methods are used. The results and methods will clearly be most powerful when these upper and lower bounds are of comparable order.

In section 2 a variety of norms will be considered and some lower bounds on the achievable error are derived. These are generally in terms of the singular values of Hankel operators. Section 3 then gives upper bounds on several norms when a variety of approximation methods are used. The results of sections 2 and 3 show that the singular values of Hankel operators give very useful information on the achievable accuracy and hence in section 4 the rate of decrease of these singular values is analysed in terms of the smoothness properties of h(t). Finally section 5 considers closing the gap between the upper and lower bounds when the L^∞ - norm of $G - \hat{G}$ is used. The results of sections 2-4 have been developed in detail elsewhere whereas the calculations of section 5 are thought to be new.

2. Error norms and lower bounds

Let the error transfer function be $E(s) := G(s) - \hat{G}(s)$ with inverse Laplace transform $e(t) = h(t) - \hat{h}(t)$. The achievable accuracy is connected with the Hankel operator, Γ_G, given by,

$$(\Gamma_G u)(t) = \int_0^\infty h(t+\tau)\, u(\tau)\, d\tau$$

whose ordered singular values will be denoted $\sigma_i(G)$, $i \geq 1$, and called the Hankel singular values of G. We will consider the following induced norms on the error system:

$$\| E \|_\infty := \sup_\omega \sigma_{max}(E(j\omega))$$

$$= \sup_{\| U(j\omega) \|_2 = 1} \| E(j\omega)\ U(j\omega) \|_2$$

$$\| e \|_1 := \int_0^\infty \sigma_{max}\ (e(t))\, dt$$

$$= \sup_{\| u \|_\infty \leq 1} \| e * u \|_\infty$$

$$\| e \|_2 := \left\{ \int_0^\infty \text{trace}\ (e^*(t)\ e(t))\ dt \right\}^{\frac{1}{2}}$$

$$= \| E\ (j\omega) \|_2$$

$$= \sup_{\| u \|_2 \leq 1} \| e * u \|_\infty$$

An advantage of the H^∞-norm is that robustness properties of closed-loop systems can be analysed [Z],[F]. A further set of norms relating to the error system's Hankel operator are,

Hankel norm, $\| E \|_H := \sigma_{max}\ (\Gamma_E) = \sigma_1(E)$

nuclear norm, $\| E \|_N := \sum_i \sigma_i\ (E)$

Hilbert-Schmidt norm, $\| \Gamma_E \|_{HS} := \left\{ \sum_i \sigma_i^2\ (E) \right\}^{\frac{1}{2}}$

$$= \left\{ \int_0^\infty t\ \text{trace}\ (e^*(t)\ e(t))\ dt \right\}^{\frac{1}{2}}$$

Finally to reflect the relative importance of different frequency ranges the frequency weighted L^∞ norm can be used, i.e.

$\| W_1 E\ W_2 \|_\infty$ where

W_1^* and W_2^* are chosen to be outer functions.

It can be shown that all these norms are invariant under a rescaling of the frequency axis ($\omega \to \alpha\omega$), except $\| e \|_2$ which will be scaled by $\alpha^{-\frac{1}{2}}$. A useful set of inequalities is

$$\| E \|_H \leq \| E \|_\infty \leq \| e \|_1 \leq 2\| E \|_N, \tag{2.1}$$

the first two being standard and a proof of the third is in [GCP]. The Hankel operator therefore defines both upper and lower bounds on the other two and the fact that $\Gamma_{\hat{G}}$ is of rank k (= McMillan degree of $\hat{G}$) can be exploited to obtain lower bounds. It is shown in [Mi] (see also [Gl]) that

$$\| \Gamma_G - \Gamma_{\hat{G}} \| \geq \| \operatorname{diag} (\sigma_{k+1}(G), \sigma_{k+2}(G), \ldots) \|$$

for any unitarily invariant norm. Hence

$$\| e \|_1 \geq \| E \|_\infty \geq \| E \|_H \geq \sigma_{k+1}(G)$$

$$\| E \|_N \geq \sum_{i \geq 1} \sigma_{k+i}(G)$$

$$\| \Gamma_E \|_{HS} \geq \left\{ \sum_{i \geq 1} \sigma^2_{k+i}(G) \right\}^{\frac{1}{2}}.$$

For $W_1 E W_2$ note that the McMillan degree of the stable part of $W_1 \hat{G} W_2$ will be no greater than k and hence [LA]

$$\| W_1 E W_2 \|_\infty = \| W_1 G W_2 - W_1 \hat{G} W_2 \|_\infty$$

$$\geq \sigma_{k+1} (\text{stable part of } W_1 G W_2).$$

3. Upper Bounds on Error Norms

3.1 Optimal Hankel-norm

Since the celebrated work of [AAK] the optimal Hankel-norm method has received much attention (see [Gl] and the references therein). The following result is proven in [GCP].

<u>Theorem 3.1</u>

Let $h \in L^1 \cap L^2$ be real, $\| \Gamma_G \|_N < \infty$, and $\sigma_i(G)$ be distinct, then there exists $\hat{h}(t)$, whose Laplace transform, $\hat{G}(s)$, has McMillan degree k such that

(a) $\| G - \hat{G} \|_H = \sigma_{k+1}(G)$

(b) there exists F(s) analytic in Re(s) < 0 such that

$$\| G - \hat{G} - F \|_\infty = \sigma_{k+1}(G)$$

(c) there exists a constant D_o such that

$$\| G - \hat{G} - D_o \|_\infty \leq M_k$$

where $M_k := \sum_{i \geq 1} \sigma_{k+i}(G)$

(d) $\| h - \hat{h} \|_1 \leq 4k\, \sigma_{k+1}(G) + 2M_k$

(e) $\| G - \hat{G} \|_N \leq 2k\, \sigma_{k+1}(G) + M_k$

(f) $\| \Gamma_G - \Gamma_G \|_{HS}^2 \leq 2k\, \sigma_{k+1}^2(G) + \sum_{i \geq 1} \sigma_{k+i}^2(G)$

The proof of these results relies on bounds on $\sigma_i(G-\hat{G})$ and $\sigma_i(F^*)$ and the inequalities of (2.1). It is seen that although only the Hankel-norm is minimized the other norms are all bounded by terms of similar magnitude to the lower bounds of section 2. Note [Mi] that in minimizing $\| \Gamma_G - X \|$ the lower bounds could all be achieved by a single X of rank k but not necessarily a Hankel operator; however the above shows that there exists a Hankel operator, X, satisfying the lower bound on the Hankel-norm. The technical assumptions of the Theorem on h(t) are probably unnecessarily restrictive.

It appears that this approximation scheme gives very close to optimal errors for the H^∞-norm. This is clearly the case if $\sigma_i(G)$ decrease rapidly and this is further supported by results in [T] for approximation of a scalar function f(z) on a disc of radius ε. Then as $\varepsilon \to o$ let r_ε be the optimal rational approximant of degree k in L^∞ and $\hat{r}_\varepsilon$ be as above (which he refers to as a CF method based on the work of Caratheodory and

Fejér). It is shown that

$$\| f - r_\varepsilon \|_\infty = O(\varepsilon^{2k+1}) \text{ whereas } \| r_\varepsilon - \hat{r}_\varepsilon \|_\infty = O(\varepsilon^{4k+3})$$

and hence $\hat{r}_\varepsilon$ is very nearly a best approximation in this asymptotic sense .

Finally it is remarked that the value of $\| G - \hat{G} \|_2$ may be arbitrarily poor since the method will tend to make the error small over a very wide frequency range.

3.2 Truncated Balanced Realizations

A balanced state-space realization is one where the controllability and observability Gramians are equal and diagonal, and the diagonal entries will in fact be the Hankel singular values. This was introduced by [Mo] with stability and other properties of truncations of such systems proved in [PS] and an L^∞-upper bound was shown in [E] (see also [Gl]). These results were all for finite dimensional systems and truncations of infinite dimensional continuous time systems are considered in [GCP] (although for technical reasons using the output normal realization i.e. observability Gramian = identity and controllabiliy Gramian is diagonal). The following Theorem gives upper bounds on the errors for this method.

Theorem 3.2 [GCP]

Under the same assumptions on G and h as in Theorem 3.1, let $\hat{G}(s)$ be the transfer function of an output normal realization of G truncated to k states, and $\hat{h}(t)$ corresponding impulse response. Defining

$$M_i = \sigma_{i+1}(G) + \sigma_{i+2}(G) + \ldots\ldots$$

then the following bounds are satisfied.

(a) $\| G - \hat{G} \|_\infty \le 2 M_k$

(b) $\frac{1}{2}\| h - \hat{h} \|_1 \le \| G - \hat{G} \|_N$

$$\le 2(i\, \sigma_{i+1}(G) + M_i)$$

where i is the largest integer such that

$$\sigma_{i+1}(G) \geq 2\sqrt{2}\, M_k .$$

(c) $$\| h - \hat{h} \|_2 \leq \min_i \left[2\sqrt{2}\, M_k \sum_{j=1}^{i} |w_j(0)| + 2 \left\{ \sum_{j>i} \sigma_j^2(G) |w_j(0)|^2 \right\}^{\frac{1}{2}} \right]$$

where $w_j(0)$ is the initial value of the j-th Schmidt vector of Γ_G.

The L^∞-error bound is just a factor of two worse than that of the optimal Hankel-norm method. The bounds of part (b) are significantly worse than those of theorem 3.2. There is however an upper bound on the L^2-error which tends to zero as $k \to \infty$ but might be substantially worse than optimal, for example when L^2- and L^∞-approximation are incompatible. The derivation of parts (b) and (c) both rely on bounds on the L^2-errors of the Schmidt vectors of the truncated system.

This method is seen to be a good compromise between different norms and this is confirmed by our numerical experience.

3.3 Modal Expansions

If a transfer function has the modal expansion

$$G(s) = \sum_{i=1}^{\infty} \frac{a_i (\mathrm{Re}(s_i))}{(s-s_i)}$$

then the truncation,

$$\hat{G}(s) = \sum_{i=1}^{k} \frac{a_i \; \mathrm{Re}(s_i)}{(s-s_i)}$$

is often a good approximation. Upper bounds on any error norm are easily computed as

$$\| G - \hat{G} \| \leq \sum_{i>k} \| \frac{a_i \, Re(s_i)}{s-s_i} \|$$

The disadvantage of the method is that the approximate system poles are forced to be a subset of the original system poles and that when the poles are not clearly divided into dominant ones and the others the errors may be much larger than is achievable.

3.4 Frequency Weighted Methods

A general method for frequency weighted approximation is derived in [LA] that exactly minimize

$$\| [W_1 (G-G) W_2]_c \|_H$$

where $[.]_c$ denotes the causal part of the transfer function. Upper bounds on the resulting L^∞ error are given in [A] in terms of the neglected singular values of $[W_1 \; G \; W_2]_c$ and a condition number on $W_1 W_2$. However the errors may be substantially larger than is possible by other methods for some problems. Two modifications to this scheme for particular choices of $W_1 W_2$ have been developed in [HG] and [G2]. The first is for $W_2 = (s-\beta)/(s-\alpha)$, $W_1 = I$, and it is shown that there exists $\hat{G}$ of McMillan degree k, such that,

$$\sigma_{k+1} \leq \| (G - \hat{G})(s - \beta)/(s - \alpha) \|_\infty$$

$$\leq (1 + \left|\frac{\alpha-\beta}{\alpha+\beta}\right|)(\sigma_{k+1} + \sigma_{k+2} + \ldots + \sigma_n)$$

where $\sigma_i = \sigma_i([W_1 \; G \; W_2]_c)$

The second special case is to consider the relative error i.e. let W be an outer spectral factor satisfying

$$G^* G = W \, W^*$$

and let $W_1 = I$, $W_2 = W^{*-1}$. Then defining $\nu_i = \sigma_i([GW^{*-1}]_c)$

[G2] shows there exists $\hat{G}$ such that

$$\nu_{k+1} \leq \| (G - \hat{G})W^{\alpha-1} \|_\infty \leq \prod_{i\geq 1} (1 + \nu_{k+i}) - 1$$

$$\simeq \sum_{i\geq 1} \nu_{k+i} \quad \text{for} \quad \nu_{k+i} \text{ small.}$$

In both these special cases the upper bounds are found to be of similar magnitude to the unachievable lower bounds. The method of [G2] bas also been applied in [GJ] to approximating spectral density functions by approximating the spectral factor. The general problem of frequency weighted approximation is an area of current research.

4. Behaviour of the Singular Values

When one considers the problem of approximating an infinte-rank Hankel operator $\Gamma : L_2(o,\infty) \to L_2(o,\infty)$, given by

$$(\Gamma u)(t) = \int_0^\infty h(t + \tau)u(\tau)\, d\tau \tag{4.1}$$

it is desirable to obtain information on the asymptotic behaviour of the singular values $(\sigma_i)_{i=1}^\infty$ of Γ, since estimates for the error are expressible in terms of these. For simplicity we shall treat only the SISO case here: for general

$$\Gamma: \ L_2(o,\infty;\ \mathbb{C}^m) \to L_2(o,\infty;\ \mathbb{C}^p)$$

analogous results hold. Likewise we shall assume h to be real.

Consider first the problem of minimising $\| G(s) - \hat{G}(s) \|_\infty$ over $\hat{G}$ rational, degree k, where G(s) is the Laplace transform of h(t), and is in $H_\infty(\text{Re } s>0)$ if $h(t) \in L_1(o,\infty)$. There is a simple lower bound, σ_{k+1}, which follows directly since $\hat{G}$ determines a rank-k operator. Conversely writing $M_k = \sigma_{k+1} + \sigma_{k+2} + \ldots$, results from [GCP] imply that one can obtain an H_∞ error of at most $2M_k$ by truncating an output normal realization, and that by using an optimal Hankel-norm approximation this error can be halved and further one can approximate h in L_1 norm to within at most $4k\sigma_{k+1} + 2M_k$. Similar results hold for discrete time systems, using a realization due to N.J. Young [Y], but we shall not discuss these here.

These estimates require Γ to be nuclear and we recall first a few properties of nuclear systems. By an abuse of language we shall say that G(s) is nuclear of the corresponding operator Γ is nuclear.

For strictly proper transfer functions G(s), the results of Peller [P] state that G(s) is nuclear if and only if

$$\iint_{\mathrm{Re}\,s>0} |G''(s)| < \infty \tag{4.2}$$

Thus for delay systems with $G(s) = \left[e^{-sT} \frac{p(s)}{q(s)} \right]_{\text{stable}}$

it follows that Γ is nuclear if and only if deg q $\geq$ (deg p) + 2, (p and q being polynoials). The corresponding result for general

$$\frac{\sum p_i(s)\, e^{-sT_i}}{\sum q_j(s)\, e^{-sU_j}}$$ is given in [PGZC].

Futher, the results of Coifman and Rochberg [CR] assert that G(s) is nuclear if and only if

$$G(s) = \sum_{i=1}^{\infty} \frac{a_i (\mathrm{Re}\, s_i)}{s - s_i} \tag{4.3}$$

in Re s > 0 for some scalars a_i with $\sum |a_i| < \infty$ and some complex numbers s_i with Re $s_i < 0$. It is important to remark that this expansion is only valid in Re s > 0 (but converges in H_∞). A quantitative version of (4.2) and (4.3) (for the disc) was given by Bonsall and Walsh [BW]. Passing to the half plane by a conformal transformation, and using these estimates in Corollary (2.1) of [GCP] we see that for a nuclear system we have not only that $\| h \|_1 \leq 2 \| \Gamma \|_N$ but also that h(t) is equal almost everywhere to a function j(t) such that

(i) $j(t)$ is continuous except possibly at zero.

(ii) $|j(t)| \le \frac{16}{e\pi} \| \Gamma \|_N / t$.

Conversely the results of Howland [H] show that a smooth impulse response $h(t)$ with a rapid decay at ∞ determines a nuclear Hankel operator : Suppose that $k(t) = h^1(t)$ exists as an L_2 function (i.e. h is the integral of k) and that

$$\int_{t=0}^{\infty} t^{\frac{1}{2}} \left(\int_{\tau=t}^{\infty} |k(\tau)|^2 d\tau \right)^{\frac{1}{2}} dt < \infty \qquad (4.4)$$

then $(\Gamma u)(r) = \int_0^{\infty} \left(\int_0^t u(\tau)\, d\tau \right) k(t+r)\, dt$ and Γ is nuclear,

with the formula in (4.4) providing an upper bound for its nuclear norm.

This result is used in [GLP] to obtain various quantitative results on the decay of (σ_i) for systems with smooth impulse responses, of which we quote two here.

Theorem 4.1

Suppose that $h(t) = 0$ for all $t \ge A$, that $r \ge 1$, that $h, h^1, \ldots, h^{(r)}$ are locally absolutely continuous and $h^{(r+1)}$ exists in L_∞.

$$\text{Then} \quad M_{N(r+1)} \le \frac{A^{r+2}}{(2N)^{r+\frac{1}{2}}} \frac{\| h^{(r+1)} \|_\infty}{r!\,(r+\frac{3}{2})} \qquad (4.5)$$

for $N \ge 1$ and so $\sigma_n = O(n^{-\frac{3}{2}-r})$ as $n \to \infty$

Theorem 4.2

Suppose that $h(t) = 0$ for all $t \ge A$ and h^1 exists in L_2.

$$\text{Then} \quad M_N < \frac{\sqrt{2}}{3} A^{\frac{3}{2}} \| h^1 \|_2 N^{-\frac{1}{2}} \qquad (4.6)$$

for $N \ge 1$ and so $\sigma_n = O(n^{-\frac{3}{2}})$ as $n \to \infty$.

A similar result is given by Reade [R] in a more general context but without the numerical estimates (which we shall require later).

In [GLP] it is also shown that one can replace the hypothesis in Theorem (4.1) of h(t) having compact support by one that $h(t),\dots,h^{(r+1)}(t)$ decay to zero as $t\to\infty$ faster than any power of t: the conclusion is now that

$$\sigma_n = O(n^{-(r+\frac{3}{2})+\varepsilon}) \quad \text{for any} \quad \varepsilon>0. \tag{4.7}$$

We now apply these results to analyse the singular values of delay systems. The basic building blocks here are functions

$$G(s) = \left[\frac{e^{-sT}}{s^m}\right]_{\text{stable}} , \quad m\geq 1 \quad , \text{ corresponding to}$$

$$h(t) = \begin{cases} (1-\frac{t}{T})^{m-1} & o \leq t \leq T \\ 0 & t \geq T \end{cases}$$

These can be analysed explicitly (see, e.g. [GLP]) and for this case one has

$n^m \sigma_n \to \left(\frac{T}{\pi}\right)^m$. Thus a general strictly proper transfer function of the form $\left[e^{-sT}\frac{p(s)}{q(s)}\right]_{\text{stable}}$, with

$$p(s) = a_o + a_1 s + \dots a_p s^p \ , \ q(s) = b_o + \dots + b_q s^q$$

can be written (assuming a_p, $b_q \neq o$), as

$$G(s) = \left[\frac{a_p}{b_q}\frac{e^{-sT}}{s^{q-p}}\right]_{\text{stable}} + \text{Remainder}$$

$$= G_o(s) + G_1(s), \quad \text{say.}$$

Now $n^m \sigma_n(G_o) \to \left(\frac{T}{\pi}\right)^m \left|\frac{a_p}{b_q}\right|$, where $m = q - p$, by the above remarks, and $n^m \sigma_n(G_1) \to 0$ (by a calculation using (4.7)). From the theorem of Ky Fan (see, e.g. [GK]) it follows that

$$n^m \sigma_n(G) \to \left(\frac{T}{\pi}\right)^m \left|\frac{a_p}{b_q}\right| .$$

A similar calculation is performed in the time domain in [GLP]. For illustration, we consider

$$G(s) = \left[\frac{e^{-s\alpha}}{s + \frac{\pi}{2} e^{-s}}\right]_{\text{stable}} , \quad 0 \le \alpha < 1,$$

for which a partial fraction expansion is analysed in [PGZC] and shown to give an H_∞ error of order $(\log n)/n^{1-\alpha}$.

$$\text{For } \alpha = 0,\ G(s) = \left[\frac{1}{s} - \frac{\left(\frac{\pi}{2}\right) e^{-s}}{s(s + \frac{\pi}{2} e^{-s})}\right]_{\text{stable}}$$

$$= \left[\frac{1}{s} - \frac{\pi}{2}\frac{e^{-s}}{s^2} + \frac{\left(\frac{\pi}{2}\right)^2 e^{-2s}}{s^2(s + \frac{\pi}{2} e^{-s})}\right]_{\text{stable}}$$

and the singular values are asymptotic to those of $\left[\frac{\pi}{2}\frac{e^{-s}}{s^2}\right]_{\text{stable}}$,

namely $n^2\sigma_n \to \frac{1}{2\pi}$, and the system is nuclear.

For $\alpha > 0$, $G(s) = \left[\frac{e^{-\alpha s}}{s} - \frac{\frac{\pi}{2} e^{-s(1+\alpha)}}{s(s + \frac{\pi}{2} e^{-s})}\right]_{\text{stable}}$ and the

singular values are asymptotic to those of $\left[\frac{e^{-s}}{s}\right]_{\text{stable}}$, namely $n\,\sigma_n \to \frac{\alpha}{\pi}$, and the system is not nuclear.

For more complicated delays, eg

$$G(s) = \left[\frac{a_1 e^{-sT_1} + a_2 e^{-sT_2}}{s}\right]_{\text{stable}}$$

where the impulse response has two discontinuities, there appears to be no simple formula for (σ_n) in general, although, since $\sigma_n(\Gamma) \geq \sigma_n(\Gamma_r)$, where Γ_r is the Hankel operator with kernel $h_r(t) = h(t+r)$ (because Γ_r is just Γ followed by a contraction) it follows that one still has, for some $A,B>0$, that $\frac{A}{n} \leq \sigma_n \leq \frac{B}{n}$ in this case. Similar arguments apply to more complicated delay systems.

In the next section we show how to obtain improved H_∞-and L_1-upper bounds in model reduction - this is of particular importance for non-nuclear systems but can be used more generally.

5. Improved error bounds in rational approximation

For a Hankel operator Γ defined as in (4.1), with singular values (σ_i) and transfer function $G(s)$, the Laplace transform of $h(t)$, we saw in the previous section that model reduction by means of truncated realizations or Hankel-norm approximation performs satisfactorily if Γ is nuclear. Since many non-nuclear systems can be decomposed into the sum of a simple non-nuclear part and a nuclear part (which is approximable), it is of interest to reconsider the elementary building blocks referred to in section 4.

$$\text{Let}\quad h_1(t) = \begin{cases} 1 & o \leq t < 1 \\ 0 & t \leq 1 \end{cases} \tag{5.1}$$

be the step function, with Laplace transform $G_1(s) = \frac{1-e^{-s}}{s}$.

In this case the singular values of Γ can be calculated explicitly (see, e.g. [GLP]) and satisfy $\sigma_n = \frac{2}{\pi(2n-1)}$.

Our strategy for approximating such a delay system will involve an intermediate step, the preliminary smoothing impulse response function, after which truncated realization or Hankel norm techniques can be used.

Theorem 5.1

Let $\varepsilon > o$ and an integer $r \geq 1$ be given. Then there exists a function $j(t)$ $(t \geq o)$ such that $\| h_1 - j \|_{L_1} \leq \varepsilon$, $j(t) = o$ $(t \geq 1 + \varepsilon)$ and $j, \ldots, j^{(r)}$ are locally absolutely continuous and $j^{(r+1)}$ exists as an L_∞ function with $\| j^{(r+1)} \|_\infty \leq C_r \varepsilon^{-1-r}$, where C_r is independent of ε.

Proof

We give two possible ways of constructing $j(t)$, each of which has applications to other functions.

Firstly, one can use a polynomial spline approximation. Define

$$j_1(x) = \begin{cases} 1 & o \leq x \leq 1 \\ \int_{x-1}^{\varepsilon} (t-\varepsilon)^r t^r \, dt \Big/ \int_o^{\varepsilon} (t-\varepsilon)^r t^r \, dt & 1 \leq x \leq 1 + \varepsilon \\ 0 & x \geq 1 + \varepsilon \end{cases} \tag{5.2}$$

Clearly $o \leq j_1 \leq 1$ and $j_1 = h_1$ except on $[1, 1+\varepsilon]$, so that $\| j_1 - h_1 \|_1 \leq \varepsilon$. Moreover $j_1(x)$ is a polynomial of degree $(2r+1)$ on $[1, 1+\varepsilon]$, $j_1^{(r)}$ is continuous (and $j_1^{(r)}(1) = j_1^{(r)}(1+\varepsilon) = 0$), and $j_1^{(r+1)}$ is zero except on $[1, 1+\varepsilon]$, where

$$j_1^{(r+1)}(1+x) = - \frac{d^r}{dx^r} ((x-\varepsilon)^r x^r) \Big/ \int_0^{\varepsilon} (t-\varepsilon)^r t^r \, dt. \tag{5.3}$$

So by Leibnitz's formula, and estimating the denominator of

(5.3) as at least $\left(\frac{2\varepsilon}{3}\right)\left(\frac{\varepsilon}{3}\right)^r \left(\frac{2\varepsilon}{3}\right)^r$ (considering the middle third of the range), one has

$$\| j_1^{(r+1)} \|_\infty \leq 2^r \left(\frac{r!}{\left[\frac{r}{2}\right]!}\right)^2 \varepsilon^r / \left(\frac{\varepsilon}{3}\right)^r \left(\frac{2\varepsilon}{3}\right)^{r+1} = C_r \varepsilon^{-1-r} \qquad (5.4)$$

Alternatively, one may obtain an approximating function by passing h_1 through an anticausal low pass filter, that is, taking the stable part of

$\dfrac{G_1(s)}{(1-sT)^{r+1}}$ (in fact one obtains slightly sharper estimates by using $\dfrac{G_1(s)\, e^{-(r+1)sT}}{(1-sT)^{r+1}}$ although the algebra is more complicated).

Equivalently, we define $f_j(t)$ inductively by $f_o(t) = h_1(t)$

$$\text{and, for } j \geq o, \quad f_{j+1}(t) = \begin{cases} \displaystyle\int_{x=t}^{1} f_j(x) \left(\frac{1}{T}\, e^{(t-x)/T}\right) dx & (o \leq t \leq 1) \\ o & t \geq 1 \end{cases} \qquad (5.5)$$

It is easily verified that

$$f_{n+1}(t) = 1 - e^{-w}\left(1+w+\ldots+\frac{w^n}{n!}\right) \quad \text{where} \quad w = \frac{1-t}{T}$$

$$\text{for} \quad n \geq o \qquad o \leq t \leq 1. \qquad (5.6)$$

Thus $f^1_{r+1}(t) = e^{-w}\dfrac{w^r}{r!}\left(-\dfrac{1}{T}\right)$ and hence

$f^1_{r+1}(1) = \ldots = f^{(r)}_{r+1}(1) = 0$, so $f^{(r)}_{r+1}$ is continuous.

Thus $f_{r+1}^{(r+1)}(t) = \left(-\frac{1}{T}\right)^{r+1} \left(\frac{e^{-w}w^r}{r!}\right)^{(r)}$

$$= \left(-\frac{1}{T}\right)^{r+1} \sum_{i=o}^{r} \binom{r}{i} (-1)^i e^{-w} \frac{w^i}{i!}$$

But $\| e^{-w}w^i \|_\infty = e^{-i}i^i$ for all i, and so

$$\| f_{r+1}^{(r+1)} \|_\infty \leq \frac{1}{T^{r+1}} \sum_{i=o}^{r} \binom{r}{i} e^{-i}i^i/i! \qquad (5.7)$$

Also $\| f_{r+1} - h_1 \|_{L_1} = T \int_0^{1/T} e^{-w}(1+w+..+ \frac{w^r}{r!})\, dw$,

since $(o \leq f_{r+1} \leq h_1) \leq T \int_0^\infty e^{-w}(1+w+..+ \frac{w^r}{r!})\, dw$

$$= T(1+1+ \ldots +1) = (r+1)T \qquad (5.8)$$

Thus taking $T = \frac{\varepsilon}{r+1}$, f_{r+1} provides another approximant satisfying the conditions of the theorem.

Similarly for a simple derivative discontinuity, such as

$$h_2(t) = \begin{cases} 1-t & o \leq t \leq 1 \\ 0 & t \geq 1 \end{cases} \qquad (5.9)$$

corresponding to $G_2(s) = \left(\frac{-e^{-s}}{s^2}\right)_{\text{stable}}$ one can find smooth approximants j_r with $\| h_2 - j_r \|_{L1} = O(\varepsilon^2)$ and $\| j_r^{(r+1)} \|_\infty = O(\varepsilon^{-r})$, for example by altering h_2 solely on the interval $[1-\varepsilon, 1+\varepsilon]$.

Corollary 5.1

Let $\varepsilon > o$ be given. Then, with h_1 and h_2 defined as in (5.1) and (5.2) and G_1 and G_2 respectively their Laplace transforms,

$$\frac{1}{\pi k} \le \inf \{ \| G_1 - \tilde{G}_1 \|_{H_\infty} : \deg \tilde{G}_1 \le k \}$$

$$\le \inf \{ \| h_1 - \tilde{h}_1 \|_{L_1} : \deg \tilde{h}_1 \le k \} = O(k^{-1+\varepsilon}) \tag{5.10}$$

and $$\frac{1}{\pi 2_k 2} \le \inf \{ \| G_2 - \tilde{G}_2 \|_{H_\infty} : \deg \tilde{G}_2 \le k \}$$

$$\le \inf \{ \| h_2 - \tilde{h}_2 \|_{L_1} : \deg \tilde{h}_2 \le k \} = O(k^{-2+\varepsilon}) \tag{5.11}$$

Proof

In each of (5.10) and (5.11) the left most inequalities follow from formulae for $\sigma_k(G_1)$ and $\sigma_k(G_2)$, obtained e.g. in [GLP]. The middle inequalities follow because the H^∞ norm of a function is always at most the L_1 norm of its inverse Laplace transform.

For the right most part of (5.10), select $r \ge 1$ such that $\frac{r+\frac{1}{2}}{r+2} \ge 1 - \varepsilon$, let $k \ge 1$ and $\delta = k^{-(r+\frac{1}{2})/(r+2)}$. By Theorem 5.1 and Theorem 4.1 one can find $j_1 \varepsilon L_1$ with $\| h_1 - j_1 \|_{L_1} \le \delta$ and $\sigma_k(j_1) \le A_r k^{-\frac{3}{2} -r} \delta^{-1-r}$. A_r independent of k. Thus there exists an optimal degree-k Hankel approximation $\tilde{h}_1$ to j_1 with $\| \tilde{h}_1 - j_1 \|_{L_1} \le B_r \delta^{-1-r} k^{-\frac{1}{2}-r}$, B_r independent of k. So

$$\| \tilde{h}_1 - h_1 \|_{L_1} \le \delta + B_r \delta^{-1-r} k^{-\frac{1}{2}-r}$$

$$= O(k^{-(r+\frac{1}{2})/(r+2)}) = O(k^{-1+\varepsilon})$$

A similar procedure applied to h_2 produces $j_2 \varepsilon L_1$ with $\| h_2 - j_2 \|_{L_1} = O(\delta^2)$ and a rank-k approximation $\tilde{h}_2$ to j_2 with

$\| \tilde{h}_2 - j \|_{L_1} = O(\delta^{-r} k^{-\frac{1}{2}-r})$; thus if $\delta = k^{-(r+\frac{1}{2})/(r+2)}$ the error is $O(k^{-(2r+1)/(r+2)})$, which can be made $O(k^{-2+\varepsilon})$ by suitable choice of r.

Using Corollary 5.1 and the technique of decomposing systems into a sum of building blocks plus a smooth remander term it follows, for example, that any piecewise smooth (ie L^∞ second derivative) h(t) with finitely many step discontinuities and faster than polynomial decay at ∞ determines a Hankel operator with singular values $\sigma_k \geq \frac{A}{k}$ (for some A>o) but can be approximated in H_∞ and L_1 norm to within $B_\varepsilon k^{-1+\varepsilon}$ for any ε>o. This is typical of many delay systems and there are of course analogues for smoother functions h(t) (and extensions of Corollary (5.1).

By using the detailed estimates of Theorem 4.1, selecting r to depend on k, one can deduce a strengthening of Corollary 5.1. We conclude by performing such a calculation.

Theorem 5.2

Let H(t) be piecewise twice differentiable (with absolutely continuous derivative and L^∞ second derivative) with finitely many simple jump discontinuities (at least one) such that as $t \to \infty$ h(t), $h^1(t)$, h"(t) are all $o(t^{-\lambda})$ for all λ>o. Let G(s) be its Laplace transform. Then there exist A,B>o such that

$$Ak^{-1} \leq \inf \{ \| G-\tilde{G} \|_{H\infty} : \deg \tilde{G} \leq k \}$$

$$\leq \inf \{ \| h-\tilde{h} \|_{L_i} : \deg \tilde{h} \leq k \} \leq B(\log k)/k.$$

for all $k \geq 2$.

Proof

We may write $h(t) = h_1(t) + \ldots + h_p(t) + h_{p+1}(t)$, where $h_1, \ldots, h_p$ are simple step functions of the form $a_i \chi_{[o,T_i]}$ and h_{p+1} is smooth, so that $\sigma_k(h_{p+1}) = O(k^{-\frac{5}{2}+\varepsilon})$ for any ε>o (by 4.7)). Since one can approximate h_{p+1} to within $O(k^{-\frac{3}{2}+\varepsilon})$ by a Hankel-norm technique, it remains to show that one can approximate a step function (without loss of generality $h_1(t)$ as in (5.1)) to within $O(\frac{\log k}{k})$. (The fact that $\sigma_k(h) \geq \frac{A}{k}$ follows from the arguments concluding section 4).

By (5.7) and (4.5),

$$M_{N(r+1)}(f_{r+1}) \leq \frac{2^r}{T^{r+1}} \frac{1}{(2N)^{r+\frac{1}{2}}} \frac{1}{r!(r+\frac{3}{2})}$$, taking $A = 1$, since

$e^{-i} i^i / i! \leq 1$ for all i.

Taking an optimal Hankel-norm approximant to f_{r+1}, we see that

$$\text{for } N \geq 1,\ \inf \{ \| h_1 - \tilde{h}_1 \|_{L_1} : \deg \tilde{h}_1 \leq N(r+1) \} \leq 4N(r+1)\sigma_{N(r+1)+1}(f_{r+1})$$

$$+ 2M_{N(r+1)}(f_{r+1}) + \| h_1 - f_{r+1} \|_{L_1} \qquad (5.12)$$

If p is even, $4p\sigma_{p+1} + 2M_p \leq \frac{4p}{(p/2)} M_{p/2} + 2M_p$, since

$$M_{p/2} \geq \sigma_{p/2+1} + \dots + \sigma_{p+1} \geq \frac{p}{2} \sigma_{p+1} ,$$

Hence for N even, (5.12) is at most

$$\frac{8}{T^{r+1}} 2^r \frac{1}{N^{r+\frac{1}{2}}} \frac{1}{r!(r+\frac{3}{2})} + \frac{2}{T^{r+1}} \frac{2^r}{(2N)^{r+\frac{1}{2}}} \frac{1}{r!(r+\frac{3}{2})} + (r+1)T \qquad (5.13)$$

with a similar expression for N odd.

For fixed N, r we choose T to minimise this bound, so

$$T^{r+2} = \frac{1}{N^{r+\frac{1}{2}}} \frac{2^{r+3} + \sqrt{2}}{r!(r+\frac{3}{2})}$$, and our upper bound in (5.13) is at most

$$(r+2)N^{-(r+\frac{1}{2})/(r+2)} \left(\frac{2^{r+4}}{r!(r+\frac{3}{2})}\right)^{\frac{1}{r+2}} \quad \text{as } 2^{r+3} + \sqrt{2} \le 2^{r+4},$$

i.e. at most $C_1 N^{-(r+\frac{1}{2})/(r+2)}$ for some C_1 indep endent of N and r.

Thus one can approximate by a rank N(r+1) operator to within an L_1 error of

$$\frac{C_2}{N^r} \exp(\frac{3}{2r} \log N + \log r) \text{ ; thus of } r = \left[\frac{3}{2} \log N\right], \text{ the error}$$

is of order $\frac{C_2}{N^r} \exp(1 + \log \frac{3}{2} + \log \log N)$, which is at most

$$\frac{C_3 \log k}{k}, \quad \text{where } k = [Nr].$$

Clearly this implies the result for general k and suggests that good rational approximations can often be found by smoothing h to obtain a function with O(log k) derivatives and then taking either a truncated realization or an optimal Hankel-norm approximation to the smoothed function. It is not known whether the optimal Hankel-norm method without smoothing would nevertheless achieve the same order of error.

References

[A] B.D.O. Anderson, "Weighted Hankel-norm approximation calculation of bounds", Systems and control letters 7 , 247-255, 1986.

[AAK] V.M. Adamjan, D.Z. Arov and, M.G. Krein, "Analytic properties of Schmidt pairs for a Hankel operator and the generalised Schur-Takagi problem", Math USSR Sbornik, 15, 31-73, 1971.

[BW] F.F. Bonsall and D. Walsh, "Symbols for trace class Hankel operators with good estimates for norms", Glasgow Math J. 28, 47-54, 1986.

[CR] R.R. Coifman and R. Rochberg, "Representation theorems for holomorphic and harmonic functions in L^p", Asterisque 77, 11-66, 1980.

[E] D. Enns, "Model reduction for control system design", Ph.D dissertation, Dept of Aero. and Astro., Stanford Univ, 1984.

[F] B.A. Francis, "A course in H_∞-control theory", Springer-Verlag, Lecture notes in Control and Information Sciences, 88, 1987.

[G1] K. Glover, "All optimal Hankel-norm approximations of linear multivariable systems and their L^∞-error bounds", Int. J. Control, 39, 1115-1193, 1984.

[G2] K. Glover, "Multiplicative approximations of linear multivariable systems with L^∞ error bounds", Proc. Amer, Control Conf., Seattle, June, 1986.

[G3] K. Glover, "Model reduction: a tutorial on Hankel-norm and related methods", Proc. IFAC World Congress, July 1987.

[GCP] K. Glover, R.F. Curtain, J.R. Partington, "Realisation and approximation of linear infinite dimensional systems with error bounds", Univ of Cambridge, Dept of Engineering Report CUED/F-CAMS/TR258, 1986.

[GJ] K. Glover and E.A. Jonckheere, "A comparison of two Hankel-norm methods for approximating spectra", MTNS-85 Stockholm, Sweden, June 1985. also in Modelling, Identification and Robust Control, C.I. Byrnes and A. Lindquist (Eds.) pp 297-306, North Holland 1986.

[GK] I.C. Gohberg and M.G. Krein, "Introduction to the theory of linear nonself adjoint operators in Hilbert space", Transl. Math Monographs, 18, Amer. Math Soc. (Providence, 1969).

[GLP] K. Glover, J. Lam and J.R. Partington, "Balanced realization and Hankel-norm approximation of systems involving delays", Proc. IEEE Conference on Decision and Control, Athens, Dec 1986.

[H] J.S. Howland, "On trace class Hankel operators", Quart J. Math 22, 1971, 147-159.

[HG] S. Hung and K. Glover, "Optimal Hankel-norm approximations of stable systems with first order stable weighting functions", Systems and Control Letters, 1986.

[LA] G.A. Latham and B.D.O. Anderson," Frequency weighted optimal Hankel norm approximations of stable transfer functions", Systems and Control Letters, 5, 229-236, 1985.

[Mi] L. Mirsky, "Symmetric gauge functions and unitarily invariant norms", Q. Journal Math. (2) 11, 50-59, 1960.

[Mo] B.C. Moore, "Principal component analysis in linear systems: controllability, observability and model reduction", IEEE Trans Aut. Control, AC-26, 17-32, 1981.

[P] V.V. Peller, "Hankel operators of class C_p and their applications (rational approximations, Gaussian processes, the problem of majorizing operators)", Math. USSR Sbornok 41, 1982, 443-479.

[PGZC] J.R. Partington, K. Glover, H. Zwart and R.F. Curtain, "L_∞-approximation and nuclearity of delay systems", in preparation.

[PS] L. Pernebo and L.M. Silverman, "Model reduction via balanced state-space representation", IEFE Trans. Aut. Control, AC-27, 382-387, 1982.

[R] J.B. Reade, "Eigenvalues of smooth kernels", Proc. Camb. Philos. Soc. 95, 1984, 135-140.

[T] L.N. Trefethan, "Rational Chebyshev approximation on the unit disk", Numer. Math. 37, 297-320, 1981.

[Y] N.J. Young, "Balanced realizations in infinite dimensions", preprint.

[Z] G. Zames, "Feedback and optimal sensitivity: model reference transformations, multiplicative seminorms, and approximate inverses", IEEE Trans. Aut. Cont., AC-23, 301-320, 1981.

Recent and New Results in Rational L^2-Approximation

Laurent Baratchart
Institut National de Recherche en Informatique et Automatique
Avenue E. Hughes Sophia-Antipolis
06560 Valbonne (France)

1. Introduction

The problem under consideration in this paper is the best approximation of a stable linear constant dynamical system by one whose Mac-Millan degree does not exceed a prescribed n. This problem can be formulated in many ways, according to the criterion which is used. In particular, striking results have been obtained concerning the Hankel norm, and a bibliographical account can be found in (KL, 81) and (Gl, 84). The criterion here will be the l^2 norm of the discrete transfer function, or equivalently the $L^2(\frac{d\omega}{1+\omega^2})$ norm of the continuous time transfer function. This problem has already been examined in the scalar case in (Ro, 78) (Ruc, 78) (De, 80) (Du, 73) and in the multivariable case in (Ba, 82) (Bs, 85) (Ba, *). In fact, the present paper can be considered as a sequel to (Ba, *), to which we refer on several occasions. Thereby, it does not really need a new introduction. Let us mention, however, that while studying qualitative generic properties as in (Ba, *), we focuse mainly on uniqueness, and give an account of the case of finite sequences. Though we do not develop here effective applications of the differential theory, we hope our results will help clarifying the situation.

2. Statement of the problem.

Throughout, the word "system" means "linear constant causal dynamical system". We define the *order* of a proper rational matrix to be its Mac-Millan degree, that is the dimension of the state-space in its minimal realizations (KFA, 69). The discrete time version of the l^2-approximation problem can be stated as follows.
Let $F = \sum_{k=0}^{\infty} F_k z^{-k}$ be the (possibly non-rational) transfer-function of a m-input p-output discrete system. The F_k's are thus real $p \times m$ matrices. Let F_k have entries $(F_k^{i,j})$, and put $\|F_k\|^2 = \sum_{i,j} (F_k^{i,j})^2$. Assume that F is l^2-stable, that is $\sum_k \|F_k\|^2 < \infty$. Now choose an integer $n \geq 1$.
Our problem is to find a finite-dimensional stable discrete system, whose order is at most n, and whose transfer-function $H = \sum_{k=0}^{\infty} H_k z^{-k}$ is such that $\sum_k \|F_k - H_k\|^2$ is as small as possible.

It is convenient to settle this in the classical framework of real Hardy spaces (e.g. (Rud, 66)). Let T be the unit circle. Let U (resp. $\bar{U}$)be the open (resp. closed) unit disk. We define H_2^- (resp. H_2^+) to be the *real* Hilbert space of functions g holomorphic in the complement of $\bar{U}$ (resp. in U), vanishing at infinity in the case of H_2^-, and such that:

$$g(z) = \sum_{k=1}^{\infty} g_k z^{-k} \quad \left(\text{resp. } \sum_{k=0}^{\infty} g_k z^k\right)$$

NATO ASI Series, Vol. F34
Modelling, Robustness and Sensitivity Reduction
in Control Systems. Edited by R. F. Curtain

with $g_k \in \mathbf{R}$ and $\|g\|^2 = \sum_k g_k^2 < \infty$. The orthogonal sum $H_2^- \oplus H_2^+$ is denoted by H_2. To each $g \in H_2^-$ (resp. H_2^+), we can associate $g^* \in L^2(T)$ by putting:

$$g^* = \sum_{k=1}^{\infty} g_k e^{-ik\theta} \quad \left(\text{resp. } g^* = \sum_{k=0}^{\infty} g_k e^{ik\theta}\right)$$

This allows us to rewrite the scalar product in H_2 as:

$$< f, g >= \frac{1}{2\pi} \int_0^{2\pi} f^*(e^{i\theta}) \, \overline{g^*(e^{i\theta})} \, d\theta \tag{1}$$

It can be shown (Rud, 66) that g^* is equal almost everywhere to the radial limit of g, and that g is the Cauchy integral of g^*, so that we often identify g and g^* unless it is necessary to keep the distinction. We now consider the space E of $p \times m$ matrices with entries in H_2. If $G = (g_{i,j})$ lies in E, the square of its norm will be $\sum_{i,j} \|g_{i,j}\|^2$. In this way, E becomes a real Hilbert space whose norm and scalar product will be denoted like those of H_2. We also define $E^- = (H_2^-)^{p \times m}$, $E^+ = (H_2^+)^{p \times m}$. When M is a matrix, we denote its tranpose by ${}^tM$ and its trace by $\operatorname{Tr} M$. If F and G lie in E, (1) gives rise to:

$$< F, G >= \frac{1}{2\pi} \operatorname{Tr} \int_0^{2\pi} {}^tF(e^{i\theta}) \, \overline{G(e^{i\theta})} \, d\theta$$

Taking into account the fact that F and G have real Fourier coefficients, and that $\overline{z} = z^{-1}$ on T, this may be converted into a line integral:

$$< F, G >= \frac{1}{2i\pi} \operatorname{Tr} \int_T {}^tF(z) \, G(\frac{1}{z}) \, \frac{dz}{z} \tag{2}$$

Now, the transfer function H of a ***strictly proper*** finite dimensional discrete system lies in E^- if and only if it is stable , that is if and only if its poles are in U. Moreover, it is easily seen that its development as a convergent power series in z^{-1} coincides with the formal "long division". Hence, the set S_n^- of stable strictly proper m-input p-output transfer-functions of order at most n is naturally included in E^-. We shall denote by Σ_n^- the subset of transfer-functions of order n. If Σ_n^- is endowed with its usual differential structure, the above identification is an embedding (Ba, *).

Now we come back to our approximation problem as stated at the beginning of the paragraph. If H is a best approximant of F, we must have $H_0 = F_0$ since the constant term does not affect the order. Therefore, the question reduces to:

Given $F \in E^-$, and $n \geq 1$, find $H^0 \in S_n^-$ such that: $\|F - H^0\| = \inf_{H \in S_n^-} \|F - H\|$

This framework allows one to formulate an analoguous problem for continuous time systems: let $\varphi : \mathbf{C} \to \mathbf{C}$ be the linear fractional transformation $z \to \frac{z+1}{z-1}$. If F is a $p \times m$ matrix function, holomorphic in a neighborhood of the closed right-half plane including ∞ and satisfying $F(-i\omega) = \overline{F(i\omega)}$, then $F \circ \varphi \in \mathbf{R}^{p \times m} \oplus E^-$, and if H^0 is a best approximant of $F \circ \varphi$ in the above sense, it is easy to check that $H^0 \circ \varphi$ is a transfer function of order at most n with poles in the left half-plane that minimizes $\|F-H\|_c$, where $\|.\|_c$ is defined by:

$$\|G\|_c^2 = \mathrm{Tr}\int_{-\infty}^{+\infty} {}^tG(-i\omega)G(i\omega)\frac{d\omega}{1+\omega^2}$$

In the sequel, we shall only consider the discrete time problem. Consequently, in addition to our previous conventions, "system" means now "discrete system", and every transfer function will be strictly proper.

There are several reasons why one should like to put additional restrictions on F. For instance, F will often be a finite sequence in practice. Some more technical reasons will be discussed in the present paper. But we give a word of warning about this. Some properties of the approximants proved in the present context hold only for generic F (see the precise definition below). These properties cannot be readily specialized to finite sequences, because those are highly nongeneric in E^-. Therefore, a separate treatment is needed, which is usually simpler at least when the result depends on transversality theory. Proposition 4 is an illustration of this. There are also, of course, special features of the finite case, and some of them are examined in this paper.

3. Existence and order of the approximants

The existence of best l^2-approximants is established in (Ba, 82)(Ba, *) and the proof can be adapted to the l^p-case. We shall not repeat the argument but we state the result.

Theorem 1. *For any $F \in E^-$, there exists $H^0 \in S_n^-$ such that $\|F-H^0\| = \inf_{H\in S_n^-} \|F-H\|$*

We shall need in the sequel the notion of local best approximant with respect to S_n^- (resp. Σ_n^-), that is a member H^0 of S_n^- (resp. Σ_n^-) that minimizes $\|F-H\|$ in a neighborhood of itself in S_n^- (resp. Σ_n^-). We pay special attention to these points, because they are those which can be obtained by differential methods. We now put under local form a result from (Ba, 82). Let $E_{i,j}$ denote the $p\times m$ matrix with entry "1" at the meeting of the ith row and j^{th} column, and "0" anywhere else. For any real numbers a, b, we define $H_{a,b}^{i,j}$ to be the rational matrix $\frac{a}{z-b}E_{i,j}$.

Proposition 1. *Let F be in E^-. If $F \notin S_{n-1}^-$, any local best approximant of F with respect to S_n^- lies in Σ_n^-, and any local best approximant with respect to Σ_n^- is a local best approximant with respect to S_n^-. If H is a local best approximant of F, we have $\|H\|^2 + \|F-H\|^2 - \|F\|^2$.*

Note: if $F \in S_n^-$ of course, there is a unique best approximant, namely F itself.

Proof: Let $H \in S_{n-1}^-$ be a local best approximant of F, and put $G = F - H$. Call $(G_{i,j})$ the entries of G. For every i, j, and any real numbers a, b, with $|b| < 1$, we have $H + H_{a,b}^{i,j} \in S_n^-$. Given such a b, we get for sufficiently small a

$$\|G + H_{a,b}^{i,j}\|^2 \geq \|G\|^2$$

Expanding, we find that

$$2 < G_{i,j}\,, \frac{a}{z-b} > + < \frac{a}{z-b}\,, \frac{a}{z-b} > \geq 0$$

Taking a very small, we see that $< G_{i,j}\,, \frac{1}{z-b} >$ must be zero. Using (2), this yields:

$$\frac{1}{2i\pi}\int_T z^{-1} G_{i,j}(z^{-1})\,\frac{dz}{z-b} = 0$$

Now, $z^{-1}G_{i,j}(z^{-1})$ lies in H_2^+ (recall that $G_{i,j}(\infty) = 0$). Hence, by Cauchy's formula, $G_{i,j}(b^{-1}) = 0$ provided b is real and $|b| < 1$. By analyticity, $G_{i,j}$ must be zero. Since i,j were arbitrary , we conclude that $F = H$, thereby proving the first assertion.
As to the second, we first let the reader check (using for instance Hautus's test) that if $H \in \Sigma_{n-1}^-$, $H + H_{a,b}^{i,j}$ belongs to Σ_n^- provided $a \neq 0$ and b is not a pole of H. Now let H^0 be a best local approximant with respect to Σ_n^-. There exist $\epsilon > 0$ such that

$$H \in \Sigma_n^- \text{ and } \|H - H^0\| < \epsilon \quad \Longrightarrow \quad \|F - H\| \geq \|F - H^0\|$$

Let $G \in \Sigma_{n-1}^-$ satisfy $\|G - H^0\| < \epsilon$, and let b be different from a pole of G. Taking $a \neq 0$ but small enough, we see that

$$\|F - G - H_{a,b}^{i,j}\| \geq \|F - H^0\|$$

Taking a very small, we get

$$G \in \Sigma_{n-1}^- \text{ and } \|G - H^0\| < \epsilon \quad \Longrightarrow \quad \|F - G\| \geq \|F - H^0\|$$

Proceeding inductively shows that H^0 is a local best approximant in S_n^-. Finally, if $H \in S_n^-$, the vector space $\mathcal{W} = \{\lambda H;\ \lambda \in \mathbf{R}\}$ is included in E^-. If in addition H is a local best approximant, it must be the orthogonal projection of F on $\mathcal{W}$. Q.E.D.

We close this paragraph by stating without proof a result borrowed from (Ba, *), concerning the continuity of best approximants.

Proposition 2. *Suppose $F^0 \notin S_{n-1}^-$ and let $\mathcal{O} \subset \Sigma_n^-$ be an open set containing every best approximant of F^0. If F is sufficiently close to F_0, $\mathcal{O}$ contains every best approximant of F.*

4. The question of uniqueness

One can raise the question wether a best approximant is unique. As we shall see, the question is not so well posed in this form. First of all, let us show by an example that the answer is negative.
Observe that the norm is invariant under the transformation $z \to -z$. Choose F to be a non rational function of z^{-2} only (for instance, if $m = p = 1$, one can choose $F = e^{\frac{1}{z^2}} - 1$). If $H^0(z)$ is a best approximant, so is $H^0(-z)$, and they are of order n by Proposition 1. It is easily seen that they cannot coincide if n is odd.

In order to study the question further, we introduce the notion of genericity. We shall say that a property depending on $F \in E^{-}$ holds for "almost all F", or "generically", if it holds whenever F belongs to some residual set in E^{-}. Recall a residual set is a countable intersection of open dense subsets, and is dense in any complete metric space by Baire's theorem. We say that a property is strongly generic if the residual set contains a dense open set. When we deal with finite sequences of fixed length N, generic means everywhere except on a set of measure zero.
It is shown in (Ba, *) that the finiteness of the number of best approximants is a strongly generic property. We now strengthen this result.

Theorem 2. *The uniqueness of a best approximant is a strongly generic property.*

Proof: First observe that S_{n-1}^{-} is closed in E^{-}. Let $F^0 \notin S_{n-1}^{-}$ have a unique nondegenerate best approximant H^0. Then $H^0 \in \Sigma_n^{-}$ by Proposition 1. By the implicit function theorem, there exists neighborhoods $\mathcal{V}_1 \subset E^{-}$ and $\mathcal{V}_2 \subset \Sigma_n^{-}$ of F^0 and H^0 such that if we define

$$\Gamma : \mathcal{V}_1 \times \mathcal{V}_2 \to \mathbf{R} \text{ by } \Gamma(F,H) = \|F - H\|^2$$

the partial map $\Gamma_F : \mathcal{V}_2 \to \mathbf{R}$ has a unique critical point $H(F)$ for each F, which is nondegenerate. Moreover, $H(F)$ is a smooth function of F. Schrinking $\mathcal{V}_1$ if necessary, we see from Proposition 2 that all best approximants of F must belong to $\mathcal{V}_2$, so that there can be only one. Hence, the set $\mathcal{O}_1$ of $F \in E^{-}$ such that $F \notin S_{n-1}^{-}$ and F has a unique nondegenerate best approximant is open. From (Ba, *), we know there is a dense open set $\mathcal{O}_2 \subset E^{-}$, such that $\mathcal{O}_2 \cap S_{n-1}^{-} = \emptyset$, whose members have a finite number of best approximants, all of which are nondegenerate. We have $\mathcal{O}_1 \subset \mathcal{O}_2$, and we have to prove that $\mathcal{O}_1$ is dense in $\mathcal{O}_2$.
Let $F^0 \in \mathcal{O}_2 - \mathcal{O}_1$, and $H^{0,i}$ $i \in \{1, ..., k\}$ its best approximants (it is important here that they are finite in number). Using the implicit function theorem as above, we can define k functions $H^i(F)$ in a neighborhood $\mathcal{V}$ of F^0. Schrinking $\mathcal{V}$ if necessary, we can assume that the sets $H^i(\mathcal{V})$ are disjoints, and by Proposition 2 that all best approximants of $F \in \mathcal{V}$ are among the $H^i(F)$. Consider the map

$$\Lambda_i : \mathcal{V} \to \mathbf{R} \text{ given by } F \to \|F - H^i(F)\|^2$$

Its derivative at F is the linear map

$$D\Lambda^i(F) : E^{-} \to \mathbf{R} \quad \text{defined by} \quad u \to 2 < F - H^i(F), u - DH^i(F).u > \qquad (3)$$

Since $H^i(F)$ is a critical point of Γ_F, $F - H^i(F)$ is orthogonal to the tangent space of Σ_n^{-} at $H^i(F)$. Therefore, (3) reduces to

$$u \to 2 < F - H^i(F), u >$$

This shows that the Λ^i have distincts derivatives, so the set on which two of them coincide is closed with empty interior in $\mathcal{V}$. In particular, in any neighborhood of F^0, there is a point F where all $\Lambda^i(F)$ are distincts, hence one of the $H^i(F)$ is the unique best approximant. Q.E.D.

The analysis of *local* best approximants is much harder, and we will not be able to tell so much about them. Using Proposition 1, it is easy to adapt the proofs of (Ba, *) to get a weak local result as follows. Given $F \in E^{-}$, denote by M_F^n the minimum of Γ_F on S_n^{-}.

Proposition 3. *Given $\epsilon > 0$, there is generically a finite number of local best approximants H of F such that $\|F - H\| < M_F^{n-1} - \epsilon$.*

To our knowledge, it is not known wether the number of local best approximants is generically finite. But it is rather puzzling that there can be no generic bound on this number. In fact, it is enough to exhibit for each $k \in \mathbf{N}$ a $H \in E^-$ who has at least k nondegenerate local best approximants, because then, using again the implicit function theorem and the local invariance of the signature of the Hessian, any member of E^- sufficiently close to F will share this property. It can be shown (but we shall not give the argument here) that for $n = m = p = 1$, the function $z \to sin\frac{1}{1+\epsilon-z}$ has an arbitrarily large number of nondegenerate local best approximants when $\epsilon \to 0$.

5. The case of finite sequences

In contrast with the above, we will show that when F is a finite sequence, all generic questions of finiteness admit rather easy positive answers.
We denote by $\mathcal{F}_N$ the subspace of finite sequences of length N in E^-, that is the space of functions of the form $\sum_{i=1}^{N} F_i z^{-i}$. The induced topology is of course that of $\mathbf{R}^{N\times m\times p}$, and the entries of the F_i's are obvious coordinates. We endow Σ_n^- with its usual manifold structure (HK, 74), where the domains of charts are indexed by nice selections, and the corresponding coordinates are the nonstructural coefficients of the nice realization.
Now, if $H \in \Sigma_n^-$ belongs to a chart as above, the long division of H yields Fourier coefficients that are polynomials in the coordinates of the chart, because the denominator is monic. Since the Fourier coefficients F_i of $F \in \mathcal{F}_N$ are zero except for N of them, it becomes appearent that $< F - H, F - H >$ is itself a polynomial in the coordinates of F and H. This observation leads to the following result.

Proposition 4. *If $N > 2n$, the property that critical points of Γ_F are finite in number and nondegenerate is strongly generic in $\mathcal{F}_N$.*

Proof: since a real algebraic variety has but a finite number of connected components, and since there are finitely many charts as above, the set of critical points will be finite as soon as its dimension is zero. Since the condition that critical points are nondegenerate is semi algebraic in F (Tarski's principle (Se, 54)), it will be strongly generic as soon as it is generic (CR, 79). Using the usual version of the transversality theorem (Hi, 76), the genericity of this condition can be established as in ((Ba, *) Lemma 5 and Prop. 3), with the only difference in the proof of the Lemma (loc. cit.) that one has first to check that the dimension of the projection on $\mathcal{F}_N$ of the tangent space to Σ_n^- (at any point) has still dimension $2n$. The hypothesis $N > 2n$ is then needed to find an element of $\mathcal{F}_N$ having prescribed scalar products with $2n$ independent vectors of $\mathcal{F}_N$. Q.E.D.

Note that in the case of finite sequences, the entire problem could be theoretically solved *via* elimination theory (Se, 54). Needless to say, the amount of computations prohibits this approach from a practical point of view. A bound on the number of critical points is easily derived from Bezout's theorem, but it is desperately large. It follows from what has been said in the preceding paragraph that any bound which is independant of F must become infinite with N, and this is not satisfactory since finite sequences are dense in E^-, while the set of best approximants is continuous by Proposition 2. No bound depending on F is known to us, except in special cases.

6. Concluding remarks

As we already mentionned in the introduction, this paper should be considered as a sequel to (Ba, *). In particular, it does not apply any of the preceeding considerations to the effective resolution of the problem *via* differential methods. Nevertheless, we have tried to make appearent that the Hardy space may sometimes exhibit strange behaviour while the set of finite sequences does not. This is related to the possibility of extending $< F, H >$ smoothly on the boundary of Σ_n^-. Though we shall not treat this here, we express our point of view that some intermediate functional spaces are more suitable frameworks from a differential viewpoint.

Aknowledgement The example mentionned at the end of Paragraph 4 is joint work with M. Olivi.

Bibliography

(Ba, 82) Baratchart L. "Une structure différentielle pour certaines classes de systèmes, application à l'approximation L^2" Thèse de docteur-Ingénieur E.N.S.M.P Paris.

(Ba, 84) Baratchart L. "On the parametrization of linear constant systems" SIAM J. cont. & opt., vol.23, n^0 5.

(Ba, *) Baratchart L., "Existence and generic properties for L^2-approximants of linear systems" to appear in I.M.A. Journal.

(BS, 85) Baratchart L., Steer S."Rosencher type equations for L^2-approximation of linear constant systems" Proc. 24th C.D.C., Fort Lauderdale, FL.

(CR, 79) Coste M., Roy M. F. "Topologies for real algebraic geometry" in "Topos theoretic methods in geometry", A. Kock ed., Arhus Universiteit, pp. 29-100.

(De, 80) Della Dora J. "Contribution à l'approximation de fonctions de la variable complexe au sens de Hermite-Padé et de Hardy" Thèse d'état Univ. Scient. & Médicale de Grenoble.

(Du, 73) Duc-Jacquet M. "Approximation des fonctionelles linéaires sur les espaces Hilbertiens à noyaux reproduisants" Thèse d'état Univ. Scient. & Médicale de Grenoble.

(Gl, 84) Glover K. "All optimal Hankel norm approximations of linear multivariable systems and their L^∞ bounds" Int. J. Cont. vol.39, n^0 6 pp. 1115-1195.

(Hi, 76) Hirsch M.W. "Differential Topology" Graduate Texts in Math. Springer-Verlag New-York.

(HK, 74) Hazewinkel M., Kalman R.E. "Moduli and canonical forms for linear systems" Report n^0 7504 Econometric Institute Erasmus Univ. Rotterdam.

(KFA, 69) Kalman R. E., Falb P. L., Arbib M. A. "Topics in mathematical system theory" Mc. Graw-Hill, New York.

(KL, 81) Kung S., Lin D. "Optimal Hankel norm model reduction: multivariable systems" IEEE trans. Aut. Cont. *vol.26*, n^0 4, pp. 832-854.

(Ro, 78) Rosencher E. "Approximation rationelle des filtres à 1 ou 2 indices: une approche Hilbertienne" Thèse de docteur-Ingénieur. Univ. Paris IX-Dauphine Paris.

(Ruc, 78) Ruckebush G. "Sur l'approximation rationelle des filtres" Rapport n^0 35 C.M.A. Ecole Polytechnique.

(Rud, 66) Rudin W. "Real and complex analysis" Mc. Graw-Hill New York.

(Se, 54) Seidenberg A. "A new decision method for elementary algebra" Ann. of Math. 60, pp. 365-374.

Design Examples Using μ-Synthesis: Space Shuttle Lateral Axis FCS During Reentry

John Doyle[1], Kathryn Lenz[2], Andy Packard[3]
Honeywell Inc.,
Systems and Research Center
Minneapolis, Minnesota

SPACE SHUTTLE ORBITER

The Authors are also affiliated with 1) California Institute of Technology 2) University of Minnesota 3) University of California

NATO ASI Series, Vol. F34
Modelling, Robustness and Sensitivity Reduction in Control Systems. Edited by R. F. Curtain

Abstract

This paper studies the application of Structured Singular Values (SSV or μ) for analysis and synthesis of the Space Shuttle lateral axis flight control system (FCS) during reentry. Comparisons are made of the conventional FCS with alternatives based on H_∞-optimal control and μ-synthesis. While this is a fairly standard FCS problem in most respects, the aircraft model is highly uncertain due to the poorly known aerodynamic characteristics (e.g. aero coefficients). The problem as formulated is particularly interesting and challenging because the uncertainty is large and highly structured.

1. Introduction

During reentry the Shuttle FCS is in automatic mode using a series of S-turns to reduce speed below Mach 1. The flight condition we will consider is at Mach .9, just prior to the heading alignment circle (HAC), which lines the Shuttle up on the runway for landing. Potential robustness problems were found at this flight condition in a previous study ([M1],[M2]) done at Honeywell's Systems and Research Center (SRC) for the Space and Strategic Avionics Division (SSAvD), who are responsible for validation of the Shuttle FCS.

The SRC study was a preliminary investigation of the use of μ in analyzing robustness of the Shuttle FCS, where the dominant uncertainty is modeled as large parameter variations in 9 key aerodynamic coefficients. SSAvD is now using μ to augment conventional analysis, which essentially involves trial and error using coefficient combinations known to produce problems. The potential advantage in using μ is that it is faster and more reliable than trying to search the high dimensional parameter space for bad coefficient values. Furthermore, μ-analysis [D1] can be combined with H_∞-optimal control methods [F1] to produce a synthesis method, called μ-synthesis [D3], which provides H_∞-performance in the presence of structured uncertainty.

This paper reports on a study at SRC using μ-synthesis to redesign the flight control laws. The objective was to mimic the performance characteristics of the existing FCS (referred to as BrandX throughout this paper), while providing this performance for a wider range of uncertainty. The resulting controller, referred to as Musyn, thus has better robust performance.

The problem was simplified to some extent to focus attention on the dominant features that were found to be the most significant problems in the actual system. The performance objective of the FCS is to execute bank commands with turn coordination in the presence of gust disturbances using aileron (actually differential elevon) and rudder (the yaw jets are turned off at Mach 1). Sensor noise, large uncertainty in the the aerodynamic coefficients, penalties on actuator magnitude, rate, and acceleration, and delays to represent effects of sampling were included. The major neglected practical issues are the effects of vehicle flexibility and nonlinearities. While these are important and significantly complicate the final design, they do not change the results in any qualitative way. It is important to emphasize that this brief study is intended only to illustrate the use of μ and is not a definitive treatment of the Shuttle FCS.

The paper is organized into 6 sections. Sections 2 and 3 briefly review μ analysis and synthesis and Section 4 describes the problem formulation. The review is minimal, covering only those methods that were actually used, and in no way claims to be a review of the robust control theory field. Section 5 begins with an analysis of BrandX along with an H_∞ controller that neglects uncertainty and gives slightly better performance but essentially no robustness. The Musyn design dramatically improves robust performance with only a slight loss of nominal performance. Comparison are made using μ-analysis and time responses. Section 6 considers the issue of controller order and reduction and Section 7 has conclusions. An appendix is included with realizations of the aircraft model, the BrandX controller, and a reduced-order μ-synthesis controller. Enough data is included in this paper so that, at least in principle, all results could be reproduced.

2. Analysis Review

This section will very briefly review the basic frequency-domain methods for analyzing the performance and robustness properties of feedback systems using μ ([D1],[D3],[D4],[M1]). The general framework to be used in this paper is illustrated in the diagram in Figure 1a. Any linear interconnection of inputs, outputs, commands, perturbations, and a controller can be rearranged to match this diagram. For the purpose of analysis the controller can be viewed as just another system component and the diagram reduces to that in figure 1b. The uncertainty in v and Δ as well as the performance specifications on e are assumed to be normalized to 1. This requires that all weighting functions and scalings have been absorbed into the interconnection structure G. We will consider performance objectives expressed in terms of $\|G_{22}\|_\infty = \sup_\omega \bar{\sigma}(G_{22}(j\omega))$. Robust stability for unstructured uncertainty (only $\bar{\sigma}(\Delta) < 1$ is known) depends on $\|G_{11}\|_\infty$. Unfortunately, norm bounds are inadequate in dealing with robust performance and realistic models of plant uncertainty involving structure. Therefore, more complicated mathematical objects involving μ are required.

To begin with, assume that Δ belongs to a set like

$$\underline{\Delta} = \{ \, diag(\Delta_1,\Delta_2, \ldots ,\Delta_n) \, \} \quad \text{or} \quad \underline{B\Delta} = \{\Delta \in \underline{\Delta} \mid \bar{\sigma}(\Delta) < 1\}. \tag{2.1}$$

The function μ has the properties $\mu(\alpha M) = |\alpha| \, \mu(M)$ and

$$det(I - M\Delta) \neq 0 \quad \forall \;\; \Delta \in \underline{B\Delta} \quad \textbf{iff} \quad \mu(M) \leq 1. \tag{2.2}$$

Obviously, μ is a function of M which depends on the *structure* of $\underline{\Delta}$. For this informal discussion just keep this fact in mind since the structure will always be clear from context. Let

$$\underline{U} \;=\; \{diag(U_1,U_2,\ldots,U_n) \mid U_i^*U_i = I\} \tag{2.3}$$

$$\underline{D} \;=\; \{diag(d_1I,d_2I,\ldots,d_nI) \mid d_i \in \mathbb{R}+\} \tag{2.4}$$

where the sets $\underline{U}$ and $\underline{D}$ match the structure of $\underline{\Delta}$. Note that the $\underline{U}$ and $\underline{D}$ leave $\underline{\Delta}$ invariant in the sense that $\Delta \in \underline{\Delta}$, $U \in \underline{U}$ and $D \in \underline{D}$ implies that $\bar{\sigma}(\Delta U) = \bar{\sigma}(U\Delta) = \bar{\sigma}(\Delta)$ and $D\Delta D^{-1} = \Delta$. The sets $\underline{U}$ and $\underline{D}$ can be used to obtain the bounds

$$\max_{U \in \underline{U}} \rho(MU) \;\le\; \mu(M) \;\le\; \inf_{D \in \underline{D}} \bar{\sigma}(DMD^{-1}) \tag{2.5}$$

where ρ denotes the spectral radius and $\bar{\sigma}$ denotes the maximum singular value.

The key theorems about μ show that the lower bound is always an equality and the upper bound is an equality when $n \le 3$. Unfortunately, the optimization problem implied by the lower bound has multiple local maxima so it does not immediately yield a reliable computational approach. Although $\bar{\sigma}(DMD^{-1})$ is convex in $\ln(D)$ so that the infimum can be found by search over $n-1$ real parameters, the infimum is not necessarily equal to μ (i.e., an example of strict inequality has been found for $n = 4$). On the other hand, extensive experimentation indicates that the upper bound may be close to μ in general, although this has not been proven. The worst case ratio of lower over upper bound found so far is .85. For all the cases in this paper, including those with 4 blocks, μ is equal to the upper bound.

Another important aspect of the upper bound is that μ may be viewed as $\bar{\sigma}$ with scaling. Thus the general synthesis methods developed for H_∞-optimization may be applied, via scalings, to optimize μ. This will be discussed further in the synthesis section. Also, μ as described above applies only to square blocks, but is easily extended to handle both nonsquare and repeated blocks.

The importance of μ for studying robustness of feedback systems is due to the following two theorems, which characterize in terms of μ the robust stability and robust performance of a system in the presence of structured uncertainty.

Theorem RS (Robust Stability)

$$\mathbf{F}_u(G,\Delta) \text{ stable } \;\forall \Delta \in \underline{B\Delta} \;\textbf{ iff }\; \sup_{\omega} \mu(G_{11}(j\omega)) \;\le\; 1$$

Theorem RP (Robust Performance)

$$\mathbf{F}_u(G,\Delta) \text{ stable and } \|\mathbf{F}_u(G,\Delta)\|_\infty \;\le\; 1 \;\forall \Delta \in \underline{B\Delta}$$

$$\textbf{iff } \; \sup_{\omega} \mu(G(j\omega)) \;\le\; 1$$

(where μ in Theorem RP is computed with respect to the structure $\underline{\tilde{\Delta}} = \{\, diag(\Delta,\Delta_{n+1}) \mid \Delta \in \underline{\Delta}\}$).

Figure 1a General Framework

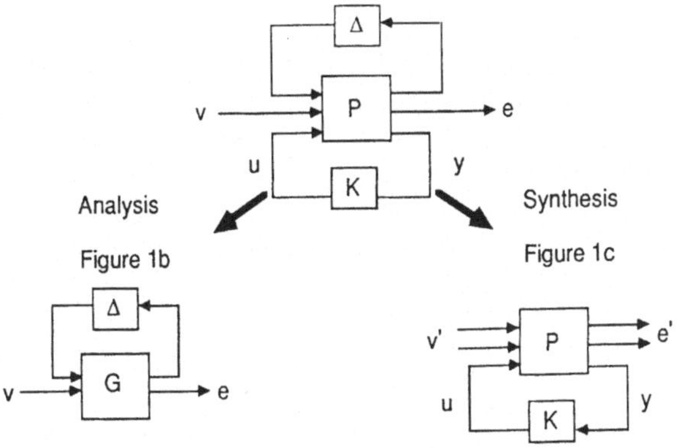

$$F_u(G,\Delta) = \left[G_{22} + G_{21}\,\Delta\,(I - G_{11}\Delta)^{-1}G_{21}\right]$$

$$F_l(P,K) = \left[P_{11} + P_{12}K\,(I - P_{22}K)^{-1}P_{21}\right]$$

Figure 1 General Framework, Analysis and Synthesis

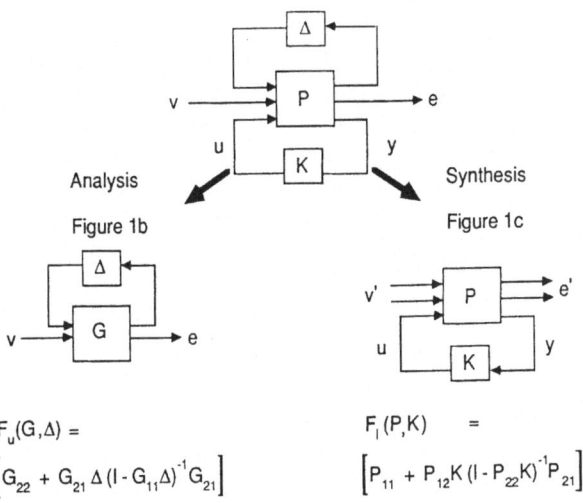

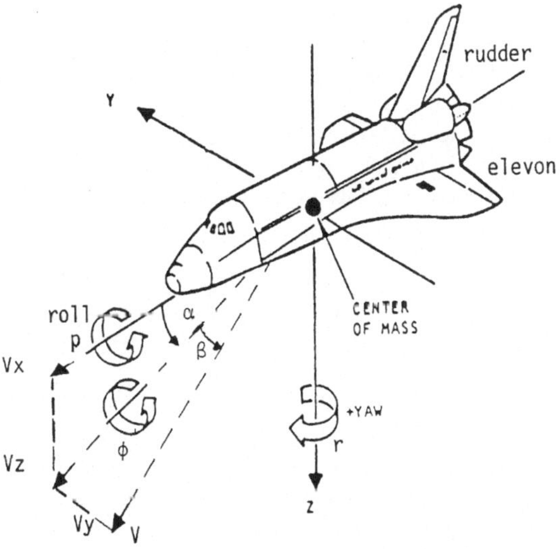

Figure 2 Rigid Body State Variables and Measurements

3. Synthesis Review

The basic framework for the general H_∞ - optimal control problem ([D2],[D3],[C2],[F1]) is shown in figure 1c. For a review of H_∞-theory, see [F1]. The objective is to find a stabilizing K which minimizes $\|\mathbf{F}_l(P,K)\|_\infty$. The first step is to find J such that $\mathbf{F}_l(P,\mathbf{F}_l(J,Q)) = \mathbf{F}_l(T,Q) = T_{11}-NQ\tilde{N} \in RH_\infty$ is stable and affine for any $Q \in RH_\infty$. We are interested in a particular J which results in N and $\tilde{N}$ being inner and co-inner respectively. That is, $N^*N = I$ and $\tilde{N}\tilde{N}^* = I$. This requires a coprime factorization with inner numerator [C1]. In addition, we require $N_\perp$ and $\tilde{N}_\perp$ inner so that $\begin{bmatrix} N & N_\perp \end{bmatrix}$ and $\begin{bmatrix} \tilde{N} \\ \tilde{N}_\perp \end{bmatrix}$ are *square* and inner. Then

$$\left\| T_{11}-NQ\tilde{N} \right\|_\infty = \left\| \begin{bmatrix} N & N_\perp \end{bmatrix}^* [T_{11}-NQ\tilde{N}] \begin{bmatrix} \tilde{N} \\ \tilde{N}_\perp \end{bmatrix}^* \right\|_\infty = \left\| \begin{bmatrix} R_{11}-Q & R_{12} \\ R_{21} & R_{22} \end{bmatrix} \right\|_\infty . \tag{3.1}$$

The standard approach to minimizing (3.1) over Q involves the so-called γ-iteration, which is computationally intensive. The alternative used in this paper is to simply choose Q to minimize $\|R_{11}-Q\|_\infty$. This provides a good approximation and is relatively cheap computationally [C2].

The μ-analysis and H_∞-synthesis methods combine to produce μ-synthesis. Recall that μ may be obtained by scaling and applying $\|\bullet\|_\infty$, so that a reasonable approach is to "solve"

$$\min_{K,D} \|D\mathbf{F}_l(P,K)D^{-1}\|_\infty \tag{3.2}$$

by iteratively solving for K and D. With either K or D fixed, the global optimum in the other variable may be found using the μ and H_∞ solutions described previously. Unfortunately, this iterative scheme is not guaranteed to find the global optimum of (3.2). Nevertheless, the approach appears promising and substantial progress is being made in developing methods to obtain the global optimum [D4]. The actual implementation computes D as a function of frequency and then fits it with rational functions for use in the H_∞-step.

4. Problem Description

The performance objective of the Musyn FCS is to mimic the BrandX FCS nominally but with better robustness to uncertainty. Since BrandX was not designed by H_∞-techniques, and since H_∞-performance objectives only make practical sense when they include meaningful variables and weights, it is necessary to carefully reinterpret the BrandX performance in terms of weighted H_∞- performance objectives. Fortunately, the mathematical properties of H_∞ make this process relatively easy. Besides, the performance specifications for a typical FCS translate fairly naturally into the H_∞-context. Based on consultation with engineers familiar with the Shuttle FCS each disturbance, command, noise, error, and actuator variable was given simple, reasonable weights. These weights were then adjusted until each variable made a nearly equal contribution to the $\|\bullet\|_\infty$ norm for the BrandX closed loop system. This approach finesses the problem of selecting weighted H_∞-performance objectives exclusively from physical considerations, an issue which will not be considered in this paper. Because flexible effects have been neglected in the problem formulation, the BrandX controller was simplified by removing bending mode filters.

The 4-state rigid body aircraft model has state variables and measurements

$$x_{state} = \begin{bmatrix} \beta \\ p \\ r \\ \phi \end{bmatrix} = \begin{bmatrix} \text{sideslip angle} \\ \text{roll rate} \\ \text{yaw rate} \\ \text{bank angle} \end{bmatrix} \quad \text{and} \quad y_{meas} = \begin{bmatrix} p \\ r \\ n_y \\ \phi \end{bmatrix} \tag{4.1}$$

where n_y is lateral acceleration. See Figure 2 for definitions of the variables. Angle of attack is denoted by α and V is the velocity vector.

The units used throughout the paper are *rad/s* for p and r, *ft/s*^{2} for n_y, *ft/s* for the gust, and *rad* for ϕ except in the plots where *deg* and *deg/s* replace *rad* and *rad/sec*. Each measurement is corrupted by additive sensor noise which becomes more severe with increasing frequency. Since p and r are both measured with comparable gyroscopes, their sensor noise weights are assumed to be identical and equal to $3\times10^{-3}(1+s/.01)/(1+s/.5)$. The measurement for ϕ is obtained from a navigation package at a reduced sample rate so its weight of $7\times10^{-3}(1+s/.01)/(1+s/2)$ was chosen to be relatively large in mid to high frequencies. An alternative scheme would have been to introduce a frequency-dependent perturbation to reflect the effects of sampling. The weight for the n_y accelerometer is $.25(1+s/.05)/(1+s/10)$. Bode magnitude plots of the sensor noise weighting filters are shown in Figure 3a. Recall that these weights were chosen consistent with the BrandX control design but may not accurately represent realistic noise characteristics of the Shuttle sensors.

The additional external inputs are the command in ϕ and a lateral gust disturbance weighted by $.5(1+s/2)/(1+s/.5)$ and $30(1+s/2)/(1+s)$, respectively. Performance is described in terms of

$$e_{perf} = W_{perf} \begin{bmatrix} n_y \\ r_p \\ \phi - \phi_{ideal} \end{bmatrix} \tag{4.2}$$

where n_y and $r_p = r - .037\phi$, the error from nominal turn rate are regulated to provide turn coordination. The ϕ_{ideal} is generated by an "ideal" model response $1/(1+2\zeta(s/\omega)+(s/\omega)^2)$ with $\omega = 1.2$ *rad/s* and $\zeta = .7$. An "ideal" turn would produce $\phi = \phi_{ideal}$ with no sensed acceleration and no turn rate error ($n_y = r_p = 0$). Of course, the vehicle physics prevents such an ideal maneuver and a good control system seeks to approach the ideal. In most conventional lateral axis control designs, n_y and r_p are blended to form a single turn coordination variable, because from a loop-shaping perspective it is

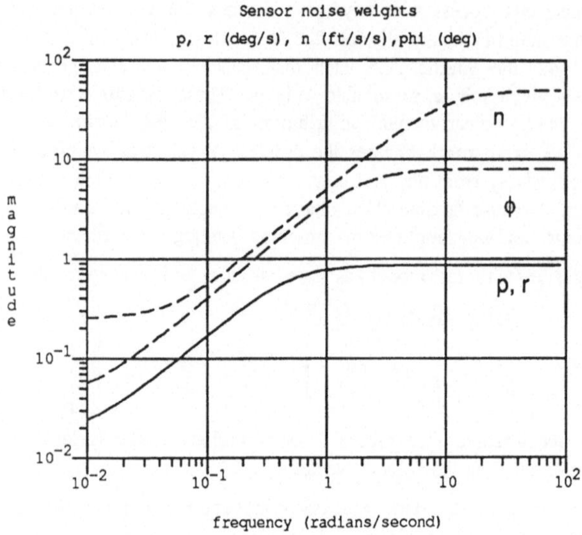

Figure 3a Sensor Noise Weighting Filter

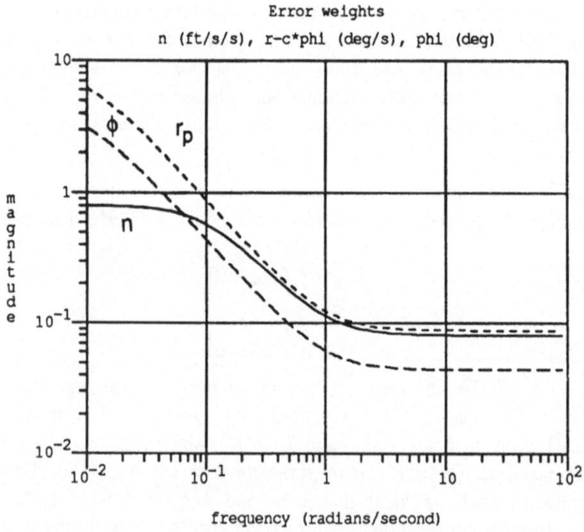

Figure 3 Performance Weights

easier to work with two instead of three performance variables to match the two inputs. Since we will not be using loop-shaping in this paper such "squaring down" is unnecessary. The W_{perf} performance weights are plotted in Figure 3. The overall shape of the weights indicates our desire to provide good performance in the low to mid frequency range. Frequencies below .01 rad are generally neglected since signals in this range are too slow to have an impact on the Shuttle reentry and landing. The performance weights are

$$W_{perf} = diag(.8\frac{(1+s)}{(1+s/.1)},\ 500\frac{(1+s)}{(1+s/.01)},\ 250\frac{(1+s)}{(1+s/.01)}).$$

The actuator models are second-order lags of the form $1/(1+2\zeta(s/\omega)+(s/\omega)^2)$ with $\omega = 21\ rad/s$ and $\zeta = .75$ for the rudder and $\omega = 14\ rad/s$ and $\zeta = .72$ for the elevon. To reflect practical saturation considerations actuator position, rate, and acceleration (in radians and seconds) are weighted by (2,.2,.009) respectively for rudder and (4,1,.005) for elevon. A second-order delay approximation of $(1-2\zeta(s/\omega)+(s/\omega)^2)/(1+2\zeta(s/\omega)+(s/\omega)^2)$, $\omega = 173$, $\zeta = .866$ was also included in each actuator to model the effects of the digital implementation of the controller. Although such a model is simplistic, experience has shown that it is entirely adequate for this type of study.

The major uncertainty in this problem is in the aerodynamic coefficients. These coefficients are standard aerodynamic parameters which express incremental forces and torques generated by incremental changes in sideslip, aileron, and rudder angles. Thus

$$\begin{bmatrix} \text{side force} \\ \text{yawing moment} \\ \text{rolling moment} \end{bmatrix} \propto \begin{bmatrix} c_{y\beta} & c_{ya} & c_{yr} \\ c_{n\beta} & c_{na} & c_{nr} \\ c_{l\beta} & c_{la} & c_{lr} \end{bmatrix} \begin{bmatrix} \text{sideslip angle} \\ \text{aileron angle} \\ \text{rudder angle} \end{bmatrix} \tag{4.3}$$

The coefficients $c_{\bullet\bullet}$ are typically estimated from theoretical predictions, numerical calculations, and experiments in wind tunnels and/or flight tests. The Shuttle at Mach .9 is in a transonic regime involving a mixture of subsonic and supersonic flows. Neither the theoretical, computational, or wind tunnel techniques are particularly accurate at this flight condition, so with extremely limited flight data the coefficient uncertainty for the Shuttle is unusually large.

Uncertainty is modeled by representing each coefficient by a nominal value plus a perturbation. In terms of the coefficient matrix in (4.3), the perturbation may be written as

$$\begin{bmatrix} r_{y\beta}\delta_{y\beta} & r_{ya}\delta_{ya} & r_{yr}\delta_{yr} \\ r_{n\beta}\delta_{n\beta} & r_{na}\delta_{na} & r_{nr}\delta_{nr} \\ r_{l\beta}\delta_{l\beta} & r_{la}\delta_{la} & r_{lr}\delta_{lr} \end{bmatrix} = \begin{bmatrix} R_{\bullet\beta} & R_{\bullet a} & R_{\bullet r} \end{bmatrix} \begin{bmatrix} diag(\delta_{\bullet\beta}, \delta_{\bullet a}, \delta_{\bullet r}) \end{bmatrix}$$

Each 3×1 vector $\delta_{\bullet\beta}$, $\delta_{\bullet a}$, $\delta_{\bullet r}$ is assumed to be of Euclidean norm 1, so our perturbation matrix is 9×3. The groupings on the perturbations is motivated by expected correlations between the uncertainties. Alternatively, we could, of course, ignore this and use a diagonal 9×9 perturbation. The R weightings are $R_{\bullet\beta} = diag(\ 2.194, -1.517, -.7180\ )$,

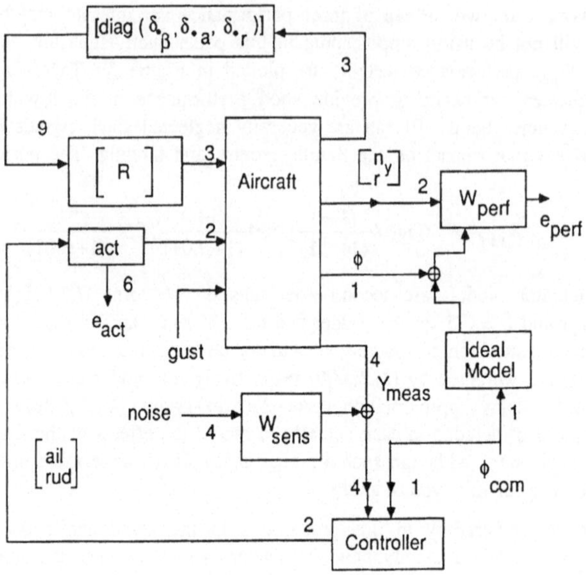

Figure 4 Block Diagram

$R_{\bullet a} = diag(-1.327, 1.347, .5185)$, $R_{\bullet r} = diag(-.3656, .8667, .2393)$, which are (conservative) current estimates of the size of the corresponding aero coefficient. The signs are simply arbitrary choices.

One conventional way to view the δ's is as fixed but unknown real parameters. This assumes that the rigid body dynamics are perfectly described by one 4th order model, but we simply do not know a priori which one. An alternative view is that since the coefficients represent the generation of aerodynamic forces and moments, they are actually themselves dynamical systems. Furthermore, they depend in complicated, nonlinear ways on quantities which are time-varying. We will not try to resolve this issue here but simply point out that these two views lead to apparently quite different uncertainty models. Roughly speaking, the former constrains the δ's to be real while the latter would suggest that they be complex with possibly frequency-dependent magnitude bounds. Some compromise would probably be appropriate. Since our μ-based methods are inherently complex, we will take the conservative approach and treat the δ's as complex. We have relatively crude extensions to μ which treat real perturbations and will show that for this problem the complex assumption is only slightly conservative. This allows us to temporarily avoid resolving the tricky issue regarding the appropriate way to view the coefficient uncertainty.

Figure 4 shows the block diagram that includes all the features discussed above. It is clearly an example of Figure 1a, with e including e_{perf} and e_{act}, v including ϕ_{com}, gust, and sensor noise, and y including the measured outputs and ϕ_{com}. The dimensions of e, v, y, u, P, and Δ are 9, 6, 5, and 2, 17×17, and 9×3, respectively. State space models for the aircraft and the BrandX controller are included in the appendix.

5. Comparisons of Designs

The μ plots of robustness and weighted performance for the BrandX design are shown in Figure 5. Referring to Figure 1b, the dashed plots are, from the top, $\mu(G)$ for robust performance, $\mu(G_{11})$ for robust stability, and $\bar{\sigma}(G_{22})$ for nominal performance. Note that the weights chosen for nominal performance make $\|G_{22}\|_\infty = .5$. If $\mu(G)$ were less than 1 then performance would only degrade by a factor of 2 for this level of uncertainty. Unfortunately, BrandX is unstable for the assumed uncertainty level because $\mu(G_{11}) > 1$ at $\omega \approx 1.5$. To give some idea of the sensitivity to the assumption that the δ's are complex, compare with the solid line which gives a lower bound for "real μ" for G_{11}. This lower bound was computed using two different programs (by M. Elgersma of SRC and M.K. Fan of U. of Maryland) which search for destabilizing real perturbations. These programs currently require that the Δ have only scalar blocks so each δ is assumed bounded in magnitude by 1. Note that since this lower bound is comparable to the complex $\mu(G_{11})$, we need not be particularly concerned about our assumptions on the δ's.

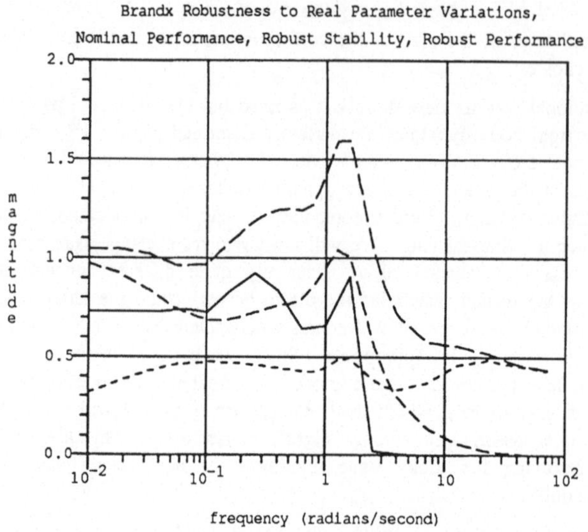

Figure 5 Robustness and Weighted Performance for the Brand X Design

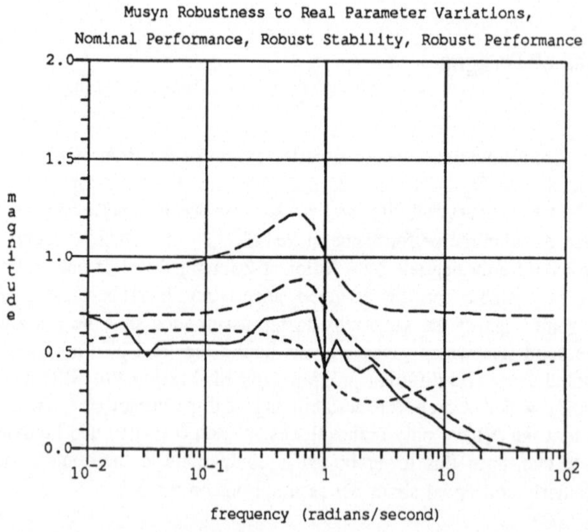

Figure 6 Robustness and Weighted Performance for the Musyn Design

The corresponding plots for the Musyn design are shown in Figure 6. Robust stability and performance are improved at the expense of a slight degradation in nominal performance. Note that we now have robust stability for the assumed perturbations, but robust performance is not quite as desired. This means that the performance in the presence of some perturbations in the allowed set degrades slightly beyond the a priori target. This is of no great consequence as the absolute level of the performance specification is somewhat arbitrary. The weights were chosen to roughly reflect desired performance characteristics, not absolute specifications. It is really only meaningful as a relative measure to compare alternatives.

It is interesting to consider a controller designed using H_∞-optimization of G_{22}, ignoring the coefficient uncertainty. This optimizes nominal performance with no consideration of robustness with respect to uncertainty in the aero coefficients. The nominal performance ($\overline{\sigma}(G_{22})$) of this controller, which we'll call Hinf, is plotted along with nominal performance of BrandX in Figure 7. Since Hinf is obtained by approximating the H_∞-optimal controller, it does not display the characteristic flat $\overline{\sigma}$ of theoretical H_∞-optimal designs, even though its norm is close. In some sense this so-called 1-block approximation is more appealing than an optimal controller, which would produce a constant $\overline{\sigma}$ slightly lower than the peak of the Hinf design. The approximation is much easier to compute and actually results in a lower $\overline{\sigma}$ over most of the frequency range.

The closeness of the BrandX and Hinf plots suggests two important points. First, the weight selection procedure was reasonably successful in capturing the BrandX performance. Had there been any "slack" in the weights (e.g. inadequate penalty on actuators or turn coordination or too small sensor noise), the Hinf design would have taken advantage of this to produce a much smaller norm. Secondly, BrandX is quite outstanding when viewed from this perspective. If the BrandX and Hinf designs were not close it would be difficult to determine what was the cause and consequently it would be difficult to proceed with confidence that the BrandX controller performance was captured in the H_∞-context.

The $\mu(G_{11})$ plot of robust stability for Hinf in Figure 8 shows that it is destabilized by even tiny perturbations. This is not too surprising since robustness was not designed for. On the other hand, it is not easy to construct physically motivated examples such as this with sensor noise, penalties on actuator signals, disturbances, and commands which produce an H_∞-controller with such poor robustness characteristics. One reason for the robustness problem is that the uncertainty enters internal to the plant and not at the inputs or outputs where signals are already penalized in the H_∞-design.

The μ-plots may seem a bit mysterious to the uninitiated, and may actually obscure the important issue of robust performance. To get another view of these designs, consider Figure 9 which plots

$$\max_{\|\Delta\|_\infty \leq \delta} \|F_u(G,\Delta)\|_\infty \quad \text{vs.} \quad \delta$$

where the maximum is taken over $\Delta \in \underline{\Delta}$. That is, the worst case performance over all $\|\Delta\|_\infty \leq \delta$ is plotted vs. δ. Nominal performance is at $\delta = 0$ and there is a vertical asymptote at the δ_u where the system goes unstable for some $\|\Delta\|_\infty = \delta_u$ (i.e. $\delta_u = 1/\mu(G_{11})$).

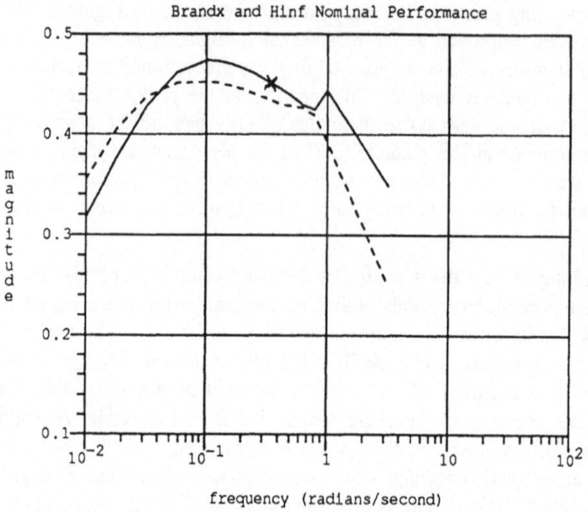

FIGURE 7

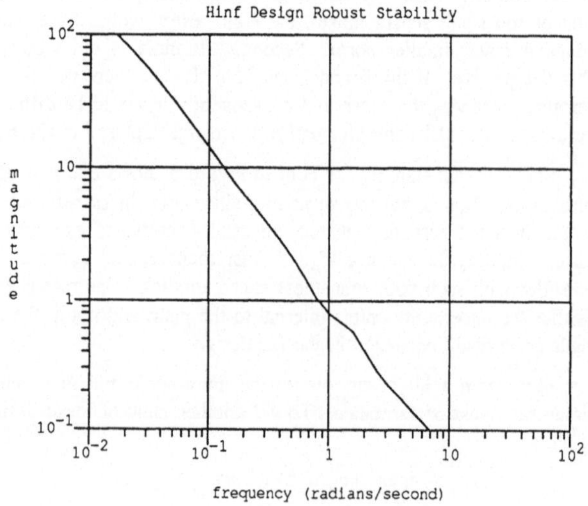

FIGURE 8

The symbols are $\times$ for BrandX, $\Box$ for Musyn, and $\bigcirc$ for an additional μ-synthesis controller designed with the perturbation weight reduced from 1 to .25. An exercise for aficianados: figure out how Figure 9 was made (hint: scale and compute μ). Note that $\Box$ has substantially better robust performance than BrandX but with slightly poorer nominal performance. $\bigcirc$ is a compromise that beats BrandX nominally with less robust performance than $\Box$. Hinf was not plotted since its robustness is so pathetic that $\delta_u \approx 0$. For comparison, the μ plots for $\bigcirc$ that correspond to Figures 5 and 6 are in Figure 10. Figure 9 illustrates an important feature of the μ-based methods; they provide for systematic exploration of the tradeoffs between performance and robustness. Since performance specifications are rarely fixed and inviolate and uncertainty bounds cannot be determined exactly, the engineer typically must compare several alternative control strategies and consider exactly these type of tradeoffs.

Some time domain comparisons of BrandX and Musyn are plotted in Figures 11 through 13. Additional analysis of Musyn revealed that response to the ϕ command is its most serious performance problem (the most significant contributor to $\|G_{22}\|$ or $\|e\|$), relative to both BrandX's command response and to the other aspects of Musyn's performance. To examine this worst-case behavior of Musyn in the time domain a step command of 28.6 deg (= .5 rad) in ϕ was chosen. Plots are shown of the response of ϕ, the turn coordination variables n_y and r_p, and the surface deflections for BrandX and Musyn in both nominal and perturbed conditions.

The nominal response for the BrandX and Musyn controllers is very similar. This is further confirmation that the H_∞- performance objective accurately captures the characteristics of BrandX. The magnitude of the errors and the actuator activity is slightly smaller for Musyn, but BrandX settles faster. This is consistent with the nominal frequency domain properties of the two controllers where Musyn was larger at low frequencies and BrandX was larger at high frequencies.

The first perturbation

$$\Delta_1 = \begin{bmatrix} \delta_{\bullet\beta} & \delta_{\bullet a} & \delta_{\bullet r} \end{bmatrix} = \begin{bmatrix} 0 & 0 & 0 \\ 1.12 & 1.12 & .93 \\ 1.12 & -1.12 & 1.12 \end{bmatrix}$$

is one for which BrandX is almost neutrally stable. The highly oscillatory response of BrandX is unacceptable, while Musyn degrades only slightly from the nominal. The second perturbation

$$\Delta_2 = \begin{bmatrix} \delta_{\bullet\beta} & \delta_{\bullet a} & \delta_{\bullet r} \end{bmatrix} = \begin{bmatrix} -1 & .645 & 1 \\ -1 & -1 & 1 \\ -1 & -1 & 1 \end{bmatrix}$$

was found to be the worst for Musyn. Although the errors and actuator inputs remain small in magnitude, they are settling more slowly than the nominal. The BrandX controller is unstable here.

The time responses suggest the same conclusions as the μ analysis, that the nominal performance of the BrandX and Musyn controllers are quite similar, but the robustness

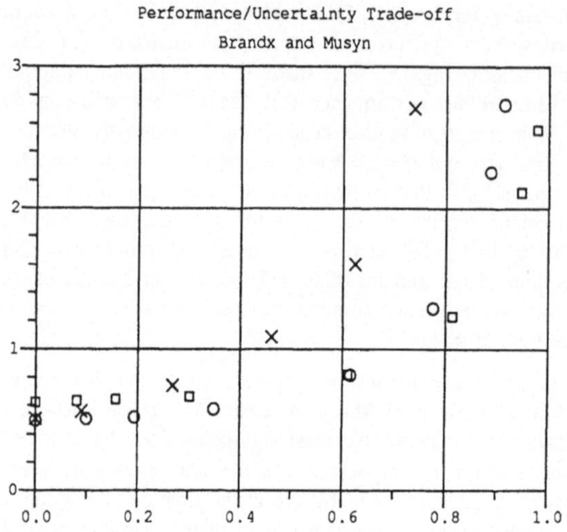

FIGURE 9 Comparison of $\max_{\|\Delta\|_\infty \le \delta} \|F_u(G, \Delta)\|_\infty$ vs. δ

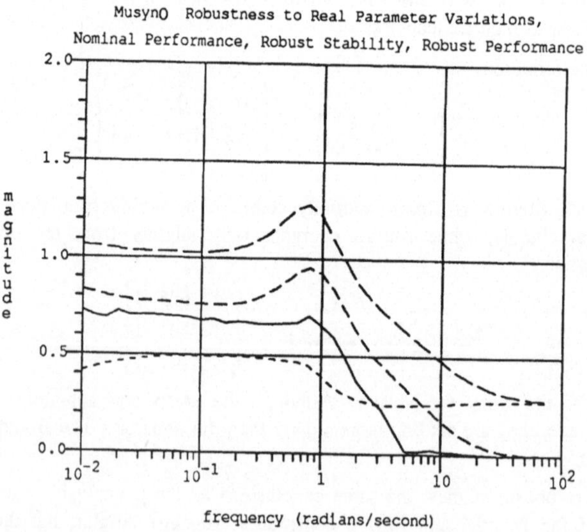

FIGURE 10

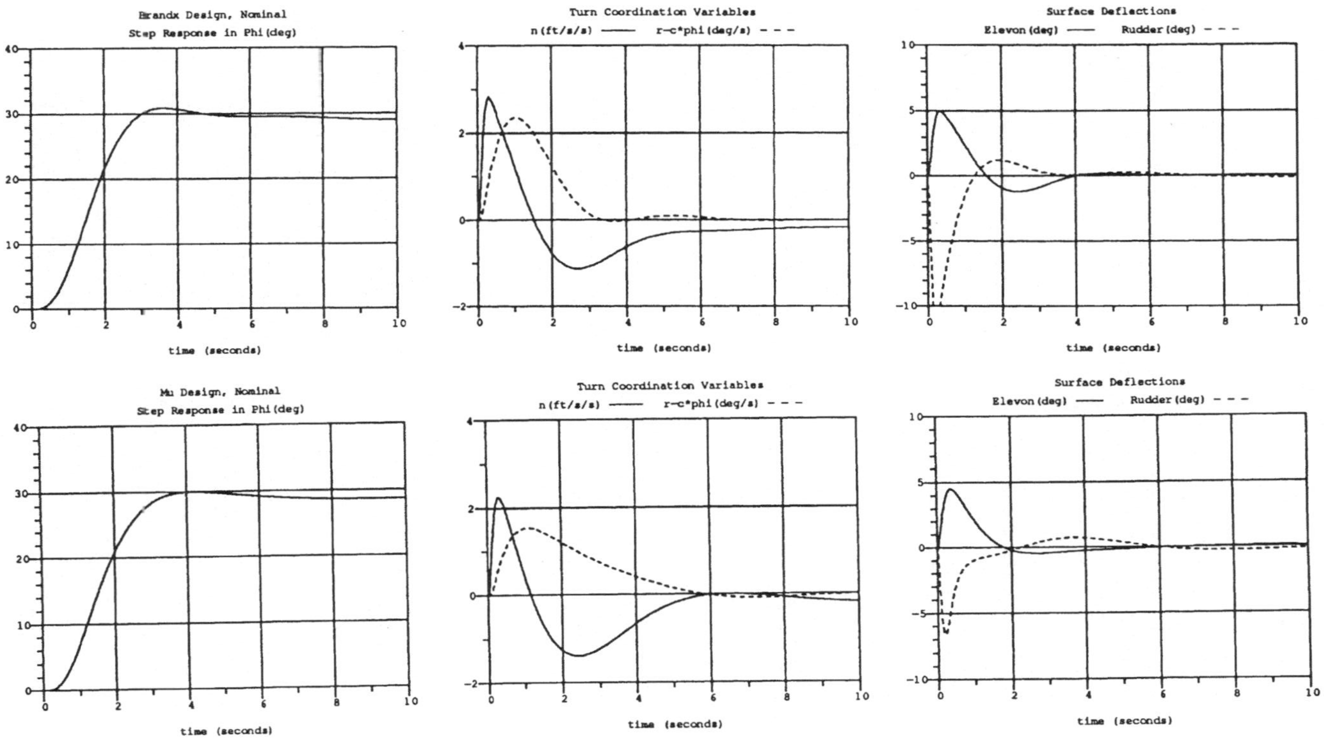

FIGURE 11 Step Responses for Nominal BrandX and Musyn Designs

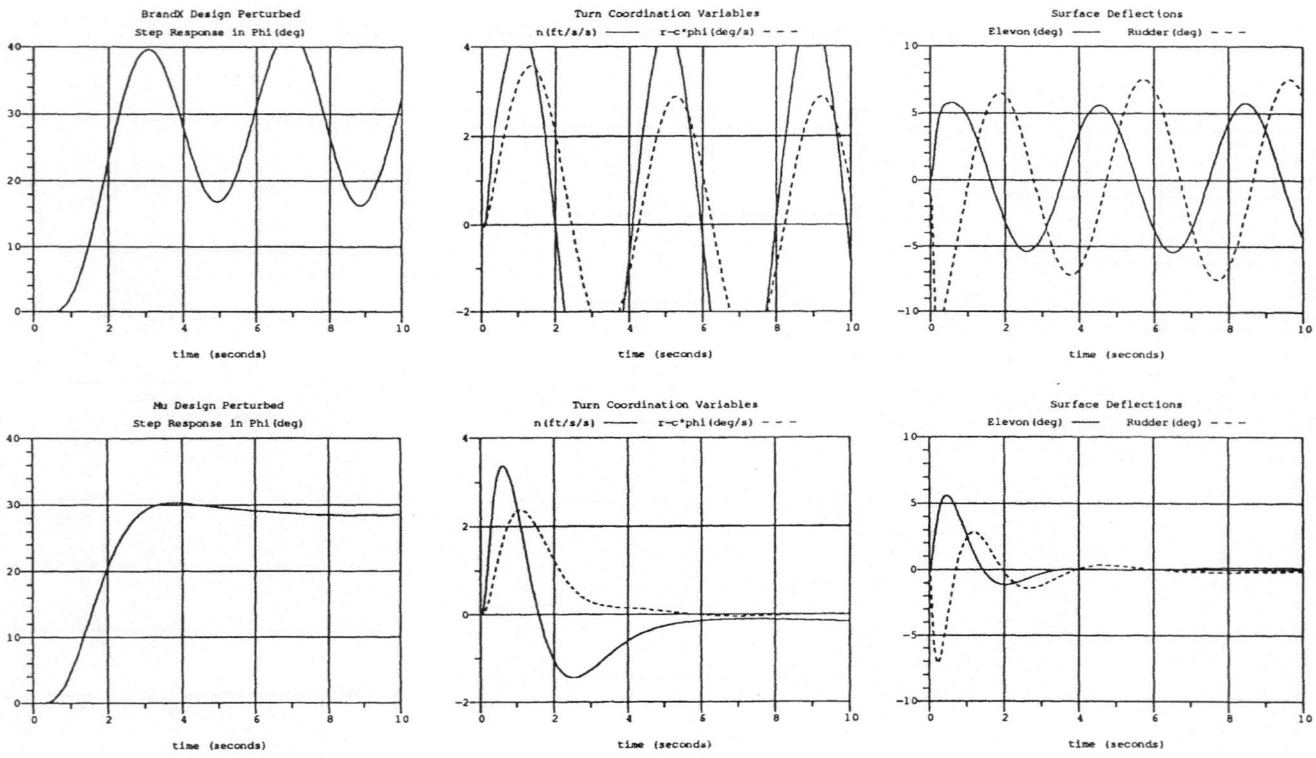

FIGURE 12 Step Responses for BrandX and Musyn Designs Perturbed by $\Delta 1$

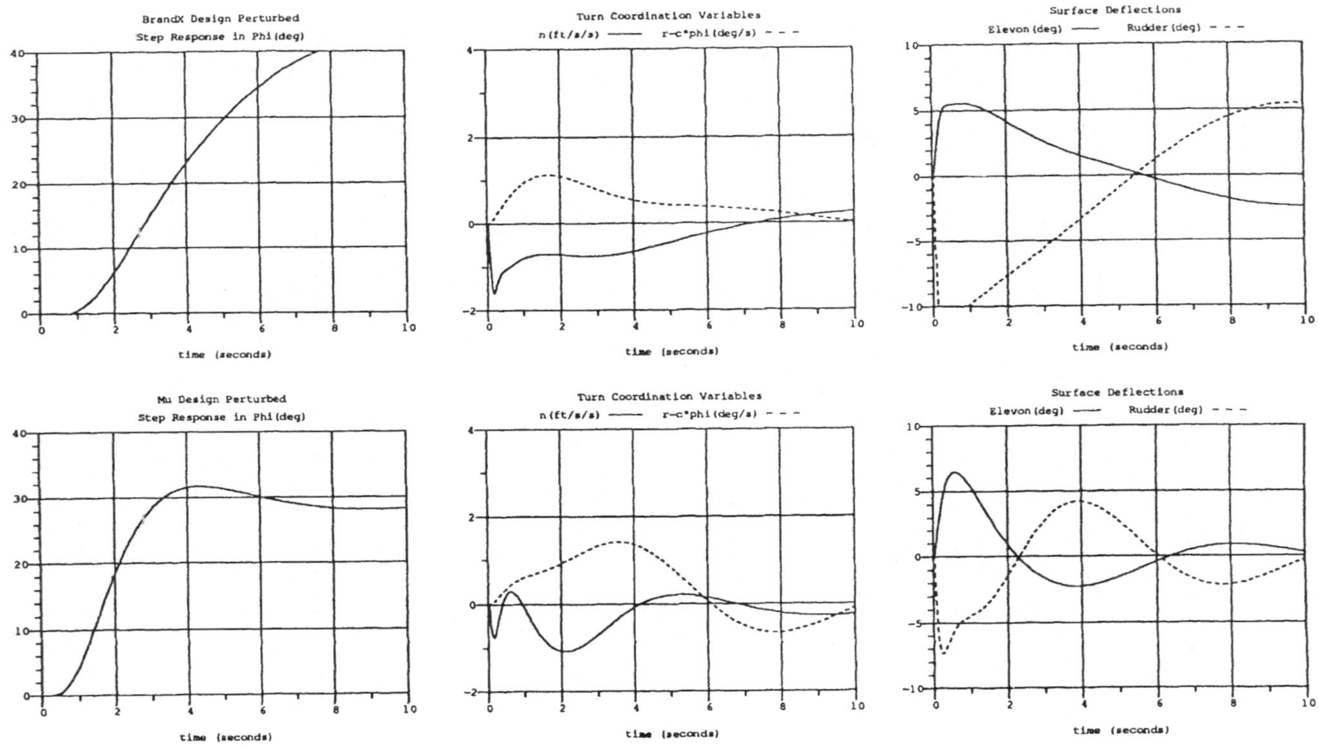

FIGURE 13 Step Responses for BrandX and Musyn Perturbed by Δ2

characteristics are dramatically different. The contrast with Hinf would be even more dramatic. Of course, these time responses are only intended to be illustrative. No definitive conclusions of the sort provided by μ can be reached on the basis of a few time responses. On the other hand, since μ is fundamentally a frequency-domain analysis tool, its only direct implications for the time domain are in terms of L_2 or sinusoidal response. Although this will usually be adequate in practice, frequency domain advocates cannot take the time domain completely for granted.

One issue of potential concern for experienced flight control engineers is the robustness characteristics of Musyn with respect to conventional measures such as gain and phase margins, or their appropriate multivariable generalizations. This was intentionally neglected in the synthesis phase. The philosophy taken here is that robustness should be obtained with respect to uncertainty actually expected to occur, as opposed to some arbitrary standard imposed on all systems. On the other hand, we might expect robustness with respect to the aerodynamic uncertainty in surface effectiveness to produce good margins. It is comforting to report that Musyn has good conventional margins, both when considered for single loop variations and for independent variations in all inputs and outputs simultaneously.

6. Controller Order and Reduction

High controller state order is a potentially annoying problem associated with H_∞ and μ-synthesis. H_∞ optimal controllers are usually at least the order (and often twice) of the interconnection structure, which includes not only the plant but also all the weights used to set up the interconnection structure. With the rigid body aircraft model, actuator models, delay approximation, ideal model response, and weights on sensor noise, performance errors, gust, and command, the nominal interconnection structure used in this paper for the H_∞ design has 23 states. The interconnection structure for μ-synthesis contains 47 states which includes an additional 12 elements for the D scalings, each with 2 states.

The μ-synthesis software used for this example produced an initial controller with 53 states. This controller was internally balanced and truncated [E1] to 13 states with no noticeable degradation in performance. This is the controller actually described and analyzed in this paper. An ad hoc weighted model reduction scheme inspired by discussions with Enns and Glover, but tailored to the μ-synthesis framework produced an 8 state controller with only moderate degradation over those with 13 or 53 states. Details on this model reduction scheme will be presented at the workshop preceding the 1987 ACC.

All the μ plots and time responses for this 8 state controller are reproduced in Figures 14 through 17 for comparison with the previous results. A realization is included in the appendix. The BrandX controller has a mere 3 states and is also included in the appendix. It would seem to be a challenging problem to come up with a 3 state controller that improves on BrandX. It is encouraging that a reasonably low order controller was found

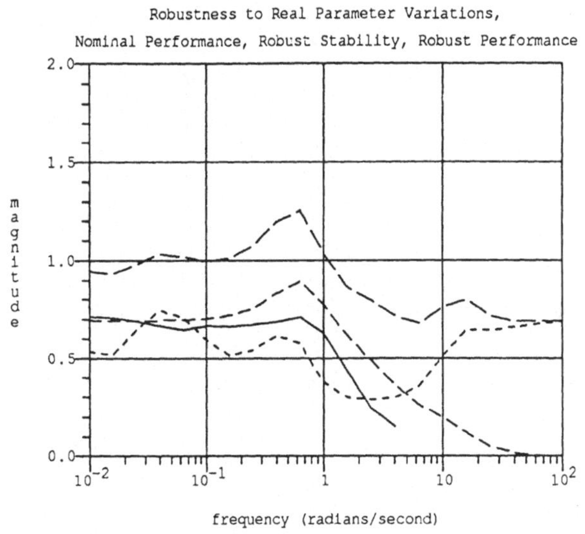

FIGURE 14 Mu Plots for System with 8th Order Controller

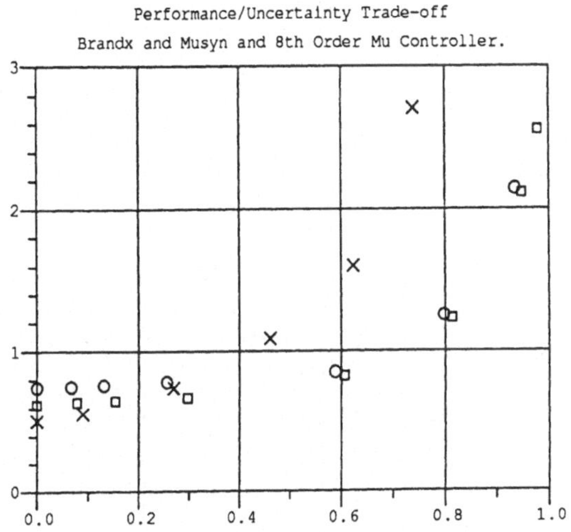

FIGURE 15 Comparison of $\max_{\|\Delta\|_\infty \leq \delta} \|F_u(G, \Delta)\|_\infty$ vs. δ

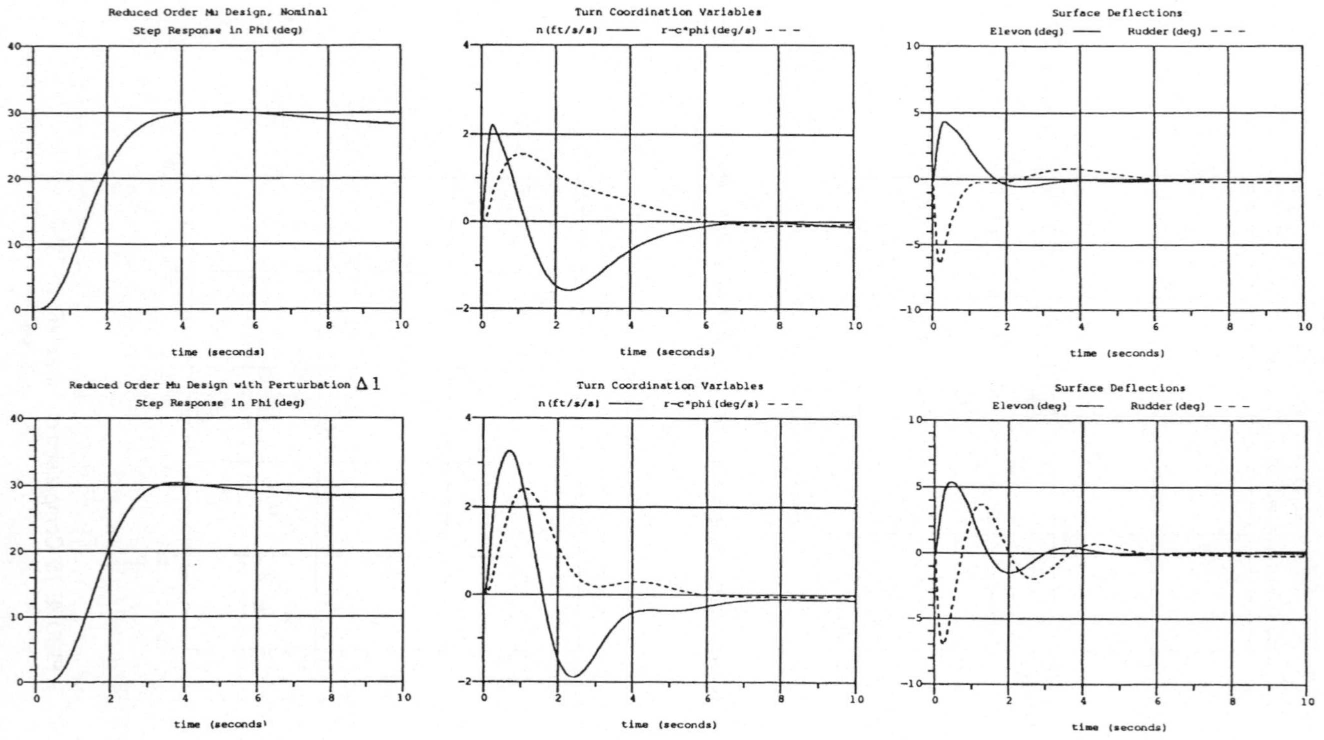

FIGURE 16 Step Responses for 8th Order Control Nominal and Perturbed by Δ1.

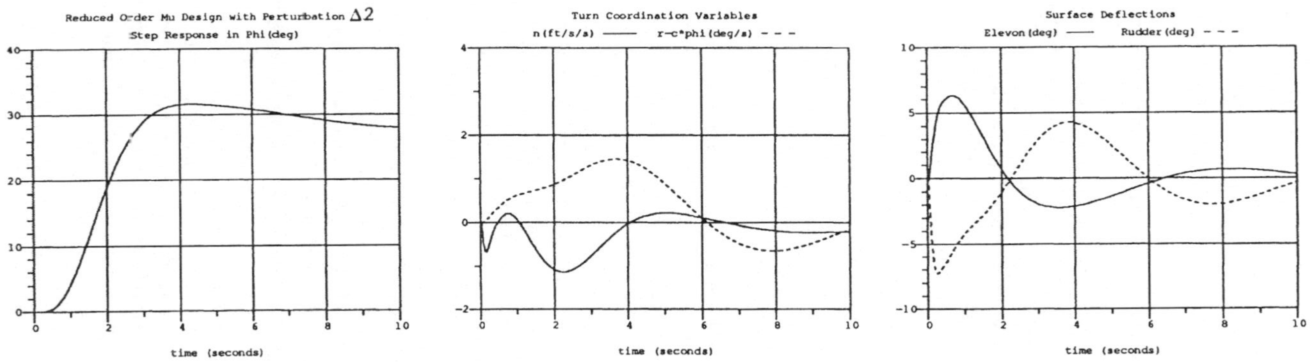

FIGURE 17 Step Responses for 8th Order Control Design Perturbed by Δ2

even though the model reduction schemes used are rather primitive compared to the other analysis and synthesis methods described in this paper. However, controller order continues to be a concern in μ-synthesis. We expect more systematic approaches based on the ad-hoc scheme used to obtain the 8 state controller to relieve this situation somewhat, but not entirely.

With the advent of more powerful computers which can in principal implement high-order controllers, the issue of controller order may seem moot. Unfortunately, it is not controller order per se that creates implementation problems but the difficulty of gain-scheduling complicated linear controllers over a wide range of flight conditions. Although the μ-based techniques described here might be used to produce robust full-range linear controllers, they would necessarily have degraded performance. A more appealing research direction might be to directly address the scheduling problem with nonlinear control. Some extensions to the techniques used here seem promising and are currently being evaluated on simplified flight control problems.

7. Conclusions

It is tempting to make wild claims, but important not to interpret the results in this paper too broadly. These results are extremely encouraging, and this study is certainly a success in demonstrating the applicability of μ to FCS design. Nevertheless, we must be cautious when drawing conclusions about the applicability of μ in general or about the relevance of this study to the Shuttle FCS.

Clearly, μ is a very powerful and promising tool, if only for analysis. Just the few plots shown in this paper yield important information about the performance and robustness of the controllers. Computation of μ has progressed to the point where it approaches that of singular values and eigenvalues in cost and reliability. While μ-synthesis is also very promising, it is highly experimental and will require additional study, application, and exposition before it can become a practical methodology. Some of the issues that must be addressed more thoroughly are the *D–K* iteration, real vs. complex perturbations, uncertainty modeling, weight selection, controller order, time-domain properties, and software and numerics. The Shuttle FCS problem as considered here was ideal in that none of these potentially serious problems presented any difficulties.

The issues addressed in this paper are typical for the design of a FCS, but much more careful and detailed study of the results would be required before making any serious conclusions about the Shuttle FCS. Even with immediate access to FCS experts and Shuttle data, in a brief study it is easy to overlook critical features of the problem. These concerns not withstanding, we have seen that μ allows an engineer, in a systematic and reliable way, to explore tradeoffs and design for robustness wherever she believes it is significant. We view μ as the fundamental analytical tool at this time for treating performance and robustness in control systems.

The problem chosen was a challenging one because of its complexity and large, structured uncertainty. This is exactly the type of problem for which we would expect μ to show the greatest benefit. In contrast, μ would have little impact on most SISO and many simpler MIMO problems. We expect that many more aerospace and process control problems will exhibit this level of complexity. See [S1] for a similar study on a process control problem.

Acknowledgements

The authors would like to thank the many people who contributed time and ideas to this paper, particularly B. Morton, R. McAfoos, D. Enns, G. Stein, and C.R. Stone, all of Honeywell SRC. This work has been supported by Honeywell IRAD Funding, ONR Research Grant N00014-82-C-0157, AFOSR Research Grant F49620-86-C-0001, and the National Science Foundation Grant ECS-8451519.

References

[C1] C.C. Chu and J.C. Doyle, "On inner-outer and spectral factorizations", *IEEE CDC*, Las Vegas, NV, 1984.

[C2] C.C. Chu, J.C. Doyle, and E.B. Lee, "The general distance problem in H_∞-optimal control", *Int. J. of Control*, 1986, Vol. 44, No.2.

[D1] J.C. Doyle, "Analysis of feedback systems with structured uncertainty", *IEE Proceedings*, Part D, No. 6, Nov., 1982.

[D2] J.C. Doyle, "Synthesis of robust controllers and filters", *IEEE CDC*, San Antonio, TX, 1983.

[D3] J.C. Doyle, "*Lectures Notes*", 1984 ONR/Honeywell Workshop on Advances in Multivariable Control, Oct. 8-10, 1984, Mpls., MN.

[D4] J.C. Doyle, "Structured uncertainty in control system design", *IEEE CDC*, Fort Lauderdale, FL, 1985.

[E1] D.F. Enns, "Model reduction for control system design", Ph.D. Dissertation, Stanford University, 1984.

[F1] B.A. Francis and J.C. Doyle, "Control theory with an H_∞-optimality criterion", *SIAM J. of Control*, 1986.

[M1] B.G. Morton and R.M. McAfoos, "A μ-test for real-parameter variations," *ACC Proceedings*, 1985.

[M2]B.G. Morton and R.M. McAfoos, "New applications fo μ to real-parameter variation problems", *IEEE CDC*, Fort Lauderdale, FL, 1985.

[S1] S. Skogestad and M. Morari, "Control of ill-conditioned plants: high purity distillation columns", *AIChE Annual Meeting*, Miami Beach, FL, Nov., 1986.

Appendix: Realizations of Aircraft, BrandX Controller and 8th Order Mu-Controller

Aircraft Model

$$x_{state} = \begin{bmatrix} \beta \\ p \\ r \\ \phi \end{bmatrix} \quad y = \begin{bmatrix} c_\beta \\ c_a \\ c_r \\ y_{meas} \end{bmatrix} \quad u = \begin{bmatrix} c_y \\ c_n \\ c_l \\ u_a \\ u_r \\ gust \end{bmatrix}$$

```
Matrix : aircraft
outputs 7   inputs 6   states 4
```

	x1	x2	x3	x4	u1	u2
x1	-9.460e-02	1.409e-01	-9.900e-01	3.637e-02	1.275e-05	0.000e+00
x2	-3.595e+00	-4.284e-01	2.809e-01	0.000e+00	0.000e+00	-3.114e-05
x3	3.950e-01	-1.263e-02	-8.142e-02	0.000e+00	0.000e+00	-1.905e-04
x4	0.000e+00	1.000e+00	-1.405e-01	0.000e+00	0.000e+00	0.000e+00
y1	1.000e+00	0.000e+00	0.000e+00	0.000e+00	0.000e+00	0.000e+00
y2	0.000e+00	0.000e+00	0.000e+00	0.000e+00	0.000e+00	0.000e+00
y3	0.000e+00	0.000e+00	0.000e+00	0.000e+00	0.000e+00	0.000e+00
y4	0.000e+00	1.000e+00	0.000e+00	0.000e+00	0.000e+00	0.000e+00
y5	0.000e+00	0.000e+00	1.000e+00	0.000e+00	0.000e+00	0.000e+00
y6	-6.804e+01	-1.744e+00	-4.058e+00	-3.720e-05	1.111e-02	-1.111e-02
y7	0.000e+00	0.000e+00	0.000e+00	1.000e+00	0.000e+00	0.000e+00

	u3	u4	u5	u6
x1	0.000e+00	-1.240e-02	1.023e-02	-1.086e-04
x2	-3.117e-03	6.571e+00	1.256e+00	-4.126e-03
x3	-6.443e-05	3.783e-01	-2.560e-01	4.533e-04
x4	0.000e+00	0.000e+00	0.000e+00	0.000e+00
y1	0.000e+00	0.000e+00	0.000e+00	1.148e-03
y2	0.000e+00	1.000e+00	0.000e+00	0.000e+00
y3	0.000e+00	0.000e+00	1.000e+00	0.000e+00
y4	0.000e+00	0.000e+00	0.000e+00	0.000e+00
y5	0.000e+00	0.000e+00	0.000e+00	0.000e+00
y6	-1.111e-02	2.667e+01	-2.952e+00	-7.810e-02
y7	0.000e+00	0.000e+00	0.000e+00	0.000e+00

Brandx Controller

$$y = \begin{bmatrix} u_a \\ u_r \end{bmatrix} \quad u = \begin{bmatrix} \phi_{com} \\ Y_{meas} \end{bmatrix}$$

```
Matrix : Brandx control
outputs 2   inputs 5   states 3
```

	x1	x2	x3	u1	u2	u3
x1	-1.000e-05	0.000e+00	0.000e+00	1.662e-01	-1.990e-01	1.033e-02
x2	0.000e+00	-1.000e-05	0.000e+00	-4.418e-02	-3.232e-02	5.960e-01
x3	0.000e+00	0.000e+00	-1.250e+00	-4.899e-09	7.652e-09	-1.286e-08
y1	2.936e-01	1.847e-01	5.026e-13	1.615e-01	-2.560e-01	4.496e-01
y2	-9.253e-02	5.697e-01	-1.128e-01	-4.037e-01	-6.507e-12	3.371e+00

	u4	u5
x1	-9.109e-04	-1.656e-01
x2	1.448e-03	2.258e-02
x3	-9.824e-02	5.329e-09
y1	1.862e-14	-1.767e-01
y2	-5.389e-04	2.806e-01

8th Order Mu-Controller

$$y = \begin{bmatrix} u_a \\ u_r \end{bmatrix} \qquad u = \begin{bmatrix} \phi_{com} \\ Y_{meas} \end{bmatrix}$$

```
Matrix : mu-control
outputs 2   inputs 5   states 8
```

	x1	x2	x3	x4	x5	x6
x1	-8.742e-03	1.384e-03	0.000e+00	0.000e+00	0.000e+00	0.000e+00
x2	-1.384e-03	-8.742e-03	0.000e+00	0.000e+00	0.000e+00	0.000e+00
x3	0.000e+00	0.000e+00	-3.145e-02	0.000e+00	0.000e+00	0.000e+00
x4	0.000e+00	0.000e+00	0.000e+00	-4.540e-01	0.000e+00	0.000e+00
x5	0.000e+00	0.000e+00	0.000e+00	0.000e+00	-9.139e-01	5.785e-01
x6	0.000e+00	0.000e+00	0.000e+00	0.000e+00	-5.785e-01	-9.139e-01
x7	0.000e+00	0.000e+00	0.000e+00	0.000e+00	0.000e+00	0.000e+00
x8	0.000e+00	0.000e+00	0.000e+00	0.000e+00	0.000e+00	0.000e+00
y1	3.330e-03	1.340e-03	-3.350e-03	-5.788e-02	-1.648e-02	5.765e-02
y2	-2.922e-03	-4.244e-03	9.938e-04	-9.349e-03	6.275e-02	1.367e-02

	x7	x8	u1	u2	u3	u4
x1	0.000e+00	0.000e+00	-6.960e+00	2.143e+01	1.623e+00	-1.081e-02
x2	0.000e+00	0.000e+00	3.713e+00	-1.642e+01	2.715e+01	-2.625e-02
x3	0.000e+00	0.000e+00	-2.268e-01	4.799e-01	1.865e-01	-2.384e-02
x4	0.000e+00	0.000e+00	-9.388e-01	-6.945e+00	-9.161e+00	3.494e-02
x5	0.000e+00	0.000e+00	4.058e+00	-5.252e+00	3.559e+01	-1.085e-01
x6	0.000e+00	0.000e+00	-7.634e+00	-1.546e+00	1.184e+01	3.307e-02
x7	-6.456e+00	0.000e+00	2.417e+00	1.002e+01	-7.948e+01	7.225e-03
x8	0.000e+00	-3.128e+01	7.237e+00	-1.673e+02	-7.383e+01	-8.628e-03
y1	1.226e-02	2.122e-01	-1.530e-01	1.404e+00	1.568e+00	2.305e-04
y2	2.414e-01	-4.140e-01	2.697e-01	-2.475e+00	-2.764e+00	-4.063e-04

	u5
x1	6.986e+00
x2	-4.737e+00
x3	4.449e-01
x4	-3.862e+00
x5	4.393e-01
x6	-2.986e+00
x7	-2.375e-01
x8	-5.137e-01
y1	6.277e-03
y2	-1.106e-02

A Necessary and Sufficient Condition for Robustness of Stability Under Known Additive Perturbation

A.I.G. Vardulakis
Department of Mathematics
Faculty of Science
Aristotle University of Thessaloniki
Thessaloniki
Greece

Abstract

A necessary and sufficient condition is derived for the robustness of stability of a unity feedback system involving a strictly proper plant P_o and a proper stabilizing compensator C_o under the assumption that the plant is perturbed to $P_o+\Delta P_o$ where ΔP_o is a known strictly proper rational matrix.

Notation

$\mathbb{R}$		the field of real numbers
$\mathbb{C}$		the field of complex numbers
$\mathbb{C}_o^-$	:=	$\{ s \in \mathbb{C},\ Re(s) < 0 \}$
$\mathbb{C}_+$	:=	$\{ s \in \mathbb{C},\ Re(s) \geq 0 \}$
$\bar{\mathbb{C}}_+$	:=	$\mathbb{C}_+ \cup \{\infty\}$
$\mathbb{R}(s)$		the field of rational functions
$\mathbb{R}_{pr}(s)$		the ring of proper rational functions
Ω		any region in $\mathbb{C}$ which is symmetrically located with respect to $\mathbb{R}$ and which excludes at least one point $-\alpha \in \mathbb{R}$ $(\alpha > 0)$
$\underline{\Omega}$	:=	$\Omega \cup \{\infty\}$
$\mathbb{R}_{\underline{\Omega}}(s)$		the Euclidean ring of rational functions $t(s) \in \mathbb{R}(s)$ with no poles inside $\underline{\Omega} = \Omega \cup \{\infty\}$, i.e. of proper and Ω-stable rational functions

If k is a set then $k^{p\times m}$ denotes the set of $p\times m$ matrices with elements in k.

NATO ASI Series, Vol. F34
Modelling, Robustness and Sensitivity Reduction in Control Systems. Edited by R. F. Curtain

Introduction

Let $\Sigma(P_o)$ be a linear multivariable system which is free of unstable hidden modes and whose input-output behaviour is described by a p×m proper (strictly proper) rational transfer function matrix P_o ("the plant"). Consider now the closed-loop unity feedback system $\Sigma_{cl}(P_o,C_o)$ of fig. 1 which involves a "stabilizing compensator" C_o in the feedforward path such that $\Sigma_{cl}(P_o,C_o)$ is internally assymptoticaly stable and the closed-loop transfer function matrix

$$H_{yu}(P_o,C_o) : \begin{bmatrix} u_1 \\ u_2 \end{bmatrix} \to \begin{bmatrix} y_1 \\ y_2 \end{bmatrix} \tag{1}$$

has arbitrary desired poles in $Re(s) < 0$. Let now that the "nominal plant" P_o is perturbed to $P_o + \Delta P_o =: P$ where ΔP_o is a known p×m proper rational matrix. In this paper we give a simple necessary and sufficient condition that has to be sutisfied by ΔP_o so that if the additevely perturbed system $\Sigma(P_o + \Delta P_o)$ is also free of unstable hidden modes, the closed loop system $\Sigma_{cl}(P_o + \Delta P_o, C_o)$ of fig. 2 is also internally assymptoticaly stable.

The set of plants stabilizable by a compensator which stabilizes a nominal plant P_o

Let $\Sigma(P_o)$ be a linear, time invariant multivariable system which is free of unstable hidden modes and whose (nominal) transfer function matrix ("the plant") is $P_o \in \mathbb{R}^{p\times m}_{pr}(s)$ and let $P_o = A_1^{-1}B_1 = B_2A_2^{-1}$ with $A_1 \in \mathbb{R}^{p\times p}_{\underline{\Omega}}(s)$, $B_1 \in \mathbb{R}^{p\times m}_{\underline{\Omega}}(s)$; $B_2 \in \mathbb{R}^{p\times m}_{\underline{\Omega}}(s)$, $A_2 \in \mathbb{R}^{m\times m}_{\underline{\Omega}}(s)$ be respectivelly any left and right coprime in $\underline{\Omega}$ "fractional representations" of P_o i.e. let the matrices $[A_1, B_1]$ and $\begin{bmatrix} A_2 \\ B_2 \end{bmatrix}$ be respectively $\underline{\Omega}$-right and left invertible. It is then known [1]-[6] that another system $\Sigma'(C_o)$ (which is also free of unstable hidden modes) with transfer function $C_o \in \mathbb{R}^{m\times p}(s)$ is a "stabilizing compensator for $\Sigma(P_o)$ such that the unity feedback closed-loop system denoted by $\Sigma_{cl}(P_o,C_o)$ of fig. 1 is internally stable (i.e. all its modes lie in $\mathbb{C}_o^-$) iff $|I_p + P_oC_o| = |I_m + C_oP_o| \neq 0$ for all $s \in \mathbb{C} \cup \{\infty\}$ and the transfer function matrix $H_{yu}(P_o,C_o)$ is an element of $\mathbb{R}^{(p+m)\times(p+m)}_{\underline{\Omega}}(s)$ where $\underline{\Omega} \equiv \bar{\mathbb{C}}_+$.

It is also well known that if C_o is a stabilizing compensator for $\Sigma(P_o)$ then C_o has unique left and right coprime in $\underline{\Omega}$ fractional representations

$$C_o = D_1^{-1}N_1 = N_2D_2^{-1} \tag{2}$$

with $D_1 \in \mathbb{R}_{\underline{\Omega}}^{m\times m}(s)$, $N_1 \in \mathbb{R}_{\underline{\Omega}}^{m\times p}(s)$; $N_2 \in \mathbb{R}_{\underline{\Omega}}^{m\times p}(s)$, $D_2 \in \mathbb{R}_{\underline{\Omega}}^{p\times p}(s)$ satisfying the Bezout identities: [1]-[6]

$$\begin{bmatrix} D_1 & N_1 \\ -B_1 & A_1 \end{bmatrix} \begin{bmatrix} A_2 & -N_2 \\ B_2 & D_2 \end{bmatrix} = \begin{bmatrix} I_m & 0 \\ 0 & I_p \end{bmatrix} \tag{3}$$

$$\begin{bmatrix} A_2 & -N_2 \\ B_2 & D_2 \end{bmatrix} \begin{bmatrix} D_1 & N_1 \\ -B_1 & A_1 \end{bmatrix} = \begin{bmatrix} I_m & 0 \\ 0 & I_p \end{bmatrix} \tag{4}$$

Multiplying (3) on the left and right by the $\mathbb{R}_{\underline{\Omega}}(s)$-unimodular matrices $\begin{bmatrix} I_m & W \\ 0 & I_p \end{bmatrix}$ and $\begin{bmatrix} I_m & -W \\ 0 & I_p \end{bmatrix}$ respectively, where $W \in \mathbb{R}_{\underline{\Omega}}^{m\times p}(s)$ and such that $|D_1 - WB_1| \not\equiv 0$, $|D_2 - B_2W| \not\equiv 0$, we obtain the identity:

$$\begin{bmatrix} D_1-WB_1 & N_1+WA_1 \\ -B_1 & A_1 \end{bmatrix} \begin{bmatrix} A_2 & -(N_2+A_2W) \\ B_2 & D_2-B_2W \end{bmatrix} = \begin{bmatrix} I_m & 0 \\ 0 & I_p \end{bmatrix} \tag{5}$$

which due to the above results shows clearly that the set $\Phi(P_o)$ of all stabilizing compensators C_W of $\Sigma(P_o)$ is parametrized by a stabilizing compensator C_o and $W \in \mathbb{R}_{\underline{\Omega}}^{m\times p}(s)$ and is given by [1]-[6] :

$$\Phi(P_o) = \{ C_W = (D_1-WB_1)^{-1}(N_1+WA_1) = (N_2+A_2W)(D_2-B_2W)^{-1} \mid W \in \mathbb{R}_{\underline{\Omega}}^{m\times p} \text{ and such that } |D_1-WB_1| \not\equiv 0,\ |D_2-B_2W| \not\equiv 0 \} \tag{6}$$

Consider now the nominal plant P_o, a (nominal) stabilizing compensator $C_o \in \Phi(P_o)$ and denote by $\Psi(C_W)$ the set of all "plants" stabilizable by a compensator $C_W \in \Phi(P_o)$. By duality to the above characterization of $\Phi(P_o)$ we can parametrize the set $\Psi(C_W)$. We have

Proposition 1 Let $P_o = A_1^{-1}B_1 = B_2A_2^{-1} \in \mathbb{R}_{pr}^{p\times m}(s)$, $C_o = D_1^{-1}N_1 = N_2D_2^{-1} \in \Phi(P_o)$ and $C_W \in \Phi(P_o)$. Then the set $\Psi(C_W)$ of all plants stabilizable by C_W which stabilizes P_o is given by

$$\begin{aligned} \Psi(C_W) = \Psi(P_o,C_o) = \{ P_{Q,W} &= [A_1+Q(N_1+WA_1)]^{-1}[B_1-Q(D_1-WB_1)] \\ &= [B_2-(D_2-B_2W)Q][A_2+(N_2+A_2W)Q]^{-1}, \qquad (7) \\ & Q \in \mathbb{R}_{\underline{\Omega}}^{p\times m}(s),\ W \in \mathbb{R}_{\underline{\Omega}}^{m\times p}(s) \\ & |A_1+Q(N_1+WA_1)| \not\equiv 0,\ |A_2+(N_2+A_2W)Q| \not\equiv 0 \} \qquad (8) \end{aligned}$$

Proof Multiplying (5) on the left and right respectively by the $\mathbb{R}_{\underline{\Omega}}(s)$-unimodular matrices:

$\begin{bmatrix} I_m & 0 \\ Q & I_p \end{bmatrix}, \begin{bmatrix} I_m & 0 \\ -Q & I_p \end{bmatrix}$ where $Q \in \mathbb{R}_{\underline{\Omega}}^{p\times m}(s)$ and such that conditions (8) are satisfied, we obtain the Bezout identity:

$$\begin{bmatrix} D_1-WB_1 & N_1+WA_1 \\ -[B_1-Q(D_1-WB_1)] & A_1+Q(N_1+WA_1) \end{bmatrix} \begin{bmatrix} A_2+(N_2+A_2W)Q & -(N_2+A_2W) \\ B_2-(D_2-B_2W)Q & D_2-B_2W \end{bmatrix} = \begin{bmatrix} I_m & 0 \\ 0 & I_p \end{bmatrix} \quad (9)$$

which due to the above results shows that the set $\Psi(C_W)$ is given by (7) □

<u>Corollary 1</u> The set $\Psi(C_o)$ of all plants stabilizable by $C_o \in \Phi(P_o)$ is given by

$$\Psi(C_o) = \{ P_Q = (A_1+QN_1)^{-1}(B_1-QD_1) = (B_2-D_2Q)(A_2+N_2Q)^{-1} \quad (7')$$

$$Q \in \mathbb{R}_{\underline{\Omega}}^{p\times m}(s), \text{ such that } |A_1+QN_1| \not\equiv 0, \ |A_2+N_2Q| \not\equiv 0 \ \} \quad (8')$$

<u>Proof</u> Just put $W = 0$ in (7) □

We investgate now conditions under which the elements of the set $\Psi(C_o)$ are proper rational matrices.

It is known [6][7] that if P_o is proper then the matrices A_1 and A_2 are biproper and if either $Q \in \mathbb{R}_{\underline{\Omega}}^{p\times m}(s)$ and/or $C_o = D_1^{-1}N_1 = N_2D_2^{-1} \in \mathbb{R}_{\underline{\Omega}}^{m\times p}(s)$ are choosen so that the product $QN_1 \in \mathbb{R}_{\underline{\Omega}}^{p\times p}(s)$ (or $N_2Q \in \mathbb{R}_{\underline{\Omega}}^{m\times m}(s)$) is not strictly proper , then the matrices $A_1+QN_1 \in \mathbb{R}_{\underline{\Omega}}^{p\times p}(s)$ and $A_2+N_2Q \in \mathbb{R}_{\underline{\Omega}}^{m\times m}(s)$ might turn out to be non-biproper so that P_Q in (7′) turns out to be non-proper. Thus in general, and even if P_o is strictly prorer, the set $\Psi(C_o)$ will contain also non-proper "plants" P_Q. If P_o is strictly proper then the strictly proper elements P_Q of $\Psi(C_o)$ can be characterised by strictly proper Q. We state this as

<u>Proposition 2</u> Let $P_o \in \mathbb{R}_{pr}^{p\times m}(s)$ be strictly proper with $P_o = A_1^{-1}B_1 = B_2A_2^{-1}$ and let $C_o \in \mathbb{R}_{pr}^{m\times p}(s)$ with $C_o = D_1^{-1}N_1 = N_2D_2^{-1}$ such that the Bezout identities (3)-(4) are satisfied, i.e. let $C_o \in \Phi(P_o)$. Then

(i) For every $Q \in \mathbb{R}_{\underline{\Omega}}^{p\times m}(s)$ which is strictly proper

$$P_Q := (A_1+QN_1)^{-1}(B_1-QD_1) = (B_2-D_2Q)(A_2+N_2Q)^{-1} \quad (9)$$

belongs to $\Psi(C_o)$ and is strictly proper.

(ii) Every strictly proper plant $P_Q \in \Psi(C_o)$ can be expressed as in (9) for some strictly proper $Q \in \mathbb{R}_{\underline{\Omega}}^{p\times m}(s)$.

Proof (i) For every strictly proper $Q \in \mathbb{R}_{\underline{\Omega}}^{p\times m}(s)$ it follows from Proposition 1 that $P_Q \in \Psi(C_o)$. Moreover we have

$$(A_1+QN_1)(\infty) = A_1(\infty) + Q(\infty)N_1(\infty) = A_1(\infty) \tag{10}$$

$$(B_1-QD_1)(\infty) = B_1(\infty) - Q(\infty)D_1(\infty) = 0 \tag{11}$$

The above equations imply that P_Q is strictly proper.
(ii) Let P_Q be strictly proper. Then according to Proposition 1, P_Q can be written as in (9) for some $Q \in \mathbb{R}_{\underline{\Omega}}^{p\times m}(s)$. Now, our hypothesis that P_o is strictly proper implies that D_2 is biproper and since also by hypothesis P_Q is strictly proper, it follows that we must have

$$(B_2-D_2Q)(\infty) = 0 \tag{12}$$

which holds true iff Q is strictly proper. □

In order to determine Q we proceed as follows. Let $P_Q = B_Q A_Q^{-1}$ with $B_Q \in \mathbb{R}_{\underline{\Omega}}^{p\times m}(s)$, $A_Q \in \mathbb{R}_{\underline{\Omega}}^{m\times m}(s)$ right coprime in $\underline{\Omega}$. Now $P \in \Psi(C_o)$ implies that

$$D_1A_Q + N_1B_Q =: U \in \mathbb{R}_{\underline{\Omega}}^{m\times m}(s) \tag{13}$$

is an $\mathbb{R}_{\underline{\Omega}}(s)$ -unimodular matrix. Also, according to the above we must have that

$$P_Q = B_Q A_Q^{-1} = (A_1+QN_1)^{-1}(B_1-QD_1) \tag{14}$$

for some $Q \in \mathbb{R}_{\underline{\Omega}}^{p\times m}(s)$. Solving (14) with respect to Q we obtain the expression for Q as

$$Q = (B_1A_Q - A_1B_Q)\, U^{-1} \tag{15}$$

which clearly is strictly proper since (15) is a right coprime fractional representation of Q and by assumption both P_o and P_Q or equivalently B_1 and B_Q are all strictly proper.

In view of Proposition 1 we can also parametrize the set $\Theta_{yu}(P_Q,C_o)$ of all closed loop transfer function matrices $H_{yu}(P_Q,C_o) : u \to y$ corresponding to a unity feedback closed loop system $\Sigma_{cl}(P_Q,C_o)$ with plant $P_Q \in \Psi(C_o)$ and stabilizing compensator $C_o \in \Phi(P_o)$.

Corollary 2 With the above notation

$$\Theta(P_Q,C_o) = \{\, H_{yu}(P_Q,C_o) = \begin{bmatrix} (A_2+N_2Q)N_1 & (A_2+N_2Q)D_1-I_m \\ & \\ (B_2-D_2Q)N_1 & (B_2-D_2Q)D_1 \end{bmatrix} \} \tag{16}$$

where the parameter $Q \in \mathbb{R}_{\underline{\Omega}}^{p\times m}(s)$ is such that conditions (8′) are satisfied.

Proof The above follows by substituting the expresions (7′) for P_Q and (2) for C_o into the formula for $H_{yu}(P_Q,C_o)$:

$$H_{yu}(P_Q,C_o) = \begin{bmatrix} (I_p+C_oP_Q)^{-1}C_o & -C_oP_Q(I_p+C_oP_Q)^{-1} \\ P_Q(I_p+C_oP_Q)^{-1} & P_Q(I_p+C_oP_Q)^{-1} \end{bmatrix}$$

A necessary and sufficient condition for robust stability of an additively perturbed plant.

Let us now consider the closed loop unity feedback system $\Sigma_{cl}(P_o,C_o)$ of fig. 1 where $P_o \in \mathbb{R}_{pr}^{p\times m}(s)$ and $C_o \in \Phi(P_o)$ are the nominal plant and compensator and let us make the following assumptions

(i) the nominal plant P_o is strictly proper and it is additivelly perturbed to $P_o + \Delta P_o =: P$ where $\Delta P_o \in \mathbb{R}_{pr}^{p\times m}(s)$ is known and strictly proper

(ii) the additively pertutbed system $\Sigma(P_o+\Delta P_o)$ is also free of unstable hidden modes, and

(iii) $|I_p+(P_o+\Delta P_o)C_o| \neq 0$ for every $s \in \mathbb{C} \cup \{\infty\}$

If we now consider the closed loop unity feedback system $\Sigma_{cl}(P_o+\Delta P_o, C_o)$ of fig. 2 then in view of the parametrization in (7') of all "plants" stabilizable by $C_o \in \Phi(P_o)$, we can state the following

Proposition 3 Under the above assumptions and notation the following statements are equivalent:

1) $\Sigma_{cl}(P_o+\Delta P_o,C_o)$ is internally assymptoticaly stable and $H_{yu}(P_o+\Delta P_o,C_o) \in \mathbb{R}_{\underline{\Omega}}^{(m+p)\times(m+p)}(s)$

2) The additevely perturbed plant $P := P_o+\Delta P_o$ is stabilizable by $C_o \in \Phi(P_o)$

3) $P := P_o+\Delta P_o \in \Psi(C_o)$

4) There exists a $Q \in \mathbb{R}_{\underline{\Omega}}^{p\times m}(s)$ and strictly proper such that:

$$P := P_o+\Delta P_o = (A_1+QN_1)^{-1}(B_1-QD_1) = (B_2-D_2Q)(A_2+N_2Q)^{-1} \tag{17}$$

with

$$|A_1+QN_1| \neq 0, \ |A_2+N_2Q| \neq 0 \tag{18}$$

Solving (17) with respect to Q we obtain the following necessary and sufficient condition for the robust stability of an additively perturbed plant.

Theorem 1 Let $P_o = A_1^{-1}B_1 = B_2A_2^{-1} \in \mathbb{R}_{pr}^{p\times m}(s)$, $C_o = D_1^{-1}N_1 = N_2D_2^{-1} \in \mathbb{R}^{m\times p}(s)$ a pair of nominal plant and compensator so that (3) and (4) are satisfied. Then $\Sigma_{cl}(P_o+\Delta P_o,C_o)$ is internally assymptoticaly stable iff

$$Q := -D_2^{-1}[(P_o+\Delta P_o)C_o + I_p]^{-1}\Delta P_o A_2 \tag{19}$$
$$= -A_1 \Delta P_o [C_o(P_o+\Delta P_o) + I_m]^{-1} D_1^{-1} \in \mathbb{R}_{\underline{\Omega}}^{p\times m}(s)$$

Example

Consider the example in [7] where $P = \frac{100}{(s-1)(s+100)}$ and a low freeequency model of the plant is $P_o = \frac{1}{s-1}$. Obviously pure gain can stabilize P_o e.g. $C_o = 2$ is a stabilizing compensator for P_o.

The question is whether C_o will also stabilize the high freequency plant model P. Now

$$\Delta P_o = P - P_o = \frac{-s}{(s-1)(s+100)}$$

Also P_o can be written as

$$P_o = \frac{1}{s-1} = \frac{1}{s+a}[\frac{s-1}{s+a}]^{-1} = BA^{-1},\ a > 0 \text{ and } A,B \text{ coprime}$$

in $\mathbb{C}_+ \cup \{\infty\}$ fractional representation of P_o, so from (19):

$$Q = -[\frac{100}{(s-1)(s+100)}\ 2 + 1]^{-1}[\frac{-s}{(s-1)(s+100)}]\frac{s-1}{s+a}$$

$$= \frac{s(s-1)}{(s^2+99s+100)(s+a)} \in \mathbb{R}_{\underline{\Omega}}(s)$$

and thus P is also stabilizable by $C_o = 2$.

If $\Sigma(P_o)$ is (open-loop) internally stable then $P_o \in \mathbb{R}_{\underline{\Omega}}^{p\times m}(s)$ where $\underline{\Omega} \equiv \mathbb{C}_+ \cup \{\infty\}$, and we can take $B_1 = B_2 = P_o$, $A_1 = I_p$, $A_2 = I_m$. In such a case the Bezout identities (3) and (4) are satisfied for $D_1 = I_m$, $N_1 = 0_{m,p}$, $D_2 = I_p$, $N_2 = 0_{p,m}$ and thus from (6)

$$\Phi(P_o) = \{\ C_W = (I_m - WP_o)^{-1}W = W(I_p - P_oW)^{-1}\ |\ W \in \mathbb{R}_{\underline{\Omega}}^{m\times p} \text{ and}$$
$$\text{and such that } |I_m - WP_o| = |I_p - P_oW| \neq 0\ \} \tag{20}$$

isthe set of all stabilizing compensators of the proper and stable plant P_o. [6]. Noticing that due to (6) the expressions in (20) constitute left and right coprime in $\underline{\Omega}$ fractional representations of C_W and substituting these expressions for C_o into condition (19) we obtain

$$Q = -(I_p - P_oW)^{-1}[(P_o+\Delta P_o)W(I_p - P_oW)^{-1} + I_p]^{-1}$$
$$= -(\Delta P_o W + I_p)^{-1}\Delta P_o \tag{21}$$

and thus we can state

Corollary 3 Let $P_o \in \mathbb{R}_{\underline{\Omega}}^{p\times m}$, $\Omega \equiv \mathbb{C}_o^-$, and let $C_W \in \Phi(P_o)$. Let P_o be additively perturbed to $P_o + \Delta P_o =: P \in \mathbb{R}_{pr}^{p\times m}(s)$. Then $\Sigma_{cl}(P_o+\Delta P_o,\ C_W)$ is internally assymptoticaly stable iff there exist a $W \in \mathbb{R}_{\underline{\Omega}}^{m\times p}(s)$ such that

$$Q := -(\Delta P_o W + I_p)^{-1}\, \Delta P_o = -\Delta P_o (W\Delta P_o + I_m)^{-1} \quad \varepsilon \;\; \mathbb{R}_{\underline{\Omega}}^{p\times m}(s) \tag{22}$$

and

$$|\Delta P_o W + I_p| = |W\Delta P_o + I_m| \not\equiv 0$$

Now from the fact that Q in (21) gives the closed-loop transfer function matrix of the configuration in fig. 3, we can rephrase Corollary 3 as

Corollary 4 Let P_o, Ω, C_W, P as in Corollary 3. Then $\Sigma_{cl}(P_o+\Delta P_o, C_W)$ is internally assymptotically stable iff ΔP_o is stabilizable by a proper and stable compensator W.

Finally solving (22) with respect to ΔP_o we obtain

Corollary 5 Let P_o, Ω, C_W as in Corollary 3. Then any $\Delta P_o \in \mathbb{R}_{pr}^{p\times m}(s)$ such that $\Sigma_{cl}(P_o+\Delta P_o, C_W)$ is internally assymptotically stable must have right and left coprime in $\underline{\Omega}$ fractional representations given by

$$\Delta P_o = -Q(I_m+WQ)^{-1} = -(I_p+WQ)^{-1}Q \tag{23}$$

for some $W \in \mathbb{R}_{\underline{\Omega}}^{m\times p}(s)$ and $Q \in \mathbb{R}_{\underline{\Omega}}^{p\times m}(s)$.

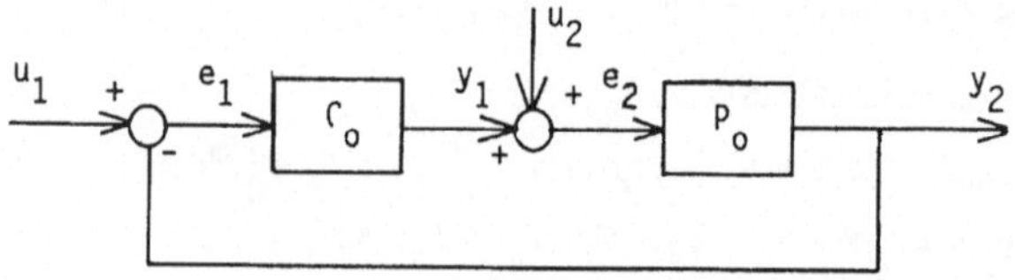

Fig. 1

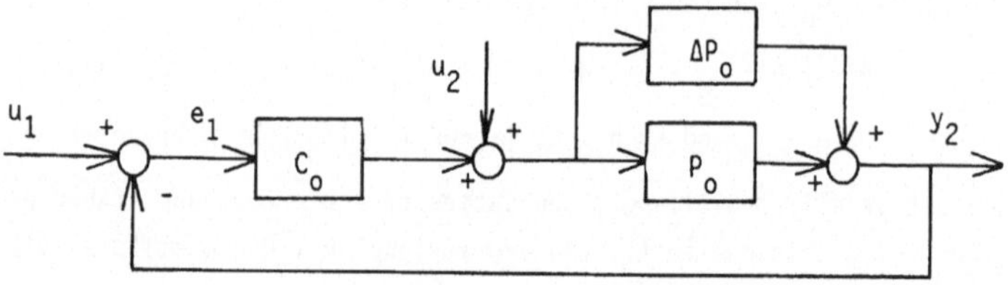

Fig. 2

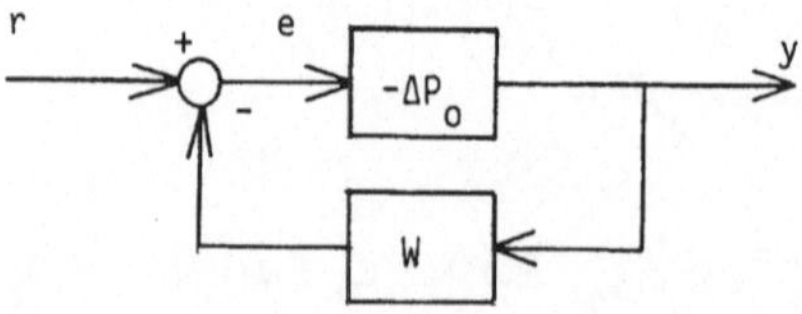

Fig. 3

References

[1] C.A.Desoer, R.W.Liu, J.W.Murray and R.Saeks, "Feedback system design: the fractional representation approach to analysis and synthesis", IEEE Trans. Auto. Control, vol. AC-25, pp399-412, 1980.

[2] M.Vidyasagar, H.Schneider and B.Francis, "Algebraic and topological aspects of feedback stabilization", IEEE Trans. Auto. Control, vol. AC -27, pp880-894, 1982.

[3] R.Saeks and J.Murray, "Feedback system design: the tracking and disturbance rejection problems", IEEE Trans. Auto. Control, vol. AC-26, pp 203-217, 1981.

[4] D.C.Youla, H.A.Jabr and J.J.Bonjorno Jr., "Modern Wiener-Hopf design of optimal controllers, Part II: the multivariable case", IEEE Trans. Auto. Control, Vol. AC-21, pp319-338, 1976.

[5] F.M.Callier and C.A.Desoer, "Multivariable Feedback Systems", Springer -Verlag, New York, 1982.

[6] M.Vidyasagar, "Control System Synthesis, A Factorization Approach", M.I.T. Press, 1985.

[7] Q.Huang and R.Liu, "Anecassary and sufficient condition for robust stability", to appear in IEEE Trans. Auto. Control.

Recursive Methods for Control Design Based on Approximate Models

David H. Owens
Department of Mathematics
University of Strathclyde
26 Richmond Street
Glasgow
G1 1XH
United Kingdom

Abstract

The use of recursive versions of fixed point theorems in robust control and model reduction is illustrated with particular note of the benefits of reducing conservatism.

1. Introduction to Recursive Methods

In many areas of feedback stability theory, it is convenient to represent the closed-loop system by an equation of the form

$$z = f(z) \tag{1}$$

where f maps a Banach space X into itself, z is a system signal of interest and the problem is formulated in such a way that the closed-loop is stable if, and only if, (1) has a solution $z \in X$. The classical example is the situation where f has finite gain λ i.e. for all $y,z \in X$

$$||f(y) - f(z)||_X \leq \lambda \; ||y-z||_X \tag{2}$$

when, with the above notation:

Theorem 1 (Contraction mapping - Holtzmann (1970))

If $\lambda < 1$, then there exists a unique solution $z \in X$ of (1). Moreover, this solution can be obtained as the formal limit of the successive approximation procedure

$$z^{(k+1)} = f(z^{(k)}) \quad , \quad k \geq 0 \tag{3}$$

for any choice of initial guess $z^{(0)} \in X$, with the bound

$$||z - z^{(1)}||_X \leq \frac{\lambda}{1-\lambda} \; ||y^{(1)} - y^{(0)}||_X \tag{4}$$

In practical terms, the condition $\lambda < 1$ can be interpreted as

NATO ASI Series, Vol. F34
Modelling, Robustness and Sensitivity Reduction in Control Systems. Edited by R. F. Curtain

a stability criterion with (4) expressing the degree of error involved with approximating z by $z^{(1)}$.

An alternative way of regarding (1) is in the recursive form

$$z = f^{(k)}(z) \tag{5}$$

where $f^{(k)}$ denotes the kth composed map of f i.e. $f^{(2)}(x) = f(f(x))$, $f^{(3)}(x) = f(f(f(x)))$, etc. A simple calculation shows that

$$||f^{(k)}(y) - f^{(k)}(z)||_X \leq \mu_k \; ||y - z||_X \tag{6}$$

where

$$\mu_k \leq \lambda^k \tag{7}$$

Simple examples indicate that f may not be a contraction whereas $f^{(k)}$ may be a contraction for some integer $k \geq 2$. The implications of this are, of course that, the contraction mapping theorem applied to (5) rather than (1) can yield stability criteria of wider applicability and produce improved estimates in the successive approximation procedure. These possibilities are the subject of this paper.

The possibilities can be illustrated in the linear case

$$z = f(z) = Az + b \tag{8}$$

with $b \in X$ and $A: X \to X$ linear and bounded. In this situation $\lambda = ||A||$ whereas equation (5) has the form

$$z = f^{(k)}(z) = A^k \, z+b+Ab+\ldots+ A^{k-1}b \tag{9}$$

from which

$$\mu_k = ||A^k||_X \leq ||A||_X^k = \lambda^k \tag{10}$$

If $r_X(A)$ denotes the spectral radius of A, the well-known formula

$$r_X(A) = \lim_{k\to\infty} ||A^k||_X^{1/k} \leq ||A||_X \tag{11}$$

yields the result:

<u>Theorem 2</u>: A necessary and sufficient condition for the existence of an integer k such that $f^{(k)}$ is a contraction, is that

$$r_X(A) < 1 \tag{12}$$

This result has an important interpretation in terms of conservatism. It is well known that, if the norm on X is replaced by any topologically equivalent norm, the induced norm $||A||$ of A will change (possibly substantially) whereas r(A) will remain unchanged. This means that contraction mapping based stability criteria can be sensitive to the technical choice of signal norm and, indeed, be extremely conservative. The interpretation of theorem 2 is that the use of the recursive form removes all excess conservatism as (12) is the best result obtainable by any norm based method! The *Principles* to be extracted from this discussion are (a) recursion improves contraction mapping methods in stability theory and (b) recursion enables an optimal norm-based stability criterion to be derived.

The natural problem arising from this is the identification of practical situations where application of the 'recursion principle' leads to computable results for computer-aided-control-systems-design. In the following sections a number of situations are identified.

2. A Simulation Based Computational Method

Consider the feedback system in $X = L_\infty(0,\infty)$

$$y = Ay+b \tag{13}$$

where A is a convolution map assumed to be generated (for simplicity) from an asymptotically stable state space system. A useful result for simulation based calculation of $||A||$ is (Owens and Chotai, 1983):

Theorem 3: With the above notation

$$||A||_{L_\infty} = N_\infty(Y) \tag{14}$$

where Y is the step response of A from zero initial conditions and the symbol $N_T(Y)$ denotes the norm Y regarded as a function of bounded variation on [0,T] i.e.

$$N_T(Y) = |Y(0+)| + \sum_{k=1}^{k^*(t)} |Y(t_k) - Y(t_{k-1})| + |Y(t) - Y(t_{k^*(t)})| \qquad (15)$$

where $0 = t_0 < t_1 < \ldots$ are the local maxima and minima of Y on the extended positive real axis and $k^*(t)$ denotes the largest integer $k \geq 0$ such that $t_k \leq t$.

The extension of the result to evaluate $||A^k||$ is

$$||A^k|| = N_\infty(Y^{(k)}) \qquad (16)$$

where $Y^{(k)}$ is the step response of A^k obtained by the recursive simulation

$$Y^{(k)} = AY^{(k-1)} \qquad (17)$$

with $Y^{(0)}$ defined to be a unit step.

One of the powerful aspects of this simple construction is (Owens and Chotai, 1984):

Theorem 4:

$$\lim_{k\to\infty} (N_\infty(Y^{(k)}))^{1/k} = r_{L_\infty(0,\infty)}(A) = ||A||_{L_2(0,\infty)} = \max_{w\geq 0} |A(iw)| \qquad (18)$$

where A(s) denotes the transfer function corresponding to A.

In effect, the result states that recursive simulation in $L_\infty(0,\infty)$ leads to an optimal L_∞ norm-based result identical to the natural L_2 norm-based result. The intriguing feature of this is that recursive simulation in the time domain leads directly to a frequency domain stability criterion!

Before moving on to applications of the above to robust and model reduction, it is of interest to note the application to the classical circle criterion (Holtzmann 1970) as described by Owens and Chotai (1984). More precisely, we consider the scalar nonlinear feedback system

$$y = G(r - N(y)) \qquad (19)$$

where N is a memoryless nonlinearity expressible in the form

$N(y) = Fy + n(y)$ where F is a scalar gain and n has incremental gain α. Equation (19) can be expressed as

$$y = L_c(r - n(y)) \tag{20}$$

where L_c is the linear closed-loop system obtained by setting $n = 0$. On the assumption that L_c is stable in the i/O sense then a sufficient condition for stability is that

$$||L_c||_{L_2(0,\infty)}\ \alpha < 1 \tag{21}$$

or, applying theorem 4, we obtain (Owens and Chotai, 1984):

Theorem 5: (The circle criterion)

The nonlinear closed-loop system is L_2 i/O stable if there exists a finite integer $k \geq 1$ such that

$$(N_\infty(Y^{(k)}))^{1/k}\ \alpha < 1 \tag{22}$$

where $Y_c^{(k)}$ denotes the unit step response of L_c^k obtained by recursive simulation.

The precise interpretation of the result is discussed in the reference. It is simply noted here that the circle-criterion is normally expressed as a graphical frequency domain interpretation of (21) whereas theorem 5 shows that it can equally well be regarded as a consequence of a recursive simulation procedure.

3. Robust Control Design Based on Reduced Order Models

For simplicity, consider a stable single-input/single-output state space system with transfer function G(s) and corresponding convolution operator G in $L_\infty(0,\infty)$. Consider the design of a unity output feedback control scheme for this process with forward path controller K(s) and demand signal r. Suppose that G is replaced by a stable reduced order model G_A of G and that K is designed to stabilize G_A. The natural problem of robustness of the design process is to construct computable conditions under which K is also guaranteed to stabilize G in the closed-loop, despite the presence of the modelling error $\Delta G = G - G_A$. The following result (Owens and

Chotai, 1983) is the basis of the analysis. Full details of the result and its multivariable generalisation can be found in this reference with further ideas given in Owens and Chotai (1987):

Theorem 6: Under the above conditions, if ΔG is known to satisfy the inequality

$$|\Delta G(iw)| \leq \Delta(w) \quad , \quad w \geq 0 \tag{23}$$

and G'K is both stabilizable and detectable for all $G' = G_A + \Delta G$ with ΔG satisfying (23), then K stabilizes the real plant G if

$$\max_{w \geq 0} \left| \frac{K(iw)}{1+K(iw)G_A(iw)} \right| \Delta(w) < 1 \tag{24}$$

This criterion has a nice graphical interpretation that is omitted here but can be found in the references. Here we concentrate on the use of recursion for computing candidates for $\Delta(w)$ using only simulation data from G. Such a procedure may be of value for high order systems where simulation is more easily undertaken than evaluation of G(s).

Let F(s) be a stable 'proper' filter and note that we can choose

$$\Delta(w) = |F^{-1}(iw)| \max_{w \geq 0} |F(iw)\Delta G(iw)| \tag{25}$$

where F^{-1} is chosen to represent our ideas (good or bad) of the frequency content in ΔG. Examination of the second factor in (25) and comparison with theorem 4 suggests that we evaluate

$$M_k = (N_\infty(E_F^{(k)}))^{1/k} \tag{26}$$

where $\{E_F^{(k)}\}$ is the sequence with $E_F^{(0)}$ equal to a unit step and $E_F^{(j)}$ obtained from $E_F^{(j-1)}$ via the simulation

$$E^{(j)} = F \, \Delta G \, E^{(j-1)} \quad , \quad j \geq 1 \tag{27}$$

Noting that, for all $k \geq 0$,

$$M_k \geq \lim_{k \to \infty} M_k = \max_{w \geq 0} |F(iw)\Delta G(iw)| \tag{28}$$

it is seen that (25) can be replaced by the *computable* error bound

$$\Delta(w) = M_k \ |F^{-1}(iw)| \tag{29}$$

which approaches the 'ideal' (25) as $k \to \infty$ but does not require evaluation of the transfer function G(s).

The practical attractions of the above procedure are its graphical nature, the ability to use simulation as a computational tool and the ability to introduce frequency shaping into the robustness analysis whilst generating a guaranteed upper bound Δ on the modelling error. The results reduce to those of Owens and Chotai (1983) if F = 1 and k = 1 and, in general, reduce the conservatism of the paper substantially. Full details can be found in Owens and Chotai (1987).

Finally, in this section, we note the following simulation based result that, by avoiding the frequency domain interpretation, generates minimally conservative criteria by using recursion (Owens and Chotai, 1987)

Theorem 7:

If K stabilizes G_A it also stabilizes G if GK is stabilizable and detectable and there exists an integer $k \geq 1$ such that

$$N_\infty(W_k^{(k)}) < 1 \tag{29}$$

where $W_k^{(0)} = E_I^{(k)}$ and $W_k^{(j)}$ is generated from $W_k^{(j-1)}$ from the simulation

$$W_k^{(j)} = (I+KG_A)^{-1}K\ W_k^{(j-1)} \tag{30}$$

$1 \leq j \leq k$, from zero initial conditions.

The proof of the result is based upon writing the closed-loop dynamics in terms of the plant input

$$u = (I+KG_A)^{-1}K\ (r - \Delta Gu) \tag{31}$$

which, in recursive form, yields the L_∞ i/O stability criterion

$$||((I+KG_A)^{-1}K\Delta G)^k||_{L_\infty} < 1 \tag{32}$$

Theorem 7 then follows from theorem 3 noting commutation of scalar convolution operators.

4. Conclusions and Other Work in the Field

The paper has attempted to motivate the use of recursive versions of the i/O equations as a tool in generalising the available stability criteria for robust control based on reduced order models. It has been demonstrated that recursion can certainly release simulation as a tool for generating modelling error bounds in robust stability criteria. Perhaps more significantly, recursion also enables the reduction of conservatism to a theoretical minimum possible for norm based methods. More details of the ideas can be found in Owens and Chotai (1986) including the derivation of 'tight' performance bounds based on recursion. Applications of recursion to non-convolution systems of a certain type can be found in Hong and Owens (1986).

5. References

Holtzmann J. (1970) Nonlinear Feedback Systems: a functional analysis approach. Prentice-Hall, Englewood Cliffs.

Hong G.S., Owens D.H. (1986) Frequency domain robust stability conditions for multi-rate predictor control schemes, IEEE Conf. on Dec. and Control, Athens, 2138-9.

Owens D.H., Chotai A. (1983) Robust control of linear dynamical systems using approximate models. Proc. IEE, 130 (D), 45-57.

Owens D.H., Chotai A. (1984) A simulation technique for checking the circle criterion. Systems and Control Letters, 4, 367-372.

Owens D.H., Chotai A. (1987) On the use of reduced order models and simulation data in control systems design. Control-Theory and Advanced Technology, to appear.

Performance-Robust Design via a multi-criteria/multi-model approach - a flight control case study[1)]

Georg Grübel and Dieter Joos
DFVLR-Institute for Flight Systems Dynamics
Oberpfaffenhofen, D-8031 Wessling

Summary

An aircraft whose centre of gravity is behind the aerodynamic centre has to be provided with "artificial stability" by an automatic control system. The performance benefits of such an aircraft are obtained at the expense of a severe safety problem in case of control system failure. Hence extreme reliability requirements have to be met for the automatic stabilization unit. For a new high-performance aircraft this problem has been approached by designing a "simplest possible" analog stabilization back-up unit. Without gain-scheduling and by using pitch rate feedback only, this back-up unit provides remarkably good flying qualities over the entire flight envelope. To achieve this result a complex "robust control" design problem had to be solved.

The performance-robust design is based on an iterative design strategy which provides the designer with a systematic decision framework to seek for a compromise which satisfies all design goals "in the best possible way". Since such a compromise is not unique, one can gain in one criterion and one

---

1) modified version of a paper presented at 14th ICAS Congress, Toulouse/France, Sept. 9-14, 1984

NATO ASI Series, Vol. F34
Modelling, Robustness and Sensitivity Reduction in Control Systems. Edited by R. F. Curtain

flight condition at the expense of other criteria and other flight conditions. Hence in order to fully exploit the performance potential of a chosen control structure one needs a design strategy which allows to consider each performance criterion and each flight condition individually while guaranteeing that simultaneously a best possible compromise is achieved. This leads to a multi-criteria/multi-model pareto-optimal design approach, to be solved in this case study by the general "design systematic with vector-valued performance index and parameter optimization" as developed by G. Kreisselmeier.

1. Introduction

Fly-by-wire technology has made it possible to define the architecture of an aircraft at the design stage without making allowances for the constraints imposed by stability requirements. Especially, the freedom to shift the centre of gravity behind the aerodynamic centre, that is making the basic aircraft unstable, allows the aircraft design to be optimized from the point of view of performance and economy [1].

The new Swedish JAS 39 multi-role high-performance aircraft is the first of a new generation of light combat aircraft to make full use of the advanced concept of artificial stability.

What has to be paid for the benefits? Guaranteed reliability of the stabilizing automatic control system is obviously a crucial demand in pursing the artificial stability concept since the safety of the aircraft and the success of its mission will depend essentially on the correct functioning of this system. How can this be achieved?

For JAS 39 the idea has been adopted that functional simplicity is something that should be required beforehand in order to properly guarantee reliability for the stabilization unit. Therefore a "simplest possible" analog back up unit has been designed for the otherwise fully digital aircraft control system. This stabilization unit by itself provides good flying qualities over the entire flight envelope without the need for gain-scheduling.

The design of such a stabilization unit which shows good performance in spite of functional simplicity, is a complex control design problem. A solution was achieved by the design procedures for 'robust control' |2|, |3| developed at the DFVLR Institute for Flight Systems Dynamics. The paper deals with the design and performance evaluation of this stabilization unit.

2. Robust Stabilization

For JAS 39, longitudinal artificial stability must be provided for subsonic flight. Especially for landing, the aircraft is highly unstable with a doubling time of 0.6 ... 0.8 seconds. The stabilization must be achieved by an automatic control system whose structure is as simple as possible. This means that

- only pitch rate q is to be used for feedback with no other aerodata or state variable measurements being employed; that is
- no control scheduling ('gain'-scheduling) as a function of the flight condition is to be used, thereby avoiding a functionally complex system which would have to be realized by software in a digital computer; instead
- an analog, that is, a hardware realization of the stabilizing control law is required with a dynamic feedback compensator and prefilter of lowest possible order and parameters independent of the flight condition. Furthermore
- as control surface for stabilization only elevon deflection δ_e is to be used, that is, the stabilization unit shall not rely on the all-movable canards as control surfaces.

Is it possible to stabilize the aircraft under such constraints, i.e. by using a simple, fixed gain, single-input single-output control loop? The answer is yes:

For control design the dynamic behaviour of the aircraft is described by the linearized equations of longitudinal motion in the form of a state equation

$$\dot{\underline{x}}(t) = [\underline{A}]_{FC}\underline{x}(t) + [\underline{b}]_{FC}\delta_{ec}(t) \tag{1}$$

with $\underline{x}$ = v_z = vertical velocity ; Fig. 1.

q = pitch rate

Θ = pitch angle

δ_e = elevon deflection.

The actuator dynamics are modeled as a first order tranfer function

$$\frac{\delta_e(s)}{\delta_{ec}(s)} = \frac{22}{s + 22} , \tag{2}$$

with δ_{ec} = commanded elevon deflection.

The flight envelope of interest is represented by 10 flight conditions FC 1 ... 10 covering a range of

$$n/\alpha = 2.4 \ldots 92.1 \text{ g/rad},$$

the aerodynamic loadfactor n/α beeing an indicator of the overall dynamic behaviour of the basic aircraft. Hence for design we have 10 different mathematical models (1), reflecting the varying dynamic behaviour of the aircraft in its operational range. The large deviations of the pitch rate step responses for FC 1 ... 10 are shown in Figure 3 as example.

We are looking for a common feedback compensator which stabilizes (1) in all flight conditions FC1 ... 10, this compensator is described by the state equation:

$$\underline{\dot{x}}_R(t) = \underline{A}_R \underline{x}_R(t) + \underline{b}_R q(t)$$
$$\delta_{ec}(t) = \underline{c}_R^T x_R + d_R q(t) + f_R \delta_P(t) \qquad (3)$$

with $\delta_P(t)$ being the pilot command input. The parameters of $\underline{A}_R$, $\underline{b}_R$, $\underline{c}_R$, d_R, f_R, are fixed and independent of the flight conditions. This we call a multi-model stabilization problem with structured uncertainty. If such a feedback compensator exists we call it a "robust" stabilizing controller, which provides stability despite large, but known, dynamic variations of the aircraft.

To see if such a robust controller exists, we first consider the pole-zero pattern of the pitch transfer function $q(s)/\delta_{ec}(s)$ for FC 1 ... 10. Three qualitatively different patterns arise as shown in Figure 2, having 0 or 1 or 2 poles in the right half s-plane. A simple root locus consideration shows that no common stable feedback compensator can be devised such that for all FC 1 ... 10 all poles can be moved into the left half complex plane since there is a zero of the transfer function at the origin. Hence from a strictly mathematical point of view no robust stabilization is possible. However, we can also deduce from a root locus consideration that "practical" stability for all flight conditions FC 1 ... 10 is possible in the sense that a remaining unstable "phugoid" pole for FC 4, 5, 6, 8 has a sufficiently small real part which yields a sufficiently large doubling time not to interfere with the pilot's short period commands. This means that this pole shall be very closely attracted by the zero at the origin. Having this in mind, a first (!) order lag-lead compensator was possible to design by the parameter space technique for pole region placement [2] which led to a robust stabilization. In fact it produced a dominant short-period pole pair satisfying performance level 3 - 2 according to the flying qualities specifications MIL 8785 B for all FC 1 ... 10, the most extreme conflicts in achievable damping ratio ζ and short-period frequency ω being between FC 2 (landing) and FC 9 (transonic flight). This was the best possible solution which was to be obtained by a systematic computer-aided tuning of the 3 parameters -static gain, 1 pole, 1 zero- of a first order compensator.

3. Performance-Robust Control

Since stabilization over the entire flight envelope can already be accomplished by a first order compensator, one can expect that more can be gained by using a higher-order compensator, thus having more poles and zeros -i.e. more design degrees of freedom- to work with. Especially, one can try to improve the flying qualities, i.e. the damping and handling behaviour.

The flying quality specifications according to MIL 8785 B are derived for aircraft which have no dynamic control. For dynamically controlled aircraft we apply these specifications as directly as possible by generalizing the underlying principles of MIL 8785 B to higher than second order models:

Stability: MIL 8785 B requires a minimum damping ratio for the short-period (dominant) pole pair, e.g. $\zeta_{SP} > 0.35$, for performance level 1.

This we generalize to

Design Goal I:

A minimum damping ratio ζ_{min} shall be achieved for all complex pole pairs, e.g.

$$\zeta \overset{!}{\geq} \zeta_{min} = 0.35 \tag{4}$$

No dominant pole pair is required.

Handling: MIL 8785 B requires that Bihrle's [4] CAP (=Control Anticipation Parameter) should be near to 1 and lie between specified bounds, e.g. for performance level 1, CAT A (CAT = category of flight condition):

$$0.28 \leq \mathrm{CAP} = \frac{\dot{q}(t=o)}{n_z(\mathrm{SP\ steady\ state})} \leq 3.6$$

where q(t) is the pitch rate response to a pilot step command input. However, since for every aircraft with actuator dynamics one has $\dot{q}(t=o) = o$, this definition of CAP is not applicable. Therefore, as proposed by D.E. Bischoff [5], we use a modified control anticipation parameter

$$\mathrm{CAP}' = \frac{\dot{q}(t)_{max}}{n_z(t_q)} \quad , \tag{5}$$

t_q being a suitable time for a "steady state" short period response $n_z(t)$, e.g. $t_q = 6$ sec. This leads to

Design Goal II:

$$0.28 \leq \lfloor \mathrm{CAP}' \overset{!}{\approx} 1 \rfloor \leq 3.6 \tag{6}$$

Control Surface Constraints: The given control command limit and control rate limit for elevon deflection have to be taken into account as constraints, yielding

Design Goal III:

$$\delta_e \leq 25^o$$

$$\dot{\delta}_e \leq 60^o/s$$

both for

- repeated max aft and forward stick commands,
- discrete gust disturbances of the form

$$w_z = 7.5(1-\cos\Omega(n/\alpha)\cdot t)\ \mathrm{m/s},\ o<t<2\Pi\Omega \tag{7}$$

$$w_z = o \quad \text{for} \quad t < o \quad \text{and} \quad t > 2\Pi/\Omega.$$

These design goals should be satisfied for all flight conditions FC 1 ... 10. However, not all flight conditions are equally important: Since an aircraft must always be able to land, special attention has to be paid to landing, i.e. FC 2.

Landing: During the design process it turned out that in this flight condition the phugoid had a considerable effect on the short period pitch rate response. Hence we have to pay attention to

Design Goal IV:

In all flight conditions, especially in FC 2, a distinct separation of the short period and phugoid motion (corresponding to a pole pair close to the origin) should be attained, i.e. the residues for the phugoid poles should be as small as possible.

For the landing task pilots ask for an especially well behaved aircraft. The requirements can be characterized by tight bounds on rise time, overshoot and settling time of the pitch rate step response [6]. This leads to

Design Goal V:

For FC 2 the pitch rate step response should be close to a prespecified "model" trajectory defined by

$$\frac{q(s)}{\delta_{p(s)}} = \frac{1/0.7\ s + 1}{1/2.25\ s^2 + \frac{2(0.8)}{1.5}\ s + 1} \tag{8}$$

This requirement has also to be satisfied for the aircraft with different external loads. This is reflected by

Design Goal VI:

For a variation of ± 25 % of the centre of gravity, the pitch rate response in landing should be as close as possible to the "model" trajectory (8).

4. A Complex Design Problem

We have 6 design goals I ... VI which cover both time domain and frequency domain specifications: Design goal I aims at pole region placement to be dealt with by eigenvalue criteria. The design goals II and III are time-response criteria which cannot be evaluated analytically and need system simulation in order to be evaluated. Design goal IV concerns the phugoid residue of the pitch transfer function which leads to a pole-zero placement criterion. The design criteria V, VI are trajectory criteria which again have to be handled by simulation.

The design goals I ... III should be satisfied equally well for the 10 aerodynamical flight conditions FC 1 ... 10. In addition for landing (FC 2) special attention has to be paid to the design goals IV ... VI, where design goal V (variation of the centre of gravity (c.g.)) requires two additional models (1) to be taken into account (models for +25 % c.g. and -25 % c.g.) besides the nominal model.

Hence we have a multi-criteria, multi-model control design problem. Analytical control design techniques are not applicable in this case since the criteria are time- and frequency criteria and hence they are of different nature. Also, no standard (scalar) optimization is possible since there are more than one performance criterion. Instead only a compromise can be sought for, which satisfies all the criteria simultaneously and "in the best possible way". Since such a compromise is not unique, one can gain in one criterion and one flight condition at the expense of other criteria and other flight conditions. Hence in order to fully exploit the ultimate potential of a given control structure one needs a design strategy which allows to control each criterion and each flight condition individually while guaranteeing that simultaneously a best possible compromise is achieved. Such a design strategy is the general design systematic using a vector performance index and parameter optimization as developed by G. Kreisselmeier [3].

This design technique has been applied for this problem and will be briefly described in section 6. The results are shown in the next section. As control structure a 3rd order compensator with 8 design degrees of freedom -static gain, 3 poles, 4 zeros- has been chosen. The mechanization of this control law is shown in Figure 12. It is a compensator in the feedback path with a simple dynamic feedforward path to additionally place one zero of the pitch rate transfer function.

5. Design Results

The controlled aircraft was analyzed with respect to the design goals and in addition some parameter sensitivity studies were made. All analysis done at DFVLR is based on the 10 linear models, whereas at SAAB the analysis of the controlled aircraft was performed with a 6 degree of freedom nonlinear model, used in a purely digital simulation, as well as in an on-ground-simulation with pilots in the loop. The model used at SAAB contains an additional body bending filter which was designed afterwards and indepently of the design of the robust controller. In the following some of these results are shown:

Design Goal I:

Figure 7 shows that a minimum damping ratio $\zeta \geq 0.35$ for all complex pole pairs could not be achieved. The time responses, Figure 4, and frequency responses, Figure 9 however indicate a sufficiently damped dynamic behaviour.

Further sensitivity analysis establishes that a sufficient stability margin is achieved. To show this, curves of constant minimum damping ratio 0.15, 0.25 and 0.35 due to relative variations in the aerodynamical parameters cm_q and $cm_{\delta e}$ are plotted in Figure 10 for FC 2 and Figure 11 for FC 9 respectively. The origin represents the nominal value for which the controller was designed, and the curves represent bounds up to which the aerodata can deviate from the nominal ones yielding the respective damping ratio. The parameter values on the 0.25 curves are the frequencies of the corresponding complex pole pairs.

Design Goal II:

For the controlled aircraft the control anticipation parameter CAP' was sought to have a value of about one. Analysis of the controlled aircraft shows relatively low values of CAP', see Figure 7. This confirms results of D.E. Bischoff [5] were for CAP' lower boundaries are suggested ($0.25 < CAP' < 1.5$) than for the original CAP of Bihrle [4].

In Figure 8 the equivalent short period frequency ω^e of a first over second order equivalent system versus n/α is plotted for all flight conditions showing the demanded range of operation of the robust controller.

These analytical characteristics for handling quality well agree with pilot opinions, as documented in Figure 13, showing the Cooper-Harper-rating for landing approach in case of several disturbance levels. The result is at least performance level 2.

Design Goal III:

For discrete gust disturbances (7) the resulting control rate $\dot{\delta}_e$ is plotted in Figure 6, showing that the rates are far below the limit (1.05 rad/s).

Design Goal IV - VI:

The result corresponding to the design goals IV-VI are shown in Figure 5. Short period and phugoid motion are sufficiently separated in the landing approach, since after 24 s simulation time no influence of the phugoid motion is visible and according to Figure 13 pilots like to fly the aircraft. The bounds on rise time, overshoot and settling time are matched for the nominal flight condition. Only for a variation of the centre of gravity of -25% the bound on the settling time is slightly failed.

6. Design Systematic for a Multi-Criteria/Multi-Model Control Problem

The design procedure used in this case study was the design systematic via performance index vector and parameter optimization as proposed in [3]. It has proven to be very useful in various practical designs and can be summarized as follows [7]:

Given a plant, and given a suitably chosen controller structure with free parameters $\underline{k} = [k_1,\ldots,k_n]$ the design problem is to determine suitable values for $\underline{k}$.

In order that the design can proceed in a systematic fashion it is necessary that all design objectives are taken care of explicitly in the design. Therefore, every design objective shall be rated quantitatively by means of a suitable mathematical performance index $J_i(\underline{k})$.

A performance index $J_i(\underline{k})$ is called suitable if $J_i(\underline{k}') < J_i(\underline{k}'')$ means that $\underline{k}'$ satisfies the i-th control objective better than $\underline{k}''$ does, and if $J_i(\underline{k})$ is a sufficiently smooth function.

For ease of notation we define the performance vector $\underline{J}(\underline{k}) = [J_1(\underline{k}),\ldots,J_L(\underline{k})]$. Moreover, for any two real vectors $\underline{x}$, $\underline{y}$ we say that $\underline{x} \leq \underline{y}$ if for all components $x_i < y_i$ and $\underline{x} \neq \underline{y}$.

Iterative Technique for a Systematic Design

The design technique is iterative, where each design iteration (the ν-th, say) comprises two steps:

Step 1: Choose a vector of design parameters $\underline{c}^\nu$ $(c_i^\nu > 0)$ such that

$$\underline{J}(\underline{k}^{\nu-1}) < \underline{c}^\nu \leq \underline{c}^{\nu-1}. \quad (9)$$

Step 2: Find $\underline{k}^\nu$ and $\alpha_o < 1$ such that

$$\underline{J}(\underline{k}^{\nu}) \leq \alpha_o \underline{c}^{\nu}. \tag{10}$$

The design iteration is initialized by using an initial guess $\underline{k}^o$ for the controller parameters and by taking $\underline{c}^o$ sufficiently large.

In step 1 of every design iteration, $\underline{c}^{\nu}$ has to be chosen where (9) provides a well defined margin. This choice determines the design direction. Step 2 then provides the margin for the next design iteration.

It is desirable that this margin be as large as possible, i.e. α_o should be as small as possible.

After the ν-th design iteration we have from (9), (10) that

$$\underline{J}(\underline{k}^{\nu}) < \underline{c}^{\nu} < \underline{c}^{\nu-1} \quad \ldots < \underline{c}^{o}, \tag{11}$$

i.e. we have a monotonically decreasing sequence of design vectors and the performance vector has become less than all of them. This establishes the systematic behaviour of the design process.

<u>Design Tool</u>

Basically, any method which is able to perform step 2 of the design iterations can serve as a design tool. Here we shall focus on a method which, in addition, tends to make α_o as small as possible.

The smallest value of α_o satisfying $\underline{J}(\underline{k}) \leq \alpha_o \underline{c}^{\nu}$ is given (as a function of $\underline{k}$) by

$$\alpha_o(\underline{k}) = \max_{1 \leq i \leq L} \{J_i(\underline{k})/c_i^{\nu}\}. \tag{12}$$

Since $\alpha_o(\underline{k})$ is not continuously differentiable everywhere, we consider a smooth, i.e. at least twice continuously differentiable approximation $\alpha(\underline{k})$ instead, where

$$\alpha(\underline{k}) = \frac{1}{\rho} \ln \sum_{i=1}^{L} \exp\{\rho J_i(\underline{k})(c_i^{\nu}\}. \tag{13}$$

and $\rho > 0$ is arbitrary. It can be verified that

$$\alpha(\underline{k}) = \alpha_o(\underline{k}) + \frac{1}{\rho} \ln \sum_{i=1}^{L} \exp\{\rho[\frac{J_i(\underline{k})}{c_i^{\nu}} - \alpha_o(\underline{k})]\}. \tag{14}$$

Hence $0 \leq \alpha - \alpha_o \leq (\ln L)/\rho$, i.e. α can be made as close to α_o as desired by choice of ρ.

The minimization of $\alpha(\underline{k})$ (for example, by applying Powell's method for function minimization without calculating derivatives) can now be used as a design tool for performing step 2 in each design iteration.

The sequence of design iterations finally terminates when the minimization of $\alpha(\underline{k})$ results in a $\underline{k}^{\nu}$ with $\alpha_o(\underline{k}^{\nu}) \geq 1$.

As in all parameter optimization problems there is a possibility of local minima. Therefore, instead of using efficient local optimization algorithms such as Powell's, one may as well use global optimization algorithms such as random search algorithms. However, the latter are known to be less efficient.

Note also that the use of optimization algorithms which assume that the function to be minimized is twice continuously differentiable (which most efficient local algorithms do), requires that the performance indices themselves must be twice continuously differentiable. This imposes a certain restriction on the mathematical formulation of performance indices.

The design procedure provides a systematic framework to take care of multiple design objectives simultaneously and individually. The design objectives can be formulated by criteria of different kind either in state space or frequency domain. Hence the design specifications problem can be tackled in the most direct way. However, the design procedure does not provide a priori guidelines what design objectives are appropriate for a particular design problem. It remains in the designer's responsibility to decide what design criteria he shall use. This decision must be based on his knowledge of control theory and the operational requirements he has to satisfy.

The design procedure provides a systematic guidance for the designer to cope with the "free-design-parameter problem": For each individual performance index J_i there is associated a free design parameter c_i which can be chosen by the designer within a well-defined margin. Typically, one starts with sufficiently large values for the design parameters and then successively reduces them so as to improve certain criteria while keeping possible deteriorations of others in tolerable bounds. Whatever choice c_i^ν in the given margin is taken, it is always guaranteed that $J_i \leq c_i^\nu$ and hence a possible degradation of the performance index J_i is well bounded. If $c_i^\nu = J_i(\underline{k}^{\nu-1})$ is chosen, then an improvement of J_i is guaranteed provided only that such an improvement is possible at all at the expense of degrading other performance indices. This allows to explore the design possibilities, and to reach a desirable trade-off between competing objectives, step by step.

In each step it is the responsibility of the designer to decide upon the design direction, i.e. which performance index shall be improved and what is a permissible expense in degrading other performance indices. The design procedure then guarantees a step by step monotonic improvement in the direction specified by the designer. The admissible margins of the design parameters which are updated in each design iteration, yield some information how easy or difficult it is to reach the individual design objectives. Hence this design procedure makes it possible to explore the system's limitations in control design.

The design approach is well suited to deal with practical controller realization constraints. This is a consequence of using parameter optimization as a design tool instead of using an analytical control synthesis technique. There is no "controller simplicity vs. plant complexity problem". The designer can specify the controller structure at will. He is free to combine control considerations with technical realizability constraints. This freedom, however, calls much at the designer's experience and physical insight into the type of system to be controlled. Furthermore, in view of the numerical convergence properties of parameter optimization, one also has to take some care in mathematically specifying the controller structure, i.e. in parameterizing the control law: There must not be redundant parameters. In addition, the free parameter to be optimized should have the same order of magnitude, i.e. proper scaling may be necessary. On the other hand, the freedom to specify a controller structure makes it possible to develop a proper structure in an iterative process: Start with a simple structure, say a P-I controller, and extend it successively by additional dynamic degrees of freedom, i.e. use a higher order controller, if the design iterations yield no further improvement with the formerly chosen structure. In this way, one can explore the trade-offs between controller simplicity and control system performance.

This design procedure is open to handle complex, nonlinear systems. There is no conceptual necessity for linearized or low order plant models. But there may be computer time limitations: In view of parameter optimization the performance indices have to be evaluated very often and hence sufficiently fast. For nonlinear systems there are usually no analytical relations for such evaluations, rather the performance indices have to be computed on the basis of system simulations. This limits the complexity of system models to be used by the speed of available computers and the efficiency of available simulation software.

References

[1] Wanner, J.C., The CCV Concept. AGARD Highlights 76/2, p. 4-10

[2] Ackermann, J., Kaesbauer, D., D-Decomposition in the Space of Feedback Gains for Arbitrary Pole Regions. Proc. VIII IFAC World Congr., Kyoto/Jap., 24.-28.08.1981, p. IV, 12-17

[3] Kreisselmeier, G., Steinhauser, R., Application of Vector Performance Optimization to a Robust Control Loop Design for a Fighter Aircraft. DFVLR FB 80-14, 1980, reprinted Int. J. Control (1983), 37, No. 2, 251-284

[4] Bihrle, W., Jr., A Handling Qualities Theory for Precise Flight Path Control. RTD Technical Report Air Force Flight Dynamics Laboratory, 3/1966

[5] Bischoff, D.E., The Definition of Short-Period Flying Qualities Characteristics via Equivalent Systems, J. Aircraft, vol. 20, No. 6, 1983, p. 494-499

[6] Fuller, S.G., et al, Design Criteria for the Future of Flight Controls. Proc. Air Force Wright-Patterson Aeronautical Labs, 7/1982, AFWAL-TR-82-3064

[7] Grübel, G., Kreisselmeier, G., Systematic Computer Aided Control Design, AGARD Lecture Series No. 128, Computer-Aided Design and Analysis of Digital Guidance and Control Systems, 9/1983, p. 8-1/8-7.

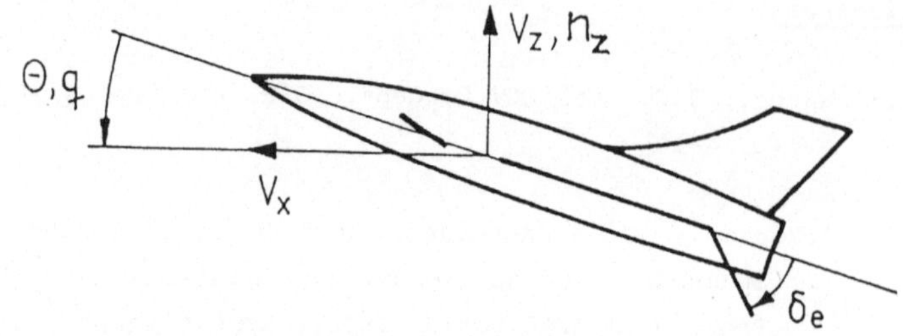

Figure 1: state variables for longitudinal motion of an aircraft.

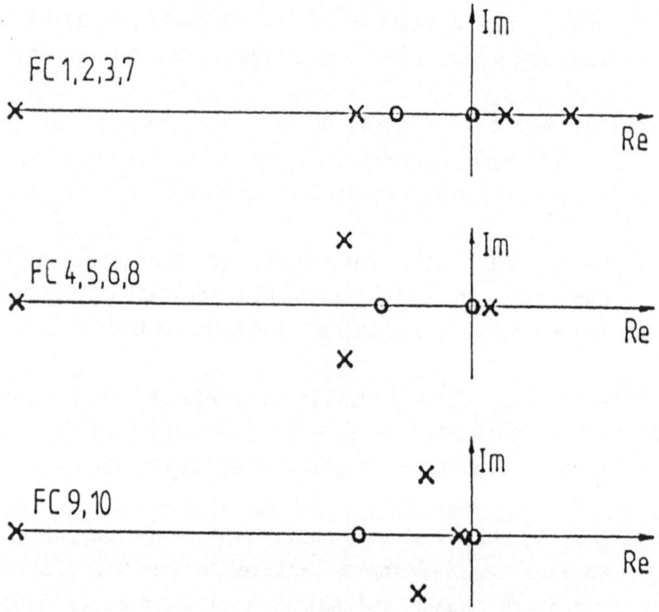

Figure 2: qualitative pole-zero pattern of pitch rate transfer function $q(s)/\delta_{ec}(s)$ for the basic aircraft for flight conditions FC 1 ... 10.

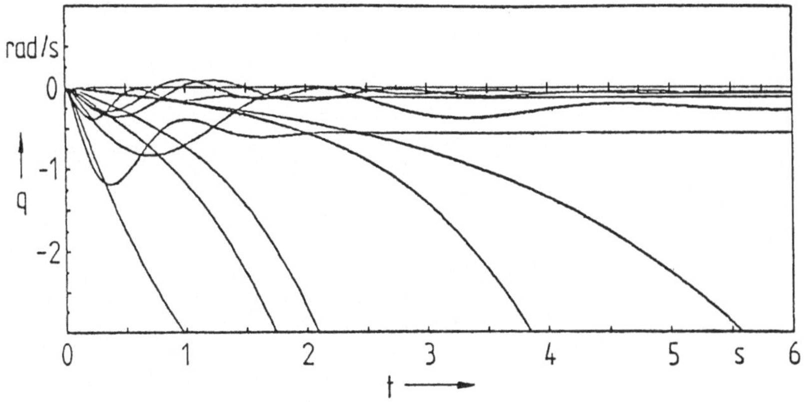

Figure 3: pitch rate step responses of the basic aircraft for flight conditions FC 1 ... 10.

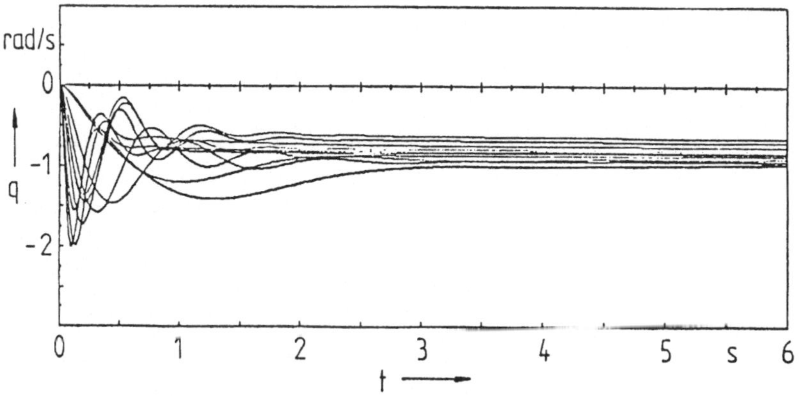

Figure 4: pitch rate step responses of the controlled aircraft with robust back-up stabilization (3rd order compensator) for flight conditions FC 1... 10.

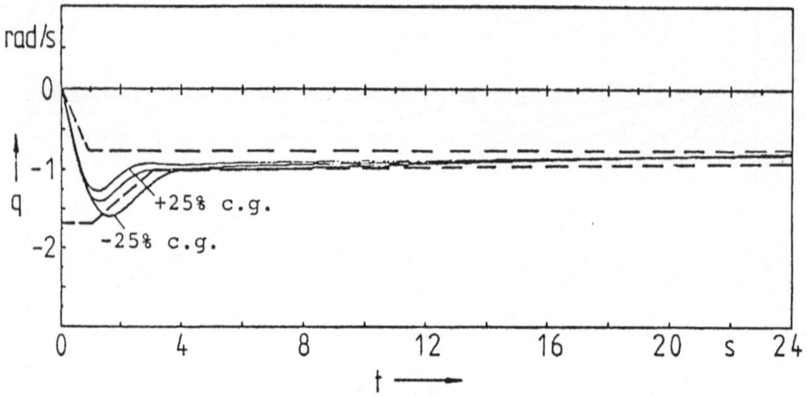

Figure 5: pitch rate step response for FC 2 (landing) whith variation of the center of gravity (c.g.) of ± 25 %.

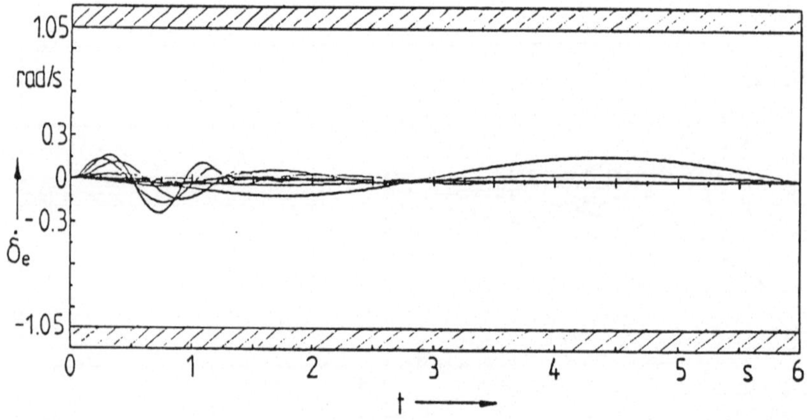

Figure 6: control surface rate due to a discrete gust model (7).

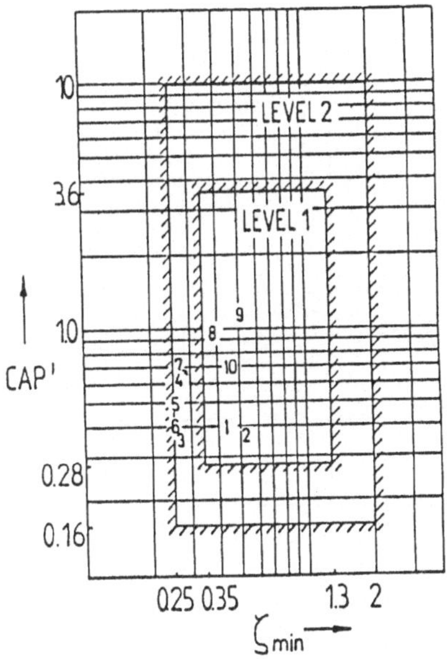

Figure 7:
Control Anticipation Parameter CAP' (8) versus minimum damping ratio, FC 1,2, Category C, FC 3 ... 10 Category A.

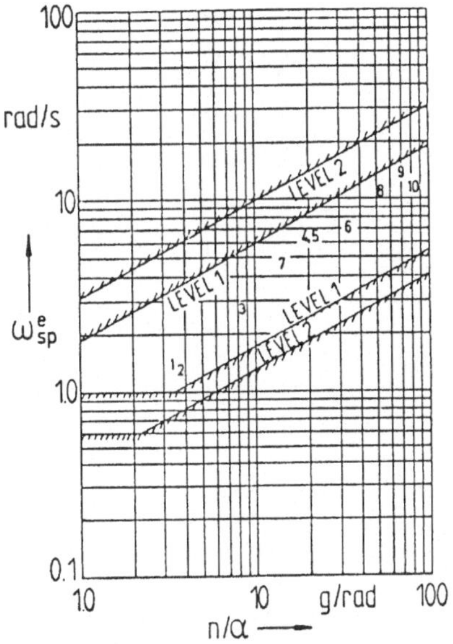

Figure 8:
Equivalent frequency ω^e versus n/α. FC 1,2 Cat. C, FC 3 ... 10 Cat. A.

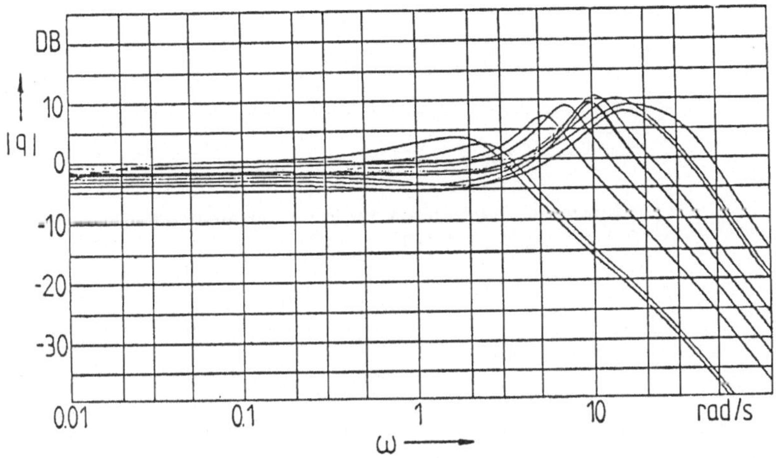

Figure 9: pitch rate frequency responses of the control augmented aircraft, FC 1,...,10.

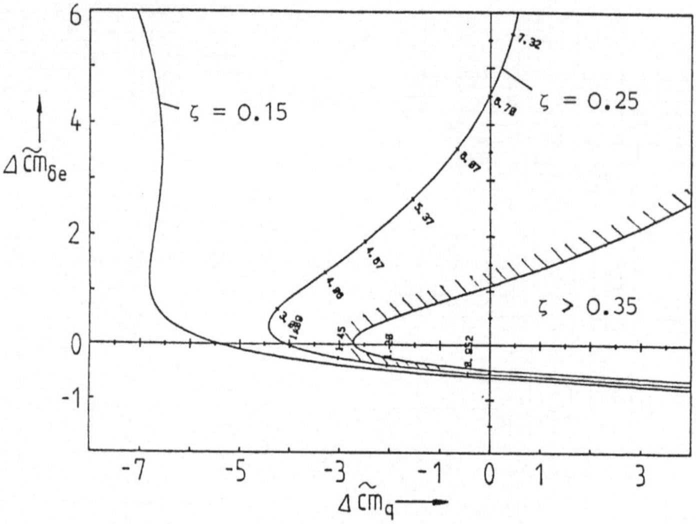

Figure 10: curves of constant minimum damping ratio ζ = 15, 0.25, 0.35 due to relative variations in cm_q and $cm_{\delta e}$ for FC 2. The parameter values indicate the frequencies of the corresponding pole pair.

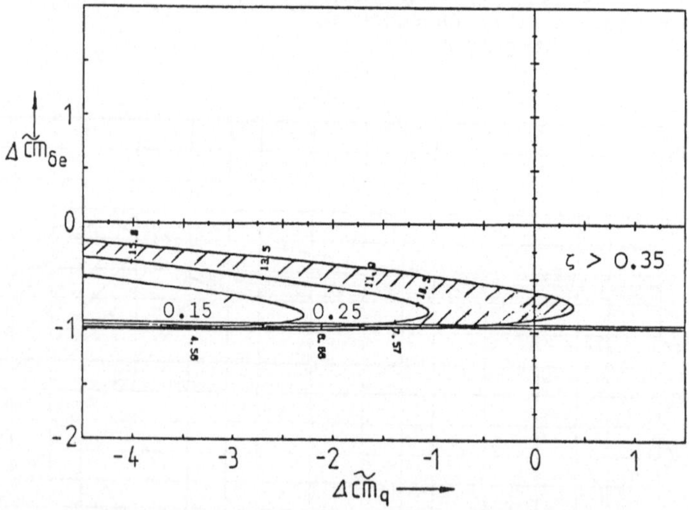

Figure 11: curves of constant minimum damping ratio ζ = 15, 0.25, 0.35 due to relative variations in cm_q and $cm_{\delta e}$ for FC 9 The parameter values indicate the frequencies of the corresponding pole pair.

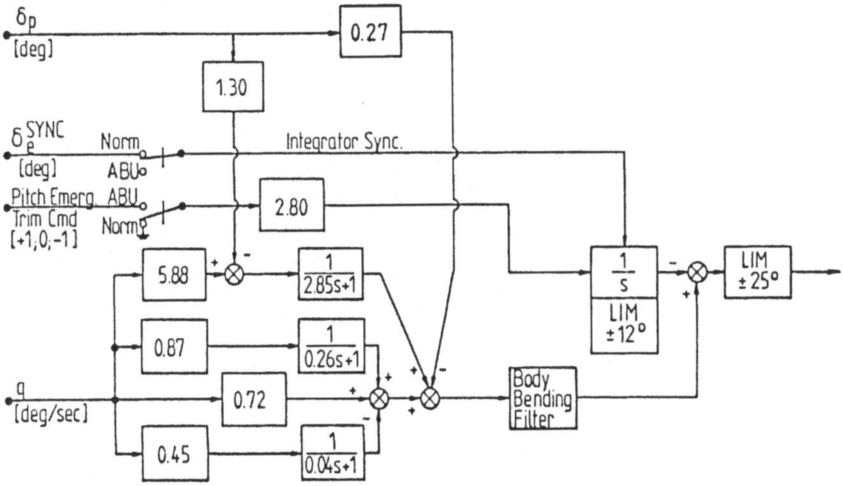

Figure 12: Mechanization of the third order robust control law including body bending filter.

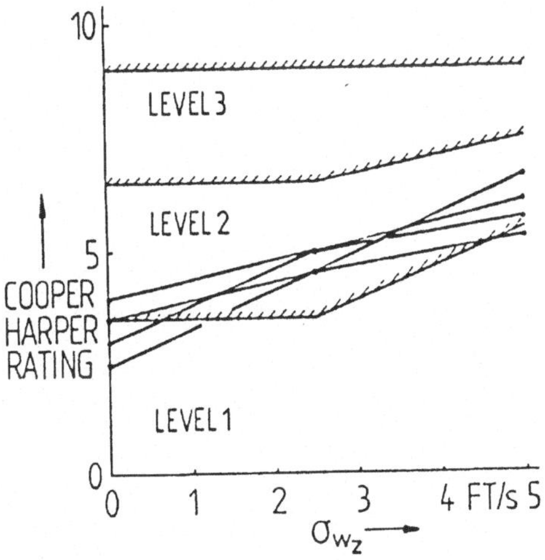

Figure 13: Cooper-Harper rating corresponding to FC 2 (landing) for several disturbance levels.

UNFALSIFIED MODELLING IN DYNAMICAL SYSTEMS

JAN C. WILLEMS

Mathematics Institute
University of Groningen
P.O. Box 800
9700 AV GRONINGEN
The Netherlands

Abstract. In this paper we will introduce the notion of the most powerful unfalsified model and derive some properties of it in the context of linear time-invariant complete systems.

1. General Principles. Assume that we are trying to model a **phenomenon.** We assume that the phenomenon can be described by attributes which belong to a set S. A **model** for the phenomenon is a subset $M \subset S$: a model claims that certain attributes are possible, others are not. In dynamical models, the attributes will be time functions - maps from a time set T to a set W (where W represents the set of 'instantaneous' values of the attributes) - and models will be specified by a subset of W^T, in other words, by what in [1] we have called the **behavior** of dynamical system. Hence the notion of a mathematical model of a dynamical phenomenon leads to the definition of a **dynamical system** as a triple $\Sigma = (T,W,\mathbf{B})$ with T the time axis, W the signal space, and $\mathbf{B} \subset W^T$ the behavior.

Assume now that we have taken measurements on the phenomenon. We will assume that each measurement corresponds to an observation of a realization of the attributes, i.e., to an element of S. Collecting these measurements will yield a subset $Z \subset S$, called the **set of measurements.**

The question which we want to consider is the following: **Assume that we have observed a set of measurements** $Z \subset S$, **which model** $M \subset S$ **should we choose?**

Assume that we have decided to select the model M from **a model class** $\mathbf{M} \subset 2^S$. We will also assume that the set of measurements Z belongs to a **set of possible set of measurements Z.** The guiding principle which we will follow in choosing an element from the model set M on the basis of

NATO ASI Series, Vol. F34
Modelling, Robustness and Sensitivity Reduction
in Control Systems. Edited by R.F. Curtain

a set of measurements Z is that we want a model which has **low complexity** (expressing the a priori appeal of the model) and **low misfit** (expressing our faith in the model as brought into evidence by the observations). We will now formalize this.

Let $\mathbf{M} \subset 2^S$ be the model class and $\mathbf{Z} \subset 2^S$ be the set of possible measurements. Now introduce two partially ordered spaces, C, the **complexity level space**, and E, the **misfit level space**, and two maps, c: $\mathbf{M} \to \mathbf{C}$, the **complexity**, and ε: $\mathbf{Z}{\times}\mathbf{M} \to \mathbf{E}$, the **misfit**. The more complex a model, the more undesirable it is. We view complexity as being related to how much a model explains: the more it explains, the more complex we will consider a model to be. The misfit is a quantitative measure which expresses in how far our model is corroborated by the measurements. The larger the misfit, the more we should distrust the model.

Let S, $\mathbf{M} \subset 2^S$, $\mathbf{Z} \subset 2^S$, c: $\mathbf{M} \to \mathbf{C}$, and ε: $\mathbf{Z}{\times}\mathbf{M} \to \mathbf{E}$ be a modelling set-up as described above. Our modelling algorithms will be based on the following principles.

(i) **Modelling with a maximal tolerated misfit.** Let $\varepsilon^{tol} \in \mathbf{E}$ and $Z \in \mathbf{Z}$ be given. We will call $M^* \in \mathbf{M}$ the **optimal approximate model within the given misfit tolerance** if:

(i) $\varepsilon(Z,M^*) \leq \varepsilon^{tol}$;

(ii) $\{M \in \mathbf{M},\ \varepsilon(Z,M) \leq \varepsilon^{tol}\} \Rightarrow \{c(M^*) \leq c(M)\}$;

(iii) $\{M \in \mathbf{M},\ \varepsilon(Z,M) \leq \varepsilon^{tol},\ c(M) = c(M^*)\} \Rightarrow \{\varepsilon(Z,M^*) \leq \varepsilon(Z,M)\}$.

The interpretation of these conditions should be clear. The first two formalize that M^* is the least complex model which fits the model within a given tolerated misfit.

(ii) **Modelling with a maximal admissible complexity.** Let $c^{adm} \in \mathbf{C}$ and $Z \in \mathbf{Z}$ be given. We will call $M^* \in \mathbf{M}$ the **optimal approximate model within the given admissible complexity** if:

(i) $c(M^*) \leq c^{adm}$;

(ii) $\{M \in \mathbf{M},\ c(M) \leq c^{adm}\} \Rightarrow \{\varepsilon(Z,M^*) \leq \varepsilon(Z,M)\}$;

(iii) $\{M \in \mathbf{M},\ c(M) \leq c^{adm},\ \varepsilon(Z,M) = \varepsilon(Z,M^*)\} \Rightarrow \{c(M^*) \leq c(M)\}$

The first two of these conditions show that M^* is the best fitting model within the given admissible complexity.

2. **Exact Modelling.** In the present paper we will concentrate on exact modelling. Exact modelling means modelling with zero tolerated misfit, formalized by means of the notion of **falsification.** Let Z be a set of measurements and M be a model. Then we will call M **unfalsified** by Z if $M \subseteq Z$.

The case of exact modelling can be formalized by means of the notion of more powerful models. We will call a model M_1 **more powerful** than M_2 if $M_1 \subseteq M_2$: the more a model forbids, the more predictive power it has, the better it is. If $\mathbf{M}$ is a model class and Z a set of measurements, then we will call $M \in \mathbf{M}$ and **undominated unfalsified** model in $\mathbf{M}$ if $Z \subseteq M$ and if $\{M' \in \mathbf{M}$ and $Z \subseteq M'\}$ implies $\{M = M'\}$. Finally, we will call M_Z^* the **most powerful unfalsified model** in $\mathbf{M}$ if $Z \subseteq M_Z^* \in \mathbf{M}$ and if $\{Z \subseteq M \in \mathbf{M}\}$ implies $\{M_Z^* \subseteq M\}$. Clearly, if M_Z^* exists, it will be unique. There is a simple condition which ensures that M_Z^* exists:

Proposition: Let S be a phenomenon, $\mathbf{M} \subset 2^S$ be a model class, and $Z \subset S$ be a set of measurements. Then M_Z^*, the most powerful unfalsified model in $\mathbf{M}$ exists if

(i) $S \in \mathbf{M}$;

(ii) for any family of models $M_\alpha \in \mathbf{M}$, $\alpha \in A$, the set $\bigcap_{\alpha \in A} M_\alpha$ also belongs to $\mathbf{M}$.

Proof: $$M_Z^* = \bigcap_{\substack{M \in \mathbf{M} \\ Z \subset M}} M.$$

□

The search for the most powerful unfalsified model as the optimal model can be formalized as follows. Let $\mathbf{C} = 2^S$, equipped with the partial order of inclusion, and view this as defining a complexity measure on $\mathbf{M}$. Define also a misfit ε: $Z \times \mathbf{M} \to \mathbf{C}$ satisfying $\varepsilon(Z,M) = 0$ if $Z \subset M$ and $\varepsilon(Z,M) > 0$ if $Z \not\subset M$. Now take $\varepsilon^{tol} = 0$. Then M_Z^*, the most powerful unfalsified model, is the optimal "approximate" model within the zero misfit "tolerance".

The conditions of the above proposition in the previous section will be satisfied in many situations encountered in practice, in particular when $S = (R^q)^Z$ and $M = \underline{\mathcal{L}}^q$, defined to be the family of linear shift invariant closed (in the topology of pointwise convergence) subspaces of $(R^q)^Z$. We state this formally.

Fact: Let Z be any set of q-dimensional time series and $M = \underline{\mathcal{L}}^q$. Then the most powerful unfalsified model, M^*, exists.

In [1] we have shown that $\underline{\mathcal{L}}^q$ allows many equivalent characterizations. In particular, it corresponds exactly to the familiar finite dimensional linear time-invariant systems. Indeed:

Let $\Sigma = (Z,R^q B)$ be a dynamical system. Then the following conditions are equivalent:

(i) $\mathbf{B} \in \underline{\mathcal{L}}^q$, i.e., **B** is a linear shift-invariant closed subspace of $(R^q)^Z$;

(ii) Σ is linear, time-invariant, and complete (see [1] for a precise definition);

(iii) $\exists$ a $g \in Z_+$ and a polynomial matrix $R(s,s^{-1}) \in R^{g\times q}[s,s^{-1}]$ such that $\mathbf{B} = \ker R(\sigma,\sigma^{-1})$ (σ denotes the left shift);

(iv) $\exists$ $f,d \in Z_+$ and polynomial matrices $R'(s,s^{-1}) \in R^{f\times g}[s,s^{-1}]$, $R''(s,s^{-1}) \in R^{f\times d}[s,s^{-1}]$ such that $\mathbf{B} = (R'(\sigma,\sigma^{-1}))^{-1} \mathrm{im}\, R''(\sigma,\sigma^{-1})$;

(v) $\exists$ $p,m \in Z_+$, $p + m = q$, a permutation matrix $\pi \in R^{q\times q}$, and polynomial matrices $P(s,s^{-1}) \in R^{p\times p}[s,s^{-1}]$, $Q(s,s^{-1}) \in R^{p\times m}[s,s^{-1}]$ with $P^{-1}(s,s^{-1})Q(s,s^{-1}) \in R_+^{p\times m}(s)$, such that $\mathbf{B} = \pi \ker[P(\sigma,\sigma^{-1}) \;\; - Q(\sigma,\sigma^{-1})]$, i.e., **B** admits a componentwise input/output representation;

(vi) $\exists$ $n \in Z_+$ such that **B** is the external behavior of a state space system $\Sigma_s = (Z,R^q,R^n,\mathbf{B}_s)$ with $\mathbf{B}_s \in \underline{\mathcal{L}}^{q+n}$ i.e., **B** has a finite dimensional linear time-invariant state realization;

(vii) $\exists$ $n,m \in Z_+$ and matrices $A \in R^{n\times m}$, $B \in R^{n\times m}$, $C \in R^{q\times n}$, $D \in R^{q\times m}$ such that $\mathbf{B} = \{\mathbf{w} | \exists\, \mathbf{x},\mathbf{v}$ such that $\sigma\mathbf{x} = Ax + Bv$, $\mathbf{w} = Cx + Dv\}$

(viii) $\exists$ $m,n, \in Z_+$, $m + p = q$, a permutation matrix $\pi \in R^{q\times q}$, and matrices $\tilde{A} \in R^{n\times n}$, $\tilde{B} \in R^{n\times m}$, $\tilde{C} \in R^{p\times n}$, $\tilde{D} \in R^{p\times m}$ such that $\pi^{-1}\mathbf{B} = \{(\mathbf{u},\mathbf{y}) | \exists \mathbf{x}$ such that $\sigma\mathbf{x} = \tilde{A}\mathbf{x} + \tilde{B}\mathbf{u}$, $\mathbf{y} = \tilde{C}\mathbf{x} + \tilde{D}\mathbf{u}\}$, i.e., **B** admits a componentwise input/state/output representation.

3. The Complexitity of Linear Systems. Above we have given many equivalent characterizations of finite dimensional linear time invariant systems. The question which we now want to consider is the following: **What is reasonable quantitative measure for the complexity of such a system?**

Let $\mathbf{C} = [0,1]^{Z_+}$ be the complexity level space, and define the complexity $\mathbf{c}$: $\underline{\mathscr{L}}^q \to \mathbf{C}$ by $\mathbf{c}_t(\mathbf{B}) := \frac{\dim \mathbf{B}_t}{q(t+1)}$, where $\mathbf{B}_t := \mathbf{B}|_{[0,t]}$. Hence $\mathbf{c}_t(\mathbf{B})$ is the relative dimension of the subspace expressing what q-dimensional time series are compatible with the behavior $\mathbf{B}$ on the interval [0,t]. We will use a total ordering on $\mathbf{C}$, which we will call the **asymptotic complexity.** Consider $\mathbf{c}(\mathbf{B})$ as defined above. Then it is easy to see that shift invariance implies $1 \geq \mathbf{c}_0(\mathbf{B}) \geq \mathbf{c}_1(\mathbf{B}) \geq \ldots \geq \mathbf{c}_t(\mathbf{B}) \geq \mathbf{c}_{t+1}(\mathbf{B}) \geq \ldots \geq 0$. Hence $c_\infty(\mathbf{B}) := \lim_{t\to\infty} \mathbf{c}_t(\mathbf{B})$ exists and $0 \leq c_\infty \leq 1$. Similarly, as follows from the next proposition, $c'_\infty(\mathbf{B}) := \lim_{t\to\infty} t(c_t(\mathbf{B}) - c_\infty(\mathbf{B}))$ exists. Note that c'_∞ can be interpreted as the derivative of $\mathbf{c}_t$ at "$t = \infty$". Let $C^{as} := [0,1]\times R_+$ be the **asymptotic complexity level space** and equip C^{as} with the lexicographic total ordering: $\{(\alpha',\beta') \geq (\alpha'',\beta'')\} :\leftrightarrow \{$either $(\alpha' > \alpha'')$ or $(\alpha' = \alpha''$ and $\beta' \geq \beta'')$. The **asymptotic complexity** on $\underline{\mathscr{L}}^q$ is defined by c_{as}: $\underline{\mathscr{L}}^q \to C^{as}$ with $\mathbf{c}_{as}(\mathbf{B}) := (c_\infty(\mathbf{B}), c'_\infty(\mathbf{B}))$. We obviously have $\{\mathbf{B}_1 \supset \mathbf{B}_2\} \to \{\mathbf{c}(\mathbf{B}_1) \geq \mathbf{c}(\mathbf{B}_2)\} \to \{\mathbf{c}_{as}(\mathbf{B}_1) \geq \mathbf{c}_{as}(\mathbf{B}_2)\}$.

Associated with any $\mathbf{B} \in \underline{\mathscr{L}}^q$ there are two important integers:

m(**B**) := the number of input variables in **B**;

and n(**B**) := the dimension of the minimal state space of **B**.

The following proposition shows that there is a very close connection between $\mathbf{c}_{as}$ and (m,n).

Proposition: Let $\mathbf{B} \in \underline{\mathscr{L}}^q$. Then

(i) for t sufficiently large, there holds:

$$\dim \mathbf{B}_t = n(\mathbf{B}) + (t+1)m(\mathbf{B})$$

(ii) $$\mathbf{c}_{as}(\mathbf{B}) = \left(\frac{m(\mathbf{B})}{q}, \frac{n(\mathbf{B})}{q}\right)$$

For proofs we refer to [1,2].

In conclusion, the asymptotic complexity is equivalent to the number of driving inputs and the number of states of a finite dimensional linear time-invariant system. If a system has more inputs than another one, then it is more complex. If it has the same number of inputs but more states, then it is also more complex.

The next proposition shows that the strong total ordering $\mathbf{c}_{as}$ is strictly related to the weak partial ordering of set inclusion, used in the definition of more powerful models and in the notion of the most powerful unfalsified model.

Proposition: 1. Let $B_1, B_2 \in \underline{\mathcal{L}}^q$, $B_1 \supseteq B_2$. Then

$$\{B_1 = B_2\} \Leftrightarrow \{\mathbf{c}(B_1) = \mathbf{c}(B_2)\} \Leftrightarrow \{\mathbf{c}_{as}(B_1) = \mathbf{c}_{as}(B_2)\}.$$

2. Let $\tilde{B} \subseteq \underline{\mathcal{L}}^q$ have the property that there exists $B^* \in \tilde{B}$ such that $B^* \subseteq B$ for all $B \in \tilde{B}$. Then

$$\{B = B^*\} \Leftrightarrow \{B \in \tilde{B} \text{ and } \mathbf{c}(B) \leq \mathbf{c}(B') \text{ for all } B \in \tilde{B}\}$$

$$\Leftrightarrow \{B \in \tilde{B} \text{ and } \mathbf{c}_{as}(B) = \min_{B' \in \tilde{B}} \mathbf{c}_{as}(B)\}$$

The above proposition can be refined for controllable systems.

In our framework the notion of controllability becomes a property of the external behavior of a system. Let $\Sigma = (\mathbb{Z}, W, B)$ be a time-invariant (i.e., $\sigma B = B$) dynamical system. Let $B^+ := B|_{[0,\infty)}$, $B^- := B|_{(-\infty,0)}$, and let ^ denote concatenation. Then Σ is said to be **controllable** if for all $w^+ \in B^+$ and $w^- \in B^-$, there exists a $T \in \mathbb{Z}_+$ and $w : [0,1,\ldots,T) \to W$ such that $w^- \wedge w \wedge w^+ \in B$. It can be shown [1,2] that for $B \in \underline{\mathcal{L}}^q$ controllability is equivalent to the existence of a controllable input/state/output representation.

Proposition: 1. Let $B_1, B_2 \in \underline{\mathcal{L}}^q$, $B_1 \supseteq B_2$ and B_1 be controllable. Then $\{c_\infty(B_1) = c_\infty(B_2)\} \Leftrightarrow \{B_1 = B_2\}$.

2. Let $\tilde{B} \subseteq \underline{\mathcal{L}}^q$ have the property that there exists a controllable $B^* \in \tilde{B}$ such that $B^* \subseteq B$ for all $B \in \tilde{B}$. Then $\{B \in \tilde{B}$, B controllable, and $c_\infty(B) = \min_{B' \in \tilde{B}} c_\infty(B')\} \Leftrightarrow \{B = B^*\}$.

We have seen that if Z is any family of time series then the unfalsified models $B \in \underline{\mathcal{L}}^q$ satisfy the conditions of $\tilde{B}$ in the first of the above two propositions. This implies that the most powerful unfalsified model is simply the unfalsified element of $\underline{\mathcal{L}}^q$ having a minimal number of inputs and, among these, the one with the minimal number of states. Consequently, if we know the number of inputs, then **"most powerful unfalsified" is equivalent to "unfalsified and minimal dimension of the state space"**. It follows from the second proposition that, if for some reason we know that the most powerful unfalsified model is controllable, then it is completely characterized by: **unfalsified, controllable, and minimal number of inputs**. It will then automatically have the minimal number of states.

4. From Time Series to State Model. We will now give an algorithm for passing from any finite set of measured vector time series $Z = \{\tilde{w}_1, \tilde{w}_2, \dots, \tilde{w}_N\}$ to this most powerful unfalsified linear time-invariant complete model Σ^*_Z. However, we can look for a specification of Σ^*_Z in many different forms.

Since they provide the most useful models in applications we prefer to look for an algorithm which passes from $\{\tilde{w}_1, \tilde{w}_2, \dots, \tilde{w}_N\}$ to a minimal i/s/o representation of Σ^*_Z. However, as we have seen, there is but a small difference between looking for a minimal i/s/o model and a minimal state/ minimal driving input model. Indeed, let $\sigma \mathbf{x} = A\mathbf{x} + B\mathbf{v}$, $\mathbf{w} = C\mathbf{x} + D\mathbf{v}$ be such a model. Then (because of minimality) ker D will be 0. Suitably choosing the components of $\mathbf{w}$ implies that this model may be written as $\sigma \mathbf{x} = A\mathbf{x} + B\mathbf{v}$, $\mathbf{w}_1 = C_1\mathbf{x} + D_1\mathbf{v}$, $\mathbf{w}_2 = C_2\mathbf{x} + D_2\mathbf{v}$, $\mathbf{w} = \mathrm{col}[\mathbf{w}_1, \mathbf{w}_2]$; with D_1 invertible. This yields as minimal i/s/o model $\sigma \mathbf{x} = (A - BD_1^{-1}C_1)\mathbf{x} + BD_1^{-1}\mathbf{u}, \mathbf{y} = (C_2 - D_1^{-1}C_1)\mathbf{x} + D_2D_1^{-1}\mathbf{u}$, $\mathbf{w} = \mathrm{col}[\mathbf{u}, \mathbf{y}]$. We will therefore concentrate on giving an algorithm for finding a minimal model in this form.

We state the problem formally:

Let $\tilde{w}_i : Z \to R^q$, $i \in N$, be a family of observed time series. The problem is to find $m, n \in Z_+$ and matrices $A \in R^{n\times n}$, $B \in R^{n\times m}$, $C \in R^{q\times n}$, and $D \in R^{q\times m}$, such that $\sigma \mathbf{x} = A\mathbf{x} + B\mathbf{v}$, $\mathbf{w} = C\mathbf{x} + D\mathbf{v}$ (DV) is a minimal state/minimal driving input representation of the most powerful unfalsified finite dimensional linear time-invariant model. Equivalently, we are looking for an unfalsified model with a minimal number of driving variables and, among these, one with a minimal number of state variables.

Our algorithm for finding (m,n) and (A,B,C,D) specifying Σ_Z^* passes via the Hankel matrix of the data. Define $\tilde{W}: Z \to R^{q \times N}$ by $\tilde{W}(t) :=$ $[\tilde{w}_1(t)\ \tilde{w}_2(t)\ \dots\ \tilde{w}_N(t)]$. Then the **Hankel matrix** $H(\tilde{W})$ is defined by

$$
\left[\begin{array}{cccccccccccc}
 & \vdots & & \vdots & \vdots & \vdots & & \vdots & \\
\dots & \tilde{W}(-2t) & \dots & \tilde{W}(-t-1) & \tilde{W}(-t) & \tilde{W}(-t+1) & \dots & \tilde{W}(0) & \dots \\
 & \vdots & & \vdots & \vdots & \vdots & & \vdots & \\
\dots & \tilde{W}(-t-1) & \dots & \tilde{W}(-2) & \tilde{W}(-1) & \tilde{W}(0) & \dots & \tilde{W}(t-1) & \dots \\
\dots & \tilde{W}(-t) & \dots & \tilde{W}(-1) & \tilde{W}(0) & \tilde{W}(1) & \dots & \tilde{W}(t) & \dots \\
\dots & \tilde{W}(-t+1) & \dots & \tilde{W}(0) & \tilde{W}(1) & \tilde{W}(2) & \dots & \tilde{W}(t+1) & \dots \\
 & \vdots & & \vdots & \vdots & \vdots & & \vdots & \\
\dots & \tilde{W}(0) & \dots & \tilde{W}(t-1) & \tilde{W}(t) & \tilde{W}(t+1) & \dots & \tilde{W}(2t) & \dots \\
 & \vdots & & \vdots & \vdots & \vdots & & \vdots &
\end{array}\right]
\begin{array}{l}
\\
\leftarrow -t\text{-th block row} \\
\\
\leftarrow -1\text{-th block row} \\
\leftarrow 0\text{-th block row} \\
\leftarrow 1\text{-th block row} \\
\\
\leftarrow t\text{-th block row} \\
\\
\end{array}
$$

Since $\tilde{W}$ will be fixed in the sequel we will denote $H(\tilde{W})$ simply as H. We will also consider the truncations H_t, H^t, and $H_{t''}^{t'}$ of H. These are defined as the submatrices of H consisting of the block rows indexed 0 to t, -t to 0, and -t' to t'', respectively.

We will now define two integers related to the linear dependence of the rows of H:

(i) per(H) = the **permanent rank** of H;

and (ii) rel(H) = the **relative rank** of H.

In order to define the permanent rank, consider the truncations H_t for $t = 0,1,2,\dots$. It is easy to see that, because of the Hankel structure, $r_t := \text{rank}\ H_{t+1} - \text{rank}\ H_t$ is a monotone nonincreasing sequence of integers, $0 \le r \le q$. Hence $\lim_{t\to\infty} r_t$ exists. We will call it the **permanent rank** of H. Note that

$$\text{per}(H) = \lim_{t\to\infty} r_t = \inf_t r_t = \lim_{t\to\infty} = \inf_t r^t$$

where $r^t := \text{rank}\ H^{t+1} - \text{rank}\ H^t$.

In order to define the relative rank of H, we will view H as a partitioned matrix

$$H = \begin{bmatrix} H^- \\ \cdots \\ H^+ \end{bmatrix}$$

with H^- consisting of the block rows of H with a negative index and H^+ consisting of the block rows of H with a nonnegative index. Now define the **relative rank** of the partitioned matrix H as

$$\mathrm{rel}(H) := \lim_{t',t''\to\infty} (\mathrm{rank}\ H^{t'} + \mathrm{rank}\ H_{t''} - \mathrm{rank}\ H^{t'}_{t''})$$

The relevance of per(H) and rel(H) follows from the following results.

Theorem: 1. $B^* \in \underline{\mathcal{L}}^q$ is the behavior of the most powerful unfalsified model if and only if B^* is unfalsified and, in addition,

$$c_\infty(B^*) = \frac{\mathrm{per}(H)}{q}$$

and $$c'_\infty(B^*) = \frac{\mathrm{rel}(H)}{q}$$

2. Let $B \in \underline{\mathcal{L}}^q$ be an unfalsified model. Then $c_\infty(B) \geq \frac{\mathrm{per}(H)}{q}$; and if $c_\infty(B) = \frac{\mathrm{per}(H)}{q}$, then $c'_\infty(B) \geq \frac{\mathrm{rel}(H)}{q}$.

We will now give an algorithm for finding the desired model (DV).

ALGORITHM:

Data: $\tilde{w}_i : \mathbb{Z} \to \mathbb{R}^q$, $i \in \mathbb{N}$.

Required: m,n and matrices (A,B,C,D) specifying Σ_Z^*, the most powerful unfalsified model in $\underline{\mathcal{L}}^q$ for the set of measurements $Z = \{\tilde{w}_1, \tilde{w}_2, \ldots, \tilde{w}_N\}$.

Step 1: Form $H(\tilde{W})$ (denoted by H) and compute rel(H).

<u>Step 2</u>: Find t', t" such that the relative rank of the partitioned matrix $H^{t'}_{t''} = \left[\frac{H^{t'}}{H_{t''}}\right]$ equals rel(H); that is, such that rank $H^{t'}$ + rank $H_{t''}$ - rank $H^{t'}_{t''}$ = rel(H). The numerical data in $H^{t'}_{t''}$ will give us enough information in order to compute the desired model.

<u>Step 3</u>: From $H^{t'}_{t''}$, derive the submatrices H'_o, H'_o, H'_2, and $\sigma H'_1$ defined as follows:

3.1 $H' = \left[\frac{H'_1}{H'_2}\right]$ is a comformably partitioned submatrix of

$$H^{t'}_{t''} = \left[\frac{H^{t'}}{H_{t''}}\right] \text{ with rank } H' = \text{rank } H^{t'}_{t''} ,$$

3.2 H'_o is the submatrix of H_o formed by the columns of H_o corresponding to those of H';

3.3 $\sigma H'_1$ is the submatrix of $H^{t'}$ formed by the same rows as H'_1 and the columns N columns to the right to those of H'_1;

Observe that

(i) dim im H'_1(mod H'_1kerH'_2) = rel(H) =: n ;

and, since $H^{t'}_o$ = col$[H^{t'}, H_o]$ has rank $H^{t'}_o$ =(t' +1) per(H) + rel(H),

that there holds

$$\text{(ii) dim im} \left[\frac{H'_1(\text{mod } H'_1 \ker H'_2)}{H'_o}\right] - \text{rel}(H) = \text{per}(H) =: m$$

H'_1(mod H'_1ker H'_2) is here viewed as a matrix with columns formed by a matrix representation of the columns of H'_2 taken modulo H'_1kerH'_2;

(iii) im $\sigma H'_1 \subseteq$ im H'_1.

Step 4: Choose a basis $(e_1,e_2,\dots,e_n)$ for the span of the columns of $H'_1(\text{mod } H'_1 \ker H'_2)$.

Choose a complementary basis $(f_1,f_2,\dots,f_m)$ such that $\left(\begin{bmatrix} e_1 \\ 0 \end{bmatrix}, \begin{bmatrix} e_2 \\ 0 \end{bmatrix}, \dots, \begin{bmatrix} e_n \\ 0 \end{bmatrix}, f_1,f_2,\dots,f_m\right)$ forms a basis for the span of the columns of

$$\left[\frac{H'_1(\text{mod } H'_1 \ker H'_2)}{H'_o}\right]$$

Step 5: Let $[\frac{X}{U}]$ be a matrix representation of the columns of

$$\left[\frac{H'_1(\text{mod } H'_1 \ker H'_2)}{H_o}\right]$$

and let σX be a matrix representation of the columns of $\sigma H'_1(\text{mod } H'_1 \ker H'_2)$.

Define M the $(n+q)\times(n+m)$ matrix such that $M[\frac{X}{U}] = [\frac{\sigma X}{H'_o}]$.

Then $M = \begin{bmatrix} A & B \\ C & D \end{bmatrix}$ defines the desired model parameters.

We refer to [1,2] for a proof of this algorithm.

5. Sufficiently Rich Inputs. We will now consider the following question: **What conditions on the input of a dynamical system does one need in order to be able to reconstruct the dynamical laws?** Specifically, assume that $B \in \underline{\mathcal{L}}^q$ and that $Z = \{\tilde{w}_1,\tilde{w}_2,\dots,\tilde{w}_N\}$ with $\tilde{w}_i \in B$ are observed responses of the system. The question thus arises: **When will B itself be the behavior of the most powerful unfalsified model in $\underline{\mathcal{L}}^q$?**

Assume that for $B \in \underline{\mathcal{L}}^q$, $W = R^q = R^m \times R^p = U\times Y$ is a partition such that $w = \text{col}(u,y)$ and u is a set of input variables, in particular $P_U B = (R^m)^Z$. Let $\tilde{w}_i = \text{col}(\tilde{u}_i,\tilde{y}_i)$. Let B^* be the most powerful unfalsified model in $\underline{\mathcal{L}}^q$ for the measurements $\{\tilde{w}_1,\tilde{w}_2,\dots,\tilde{w}_N\}$ and B^*_u be the most powerful unfalsified model in $\underline{\mathcal{L}}^m$ for the measurement $\{\tilde{u}_1,\tilde{u}_2,\dots,\tilde{u}_N\}$. Then we have

Theorem: Assume that $B \in \underline{\mathcal{L}}^q$ is controllable. Let B^* and B^*_u be as defined above. Then $\{B^* = B\} \Leftrightarrow \{\bar{B}^*_u = (R^m)^Z\}$.

Let us call an input **u**: $Z \to R^m$ **unstructured** if $B_u^* = (R^m)^Z$. Note that generic elements of $(R^m)^Z$ will be unstructured. The above theorem guarantees us that an unstructured input will be sufficiently rich so as to allow us to identify a controllable input/output system.

However, there are situation in which partial knowledge of the system allows one to deduce the system dynamics from the most powerful unfalsified model even when the inputs are structured. Let $B \in \underline{\mathcal{L}}^q$. Important invariants of B are:

1. n(B), its **McMillan degree**, defined as the minimal dimension of the state space of a state realization of B;
2. L(B), its **lag**, defined as the minimal degree of a polynomial matrix R such that $B = \ker R(\sigma)$;
3. $\ell(B)$, its **shortest structure index**, defined as the smallest t such that $B|_{[0,t]} \neq (R^q)^{(t+1)}$.

As in [1] we will call B **t-complete** if $L(B) \leq t$. It is easy to see that for any $\{\tilde{w}_1, \tilde{w}_2, \ldots, \tilde{w}_N\}$ there exists a most powerful t-complete element of $\underline{\mathcal{L}}^q$, denoted by B_t^*.

Theorem: Let B, B_t^* **and** B_u^* **be as before, and assume that B is controllable. Then** $\{\ell(B_u^*) > n(B) + L(B)\} \Rightarrow \{B_{L(B)}^* = B\}$.

The above theorem implies our earlier result since if the input is unstructured we have $\ell(B_u^*) = \infty$. In general the theorem says that (AR) relations for B_u^*, if they exist, must have a sufficiently degree so as not to interfere with the lag structure of B.

As a final item, we would like to record an application of these ideas to the notion of feedback. Let $\tilde{w} = (\tilde{u}, \tilde{y}) : Z \to R^m x R^p$ be an input/output response pair of a controllable system with $B \in \underline{\mathcal{L}}^q$. We will say that $\tilde{u}$ is **free of (linear) feedback from** B if, with $B^* \in \underline{\mathcal{L}}^q$ the most powerful unfalsified model for $\{\tilde{w}\}$ and $B_u^* \in \underline{\mathcal{L}}^m$ the most powerful unfalsified model for $\{\tilde{u}\}$, there holds $B^* = B \cap (B_u^* \oplus (R^p)^Z)$. With this notion it is possible to refine the above theorem somewhat.

Theorem: Assume that $\tilde{u}$ **in** $\tilde{w} = (\tilde{u}, \tilde{y})$ **is free of feedback from** B, **and that** B **is controllable. Then** $\{\ell(B_u^*) > L(B)\} \Rightarrow \{B_{L(B)}^* = B\}$.

6. References

[1] J.C. Willems: "From Time Series to Linear System
Part I : Finite Dimensional Linear Time Invariant systems;
Part II : Exact Modelling;
Part III : Approximate Modelling."
AUTOMATICA, Vol. 22, No. 5, pp. 561-580, 1986 (Part I)
Vol. 22, No. 6, pp. 675-694, 1986 (Part II)
Vol. 23, No. 1, 1987 (Part III).

[2] J.C. Willems: "Models for Dynamics", Submitted.

APPROXIMATIVE INVERSION OF POSITIVE MATRICES WITH APPLICATIONS TO MODELLING

P. Dewilde and Ed.F. Deprettere
Delft University of Technology
Network Theory Section
Delft, The Netherlands

Abstract

An algorithm is presented to invert a (large and full) positive matrix approximatively, based on a (generalized) band of data centered around the diagonal. The approximative inverse is also positive with its off-band elements zero. As a result, the original matrix is approximated in the maximum entropy sense with another positive matrix which coincides in the band with the original but has (implicit) off-band elements. The algorithm is an extension of the classical Schur algorithm for maximum entropy estimation, has minimal complexity and is ideally suited for pipelined implementation in a concurrent processor.

0. This research has received partial support by the Commission of the EEC under the ESPRIT 991 program and by the IOP-IC program TEL 45.009.

NATO ASI Series, Vol. F34
Modelling, Robustness and Sensitivity Reduction in Control Systems. Edited by R. F. Curtain

1. INTRODUCTION

We consider a hermitian positive definite matrix C of dimension $(n+1)^*(n+1)$:

$$C = \begin{bmatrix} C_{00} & C_{01} & \dots & C_{0n} \\ C_{10} & C_{11} & \cdots & C_{1n} \\ & & \dots & \\ C_{n0} & C_{n1} & \cdots & C_{nn} \end{bmatrix}$$

Typically n will be large, and its elements computable but not yet computed (as would be the case of a finite element method). Typically also, many C_{ij}, especially when i and j are very different, will be small, and may be "neglected". Putting them equal to zero will produce unacceptable errors (the matrix may even fail to be positive - see discussion in sect. 6). Better is to let them assume an implicit, optimal value without computing them.

Desired will be an inverse of the matrix C. It has been shown [1], see also [2], that the unknown (or uncomputed) elements in C will be optimal in the "maximum entropy" sense if their corresponding positions in C^{-1} are filled with zeros.

In this paper we present an algorithm for computing an approximate inverse C_a^{-1} of C_a, whereby a number of (properly located) elements of C_a are given, and the positions in C_a^{-1} corresponding to the unspecified elements in C_a are forced to be zero, while keeping C_a^{-1} positive definite. The algorithm will be shown to have low - even minimal - complexity for the class of problems at hand. The existence of such a solution for the case where the specified elements are on a block-diagonal was shown in [3]. Our contributions are (1) the algorithm which is recursive and allows for incremental solutions, (2) a new theory which places the problem in the context of the Schur-Levinson [4] theory, (3) a pipelined computational scheme allowing for very efficient evaluation in a parallel (concurrent) environment.

In section 2 we shall introduce a number of preliminaries which will ease the subsequent development. In section 3 we present the generalized Levinson approach to the inversion problem which was pioneered independently in [5], and [6]. In section 4 we shall deduce from it a generalized Schur algorithm which solves the approximation problem. Section 5 will be devoted to the low complexity computational scheme. In section 6 we shall consider approximation properties, and we shall show convergence of the algorithm. It will appear that the standard invariant-subspace type arguments follow through even in the more primitive environment of algebras of matrices.

2. PRELIMINARIES

With the original positive definite matrix $C = [C_{ij}]$, $i=0..n, j=0..n$, we define the diagonal matrix $D = [C_{00} + C_{11} + \cdots + C_{nn}]$ and the normalization of C

$$C_{norm} = D^{-1/2} C D^{-1/2} \tag{2.1}$$

The element of C_{norm} will be designated by lower case c's.

$$C_{norm} = \begin{bmatrix} 1 & c_{01} & c_{02} & c_{0n} \\ c_{10} & 1 & c_{12} & c_{1n} \\ & & & \\ c_{n0} & c_{n1} & & 1 \end{bmatrix} \tag{2.2}$$

with $c_{ij} = C_{ij} / \sqrt{C_{ii} C_{jj}}$

The computation of C_{norm}^{-1} is of course equivalent to the computation of C^{-1}. While this normalization is not strictly necessary, it will ease the subsequent development. Any matrix obtained as an intermediate approximation of C_{norm} will be valid of an intermediate approximation of C after denormalization with the same diagonal matrix $D^{1/2} \cdots D^{1/2}$.

We shall always suppose that the diagonal elements of C (or C_{norm}) are given. In the first four sections of this paper we shall assume that subsequent elements of C are given in the following order: parts of the first subdiagonal, the first subdiagonal, parts of the second subdiagonal, the second subdiagonal etc. It is indeed possible to follow different orders, but this issue will be postponed to section 5 in order not to cluster the main presentation with these ordering details.

The Levinson method

One key ingredient of our algorithm is a generalization of the Generalized Levinson method. Let

$$C_i^k = \begin{bmatrix} c_{ii} & c_{i\,i+1} & .. & c_{ik} \\ & & & \\ c_{ki} & & .. & c_{kk} \end{bmatrix} \tag{2.3}$$

be a principal submatrix of C_{norm} (with $c_{ii} = \cdots = c_{kk} = 1$). We shall say (for historical reasons - see [5]) that a_i^{k-i} is the "$(k-i)^{th}$ order normalized forward predictor at position i", if the following two conditions are satisfied:

$$C_i^k \, a_i^{k-i} = \begin{bmatrix} F_i^{k-i} \\ 0 \\ \\ \\ 0 \end{bmatrix} \tag{2.4}$$

$$\|a_i^{k-i}\|_{C_{norm}} = (a_i^{k-i})^* C_i^k (a_i^{k-i}) = 1 \tag{2.5}$$

These two conditions uniquely specify a_i^{k-i} because C_i^k is positive definite as a principal submatrix of the positive definite matrix C_{norm}. The normalization of a_i^{k-i} fixes a relation between F_i^{k-i} and the first element of a_i^{k-i}, $(a_i^{k-i})_i$ as follows

$$(a_i^{k-i})_i^* F_i^{k-i} = 1 \tag{2.6}$$

(Remark: because we allow extensions of all vectors and matrices with zeros (or other values), we keep indexing according to the position in a supposedly doubly infinite indexing scheme. Hence the first element of a_i^{k-i} is denoted $(a_i^{k-i})_i$ rather than $(a_i^{k-i})_0$).

Likewise, the "$(k-i)^{th}$ order normalized backward predictor" b_k^{k-i} will be defined by

$$C_i^k \, b_k^{k-i} = \begin{bmatrix} 0 \\ \\ \\ 0 \\ R_k^{k-i} \end{bmatrix} \tag{2.7}$$

with $(b_k^{k-i})^* C_i^k \, b_k^{k-i} = 1$ and hence

$$(b_k^{k-i})_k^* R_k^{k-i} = 1 \tag{2.8}$$

The central fact in the Szegö-Levinson scheme is that (a_i^{k-i}, b_k^{k-i}) can be computed recursively from $(a_i^{k-i-1}, b_k^{k-i-1})$. We have:

$$C_i^k \begin{bmatrix} a_i^{k-i-1} & 0 \\ & --- \\ --- & \\ 0 & b_k^{k-i-1} \end{bmatrix} = \begin{bmatrix} F_i^{k-i-1} & \Gamma_i^k \\ 0 & 0 \\ & \\ 0 & 0 \\ \Delta_i^k & R_k^{k-i-1} \end{bmatrix} \tag{2.9}$$

for some Δ_i^k, Γ_i^k. Using the normalization defined earlier for the a's and b's we have

$$\begin{bmatrix} a_i^{k-i-1} & 0 \\ --- & --- \\ 0 & b_k^{k-i-1} \end{bmatrix}^* C_i^k \begin{bmatrix} a_i^{k-i-1} & 0 \\ --- & --- \\ 0 & b_k^{k-i-1} \end{bmatrix} = \begin{bmatrix} 1 & \dfrac{\Gamma_i^k}{F_i^{k-i-1}} \\ \dfrac{\Delta_i^k}{R_k^{k-i-1}} & 1 \end{bmatrix} \tag{2.10}$$

This matrix is again hermitian and positive definite because C_i^k is positive definite and

$$\begin{bmatrix} a_i^{k-i-1} & 0 \\ --- & --- \\ 0 & b_k^{k-i-1} \end{bmatrix}$$

has rank not smaller than 2, since $(a_i^{k-i-1})_i$ and $(b_k^{k-i-1})_k$ are non-zero. In particular:

$$\left(\frac{\Delta_i^k}{R_k^{k-i-1}} \right) = \left(\frac{\Gamma_i^k}{F_i^{k-i-1}} \right)^* \tag{2.11}$$

Let

$$\rho_i^k = -\frac{\Gamma_i^k}{F_i^{k-i-1}} \tag{2.12}$$

be the so-called "reflection coefficient" and

$$\theta_i^k = \frac{1}{\sqrt{1-|\rho_i^k|^2}} \begin{bmatrix} 1 & (\rho_i^k) \\ (\rho_i^k)^* & 1 \end{bmatrix} \tag{2.13}$$

then postmultiplying (2.9) with θ_i^k will produce the desired effect:

$$C_i^k \left[a_i^{k-i} \mid b_k^{k-i} \right] = C_i^k \begin{bmatrix} a_i^{k-i-1} & 0 \\ --- & --- \\ 0 & b_k^{k-i-1} \end{bmatrix} \theta_i^k$$

$$= \begin{bmatrix} F_i^{k-i-1} & \Gamma_i^k \\ 0 & 0 \\ 0 & 0 \\ \Delta_i^k & R_k^{k-i-1} \end{bmatrix} \frac{1}{\sqrt{1-|\rho_i^k|^2}} \begin{bmatrix} 1 & \rho_i^k \\ (\rho_i^k)^* & 1 \end{bmatrix}$$

$$= \begin{bmatrix} F_i^{k-i-1}(\sqrt{1-|\rho_i^k|^2}) & 0 \\ 0 & 0 \\ 0 & R_k^{k-i-1}(\sqrt{1-|\rho_i^k|^2}) \end{bmatrix} = \begin{bmatrix} F_i^{k-i} & 0 \\ 0 & \\ & \\ 0 & R_k^{k-i} \end{bmatrix} \tag{2.14}$$

where the normalization on the $[a\ b]$ indeed checks out because

$$[a_i^{k-i} \mid b_k^{k-i}]^* \, C_i^k \, [a_i^{k-i} \mid b_k^{k-i}] = \begin{bmatrix} 1 & \rho_i^k \\ (\rho_i^k)^* & 1 \end{bmatrix} \tag{2.15}$$

The special form of θ_i^k which pulls the trick can nicely be motivated from geometric considerations, see [7].

Right-sided Θ-matrices

The matrix

$$\theta(\rho) = \frac{1}{\sqrt{1-|\rho|^2}} \begin{bmatrix} 1 & \rho \\ \rho^* & 1 \end{bmatrix} \tag{2.16}$$

is an elementary instance of a J-unitary matrix with

$$J = \begin{bmatrix} 1 & \\ & -1 \end{bmatrix} \tag{2.17}$$

Indeed

$$\theta^* J \theta = J \tag{2.18}$$

In the sequel we shall need a more general form based on

$$J = \begin{bmatrix} I_n & 0_n \\ 0_n & -I_n \end{bmatrix} \tag{2.19}$$

with I_n an $(n+1)*(n+1)$ unit matrix. All the elementary θ-matrices to be used in the sequel will be "embedded" in a

$$\Theta = \begin{bmatrix} \theta_{11} & \theta_{12} \\ \\ \theta_{21} & \theta_{22} \end{bmatrix}$$

in which θ_{11} and θ_{21} are lower triangular, θ_{12} and θ_{22} are upper triangular and the diagonal entires of θ_{12} and θ_{21} are zero. Such Θ-matrices will be called *right-sided.*

Theorem 1

Let Θ be a right-sided matrix. Then it has the form

$$\Theta = \begin{bmatrix} \frac{1+G^*}{2} L^{-*} & \frac{1-G}{2} M^{-1} \\ \\ \frac{1-G^*}{2} L^{-*} & \frac{1+G}{2} M^{-1} \end{bmatrix} \tag{2.20}$$

where

1. G is upper triangular, and such that $\frac{G+G^*}{2}$ is strictly positive definite.
2. L and M are upper triangular.
3. $\frac{G+G^*}{2} = LL^* = M^* M.$

PROOF We first establish that $\theta_{11} + \theta_{21}$ and $\theta_{22} + \theta_{12}$ are non-singular. For any vector $x \in C^{n+1}$ we have

$$\|(\theta_{11} + \theta_{21})x\| \geqslant |\|\theta_{11}x\| - \|\theta_{12}x\||$$

by the triangle inequality for Euclidean norms and next, because of J-unitarity

$$\|(\theta_{11} + \theta_{21})x\| \geqslant |\sqrt{\|x\|^2 + \|\theta_{12}x\|^2} - \|\theta_{12}x\||.$$

The latter expression can be zero only when $\|x\| = 0$. This shows that $(\theta_{11} + \theta_{21})$ is non-singular. A similar proof applies to $(\theta_{22} + \theta_{12})$. The rest of the assertions is now immediate in view of the hypothesis. See [5] for a similar proof in a different context.

3. RECURSIVE ORTHONORMALIZATION

A recursive orthonormalization via partial factorization of the original matrix C_{norm} can be obtained by collecting predicting vectors of increasing orders and varying positions together. This way of arranging things was first proposed in [5]. Its remarkable approximation properties will be explored later. We shall show the algorithm for one possible sequence of operations: the natural sequence. Using the previously defined notation for forward and backward predictors we begin by collecting all zero'th order predictors in one matrix:

$$[a_0^0 \; a_1^0 \; \cdots \; a_n^0 \mid b_0^0 \; b_1^0 \; \cdots \; b_n^0] \tag{3.1}$$

$$= [1_n \;\mid\; 1_n]$$

The sequence of operations is then as shown in the following scheme:

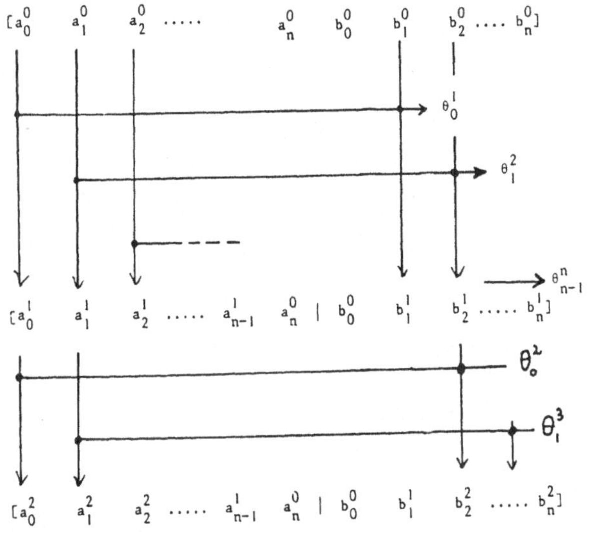

etcetera.

Expressed in matrices and writing out the second members pretty much as in (2.14) but collecting all information pertaining to a certain order together, we obtain:

First Step (initialization)

$$C[a_0^0 \; a_1^0 \; \cdots \; a_n^0 \mid b_0^0 \; b_1^0 \; \cdots \; b_n^0] =$$

$$= \begin{bmatrix} 1 & c_{01} & c_{02} & \cdots \\ c_{10} & 1 & c_{12} & \cdots \\ & & & \\ c_{n0} & c_{n1} & \cdots & \end{bmatrix} \begin{bmatrix} 1_n & \mid & 1_n \end{bmatrix} =$$

$$= \begin{bmatrix} 1 & c_{01} & c_{02} & \cdots & c_{0n} & 1 & c_{01} & c_{02} & \cdots & c_{0n} \\ c_{10} & 1 & c_{12} & \cdots & c_{1n} & c_{01} & 1 & c_{1n} & \cdots & c_{1n} \\ c_{20} & c_{21} & 1 & \cdots & c_{2n} & c_{20} & c_{21} & 1 & \cdots & c_{2n} \\ c_{n0} & & & & c_{nn} & c_{n0} & & & \cdots & c_{nn} \end{bmatrix}$$

As in (2.14), the second members will determine the reflection coefficients:

$$\rho_0^1 = -c_{01} \,;\, \rho_1^2 = -c_{12} \,;\, \cdots \,;\, \rho_{n-1}^n = -c_{n-1,n} \,.$$

The newly determined Szegö-Levinson vectors are obtained by applying appropriate elementary θ-matrices to the right side of both members of the previous equation:

$$\theta^{(1)} = \theta_0^1 \, \theta_1^2 \cdots \theta_{n-1}^n = \tag{3.2}$$

$$\left[\begin{array}{cccc|cccc}
\frac{1}{\sqrt{1-|\rho_0^1|^2}} & 0 & \cdots & 0 & 0 & \frac{\rho_0^1}{\sqrt{1-|\rho_0^1|^2}} & \cdots & 0 \\
0 & \frac{1}{\sqrt{1-|\rho_1^2|^2}} & \cdots & 0 & 0 & 0 & \cdots & 0 \\
0 & 0 & \cdots & 0 & 0 & 0 & \cdots & \frac{\rho_{n-1}}{\sqrt{1-|\rho_{n-1}^2|^2}} \\
0 & 0 & \cdots & 0 & 1 & 0 & \cdots & 0 \\
\hline
0 & 0 & \cdots & 0 & 1 & 0 & \cdots & 0 \\
\frac{(\rho_0^1)^*}{\sqrt{1-|\rho_0^1|^2}} & 0 & \cdots & 0 & 0 & \frac{1}{\sqrt{1-|\rho_0^1|^2}} & \cdots & 0 \\
0 & \frac{(\rho_{n-1}^n)^*}{\sqrt{1-|\rho_{n-1}^n|^2}} & \cdots & 0 & 0 & 0 & \cdots & \frac{1}{\sqrt{1-|\rho_{n-1}^n|^2}}
\end{array}\right]$$

which is a matrix of the right-sided type. Applying it leads to the

Second Step: $C[a_0^1 \; a_1^1 \, .. \, a_n^0 \mid b_0^0 \; b_1^1 \cdots b_n^1] =$

$$\begin{bmatrix}
1 & c_{01} & c_{02} & \cdots & c_{0n} \\
c_{10} & 1 & c_{12} & \cdots & c_{1n} \\
c_{20} & c_{21} & 1 & \cdots & c_{2n} \\
. & . & . & \cdots & . \\
c_{n0} & c_{n1} & c_{n2} & \cdots & c_{nn}
\end{bmatrix}
\left[\begin{array}{ccccc|ccccc}
a_{00}^1 & 0 & \cdots & 0 & 0 & 1 & b_{01}^1 & 0 & \cdots & 0 \\
a_{01}^1 & a_{11}^1 & \cdots & 0 & 0 & 0 & b_{11}^1 & b_{12}^1 & \cdots & 0 \\
0 & a_{21}^1 & \cdots & 0 & 0 & 0 & 0 & b_{22}^1 & \cdots & 0 \\
. & . & \cdots & . & . & \cdots & . & . & . & b_{n-1,n}^1 \\
0 & 0 & \cdots & a_{n,n-1}^1 & a_{n,n}^1 & 0 & 0 & 0 & \cdots & b_{n,n}^1
\end{array}\right]$$

$$\left[\begin{array}{ccccc|ccccc}
\gamma_{00}^1 & \gamma_{01}^1 & \gamma_{02}^1 & \cdots & \gamma_{0n}^1 & \delta_{00}^1 & 0 & \delta_{02}^1 & \cdots & \delta_{0n}^1 \\
0 & \gamma_{11}^1 & \gamma_{12}^1 & \cdots & \gamma_{1n}^1 & \delta_{10}^1 & \delta_{11}^1 & 0 & \cdots & \delta_{1n}^1 \\
\gamma_{20}^1 & 0 & \gamma_{22}^1 & \cdots & \gamma_{2n}^1 & \delta_{20}^1 & \delta_{21}^1 & \delta_{22}^1 & \cdots & \delta_{2n}^1 \\
. & . & . & \cdots & . & . & . & . & \cdots & . \\
\gamma_{n0}^1 & \gamma_{n1}^1 & \gamma_{n2}^1 & \cdots & \gamma_{nn}^1 & \delta_{n0}^1 & . & . & \cdots & \delta_{nn}^1
\end{array}\right]$$

The orthonormalization of the second order vectors will lead to:

$$\rho_0^2 = \frac{\delta_{02}^1}{\gamma_{00}^1}, \rho_1^3 = \frac{\delta_{13}^1}{\gamma_{11}^1}, \cdots, \rho_{n-2}^n = \frac{\delta_{n-2,n}^1}{\gamma_{n-2,n-2}^1}$$

and the partial θ-matrix:

$$\theta^{(2)} = \theta_0^2 \theta_1^3 \,..\, \theta_{n-2}^n = \tag{3.3}$$

$$\left[\begin{array}{ccccc|ccccc}
\frac{1}{\sqrt{1-|\rho_0^2|^2}} & 0 & \cdots & 0 & 0 & 0 & 0 & \frac{\rho_0^2}{\sqrt{1-|\rho_0^2|^2}} & \cdots & 0 \\
0 & \frac{1}{\sqrt{1-|\rho_1^3|^2}} & \cdots & 0 & 0 & 0 & 0 & 0 & \cdots & 0 \\
\cdot & \cdot & \cdots & \cdot & \cdot & \cdot & \cdot & \cdot & \cdots & \cdot \\
0 & 0 & \cdots & 1 & 0 & 0 & 0 & 0 & \cdots & 0 \\
0 & 0 & \cdots & 0 & 1 & 0 & 0 & 0 & \cdots & 0 \\
\hline
0 & 0 & \cdots & 0 & 0 & 1 & 0 & 0 & & 0 \\
0 & 0 & \cdots & 0 & 0 & 0 & 1 & 0 & \cdots & 0 \\
\frac{(\rho_0^2)^*}{\sqrt{1-|\rho_0^2|^2}} & 0 & \cdots & 0 & 0 & 0 & 0 & \frac{1}{\sqrt{1-|\rho_0^2|^2}} & \cdots & \cdot \\
\cdot & \cdot & \cdots & \cdot & \cdot & \cdot & \cdot & \cdot & \cdots & \cdot \\
0 & 0 & \cdots & 0 & 0 & 0 & 0 & 0 & \cdots & 0 \\
0 & 0 & \cdots & 0 & 0 & 0 & 0 & 0 & \cdots & \frac{1}{\sqrt{1-|\rho_{n-2}^n|^2}}
\end{array}\right]$$

which again is of the right-sided type as well as the overall θ-matrix obtained so far: $\theta^{(1)}\,\theta^{(2)} =$

$$\left[\begin{array}{cccccc|cccccc}
x & 0 & 0 & \cdots & 0 & 0 & 0 & x & x & \cdots & 0 & 0 \\
x & x & 0 & \cdots & 0 & 0 & 0 & 0 & x & \cdots & 0 & 0 \\
0 & x & x & \cdots & 0 & 0 & 0 & 0 & 0 & \cdots & 0 & 0 \\
\cdot & \cdot & \cdot & \cdots & \cdot & \cdot & \cdot & \cdot & \cdot & \cdots & \cdot & \cdot \\
0 & 0 & 0 & \cdots & x & 0 & 0 & 0 & 0 & \cdots & 0 & 0 \\
\hline
0 & 0 & 0 & \cdots & 0 & 0 & 1 & x & 0 & \cdots & 0 & 0 \\
x & 0 & 0 & \cdots & 0 & 0 & 0 & x & x & \cdots & 0 & 0 \\
\cdot & \cdot & \cdot & \cdots & \cdot & \cdot & \cdot & \cdot & \cdot & \cdots & \cdot & \cdot \\
0 & 0 & 0 & \cdots & 0 & 0 & 0 & 0 & 0 & \cdots & x & x \\
0 & 0 & 0 & \cdots & x & 0 & 0 & 0 & 0 & \cdots & 0 & x
\end{array}\right]$$

where the "x" indicate fill-ins. The resulting second member will now have the form:

$$\left[\begin{array}{ccccc|cccccc} \gamma_{00}^2 & \gamma_{01}^2 & \gamma_{02}^2 & \cdots & \gamma_{0n}^2 & \delta_{00}^2 & 0 & 0 & \delta_{03}^2 & \cdots & \delta_{0n}^2 \\ 0 & \gamma_{11}^2 & \gamma_{12}^2 & \cdots & \gamma_{1n}^2 & \delta_{10}^2 & \delta_{11}^2 & 0 & 0 & \cdots & \delta_{1n}^2 \\ 0 & 0 & \gamma_{22}^2 & \cdots & \gamma_{sn}^2 & \delta_{20}^2 & \delta_{21}^2 & \delta_{22}^2 & 0 & \cdots & \delta_{2n}^2 \\ \gamma_{30}^2 & 0 & 0 & \cdots & \gamma_{3n}^2 & . & . & . & . & \cdots & . \\ \gamma_{n0}^2 & \gamma_{n1}^2 & \gamma_{n2}^2 & \cdots & \gamma_{nn}^2 & \delta_{n0}^2 & \delta_{n1}^2 & . & . & \cdots & \delta_{nn}^2 \end{array}\right]$$

The procedure contimes recursively with a shift and an increase in order at very step. Also, the resulting θ-matrices, as well as the global θ-matrix obtained at each step will be J-unitary, right-sided.

At this point the following remarks are in order

1. All the reflection coefficients computed in the course of the procedure are strictly bounded in magnitude by one because of the Szegö-Levinson property discussed above.

2. The interaction of the columns of the subsequent Γ and Δ matrices is such that the upper-triangular part of Γ is only influenced by the strictly upper-triangular (i.e. upper-triangular minus diagonal) part of Δ and vice-versa. The same holds for the strictly lower triangular part of Γ with the lower triangular part of Δ. The latter two matrices carry only redundant information. This all important observation leads to a generalized Schur interpretation of the algorithm.

4. THE GENERALIZED SCHUR ALGORITHM

Starting out from the second remark at the end of the previous section, we rewrite the algorithm as follows:

Initial phase

$$[\Gamma_0 \mid \Delta_0] \underset{=}{\Delta} \left|\begin{array}{ccccccccccc} 1 & c_{01} & c_{02} & \cdots & c_{0n} & 0 & c_{01} & c_{02} & \cdots & c_{0n} \\ 0 & 1 & c_{12} & \cdots & c_{1n} & 0 & 0 & c_{12} & \cdots & c_{1n} \\ 0 & 0 & 1 & \cdots & c_{2n} & 0 & 0 & 0 & \cdots & c_{2n} \\ \\ 0 & 0 & 0 & \cdots & 1 & 0 & 0 & 0 & \cdots & 0 \end{array}\right| \tag{4.1}$$

First Pass

In the first pass one uses elementary θ-matrices to eliminate c_{01} in column $n+2$ against 1 in column 0 producing ρ_0^1 etcetera as in the first pass of section 3. This will produce new Γ,Δ matrices of the type:

$$[\Gamma_1 \quad \Delta_1] = \left|\begin{array}{cccccccccc} \gamma_{00}^1 & \gamma_{01}^1 & \cdots & \gamma_{0n}^1 & 0 & 0 & \delta_{02}^1 & \cdots & \delta_{0n}^1 \\ 0 & \gamma_{11}^1 & \cdots & \gamma_{1n}^1 & 0 & 0 & 0 & \cdots & \delta_{1n}^1 \\ 0 & 0 & \cdots & \gamma_{2n}^1 & 0 & 0 & 0 & \cdots & \delta_{2n}^1 \\ . & . & \cdots & . & . & . & . & \cdots & . \\ 0 & 0 & \cdots & \gamma_{nn}^1 & 0 & 0 & 0 & . & 0 \end{array}\right| \tag{4.2}$$

Recursion

Proceeding further and after a number of steps in which $\delta_{ij}^{(0)}$ is eliminated (corresponding to the determination of ρ_i^j) we will have obtained:

$$[\Gamma_k \quad \Delta_k] = \tag{4.3}$$

$$\left|\begin{array}{cccc|cccccccc}
\gamma_{00}^k & \gamma_{01}^k & \cdots & \gamma_{0n}^k & 0 & 0 & \cdots & 0 & \delta_{..}^k & \cdots & \delta_{0n}^k \\
0 & \gamma_{11}^k & \cdots & \gamma_{1n}^k & 0 & 0 & \cdots & 0 & 0 & \cdots & \delta_{1n}^k \\
0 & 0 & \cdots & \gamma_{2n}^k & 0 & 0 & \cdots & 0 & 0 & \cdots & \delta_{2n}^k \\
\cdot & \cdot & \cdots & \cdot & \cdot & \cdot & \cdots & \cdot & \cdot & \cdots & \cdot \\
\cdot & \cdot & \cdots & \cdot\cdot & 0 & 0 & \cdots & \cdot & \cdot & \cdots & 0 \\
\cdot & \cdot & \cdots & \cdot & \cdot & \cdot & \cdots & \cdot & \cdot & \cdots & \cdot \\
0 & 0 & \cdots & \gamma_n^k & 0 & 0 & \cdots & \cdot & \cdot & \cdots & 0
\end{array}\right|$$

with some $\delta_{i+l,j+l}$ zero, and where a large chunk of the matrix Δ_k has become zero (in a specific order). We have

$$[\Gamma_0 \quad \Delta_0]\theta_k = [\Gamma_k \quad \Delta_k] \tag{4.4}$$

where θ_k is J-unitary and right-sided (θ_k has a structure complementing the structure of Δ_k - see the next section).

We are now ready for our main results. Suppose that we have stopped the procedures at level k. Since θ_k is right-sided, J-unitary, it has the form:

$$\theta_k = \begin{bmatrix} \frac{1+G_k^*}{2} L_k^{-*} & \frac{1-G_k}{2} M_k^{-1} \\ \frac{1-G_k^*}{2} L_k^{-*} & \frac{1+G_k}{2} M_k^{-1} \end{bmatrix} \tag{4.5}$$

as in theorem 1. Let, on the other hand

$$G = \begin{bmatrix} 1 & 2c_{01} & \cdots & 2c_{0n} \\ 0 & 1 & \cdots & 2c_{1n} \\ & & & \\ 0 & 0 & \cdots & 1 \end{bmatrix} \tag{4.6}$$

so that

$$[\Gamma_0 \quad \Delta_0] = \left|\frac{G+1}{2} \quad \frac{G-1}{2}\right|. \tag{4.7}$$

Finally, let $\mathbf{S}_k$ be the set of pairs (i,j) with $j \geqslant i$ and such that δ_{ij}^l has been eliminated (for some l).

Theorem 2

For all $(i,j) \in \mathbf{S}_k$ we have that $(G-G_k)_{ij} = 0$.

Proof

From (4.4) and (4.7) we have that

$$\left|\frac{G+1}{2} \quad \frac{G-1}{2}\right| = [\Gamma_k \quad \Delta_k]\theta_k^{-1} \tag{4.8}$$

Inverting (4.5) and using the fact that θ_k is J-unitary:

$$\theta_k^{-1} = J\theta_k^* J = \begin{bmatrix} L_k^{-1}\dfrac{G_k+1}{2} & L_k^{-1}\dfrac{G_k-1}{2} \\ M_k^{-*}\dfrac{G_k^*-1}{2} & M_k^{-*}\dfrac{G_k^*+1}{2} \end{bmatrix} \tag{4.9}$$

and hence

$$\begin{bmatrix} \dfrac{G_k+1}{2} & \dfrac{G_k-1}{2} \end{bmatrix} = [L_k \quad 0]\theta_k^{-1} \tag{4.10}$$

It follows by subtracting (4.10) from (4.8) that

$$\left(\frac{G-G_k}{2}\right)[1 \quad 1]\theta_k = [\Gamma_k - L_k \quad \Delta_k]$$

and because of (2.20)

$$\left(\frac{G-G_k}{2}\right)[L_k^{-*} \quad M_k^{-1}] = [\Gamma_k - L_k \quad \Delta_k] \tag{4.11}$$

from which we have

$$G-G_k = 2(\Gamma_k - L_k)L_k^* \tag{4.12}$$

and finally

$$G-G_k = 2\Delta_k M_k \tag{4.13a}$$

(From (4.12) one also deduces

$$G+G_k^* = 2\Gamma_k L_k^*) \tag{4.13b}$$

M_k is upper triangular as well as Δ_k. In the product $\Delta_k M_k$. Therefore, the zero's that have been produced in Δ_k by elimination will remain in the product $\Delta_k M_k$ (see the special form of Δ_k in (4.3)).

□

Define $C_{norm}^k = \dfrac{G_k+G_k^*}{2} = L_k L_k^* = M_k^* M_k$. We see that $C_{norm} - C_{norm}^k$ vanishes for all $(i,j) \in \mathbf{S}_k$. On the other hand:

$$(C_{norm}^k)^{-1} = L_k^{-*}L_k^{-1} = A_k^* A_k$$

where A_k is the kth order forward prediction matrix obtained in the course of the orthonormalization procedure. It follows that $\mathbf{S}_k$ is precisely the support of $(C_{norm}^k)^{-1}$ i.e. the collection of indices (i,j) on which $(C_{norm}^k)^{-1}$ can have a non-zero entry. Hence C_{norm}^k is a solution of the problem:

find a matrix which:

1. *coincides with* C_{norm} *on* $\mathbf{S}_k$*;*
2. *is positive definite;*
3. *has an inverse which is zero for all index pairs* (i,j) *for which neither* (i,j) *nor* (j,i) *belongs to* $\mathbf{S}_k$*.*

The same three properties also hold for the denormalized matrices:

$$C = D^{1/2} C_{norm} D^{1/2}$$

$$C^{(k)} = D^{1/2} C_{norm}^{k} D^{1/2}$$

$$L^{(k)} = D^{1/2} L^{k}$$

since none of the properties are affected by pre- and post multiplication.

5. ORDER, DATA FLOW AND COMPLEXITY

The recursive Schur-Levinson procedure given above can conveniently be described using a graphical representation. From it follows a data flow scheme representing the algorithm. It will be described in this chapter.

We start by assigning nodes to columns (or equivalently rows) of the original matrix C in a graph representation - for $(n+1)$ elements $(n+1)$ nodes. The elements are strictly ordered, (an original ordering of rows/columns is necessary). In the course of the proceedings columns will be combined to produce higher order estimates. Each such will be represented by a node in the graph which will be characterized by a sequence of elements that have led to the given node. The construction of nodes will be subjected to three rules which will tightly control the operation, but also leave a number of degrees of freedom which can be exploited. Each node will furthermore, while the construction proceeds, be characterized by two "tokens" T or B (for "top" and "bottom") which will move down the graph (which has the form of a tree) with the highest order forward and backward predictor. The evolution of the graph for the case treated in section 3 is shown in fig. 5.1.

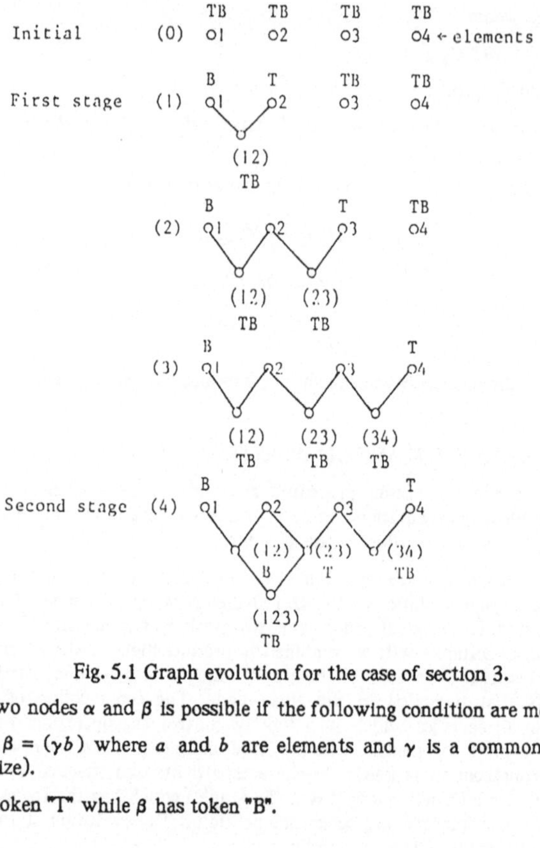

Fig. 5.1 Graph evolution for the case of section 3.

The joining of two nodes α and β is possible if the following condition are met:

1. $\alpha = (a\gamma)$, $\beta = (\gamma b)$ where a and b are elements and γ is a common piece (of arbitrary size).
2. α has the token "T" while β has token "B".
3. $a < b$.

The result of a join is a new node $(a\gamma b)$ with characteristic TB whereby α (respect. β) looses its characteristic T (respect. B).

At each intermediate stage one may stop the procedure and consider the then-obtained graph as an end result. The resulting approximation will have as forward matrix, a collection of forward vectors corresponding to T-nodes, while the backward matrix will consist of a collection of backward vectors corresponding to B-nodes. E.g. for graph (4) in fig. 4.1, the forward matrix will be

$$[a_2^2 \ a_2^1 \ a_3^1 \ a_4^0]$$

and the backward matrix:

$$[b_1^0 \ b_2^1 \ b_3^2 \ b_4^1].$$

The filling of these two matrices is as follows:

$$\begin{bmatrix} x & & & \\ x & x & & \\ x & x & x & \\ & & x & x \end{bmatrix} \quad \begin{bmatrix} x & x & x & \\ & x & x & \\ & & x & x \\ & & & x \end{bmatrix}$$

From the preceding it will be clear that the procedure can only lead to approximations of the block-band type (see next section).

To each join there also corresponds an orthonormalizing Θ matrix which takes as inputs the respective second members or Schur-vectors, properly restricted. For example, if $(\alpha_1\ \alpha_2 \cdots \alpha_{k-1})$ and $(\alpha_2 \cdots \alpha_{k-1}\, \alpha_k)$ are joined, we must at first process $\gamma_{\alpha_1\alpha_1}^{(k-2)}, \delta_{\alpha_1\alpha_k}^{(k-2)}$ to produce the reflection coefficient. Next one will have to process data that will be used by the subsequent processors in the chain. For example, $\gamma_{\alpha_1\alpha_1}^{(k-2)}$ will have been produced in the join leading to $(\alpha_1\alpha_2 \cdots \alpha_{k-1})$ while $\delta_{\alpha_1\alpha_k}^{(k-2)}$ comes from the combination of $\gamma_{\alpha_1\alpha_2}^{(k-3)}$ and $\delta_{\alpha_1\alpha_k}^{(k-3)}$ which in turn are produced by lower order combinations. The Θ-matrix that goes with each join is pictured by the symbol in fig. 5.2.

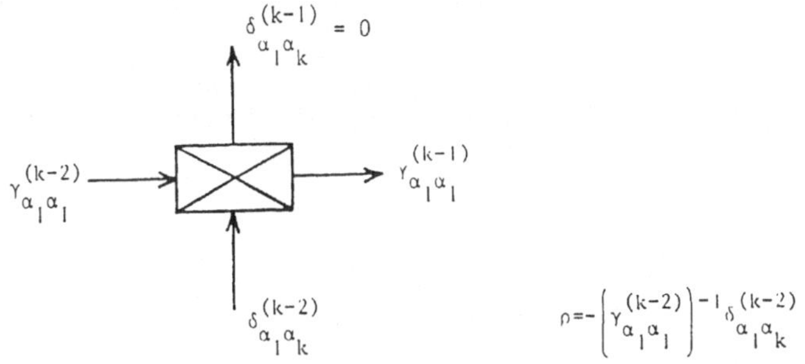

Fig. 5.2 Representation of the Θ matrix, with defining quantities for join $(\alpha_1 \ldots\ldots \alpha_k)$.

We let the γ's flow eastwards and the δ's northward. The defining relations for join $(\alpha_1 \cdots \alpha_k)$ are shown in fig. 5.2. Remarks that $\delta_{\alpha_1\alpha_k}^{(k-1)} = 0$, showing that the elimination has been properly preformed. Once the join has been defined, subsequent elements are computed as shown in fig. 5.3.

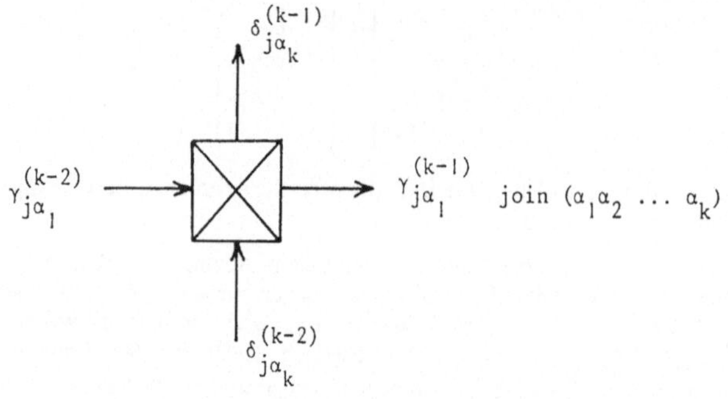

Fig. 5.3 Subsequent computations in join $(\alpha_1 \cdots \alpha_k)$.

Connecting joins requires some care. For one thing every time a join is added to the tree, new data has to be processed. For another, there is a shift occurring between the data streams at the joins, which has to be taken care of. If $(\alpha_1 \cdots \alpha_k)$ is a terminal join, it will need only the data shown in fig. 5.2. The two preceding joins will have to process new data, as shown in fig. 5.4, which shows the data needed for the new join that has been added.

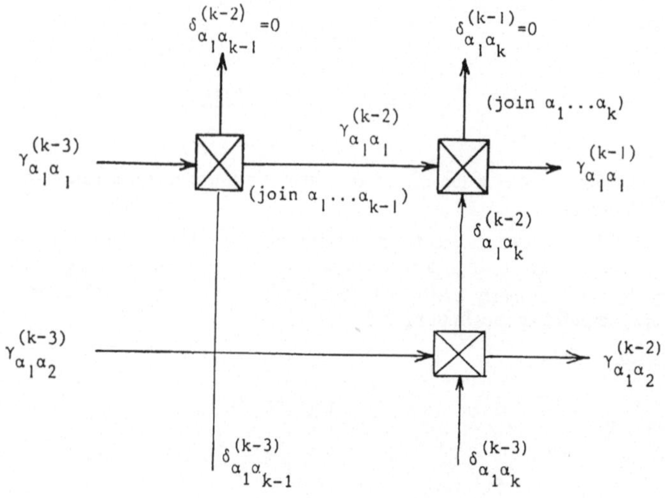

Fig. 5.4 Producing the data needed for terminal join $(\alpha_1 \cdots \alpha_k)$.

In fig. 5.4 the $(\alpha_1 \cdots \alpha_{k-1})$ join is also terminal, while $(\alpha_2 \cdots \alpha_k)$ executes a new operation. Because of that, there is a shift between the two operations. A signal flow graph representation of the three operations with correct timing is shown in fig. 5.5.

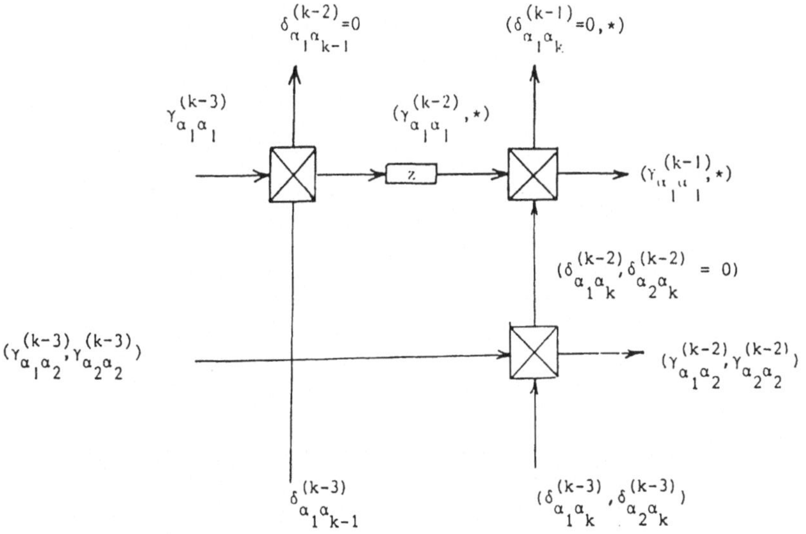

Fig. 5.5 Signal Flow Graph representation of the operations induced by join ($\alpha_1 \cdots \alpha_k$) on its predecessors. Signals are represented by vectors in reverse order (last goes in first) and '*' indicates 'don't care'.

It should be clear from the preceding that both the number of operations and of data points to be processed increases with the complexity of the graph and that the processing scheme can directly be derived from the representing graph in a one-to-one fashion. For example, the scheme resulting from the last graph of fig. 5.1 is shown in fig. 5.6.

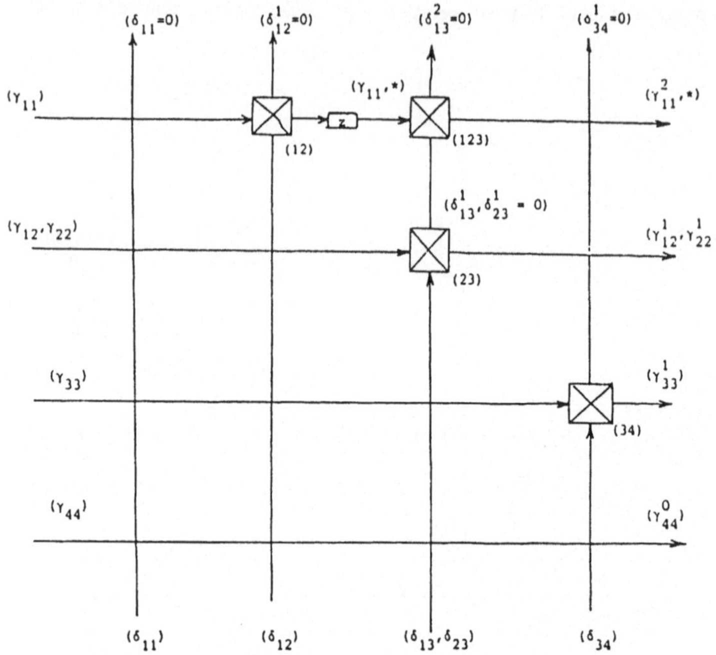

Fig. 5.6 Computing the components of the Θ-matrix for the case of fig. 5.1, second stage (dummy lines have been added for connections which are there in principle but need no computations.

Modelling

The computations described above will have produced the components of a Θ matrix with the following structure (recall section 3):

$$\Theta = \theta_{(12)} \cdot \theta_{(23)} \cdot \theta_{(24)} \cdot \theta_{(123)} \cdots$$

where each of the θ's correspond to a join. In the construction of the Θ-matrix the delays "z", which were necessary for ordering the sequence of computations, must now be dropped. Also, all θ's corresponding to the same level of joining commute and can be assembled in one matrix:

$$\Theta = \theta_{level\,1} \cdot \theta_{level\,2} \cdots$$

the higher levels becoming more sparse as the approximation progresses. For the Θ obtained in this way, we shall use the representation of fig. 5.7

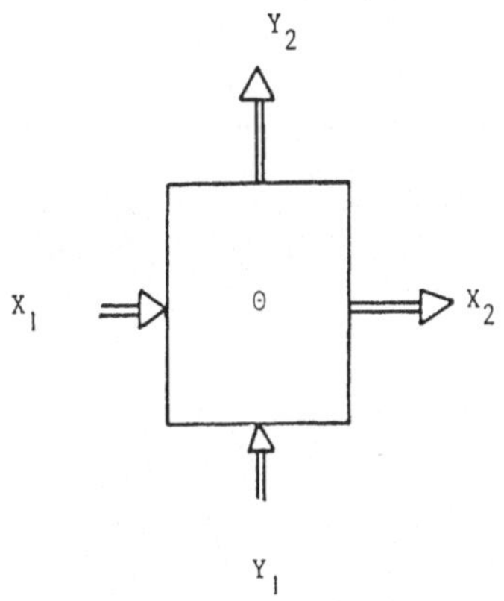

Fig. 5.7 Block representation of the resulting Θ matrix.

Fig. 5.8 shows the result for the case of fig. 5.6.

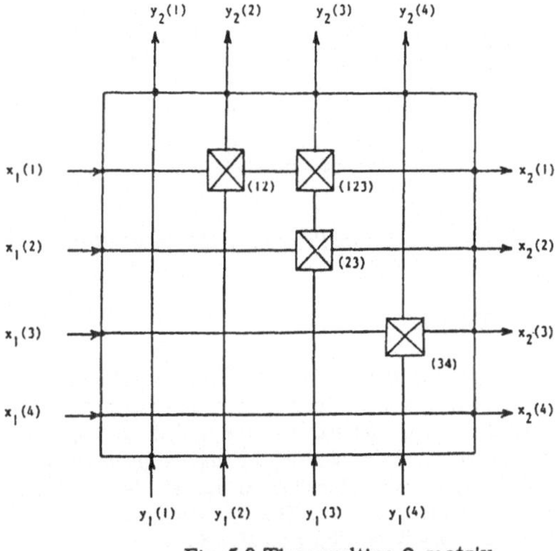

Fig. 5.8 The resulting Θ matrix.

It shall be clear now that the reasoning of section 3 generalizes and that, whatever the scheme used, Θ will be a "right-sided CSM" as described in section 2. It will thus have the form:

$$\Theta = \begin{bmatrix} \frac{G^*+1}{2}(L^{-1})^* & \frac{G-1}{2}M^{-1} \\ \frac{G^*-1}{2}(L^{-1})^* & \frac{G+1}{2}M^{-1} \end{bmatrix}$$

with L and M upper triangular and

$$\frac{G+G^*}{2} = LL^* = M^*M$$

Finally, $(L^{-1})^*$ and M^{-1} can be obtained directly from Θ by inputing the block matrix $[I \quad I]$. Usually a (row-)vector x is given (see section 6) and one will have to compute xL^{-*} and/or xM^{-1} for some vector x. This is achieved in fig. 5.9. One should remark that again, the operational complexity is proportional to the number of elements actually computed.

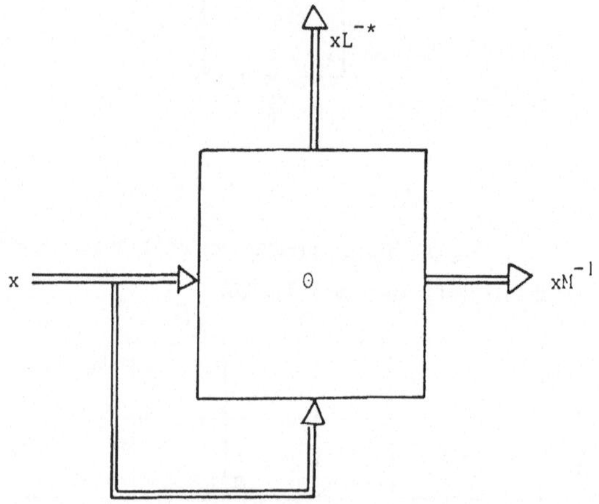

Fig. 5.9 Producing xL^{-*} and xM^{-1}.

6. APPROXIMATION THEORY

The approximation theory for the case at hand follows closely the theory for the classical Schur algorithm as discussed in [8]. Let us suppose that the algorithm has proceeded up to a stage characterized by the index set $\mathbf{S}_k$, the corresponding computation graph $\mathbf{G}_k$, and the corresponding θ_k matrix given by (4.5). The index set froms a staircase in the uppertriangular matrix. Let $\mathbf{H}_k$ be the linear space of upper triangular matrices with support on $\mathbf{S}_k$. The complementary subspace $\mathbf{M}_k$ of upper triangular matrices with support vanishing outside $\mathbf{S}_k$ is invariant with respect to pre- and postmultiplication with upper triangular matrices. The following (trivial) proposition is of key importance:

Proposition 6.1

Let $\mathbf{M}_k$ be a subspace of matrices whose support is upper staircase (i.e. if (i,j) belongs to the support, then so do $(i-k,j+l)$ for any feasible $k \geq 0, l \geq 0$), then $\mathbf{M}_k$ is invariant for pre- or postmultiplication with upper triangular matrices.

Proof: Let's treat the case of premultiplication, say with L, the other being symmetric. Any row, say i, in L contains the recipe to construct the resulting row:

take row i multiply with l_{ii}, next row $(i+1)$, multiply with $l_{i(i+1)}$ and add to the previous, etc.. The resulting row will have as support the union of the supports of row i, row$(i+1)$,... which is equal to the support of row i. Since the property holds for all such rows, the property is proven.

□

Proposition 6.1 admits an (equally trivial) converse which, for the sake of completeness we give in appendix.

We shall endow our matrix subspaces with an appropriate inner product. Let F_1 and F_2 be two square matrices of dimension $(n+1)*(n+1)$. Define

$$< F_1 . F_2 >_C = F_1 C F_2^* \tag{6.1}$$

(where we take C as C_{norm} for brevity) and

$$(F_1 . F_2)_C = \frac{1}{(n+1)} \text{trace } F_1 C F_2^* \tag{6.2}$$

It is easy to see that $(F_1 . F_1)^{½}$ is an orthodox norm on the space of matrices (it is actually the Frobenius [9] or Hilbert-Schmidt norm of $F_1 C^{½}$. Define moreover the projection operator P_0 as the operator that selects the diagonal enties:

$$(P_0 A)_{ij} = 0 \text{ if } i \neq j \text{ and } (P_0 A)_{ii} = A_{ii} . \tag{6.3}$$

Let $\mathbf{G}_k$, $\mathbf{M}_k$ and $\mathbf{L}_k$ be as in (4.5) and $C_k = (G_k + G_k)^* /2 = L_k L_k^* = M_k^* M_k$. We have the following sequence of properties:

Proposition 6.2

For all F_1 upper triangular and $F_2 \in \mathbf{H}_k$

$$P_0 < F_1 . F_2 >_C = P_0 < F_1 . F_2 >_{C_k} \tag{6.4}$$

and in particular

$$(F_1 . F_2)_C = (F_1 . F_2)_{C_k} \tag{6.5}$$

$$< F_1 . F_2 >_C - < F_1 . F_2 >_{C_k} = F_1(G - G_k)F_2^* + F_2(G^* - G_k^*)F_1^* \tag{6.6}$$

Because of (4.13a) we have:

$$F_1(G - G_k)F_2^* = 2F_1 \Delta_k M_k F_2^*$$

Because of proposition 6.1, $F_1 \Delta_k \mathbf{M}_k \in M_k$. Since F_2 is upper-triangular and has support on $\mathbf{S}_k$, the inner product of rows of $F_1 \Delta_k M_k$ with columns of like index in F_2 will vanish. The diagonal entries in the second term will vanish likewise, proving (6.4). (6.5) is now immediate after taking traces.

□

Corollary 6.2

$$\|L_k^{-1}\|_C = \left[\frac{1}{(n+1)} \text{trace } L_k^{-1} C \, L_k^{-*} \right]^{½} = 1 \tag{6.7}$$

Proof: Because of proposition 6.1 we have

$$\|L_k^{-1}\|_C^2 = \|L_k^{-1}\|_{C_k}^2 = \frac{1}{(n+1)} \text{trace } L_k^{-1} C_k L_k^{-*} = 1 \tag{6.8}$$

□

The matrix $(P_0L_k)^{-1}L_k^{-1}$ has a reproducing property which we formulate as follows.

Proposition 6.3

For any $F \in \mathbf{H}_k$, the following holds:

$$P_0 < F \,.\, (P_0L_k)^{-1}L_k^{-1} >_C = P_0F \tag{6.9}$$

Proof: L_k^{-1} is also in $\mathbf{H}_k$ and hence

$< F \,.\, (P_0L_k)^{-1}L_k^{-1} >_C = FC_kL_k^{-*}(P_0L_k)^{-1} = FL_k(P_0L_k)^{-1}$

Since both F and $L_k(P_0L_k)^{-1}$ are upper-triangular and the diagonal entries of the latter are 1, we have:

$P_0 < F \,.\, (P_0L_k)^{-1}L_k^{-1} >_C = (P_0F) \,.\, (P_0L_k)(P_0L_k)^{-1} = P_0F.$

□

The way in which $(P_0L_k)^{-1}L_k^{-1}$ approximates $(P_0L)^{-1}L^{-1}$ can now be evaluated:

Proposition 6.4

$$((P_0L)^{-1}L^{-1} - (P_0L_k)^{-1}L_k^{-1} \,.\, (P_0L)^{-1}L^{-1} - (P_0L_k)^{-1}L_k^{-1})_C \tag{6.10}$$
$$= (P_0L^{-1} \,.\, P_0L^{-1})_C - (P_0L_k^{-1} \,.\, P_0L_k^{-1})_C$$

Proof: The proof proceeds by direct evaluation:

$$<(P_0L)^{-1}L^{-1} \,.\, (P_0L)^{-1}L^{-1} >_C = (P_0L)^{-1}L^{-1}CL^{-*}(P_0L)^{-1}$$
$$= (P_0L)^{-2}$$

$$P_0 < (P_0L_k)^{-1}L_k^{-1} \,.\, (P_0L)^{-1}L^{-1} >_C = (P_0L_k)^{-2}$$

by the reproduction property of $(P_0L)^{-1}L^{-1}$,
and $<(P_0L_k)^{-1}L_k^{-1} \,.\, (P_0L_k)^{-1}L_k^{-2}>_C = (P_0L_k)^{-2}$, by proposition 6.2.

□

Let $\mathbf{T}_k$ be the projection operator onto $\mathbf{H}_k$ (w.r. to the $(...)_C$ inner product).

Proposition 6.5

$$(P_0L_k)^{-1}L_k^{-1} = T_k[(P_0L)^{-1}L^{-1}] \tag{6.11}$$

Proof: We show that the two members of the equation produce the same inner product with any member F of H_k:

$$(F \,.\, \mathbf{T}_k(P_0L)^{-1}L^{-1})_C = (F \,.\, (P_0L^{-1})L^{-1}) =$$
$$\mathit{trace}\ P_0 < F \,.\, (P_0L^{-1})L^{-1} > = \mathit{trace}\ P_0F = \mathit{trace}\ F$$

while the same holds for the first member because of proposition 6.3.

□

This leads us to the result:

Theorem 4

Of all the matrices in $\mathbf{H}_k$, $(P_0L_k)^{-1}L_k^{-1}$ is the closest to $(P_0L)^{-1}L^{-1}$ in $\|.\|_C$, with increasing k, the approximation error $\|(P_0L)^{-1}L^{-1} - (P_0L_k)^{-1}\|_c$ goes monotonely to zero. The approximation error is given by:

$$trace\,[(P_0L)^{-2} - (P_0L_k)^{-2}] \qquad (6.12)$$

Proof: The first property is equivalent to proposition (6.11). The second follows directly from the fact that the subspaces $\mathbf{H}_k$ form an increasing sequence, while the third property follows from proposition 6.4.

□

Because L and L_k are uppertriangular matrices, (6.12) is easy to evaluate in terms of the reflection coefficients obtained, the increasing *trace* $(P_0L_k)^{-2}$ giving a good indication of the accuracy.

7. DISCUSSION

In[10, 11, 12] techniques have been presented for modelling interconnect structures in VLSI circuits. They use finite elements on conductors and compute - via a Green's function - the elasticity matrix (i.e. the inverse of the capacitance matrix) G which consists of the mutual influences of the finite elements on each other. It is a matrix characteristic for the geometry. Where elements are far from each other, their influence will be small, and one will wish to neglect them. We shall see shortly that putting them equal to zero will produce incorrect results. Suppose that there are m conductors and $(n+1)$ finite elements, with $n \ll m$. The finite elements are located on conductors. Define the $m*(n+1)$ conductor-finite-element incidence matrix F, with $F_{ij} = 1$ if conductor i contains finite element j. It is easy to see that the capacitance matrix will be given by (see [10] for more detail):

$$Cap = FG^{-1}F^*$$

If we approximate G as in the previous sections we obtain a

$$G_a = L_a L_a^*$$

and

$$Cap = (F\ L_a^{-*})(L_a^{-1}\ F^*).$$

With the m rows of F represented as $F = [f_1\, f_2 \cdots f_m]^T$ it will suffice to compute $f_i L_a^{-*}$ for all i with the scheme of Fig. 5.9 - hence m passes through the Θ matrix. It follows that the overall complexity will be proportional to (1) the number of elements in the Θ-matrix squared and (2) the number of capacitances desired, i.e. a massive reduction w.r. to the original problem (and probably the smallest obtainable in general). If the method of the previous section is used, one will neglect entries in the original G by putting the corresponding entries in the inverse zero. Suppose that this happens with element i,j. Then the procedure induces implicitly a value for G_{ij} which is such that there is no *direct* capacitance between the finite elements i and j. However, there is an influence of i and j and vice versa via intermediate capacitors. If G_{ij} were part equal to zero then all capacitance between i and j would be forced to zero (short by other nodes) hence implicitly producing unphysical negative capacitors! There is a more mathematical way to view these modelling problems. Suppose that G is actually Toeplitz (corresponding to an array of finite elements), with

$$g_0 = g_{00} = g_{11} = \cdots = g_{nn} \cdot g_1 = g_{01} = g_{12} = \cdots, etc.$$

then there is a spectrum associated with G given by:

$$W(\theta) = g_0 + 2g_1\cos\theta + 2g_2\cos 2\theta + \cdots$$

The minimum and maximum of the spectrum correspond to the smallest and largest eigenvalues of G (when large enough, see e.g.[13]). Very often (i.e. in almost all cases), the minimum will be close to zero. Cuting off the tail of the sequence will then almost surely force the spectrum to become negative for some values of θ, destroying the positivity of G and resulting in an incorrect G^{-1}. We dare say that our procedure gives the only acceptable approximation and is able to compute it with minimal complexity.

In fig. 7.1 we show a small network theoretical example illustrating the type of approximations obtained. The 0th order approximation, with only node-to-ground capacitors is shown in (b). The impedance seen at each node is equal for the two networks. The 1st order approximant whereby the first off-diagonal is eliminated is shown in (c). Now, 2*2 impedance matrices between adjacent nodes are equal when (a) and (c) are compared. Finally, (d) shows the 2nd order approximant, with the third order giving the original circuit back.

Many offshoots and improvements on the technique presented can be envisaged. First of all, the connection with the classical Schur estimation theory[4, 14, 8, 15] should be clear. The approximation results presented in section 6 show a strong correspondence with the classical Beurling-Lax theory[16] and reproducing kernel theory as well. This was also remarked in a somewhat different context in [17]. Another connection is with inverse scattering theory, especially of the "layer-peeling-type"[18, 19]. The connection between the present theory and the theory of alpha-stationary processes of Lev-Ari[20] has also not yet been elucidated. Last but not least, the practiculities have also to be tested out in a realistic physical context, leading, perhaps, to a more general theory of modelling.

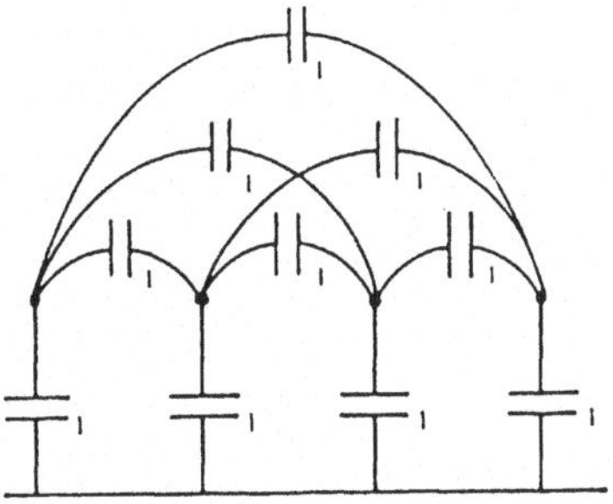

a. Original Circuit

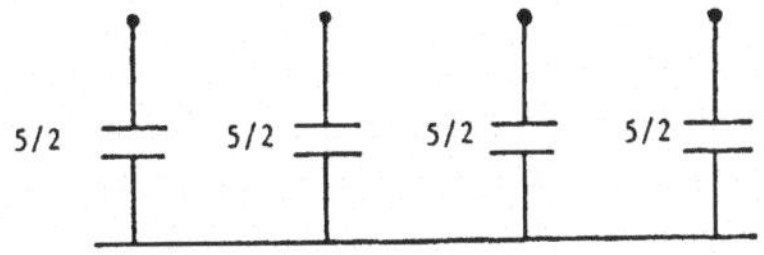

b. Zeroth order circuit

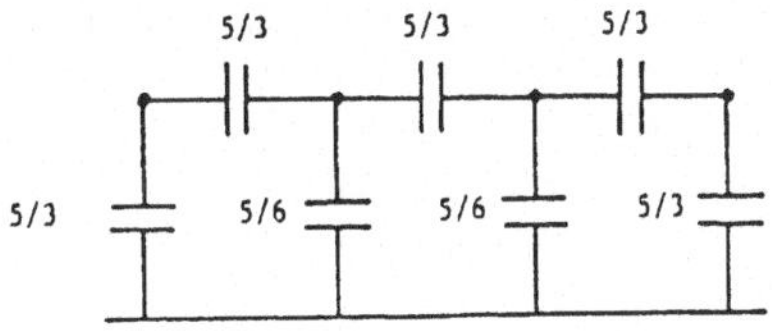

c. First order circuit

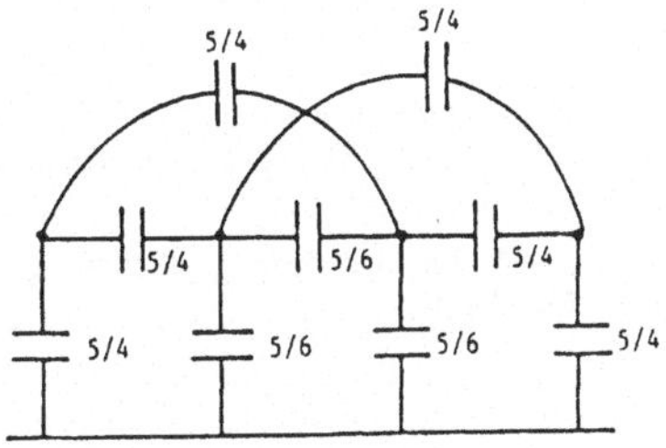

d. Second order circuit

Subsequent approximations of circuit (a), whereby off-diagonals of increasing offsets are used.

References

1. H. Lev-Ari, "Multidimensional Maximum-Entropy Covariance Extension," *Proceedings ICASSP*, pp. 21.7.1-21.7.4 (1985).

2. Y. Kamp, "Some Results on Constrained Maximum Likelihood Estimation," *Proceedings ICASSP*, pp. 27.16.1-3 (1986).

3. H. Dym and I. Gohberg, "Extensions of Band Matrices with Band Inverses," *Linear Algebra and its Applications*, pp. 1-24 Elsevier, (1981).

4. P. Dewilde, A. Vieira, and T. Kailath, "On a Generalized Szego-Levinson Realization Algorithm for Optimal Linear Predictors based on a Network Theoretic Approach," *IEEE Trans. CAS* CAS-25(9) pp. 663-675 (Sept. 1978).

5. E. Deprettere, "Mixed Form Time-Variant Lattice Recursions," pp. 545-562 in *Outils et Modèles Mathèmatiques pour l'Automatique, l'Analyse de systèmes et le traitement du signal*, CNRS, Paris (1981).

6. H. Lev-Ari and T. Kailath, "Schur and Levinson Algorithms for Nonstationary Processes," *Proceedings ICASSP*, (1981).

7. , "Deuxième Partie: Algorithmes Rapides pour les Systèmes Dynamiques Linèaires," in *Outils et Modèles Mathèmatiques pout l'Automatique, l'Analyse de Systèmes et le Traitement du Signal.*, ed. P. Dewilde, CNRS, Paris (1981).

8. P. Dewilde and H. Dym, "Lossless Chain Scattering Matrices and Optimum Linear Prediction: The Vector Case," *Intl. J. Circuit Theory and Appln.* 9 pp. 135-175 (1981).

9. J.M. Wilkinson and C. Reinsch, *Handbook for Automatic Computation II: Linear Algebra*, Springer Verlag, Berlin & New York (1971).

10. Z.Q. Ning, P. Dewilde, and F.L. Neerhof, "Calculation of Capacitance Coefficients for Multilevel Metallization Lines," *Technical Reports, Network Theory Section, Dept. of EE., Delft Univ. (to appear, IEEE Trans. on Solid State and Electron Devices, 1987)*, (1986).

11. A.E. Ruehli and P.A. Brennan, "Efficient capacitance calculations for three-dimensional multiconductor systems," *IEEE Trans. on Microwave Th. and Techn.* 21(Feb. 1973).

12. P.D. Patel, "Calculation of capacitance coefficients for a system of irregular finite conductors on a dielectric sheet," *IEEE Trans. on Microwave Theory and Techn.* 19 pp. 862-869 (Nov. 1971).

13. U. Grenander and G. Szegö, *Toeplitz forms and their applications*, University of California Press, Berkeley (1958).

14. P. Dewilde and H. Dym, "Schur Recursions, Error Formulas and Convergence of Rational Estimators for Stationary Stochastic Sequences," *IEEE Trans. Info. Theory* IT-27(4) pp. 446-461 (Jul 1981).

15. P. Dewilde and H. Dym, "Lossless Inverse Scattering, Estimation Theory and Digital Filters," *IEEE Trans. Info. Theory* IT-30(4) pp. 644-662 (July 1984).

16. D. Alpay and H. Dym, "Hilbert Spaces of Analytic Functions, Inverse Scattering and Operator Models. I," *Integral Equations and Operator Theory* 7 pp. 589-641 (1984).

17. J. Ball and J.W. Helton, "A Beurling-Lax Theorem for the Lie Group U(m,n) which contains most classical Interpolation Theory," *J. Operator Theory* 9 pp.

107-142 (1983).

18. P. Dewilde, J.T. Fokkema, and I. Widya, "Inverse Scattering and Linear Prediction, The Time Continuous Case," *Stochastic Systems: The Mathematics of Filtering and Identification and Applications*, pp. 351-382 D. Riedel Publ. House, (1981).

19. I. Widya, "Continuous Time Stochastic Modeling with Lossless Structures," Ph.D. Thesis, Delft Univ. of Technology (1982).

20. H. Lev-Ari, "Lattice Modelling of Nonstationary Processes," *Ph.D. Thesis, Stanford Univ.*, (1983).

APPENDIX: Invariant Subspaces of upper triangular matrices.

Subspaces of upper-triangular matrices which are invariant for both right and left multiplicattion with upper-triangular matrices can easily be characterized as follows. Let $\mathbf{M}$ be such a subspace. Consider the union of the supports of the first row of elements of $\mathbf{M}$. Let i_0 be the leftmost index and let e_{0i_0} be the matrix with only one non zero element, namely a 1 at the position $(0,i_0)$. e_{0i_0} belongs to $\mathbf{M}$ because it can be produced from any matrix with a non-zero entry x at position $(0,i_0)$ by permultiplication with the diagonal matrix

$$\begin{bmatrix} x^{-1} & & \\ & 0 & \\ & & 0 \end{bmatrix}$$

and postmultiplication with an upper triangular matrix which eliminates the other entries in the 0th row. If the same is done on the second row, an index i_1 is obtained, and again e_{1i_1} will belong to $\mathbf{M}$. Moreover, $i_1 \geqslant i_0$ because in the contrary case, a non-zero entry at a lower position than i_0 could have been found in the 0th row. This procedure continues recursively. Define now the chain-matrix:

$$S = e_{01_0} + e_{1i_1} + \cdots + e_{ni_n}$$

and let $\mathbf{R}$ be linear space of upper-triangular matrices. We have:

Proposition A1

Every subspace of upper-triangular matrices which is invariant under left and right multiplication with uppertriangular matrices is given by $\mathbf{R}S\mathbf{R}$.

The periodic prediction problem for cyclostationary processes - an introduction

Sergio Bittanti
Dipartimento di Elettronica
Politecnico di Milano
Piazza Leonardo da Vinci 32
20133 Milano
Italy

Abstract

With reference to discrete-time systems with periodic coefficients, two problems are considered in this paper. First, the existence of a cyclostationary process solving (in some suitable sense) the given state-space description. Second, the one step ahead prediction problem is dealt with. The two problems require the analysis of the periodic Lyapunov equation and periodic Riccati equation, for which suitable inertia theorems are used.

1. Introduction

The class of cyclostationary processes is defined as the class of wide sense stationary processes with expected value periodic of period T and autocorrelation function $\gamma(\cdot,\cdot)$ such that

$$\gamma(t+T, \tau+T) = \gamma(t,\tau) \qquad \forall t,\tau. \tag{1}$$

Here, only discrete-time processes are considered, so that $t \in Z$, $\tau \in Z$ and $T \in Z^+$, T being the process period.
The interest for these processes originates in different problems of signal processing, multirate digital filtering, time series analysis and stochastic control, see e.g. (Gardner and Franks, 1975), (Meyer and Burrus, 1975 and 1976), (Jones, 1967), (Bittanti and Guardabassi, 1986).

NATO ASI Series, Vol. F34
Modelling, Robustness and Sensitivity Reduction in Control Systems. Edited by R. F. Curtain

The main dynamic representations which can be used for the description of cyclostationary processes are: (i) linear and periodic state-space systems; (ii) difference equations where the process is seen as the output of an ARMA type model with periodic coefficients (PARMA models); (iii) impulse response characterizations.

Each of these representations enables one to describe a subclass of cyclostationary processes. Actually, the relationship between two such subclasses is far to be fully understood yet. For instance, many subtle problems can be encountered along the route of qualifying the class of linear systems which can be represented as PARMA models, (Bittanti, Bolzern and Guardabassi, 1985). In this paper, the attempt is made to pose on solid grounds a theory of optimal prediction for cyclostationary processes by means of state-space representations. For, consider the model

$$x(t+1) = A(t)\ x(t) + B(t)\ v(t) \tag{2.a}$$

$$y(t) = C(t)\ x(t) + w(t)\ , \tag{2.b}$$

where $A:Z\to R^{n\times n}$, $B:Z\to R^{n\times m}$ and $C:Z\to R^{p\times n}$ are periodic of period T:

$$A(t+T) = A(t) \quad , \quad B(t+T)=B(t) \quad , \quad C(t+T)=C(t) \quad , \quad \forall t \tag{2.c}$$

and v and w are white noises with zero expected value and identity covariance matrix. The generalization to the case when v and w have T-periodic expected value and T-periodic covariance matrix is easy. Moreover, v, w and the initial condition are independent of each other.

The objective of the analysis is twofold: First to investigate under which conditions there exists a (possibly unique) cyclostationary process of period T satisfying (2). Second, in the general framework of Kalman prediction theory, to derive the optimal periodic predictor for system (2), and point out a few major properties.

The paper is organized as follows. Some basic notions relative to periodic systems are introduced in Sect. 2. The existence of a cyclostationary process satisfying (2) is the subject of Sect.3, while the theory of optimal periodic prediction is worked out in Sect. 4. The discussion of Sections 3 and 4 evolve around the periodic Lyapunov and periodic Riccati equation, respectively. It is largely based on the findings achieved in the last decade on these equations, see the quoted references.

2. Discrete-time periodic systems - A few basic properties

2.1 Stability

The transition matrix of system (2) is given by

$$\psi(t,\tau) = A(t-1)\ A(t-2)\ \dots\ A(\tau)$$

The matrix $\phi_\tau := \psi(\tau+T,\tau)$ is called monodromy matrix at τ. Its eigenvalues, which turn out to be independent of τ, see e.g. (Bittanti, 1986), are named characteristic multipliers. The system is asymptotically stable if and only if the characteristic multipliers belong to the open unit disk.

2.2 Structural properties and canonical decomposition

Referring the interested reader to (Bittanti, 1986) for details, we will only mention here a few facts. Obviously, all the structural subspaces of (2) are periodically time-varying. However, there is a an important difference between the controllability and reconstructibility subspaces and the reachability and observability ones. The former ones have constant dimensions,whereas the latter ones may have time-varying dimensions. This is why a canonical decomposition of discrete-time periodic systems of general validity cannot be performed by means of the reachability or/and observability concepts. One can wonder whether the canonical decomposition can actually be performed in terms of controllability and reconstructibility. By a nontrivial analysis, it can be shown that the reply is affermative. In other words, there is only one decomposition of general validity for periodic systems, the one based on controllability and reconstructibility.

This conclusion enables one to define the notions of stabilizablity and detectability as follows.

Definitions

1. System (2) is stabilizable if the uncontrollable part of the system is asymptotically stable.
2. System (2) is detectable if the unreconstructible part of the system is asymptotically stable.

The structural properties can be given a spectral characterization similar to the well known one relative to the time-invariant case. For, consider the reachability Gramian

$$W_r(\tau,t) = \sum_{\tau+1}^{t} j\ \phi(t,j)B(j-1)B(j-i)'\ \ \phi(t,j)'\ , \qquad t \geq \tau.$$

Note that the forward transition matrix appears in this Gramian i.e. $\phi(t,j)$, $t \geq j$. In general, the controllability Gramian cannot be defined since it calls for the backward transition matrix, i.e. $\phi(j,t)$, $t \geq j$.

We can then state the

Stabilizability criterion

System (2) is stabilizable if and only if , at a time point t, for each nonzero eigenvalue λ of ϕ_t such that $|\lambda| \geq 1$

$$\phi_t' x = \lambda x \quad \text{and} \quad W_r(t,t+T)x = 0 \quad \text{imply} \quad x = 0.$$

An analogous characterization can also be given for reachability and controllability. By duality arguments, the corresponding criterions for detectability can be worked out.

3. State-space description of cyclostationary processes

The first and second order statistics of stochastic process $x(\cdot)$,

$$z(t) = E[x(t)]$$

$$\gamma(t,\tau) = E[(x(t)-z(t))(x(\tau)-z(\tau)')],$$

are easily obtained from (2):

$$z(t+1) = A(t)\, z(t) \tag{3}$$

$$\gamma(t,\tau) = \psi(t,\tau)\, \Gamma(\tau) \tag{4}$$

$$\Gamma(t+1) = A(t)\Gamma(t)\, A(t)' + B(t)\, B(t)', \tag{5}$$

where

$$\Gamma(t) = \gamma(t,t)$$

is the process covariance matrix.

For a T-periodic system, $\psi(t+T, \tau+T) = \psi(t,\tau)$. Hence, $\gamma(\cdot,\cdot)$ satisfies (1) whenever $\Gamma(\cdot)$ is periodic. This implies that the problem of analyzing the existence and the characteristics of the cyclostationary process solving (1) reduces to the problem of studying the existence and the characteristics of the periodic solutions of (3) and (5).

From (3) it is apparent that if no characteristic multiplier is equal to 1, (3) admits the trivial solution only.

Turn now to the difference periodic Lyapunov equation (DPLE) (4).

Since any covariance matrix is symmetric and positive semidefinite, the problem is to establish under which conditions (5) admits a symmetric T-periodic and positive semidefinite (SPPS) solution. Integrating eq.(5) over a period and imposing the periodicity constraint $\Gamma(T) = \Gamma(0)$, the following difference algebraic Lyapunov equation (DALE) :

$$Y = \phi_o T \phi_o' + W_r (0,T) \tag{6}$$

is obtained, where $Y := \Gamma(0)$. Then, three questions arise: (i) under which conditions, (6) admits a symmetric and positive semidefinite solution $\bar{Y}$? Such a solution acts as a periodic generator for (5), i.e. the solution $\bar{\Gamma}(\cdot)$ of (5) such that $\bar{\Gamma}(0) = \bar{Y}$, is T-periodic. Also, $\bar{\Gamma}(\cdot)$ is obviously symmetric at any time point. (ii) $\bar{\Gamma}(0) = \bar{Y}$ is positive semidefinite. Is $\bar{\Gamma}(t)$ positive semidefinite for any t? (iii) Does (5) admit a unique SPPS solution? Question (i) is a standard question for an algebraic Lyapunov equation, for which a well known reply is: If the system (2) is asymptotically stable, the DALE (6) admits a unique symmetric and positive semidefinite solution.

As for questions (ii) and (iii), the reply can be obtained by resorting to a suitable inertia theorem. In general, an inertia theorem for a DPLE enables one to link the number of eigenvalues of a solution lying in a half plane to the number of characteristic multipliers lying inside or outside the unit disk. A number of inertia theorems have been derived for the DPLE, see (Bittanti, Bolzern and Colaneri, 1986). Amongst them, the following Extended Inertia Theorem (EIT) holds true, for the derivation of which a major role is played by the spectral characterization of stabilizability.

Theorem 1 (DPLE EIT)

Suppose that $(A(\cdot), B(\cdot))$ is stabilizable and that there exists a T-periodic symmetric solution $\bar{\Gamma}(\cdot)$ of (5). Then, no characteristic multiplier lies on the unit circle and the number of non-negative eigenvalues of $\bar{\Gamma}(t)$ coincides, for any t, with the number of characteristic multipliers belonging to the open unit disk.

Therefore, if the system is stabilizable and there exists a T-periodic symmetric solution of the DPLE such that $\bar{\Gamma}(0) \geq 0$, then $\bar{\Gamma}(t) \geq 0$, $\forall t$.
In turn, this implies that the number of SPPS solutions of the DPLE (5) coincides with the number of positive semidefinite solution of the associate DALE (6).
Since the system stability obviously implies its stabilizability, the mentioned results can be summarized as follows.

Theorem 2

Suppose that system (2) is asymptotically stable. Then, there exists a unique cyclostationary process satisfying (2).

Another corollary of the DPLE EIT is that if system (2) is stabilizable but not asymptotically stable, then the DPLE does not admit any SPPS solution. Indeed, if the system is not asymptotically stable, either some characteristic multiplier lies on the unit circle or outside the unit circle. In the first case, as stated by the DPLE EIT, (5) does not have any T-periodic symmetric solution. In the second case, assume that - say - q characteristic multipliers are outside the unit circle.Then,in

view of the DPLE EIT, any T-periodic symmetric solution would have q eigenvalues in the left half plane, so that the solution would not be positive semidefinite.
In conclusion, if the system is stabilizable, a necessary and sufficient condition for the existence of a unique cyclostationary solution is that the system is asymptotically stable.

4. Optimal Prediction

As is well known, see e.g. (Anderson and Moore, 1971), the optimal one step ahead predictor for system (2) has an error covariance matrix $P(t) \in R^{n\times n}$ satisfying the difference periodic Riccati equation (DPRE):

$$P(t+1) = A(t)P(t)A(t)' + B(t)B(t)' -A(t)P(t)C(t)' [I+(C(t)P(t)C(t)']^{-1} C(t)P(t)A(t)' \tag{7}$$

Denoting by $\overline{P}(\cdot)$ a SPPS solution of (7), and letting

$$\overline{K}(t) = A(t)\overline{P}(t)C(t)' [I+C(t)\overline{P}(t)C(t)']^{-1} , \tag{8}$$

an optimal periodic predictor is given by

$$x(t+1|t)= A(t)\; x(t|t-1) + \overline{K}(t)e(t)$$
$$e(t) = y(t) - C(t)\; x(t|t-1).$$

The stability of this predictor is determined by the characteristic multipliers of

$$\hat{A}(t) = A(t) - \overline{K}(t)C(t).$$

These multipliers will be referred to as closed loop characteristic multipliers by short. The monodromy matrix at time τ relative to $\hat{A}(\cdot)$ is denoted by $\hat{\phi}_t$.

We will say that the periodic prediction problem has a stabilizing solution if the DPRE admits a SPPS solution and the corresponding closed-loop characteristic multipliers belong to the open unit disk.

For discrete-time systems, the periodic filtering problem was considered in the pioneer paper (Nishimura, 1971), where an extension of the Potter algorithm (Potter, 1966) to the periodic case was suggested. The theoretical analysis of the periodic solutions of the DPRE is partially reported in (Bittanti, Colaneri and De Nicolao, Dec. 1986), where the following result is derived by means of an iterative linearization technique.

Theorem 3

The periodic prediction problem admits a unique stabilizing solution if and only if system (2) is stabilizable and detectable.

A number of interesting properties of the T-periodic solutions of the DPRE can be pointed out by means of the inertia theory recently developed for the DPRE. Here, the following Extended Inertia Theorem will be mentioned.

Theorem 4 (DPRE EIT)

Suppose that system (2) is stabilizable and that the DPRE admits a T-periodic symmetric solution $\overline{P}(\cdot)$. Then, no characteristic multiplier of the associated matrix $\hat{A}(\cdot)$ lies on the unit circle and the number of nonnegative eigenvalues of $\overline{P}(t)$ coincides, for any t, with the number of closed loop characteristic multipliers belonging to the open unit disk.

From this theorem it follows that, if the system is stabilizable and a T-periodic solution of the DPRE is positive semidefinite at a given time point, it is positive semidefinite at any time point.

Hence, a SPPS solution can be computed by direct integration over one period starting from a symmetric and positive semidefinite periodic generator. In turn, it can be shown (Bittanti, Colaneri,De Nicolao,Aug. 1986), that the problem of finding a symmetric and positive semidefinite generator reduces to that of finding a symmetric and positive semidefinite solution of an *algebraic* Riccati equation. This observation supplies the basis for a computational scheme of the SPPS solution of (5). Alternatively such a solution can be computed by means of an iterative linearization technique, calling for a sequence of periodic Lyapunov equations. For more details on the computation of the SPPS solution, see (Bittanti, Colaneri and De Nicolao, Aug. 1986).

Acknowledgement

This paper is supported by Centro di Teoria dei Sistemi (C.N.R.) and MPI.

References

Anderson BDO, Moore JB (1971) Linear Optimal Control. Prentice Hall.

Bittanti S (1986) Deterministic and stochastic periodic system. in: Time Series and Linear Systems (Bittanti S ed.) Springer -Verlag, Berlin.

Bittanti S, Bolzern P, Guardabassi G (July 1985) Some critical issues on the state-representation of time-varying ARMA models. IFAC Symp. on Identification and System Par. Estimation, York: 1479-1483.

Bittanti S, Bolzern P, Colaneri P (Aug. 1986) Periodic Lyapunov and Riccati equations: some recent results on the inertia of periodic solutions. SIAM Conf. on Linear Algebra in Systems, Signals and Control, Boston.

Bittanti S, Colaneri P and De Nicolao G (Aug. 1986): Two techniques for the solution of the discrete-time periodic Riccati equation. Conf. on Linear Algebra in Signals, Systems and Control, Boston.

Bittanti S, Colaneri P and De Nicolao G (Dec. 1986): Analysis of the discrete-time periodic Riccati equation by a Kleinman procedure. 25th Conf. on Dec. and Control, Athens.

Bittanti S, Guardabassi G (Dec. 1986) Optimal periodic control and periodic systems analysis - an overview. 25th Conf. on Dec. and Control, Athens.

Gardner WA, Franks DE (1975) Characterization of cyclo-stationary random processes. IEEE Trans. Information Theory 2: 1-24.

Jones RH (1967) Time series with periodic structure. Biometrika 54: 403-407.

Meyer RA, Burrus CS (1975) A unified analysis of multirate and periodically time-varying digital filters. IEEE Trans. Circuits and Systems 22, 3:162-167.

Meyer RA, Burrus CS (1976) Design and implementation of multirate digital filters. IEEE Trans. Acoustics, Speech and Signal Processing 1:53-58.

Nishimura T (1971) Spectral factorization in periodically time-varying systems and application to navigation problems. J. Spacecraft, 9, 7:540-546.

Potter JF (1966) Matrix quadratic solutions. J. SIAM Appl. Math., 14: 496-501.

SYSTEM IDENTIFICATION: DESIGN VARIABLES AND THE DESIGN OBJECTIVE

L. Ljung
Division of Automatic Control
Department of Electrical Engineering
Linköping University
S-581 83 Linköping, Sweden

Abstract

Models and model quality are prime concerns for most design issues in control and system analysis. In this contribution we discuss how to build mathematical models that given certain constraints, are of optimum quality for a prespecified application. We then take into account the influence of both bias errors and random errors on the model. It turns out that for a fairly broad class of identification methods in the prediction error family, the optimal choices of design variables can be given in an explicit form.

1 INTRODUCTION

Building mathematical models of dynamical system involves many possibilities and choices of design variables. The particular route taken may have a substantial influence on the quality of the resulting model, and it is of course desirable to make the choices so that a model of "optimal quality" is achieved, given the constraints. A complication here is that there will typically be no "uniformly good" designs, so the model quality concept must be tied to the intended application.

A completely general treatment of this problem is no doubt difficult. In this contribution we shall formulate and solve a subproblem, where the list of possibilities have been con-

NATO ASI Series, Vol. F34
Modelling, Robustness and Sensitivity Reduction in Control Systems. Edited by R. F. Curtain

strained as follows:

- Only linear models will be considered.

- The true system will be assumed to be linear (but may be much more complex than the models considered)

- The model construction will be by system identification in a class of prediction error methods (to be precisely defined in Section 4)

- The intended model application will be defined in terms of a quadratic criterion of fit in the frequency domain.

The analysis is based on general asymptotic results given in Ljung (1985ab) and Wahlberg and Ljung (1986). Related discussions are given in Yuan and Ljung (1985), Gevers and Ljung (1986) and Ljung (1986). For a comprehensive treatment, see also Ljung (1987). For general discussions on Systems Identification, see also Goodwin and Payne (1977), Eykhoff (1974, 1981) and Åström and Eykhoff (1971).

2 PROBLEM SETUP

In this contribution we shall assume that there is a true linear system S, that generates the observed data. If $y(t)$ and $u(t)$ denote the output and the input, respectively at time instant t we thus assume that

$$y(t) = G_0(q)u(t) + v_0(t) \tag{1}$$

Here $G_0(q)$ is the transfer operator

$$G_0(q)u(t)=[\sum_{k=1}^{\infty} g_0(k)q^{-k}]u(t)=\sum_{k=1}^{\infty} g_0(k)u(t-k) \tag{2}$$

in the shift operator q

$[qu(t) = u(t+1);\ q^{-1}u(t) = u(t-1)]$.

We thus describe the system in discrete time and, for simplicity, the sampling interval is taken to be the time unit. In (1), $v_0(t)$ is an additive disturbance, which is supposed to be a stationary stochastic process with spectrum

$$\Phi_v(\omega) = \lambda_0 |H_0(e^{i\omega})|^2 \tag{3}$$

This means that $\{v_0(t)\}$ can be regarded as generated by

$$v_0(t) = H_0(q)e_0(t) \tag{4}$$

where $\{e_0(t)\}$ is white noise with variance λ_0.

For the system (1) we may generate an input $\{u(t)\}$, such that

$$\lim_{N\to\infty} \frac{1}{N} \sum_{t=1}^{N} u(t)u(t-\tau) = R_u(\tau)$$

exist for all τ, and the spectrum is

$$\Phi_u(\omega) = \sum_{\tau=-\infty}^{\infty} R_u(\tau)r^{-i\tau\omega}. \tag{5}$$

We allow the possibility of output feedback, in which case the cross spectrum between $\{u(t)\}$ and $\{e(t)\}$, $\Phi_{ue}(\omega)$, is non-zero. Thus collecting the data set

$$Z^N = \{u(1),y(1),\ldots,u(N),y(N)\} \tag{6}$$

we may proceed to estimate the transfer functions G_0 and H_0 in (1), (4). Let the result be denoted by

$$\hat{G}_N(q) \quad (=\hat{G}(q,Z^N))$$
$$\hat{H}_N(q) \quad (=\hat{H}(q,Z^N)) \tag{7}$$

We shall discuss procedures for this in Section 4.

3 MEASURES OF MODEL QUALITY

The true system and the model.

Suppose that the true system is subject to (1)- (4), i.e. that

$$y(t)=G_0(q)u(t) + H_0(q)e_0(t) \tag{8}$$

where $\{e_0(t)\}$ is white noise with variance λ_0.

For simpler notation, we shall also use

$$T_0(q) = [G_0(q) \ H_0(q)] \tag{9}$$

Suppose that we have decided upon all the design variables $\mathcal{D}$, associate with the model construction and as a result obtained the model

$$\hat{T}(q,\mathcal{D}) = [\hat{G}(q,\mathcal{D}) \quad \hat{H}(q,\mathcal{D})] \tag{10}$$

$\mathcal{D}$ will contain, among other things N, the number of collected data.

A scalar design criterion

It is of course desirable that the model $T(q,\mathcal{D})$ is close to $T_0(q)$. The difference

$$\tilde{T}(e^{i\omega},\mathcal{D}) \triangleq \hat{T}(e^{i\omega},\mathcal{D})-T_0(e^{i\omega}) \tag{11}$$

should, in other words, be small. Let us develop a formal measure of the size of $\tilde{T}$. Depending on the intended use of the model a good fit in some frequency ranges may be more important than in others. To capture this fact, we introduce a frequency weighted scalar criterion

$$J_1(\tilde{T}(\cdot,\quad)) = \int_{-\pi}^{\pi} \tilde{T}(e^{i\omega},\quad)C(\omega)\tilde{T}^T(e^{-i\omega},\quad)d\omega \tag{12}$$

where the 2×2 matrix function

$$C(\omega) = \begin{bmatrix} C_{11}(\omega) & C_{12}(\omega) \\ C_{21}(\omega) & C_{22}(\omega) \end{bmatrix} \tag{13}$$

describes the relative importance of a good fit at different frequencies as well as the relative importance of the fit in G and H, respectively. We shall generally assume that $C(\omega)$ is Hermitian, i.e. that

$$C_{21}(\omega)=\overline{C_{12}(\omega)}\ [=C_{12}(-\omega)]$$

(the last equality follows when the dependence on ω is via $e^{i\omega}$.) We shall shortly give an example of how such weighting functions can be determined.

Now, the scalar $J_1(\tilde{T}(\cdot,\mathcal{D}))$ is a random variable due to the randomness in $\hat{T}$. To obtain a realization independent quality measure, it is natural to take the expectation of J_1 and form the criterion

$$J(\mathcal{D}) = \int_{-\pi}^{\pi} E\ \tilde{T}(e^{i\omega},\mathcal{D})C(\omega)\tilde{T}^T(e^{-i\omega},\mathcal{D})d\omega =$$

$$= \int_{-\pi}^{\pi} \mathrm{tr}[\Pi(\omega,\mathcal{D})C(\omega)]d\omega \tag{14}$$

where the 2×2 matrix Π is given by

$$\Pi(\omega,\mathcal{D}) = E\ \tilde{T}^T(e^{-i\omega},\mathcal{D})\tilde{T}(e^{i\omega},\mathcal{D}) \tag{15}$$

The problem of choosing design variables can now be stated as

$$\min_{\mathcal{D}\in\Delta} \bar{J}(\mathcal{D}) \tag{16}$$

where Δ denotes the constraints associated with the design options. These will typically include a maximum number of samples, signal power constraints, not too complex numerical procedures etc. The constraints Δ could also include that certain design variables simply are not available to the user in the particular application in question.

Model application: Simulation

Suppose that the transfer function $\hat{G}$ is used to simulate the input-output part of the system with input $u^*(t)$. The model $\hat{G}(q,\mathcal{D})$ then produces the output

$$y_{\mathcal{D}}(t) = \hat{G}(q,\mathcal{D})u^*(t)$$

while the true system would give the correct output

$$y_0(t) = G_0(q)u^*(t).$$

The error signal

$$\tilde{y}_{\mathcal{D}}(t) = y_{\mathcal{D}}(t) - y_0(t) = [\hat{G}(q,\mathcal{D})-G_0(q)]u^*(t)$$

has the spectrum

$$\Phi_{\tilde{y}}(\omega,\mathcal{D}) = |\hat{G}(e^{i\omega},\mathcal{D})-G_0(e^{i\omega})|^2\Phi_u^*(\omega)$$

where $\Phi^*_u(\omega)$ is the spectrum of $\{u^*(t)\}$. This, again, is a random function, and its expectation w.r.t $\hat{G}$

$$\Psi_{\tilde{y}}(\omega,\mathcal{D}) = E\left|\hat{G}(e^{i\omega},\mathcal{D})-G_0(e^{i\omega})\right|^2\Phi^*_u(\omega) \tag{17}$$

is a measure of the average performance degradation due to errors in the model G. Note that, with (15) and

$$C(\omega) = \begin{bmatrix} \Phi^*_u(\omega) & 0 \\ 0 & 0 \end{bmatrix} \tag{18}$$

we can rewrite (17) as

$$\Psi_{\tilde{y}}(\omega,\mathcal{D}) = \text{tr}\ \Pi(\omega,\mathcal{D})C(\omega)$$

Finally, the average variance $E\tilde{y}^2(t)$ (averaged over u*(t) as well as over $\hat{G}$) will be

$$E\tilde{y}^2(t) = \bar{J}(\mathcal{D}) = \frac{1}{2\pi}\int_{-\pi}^{\pi}\Psi_{\tilde{y}}(\omega,\mathcal{D})d\omega$$

which is a special case of (14).

Model application: Pole placement

Now, assume that we want to create a servo system from a reference input r(t) to the output, such that

$$y(t) = R(q)r(t) \tag{19}$$

and that the noise spectrum is insignificant in comparison with the spectrum of r; $\Phi_r(\omega)$. A general solution to obtain

(19), in case G has no zeros outside the unit circle, is to use the regulator

$$u(t) = Q(q)r(t)-P(q)y(t) \tag{20}$$

with transfer functions Q and P such that

$$\frac{\hat{G}(q)Q(q)}{1+G(q)P(q)} = R(q)$$

Using (20) in the true system gives

$$y_{\mathcal{D}}(t) = \frac{G_0 Q}{1+G_0 P} r(t) = Rr + \left[\frac{G_0 Q}{1+G_0 P} - \frac{\hat{G}Q}{1+GP}\right] r \approx Rr + \frac{Q\hat{G}}{(1+G_0 P)^2} r$$

if $\tilde{G}$ is small. Comparing with (12)-(13) we see that this corresponds to

$$C(\omega) = \begin{bmatrix} \dfrac{|R|^2}{|G_0|^2 |1+G_0 P|^2} \Phi_r(\omega) & 0 \\ 0 & 0 \end{bmatrix} \tag{21}$$

In view of the common special cases (21) and (18) we shall in this contribution specialize from (13) to

$$C(\omega) = \begin{bmatrix} C_{11}(\omega) & 0 \\ 0 & 0 \end{bmatrix} \tag{22}$$

4 PREDICTION ERROR IDENTIFICATION METHODS

The model set

The perhaps most common approach in modern identification is to postulate that the transfer function is to be sought within a certain set:

$$\mathcal{G} = \{G(e^{i\omega},\theta) \mid \theta \in D_M\} \tag{23}$$

Here D_M typically is a subset of R^d. In order to improve the result, it is customary to also include assumptions about the disturbance spectrum $\Phi_v(\omega)$ (see (1)-(4)). It is assumed to belong to a set

$$\Phi_v(\omega) = \lambda |H(e^{i\omega},\theta)|^2; \; H(e^{i\omega},\theta) \in \mathcal{H} \tag{24}$$

$$\mathcal{H} = \{H(e^{i\omega},\theta) \mid \theta \in D_M\}.$$

This means that the system is assumed to be described as

$$y(t) = G(q,\theta)u(t) + H(q,\theta)e(t) \tag{25}$$

for some $\theta \in D_M$. Here $\{e(t)\}$ is a sequence of independent random variables with zero mean values and variances λ, and G and H are functions of the shift operator q;

$$G(q,\theta) = \sum_{k=1}^{\infty} g_k(\theta) q^{-k} \tag{26a}$$

$$H(q,\theta) = 1 + \sum_{k=1}^{\infty} h_k(\theta) q^{-k}. \tag{26b}$$

There are several ways by which the transfer functions in (25) can be parametrized. Common ones include state-space models, ARMAX models, output-error models, etc.

<u>The estimation method</u>

Given the model (25) and input-output data up to time t-1, we can determine the prediction output at time t as

$$\hat{y}(t|\theta) = (1-H^{-1}(q,\theta))y(t) + H^{-1}(q,\theta)G(q,\theta)u(t) \tag{27}$$

At time t, when y(t) has been recorded we can compute the prediction error that the model (25) led to:

$$\varepsilon(t,\theta)=y(t)-\hat{y}(t|\theta)=H^{-1}(q,\theta)\left(y(t)-G(q,\theta)u(t)\right) \tag{28}$$

We may say that the model (25) is "good" if the sequence $\varepsilon(t,\theta), t=1,2,\ldots,N$ is "small". In a very common class of identification methods, the squared sum of prediction errors is minimized to find the "best" model:

$$\hat{\theta}_N = \arg\min_{\theta\in D_M} \frac{1}{N} \sum_{t=1}^{N} \varepsilon^2(t,\theta) \tag{29}$$

With $\hat{\theta}_N$ determined in this way, the transfer function estimate becomes

$$\hat{G}_N(e^{i\omega}) = G(e^{i\omega},\hat{\theta}_N) \tag{30}$$

Among methods that can be expressed as (29) we find the "maximum likelihood method" (assuming Gaussian disturbances), the "least squares method" and others. See Ljung (1987) and Åström (1980) for further discussions.

Some extensions

It may often be worthwhile to consider a modified criterion (29) where the prediction errors $\varepsilon(t,\theta)$ (or, equivalently the input-output sequences) first are filtered through a filter $L(q)$:

$$\varepsilon_F(t,\theta) = L(q)\varepsilon(t,\theta) \tag{31}$$

This is, however, equivalent to replacing the noise model $H(q,\theta)$ by $H(q,\theta)/L(q)$. See (28) and, for a further discussion, Wahlberg and Ljung (1986). Prefiltering the data thus corresponds to selecting another noise model set.

Also the use of k-step ahead predictors in (27) might be useful. As elaborated on in Wahlberg and Ljung (1986), k-step ahead prediction methods are equivalent to replacing $H(q,\theta)$ by

$$H(q,\theta)M_k^{-1}(q,\theta) \tag{32}$$

where $M_k(q,\theta)$ are the first k terms in the Laurent expansion of $H(q,\theta)$. The use of k-step ahead predictors is thus equivalent to prefiltering $(L(q)=M_k(q,\theta))$ or to selecting another noise model set.

Design variables

Let us list the available design variables:

- $\Phi_u(\omega)$: spectrum of the extra input u in (5) (33a)
- $\Phi_{ue}(\omega)$ cross spectrum between u and e (resulting from output fedback (33b)
- $\mathcal{H}=\{H(q,\theta)\,|\,\theta\in D_\mu\}$: set of noise models. This includes, as we noted, the possibility of prefiltering with L in (31) and the use of k-step ahead predictors (see (32)).
- In this study we shall confine ourselves to fixed noise models, i.e. the set $\mathcal{H}$ is a singleton:

  $$\mathcal{H} = \{H_*(q)\} \tag{33c}$$

  The coice of H_* is however included among the design variables.

These three items will henceforth be denoted collectively by the symbol $\mathcal{D}$.

Other design variables, such as N, the number of collected data, and $\mathcal{G}$, the set of transfer function models (including the model order n) will be regarded as fixed in this study.

5 ASYMPTOTIC PROPERTIES OF THE ESTIMATED TRANSFER FUNCTIONS

Convergence

Under weak conditions it can be shown that

$$\hat{\Theta}_N \to \Theta^* = \arg\min_{\Theta \in D_M} \bar{V}(\Theta) \text{ w.p.1 as } N\to\infty \tag{34}$$

where

$$\bar{V}(\Theta) = \lim_{N\to\infty} \frac{1}{N} \sum_{t=1}^{N} E\varepsilon^2(t,\Theta) \tag{35}$$

See, e.g. Ljung (1978).

Applying Parseval's relationship to (35) gives, after some calculations (see Ljung (1987)), using also (33c)

$$\Theta^* = \arg\min_{\Theta} \int_{-\pi}^{\pi} \mathrm{tr}\left[R(\omega,\Theta)\cdot Q(\omega)\right]d\omega \tag{36}$$

with

$$R(\omega,\Theta) = \tilde{T}^T(e^{-i\omega},\Theta)\tilde{T}(e^{i\omega},\Theta) \tag{37}$$

$$Q(\omega) = \Phi_\chi(\omega)/\left|H_*(e^{i\omega})\right|^2 \tag{38}$$

$$\tilde{T}(q,\Theta) = \left[G(q,\Theta) \quad H_*(q)\right]-T_0(q) \tag{39}$$

and

$$\Phi_\chi(\omega) = \begin{bmatrix} \Phi_u(\omega) & \Phi_{ue}(\omega) \\ \Phi_{eu}(\omega) & \lambda_0 \end{bmatrix} \tag{40}$$

Note that for open loop operation ($\Phi_{ue}(\omega)\equiv 0$), this expression specializes to

$$\theta^* = \arg\min_{\theta} \int_{-\pi}^{\pi} |\tilde{G}(e^{i\omega},\theta)|^2 \cdot \frac{\Phi_u(\omega)}{|H_*(e^{i\omega})|^2} d\omega \tag{41}$$

Variance

Let

$$T^*(q) = T(q,\theta^*) \tag{42}$$

with θ^* defined as above.

Under fairly general conditions it can then be shown that

$$\sqrt{N}\,[\hat{T}_N(e^{i\omega}) - T^*(e^{i\omega})] \in AsN\,\big(0, P_n(\omega)\big) \tag{43}$$

Here (43) means that the random variable on the left converges in distribution to the normal distribution with zero mean and covariance matrix $P_n(\omega)$. Here the index n denotes the order of the model used in $T(q,\theta)$.

Results, such as (43) go back to the asymptotic normalitiy of the parameter estimate $\hat{\theta}_N$, established, e.g. in Ljung and Caines (1979). The expression for $P_n(\omega)$ is in general complicated.

For models that are parametrized as "black boxes" we have, however the following general result, Ljung (1985b).

$$\lim_{n\to\infty} \frac{1}{n} P_n(\omega) = \Phi_v(\omega)[\Phi_\chi(\omega)]^{-1} \tag{44}$$

with Φ_v and Φ_χ defined by (3) and (40), respectively.

A pragmatic interpretation

Even though the covariance of $\hat{T}_N$ need not converge (convergence in distribution does not imply convergence in L_2), we shall allow ourselves to use the result (43)-(44) in the following more suggestive version:

$$\text{Cov}\ \hat{T}_N(e^{i\omega}) \sim \frac{n}{N}\,\Phi_v(\omega)[\Phi_\chi(\omega)]^{-1} \tag{45}$$

We shall also allow the approximation

$$E\ \hat{T}_N(e^{i\omega}) \approx T^*(e^{i\omega}) \tag{46}$$

(See Ljung (1987) for justifications.)

With (45) and (46) the expression (15) can be rewritten

$$\Pi(\omega,\mathcal{D})=E\ \tilde{T}^T(e^{-i\omega},\mathcal{D})\tilde{T}(e^{i\omega},\mathcal{D})=R(\omega,\theta^*(\mathcal{D}))+\text{Cov}\ \hat{T}_N(e^{i\omega},\mathcal{D}) \tag{47}$$

where the bias contribution R was defined in (37). We have here appended the argument $(\theta^*=\theta^*(\mathcal{D}),\ \hat{\theta}_N=\hat{\theta}_N(\mathcal{D}))$ to stress the dependence on the design variables.

The criterion (14) can thus be split into a bias and a variance contribution:

$$J(\mathcal{D}) \approx J_B(\mathcal{D}) + J_P(\mathcal{D}) \tag{48}$$

where

$$J_B(\mathcal{D}) = \int_{-\pi}^{\pi} \text{tr}\ R(\omega,\theta^*(\mathcal{D}))C(\omega)d\omega \tag{49}$$

$$J_P(\mathcal{D}) = \int_{-\pi}^{\pi} \mathrm{tr}[\mathrm{Cov}\ \hat{T}_N(e^{i\omega},\mathcal{D}) \cdot C(\omega) d\omega] \approx$$

$$\approx \frac{n}{N} \int_{-\pi}^{\pi} \Phi_v(\omega) \cdot \mathrm{tr}[\Phi_\chi^{-1}(\omega,\mathcal{D}) C(\omega)] d\omega \tag{50}$$

using (44).

In the following two sections we shall discuss the minimization of these two contributions to the design criterion.

6 MINIMIZING THE BIAS CONTRIBUTION

Consider now the problem of minimizing the bias distribution, i.e.

$$\min_{\mathcal{D}\in\Delta} J_B(\mathcal{D}) \tag{51}$$

where $J_B(\mathcal{D})$ is defined by (37), (49). The function $J_B(\mathcal{D})$ depends on $\mathcal{D}$ via $\Theta^*(\mathcal{D})$. The dependence on $\mathcal{D}$ of the latter function, in turn, is defined by (36), which we write as

$$\Theta^*(\mathcal{D}) = \arg\min_{\Theta} \int_{-\pi}^{\pi} \mathrm{tr}[R(\omega,\Theta) \cdot Q(\omega,\mathcal{D})] d\omega \tag{52}$$

$R(\omega,\Theta)$ is defined by (37) and Q by (38). We have appended the argument $\mathcal{D}$ to Q, to stress hat it is made up from the design variables (33). See also Ljung (1986) and Gevers and Ljung (1986).

Comparing (52) with the minimization problem (51), (49) it is intuitively clear that the best choice of $\mathcal{D}$ should be one that makes $C(\omega)$ and $Q(\omega,\mathcal{D})$ proportional. That this is indeed the case is proven in Yuan and Ljung (1985) and Ljung (1986b). We thus have the following result.

<u>Theorem 1</u>. Consider the problem to minimize (51) (via (49) and (22)) with respect to $\mathcal{D} = \{\Phi_u(\omega),\ \Phi_{ue}(\omega),\ L(e^{i\omega}),\ H_*(e^{i\omega})\}$ (see (33)) under the assumptions (22). Then $\mathcal{D}_{opt}$ is such that

$$\Phi_{ue}^{opt}(\omega) \equiv 0 \tag{53a}$$

$$\frac{|L(e^{i\omega})|^2 \Phi_u^{opt}(\omega)}{|H_*^{opt}(e^{i\omega})|^2} = \alpha \cdot C_{11}(\omega) \tag{53b}$$

where α is any constant that makes $\mathcal{D}_{opt}$ belong to the admissible set.

□

Here are included the prefilter L in (31) as an explicit option. Notice that there are several ways of obtaining the optimal design. Any combinations of input spectrum and noise model that obey (53b) will give the optimal bias distribution. Also recall that the choice of noise model $H_*(q)$ contains the option of prediction horizon k (see (32)).

7 MINIMIZING THE VARIANCE CONTRIBUTION

<u>The problem</u>

Let us now turn to the problem

$$\min_{\mathcal{D} \in \Delta} J_p(\mathcal{D}) \tag{54}$$

where $J_p(\mathcal{D})$ is given by (50). We shall generally assume that the input power is constrained:

$$\Delta:\ \int_{-\pi}^{\pi} \Phi_u(\omega)d\omega \le \beta \tag{55}$$

Spelling out (50) gives

$$J_p(\mathcal{D}) = \int_{-\pi}^{\pi} \Psi(\omega,\mathcal{D})d\omega$$

where

$$\Psi(\omega,\mathcal{D})=\frac{\lambda C_{11}(\omega)-2\mathrm{Re}\left[C_{12}(\omega)\Phi_{eu}(\omega)\right]+C_{22}(\omega)\Phi_u(\omega)}{\lambda\Phi_u(\omega)-|\Phi_{ue}(\omega)|^2}\cdot\Phi_v(\omega) \tag{56}$$

$$\mathcal{D} = \{\Phi_u,\Phi_{ue}\} \tag{57}$$

Here we dispensed with the scaling n/N, which is immaterial.

For the case (22) we obtain the problem

$$\min_{\Phi_u,\Phi_{ue}} \int_{-\pi}^{\pi} \frac{\lambda_0 C_{11}(\omega)}{\lambda_0\Phi_u(\omega)-|\Phi_{ue}(\omega)|^2}\Phi_v(\omega)d\omega \tag{58}$$

subject to the constraint that

$$\int_{-\pi}^{\pi} \Phi_u(\omega)d\omega \leq C \tag{59}$$

From (58) and the fact that $\Phi_{ue}(\omega)$ does not enter the constraint, it follows that

$$\Phi_{ue}^{opt}(\omega) \equiv 0. \tag{60}$$

It is thus optimal to use open loop experiments, and the optimal input is easy to compute using Schwarz's inequality:

<u>Lemma 1</u> The solution to (58)-(59)

$$\Phi_u^{opt}(\omega) = \mu \cdot \sqrt{C_{11}(\omega) \cdot \Phi_v(\omega)} \tag{61}$$

where μ is a constant, adjusted so that

$$\int_{-\pi}^{\pi} \Phi_u^{opt}(\omega) d\omega = C \tag{62}$$

8 MINIMIZING THE DESIGN CRITERIA

Let us now turn to the full design criterion (13)- (16), (22) in its pragmatic form (48)-(50). Our partial results on bias- and variance-minimization then show that it in certain cases is possible to minimize the two contributions simultaneously. Then of course the full criterion is also minimized. For the case of Theorem 1 we thus have the following result.

<u>Theorem 2</u>: Consider the problem to minimize (48)-(50) with respect to

$$\mathcal{D} = \{\Phi_u(\omega),\ \Phi_{ue}(\omega),\ L(e^{i\omega}),\ H_*(e^{i\omega})\}$$

under the assumptions (22), (33), and subject to the constraint (59). Then $\mathcal{D}_{opt}$ is given by

$$\Phi_{ue}(\omega) \equiv 0$$

$$\Phi_u(\omega) = \mu_2 \sqrt{C_{11}(\omega) \cdot \Phi_v(\omega)} \tag{63}$$

$$\left| \frac{L(e^{i\omega})}{H_*(e^{i\omega})} \right|^2 = \mu_1 \sqrt{\frac{C_{11}(\omega)}{\Phi_v(\omega)}}$$

Here μ_1 is a constant, adjusted so that the left hand side has a Laurent expression that starts with a "1", and μ_2 is a constant adjusted so that the input power constraint is met.

□

Note that the freedom in the choice of noise model and prefilter is imaginary, since they always appear in the combination $L(q)/H^*(q)$ in the criterion.

The case where our prime interest is in the transfer function G is probably the most common one, and therefore the optimal design variables offered by Theorem 2 should be of interest. The only drawback with this solution may be that the choice of constant noise model may lead to more calculations in the numerical minimization of the prediction error criterion.

9 CONCLUSIONS

In this contribution we have focused our interest on the design variables that are available for the estimation of transfer functions.

We have studied the family of prediction error identification methods for the parameter estimation, and made use of some recently derived asymptotic expressions for bias and variance of the transfer function estimate. Under certain assumptions some fairly explicit advice for the choice of input spectra, feedback mechanisms, prefilters and noise models have been derived.

Acknowledgement. This work was supported by the Swedish National Board for Technical Development (STUF).

REFERENCES

Åström, K J (1980). Maximum likelihood and prediction error methods. Automatica, vol 16, pp 551-574.

Åström, K J and P. Eykhoff (1971). System identification - a survey. Automatica, vol 13, pp 457-476.

Eykhoff, P (1974). System Identification. Wiley, London.

Eykhoff, P (1981) (Ed). Trends and progress in system Identification, Pergamon Press, Oxford.

Gevers, M and L Ljung (1986). Optimal experiment design with respect to the intended model application. Automatica, vol 22, No. 5, pp 543-554.

Ljung, L (1978). Convergence analysis of parametric identification methods. IEEE Transactions on Automatic Control, vol AC-23, pp 770-783.

Ljung, L (1985a). On the estimation of transfer functions, Automatica, vol 21, pp 677-696.

Ljung, L (1985b). Asymptotic variance expressions for identified black-box transfer function models. IEEE Trans Autom. Control, vol AC-30, pp 834-844.

Ljung, L (1986). Parametric methods for identification of transfer functions of linear systems. In Advances in Control vol XXIV, (C.L. Leondes, Ed), Academic Press, N.Y. 1986, to appear.

Ljung, L (1987). System Identification - Theory for the User, Prentice-Hall, Engelwood Cliffs, to appear.

Ljung, L and P E Caines (1979). Asymptotic normality of prediction error estimation for approximate system models. Stochastics vol 3, pp 29-46.

Wahlberg, B and L Ljung (1986). Design variables for bias distribution in transfer function estimation. IEEE Trans Autom Control, vol AC-31, Feb 1986.

Yuan, Z D and L Ljung (1985). Unprejudiced optimal open loop input design for identification of transfer functions, Automatica, vol 21, pp 697-708.

APPROXIMATE MODELLING OF DETERMINISTIC SYSTEMS

C. Heij
Econometrics Institute
University of Groningen
P.O. Box 800
9700 AV Groningen
The Netherlands

Abstract

This paper contains a description of a procedure for modelling a time series by means of a (deterministic) linear system.
Given the time series, the procedure identifies the system of minimal complexity which has a prescribed predictive accuracy.

Keywords

Data modelling, time series analysis, identification, prediction, linear systems.

1. INTRODUCTION

In this paper we will describe a procedure of system identification which is inspired by the purpose of prediction. On the basis of an observed time series the procedure determines a deterministic linear system which is reasonable in the sense that the complexity of the system is relatively low and that the predictive power of the system is relatively high.

Exact definitions of the concepts "complexity" and predictive power (or "error") are given in section 3 for a class of prediction models defined in section 2. An outline of the procedure is given in section 5 and is based on the concept of canonical variables, described in section 4.

NATO ASI Series, Vol. F34
Modelling, Robustness and Sensitivity Reduction in Control Systems. Edited by R. F. Curtain

The prediction models consist of models described by autoregressive equations. Two main features of the identification procedure described in this paper are the following.

1. The time series and the identified models are deterministic. It is not assumed that the observed time series is generated by an underlying stochastic process.
2. The lags of the autoregressive equations are determined from the data. The lags are not imposed a priori.

Related literature on identification is abundant. We confine ourselves to the following references. For identification of stochastic models with a priori imposed lags we refer to Maddala [6], Theil [10]. For prediction error identification see e.g. Åström [2]. An overview of least squares prediction can be found in Kailath [4]. Within the stochastic framework the determination of lags is investigated e.g. by Akaike [1], Hannan and Quinn [3], Shibata [9]. Some work on identification on the basis of a deterministic time series with given lags can be found e.g. in Ljung [5], Makhoul [7].

The procedure described in this paper is basically inspired by Willems [11]. The main difference is that in [11] the error measure is equation error, reflecting the purpose of description, while here the error measure is prediction error, reflecting the purpose of prediction.

We finally remark that a completely different approach for modelling deterministic time series by means of models without a priori imposed lags is given by Rissanen [8].

2. PREDICTION MODELS

We will consider prediction models which can be described by a number of autoregressive equations. We give such a model the following interpretation. Some linear combinations of the variables on time $t+1$ are predicted on the basis of a number of linear combinations of these variables on times $t' \leq t$.

To make this specific we introduce the following notation. We assume we have observed a time series of length T in q variables, $w := (w(1), \ldots, w(T))$, $w(t) \in \mathbb{R}^q$. Let σ^* denote right shift $\sigma^* : (\mathbb{R}^q)^T \to (\mathbb{R}^q)^T$ defined by $(\sigma^* w)(1) := o$ and $(\sigma^* w)(t) := w(t-1)$ for $t = 2, \ldots, T$. Let $R \in \mathbb{R}^{p\times q}[\sigma^*]$ denote a pxq matrix of polynomials in σ^* and let $d := d(R)$, the maximal degree of elements of R. Let $A \in \mathbb{R}^{p\times q}$. Finally, let $\pi_t : (\mathbb{R}^q)^t \to 2^{\mathbb{R}^q}$ be a predictor, with the interpretation that on observing $w(1), \ldots, w(t)$ π_t predicts that $w(t+1)$ belongs to $\pi_t((w(1), \ldots, w(t))) \subset \mathbb{R}^q$, a subset of $\mathbb{R}^q$. We will use the notation $\hat{w}(t+1) := \pi_t((w(1), \ldots, w(t)))$.

<u>Definition</u>. The prediction model (A,R) predicts according to

$$\hat{w}(t+1) := A^{-1}\{(R(\sigma^*) w)(t)\}, \quad t = d+1, \ldots, T.$$

This has the following interpretation.

Let $R(\sigma^*) = \sum_{j=o}^{d} R_j \sigma^{*j}$, $R_j \in R^{p\times q}$.

Given the observations $w(1), \ldots, w(t)$, the model (A,R) predicts that $w(t+1)$ will satisfy $Aw(t+1) = \sum_{j=o}^{d} R_j w(t-j)$. So some linear combinations of $w(t+1)$ are predicted by means of linear combinations of $w(1), \ldots, w(t)$.

<u>Remark</u>. 1. These prediction models consist of one step ahead prediction by means of autoregressive equations. If $p = q$ and A is invertible, then the model gives point predictions. If $p < q$ then $w(t+1)$ is not uniquely predicted. This may be sensible, e.g. because some of the variables in w are free and cannot be predicted.

2. If the predictions would be exact, i.e. $Aw(t+1) = (R(\sigma^*)w)(t)$, $t = d+1, \ldots, T$, then the time series w is a (finite time) observation from a linear system with finite dimensional state space, see Willems [11]. This means that the model (A,R) on t predicts according to the assumption that $(w(t-d), \ldots, w(t+1))$ is an element of such a linear system. The model (A,R) is not meant to claim that w is generated by such a system, it only claims that one step ahead predictions can reasonably be made by it.

3. In the sequel we will make the assumption that $d \leq T-1$ and that A has full row rank p.

We call two prediction models (A_1,R_1) and (A_2,R_2) equivalent if they give identical predictions for all time series w. It is clear that (A_1,R_1) and (A_2,R_2) are equivalent if and only if $p_1 = p_2 = p$ and there exists a nonsingular $S \in \mathbb{R}^{pxp}$ such that $(A_2,R_2) = (SA_1,SR_1)$.

For prediction models we define a canonical form as follows. Let $R^{(k)}$ consist of those rows of R which have degree k-1, and let A_k consist of the corresponding rows of A, $k \geq 1$. Furthermore, let A_o denote the rows of A corresponding to the zero rows in R. For $k \geq 1$ let $R^{(k)} = \sum_{j=o}^{k-1} R_j^{(k)} \sigma^{*j}$. For $M \in \mathbb{R}^{nxq}$, let sp $M \subset \mathbb{R}^{1xq}$ denote the space spanned by the rows of M. Finally let d_i denote the degree of the i-th row of R, with $d(o) := -\infty$.

The following clearly defines a canonical form.

Definition. (A,R) is in standard prediction form if

(i) $R_{k-1}^{(k)}$ has full row rank, $k \geq 1$ (and A has full row rank p)

(ii) $d_i \leq d_{i+1}$, $i = 1, \ldots, p-1$

(iii) $\text{sp}\, A_k \perp \text{sp}\, A_l$, $k \neq l$, $k,l = 0,1, \ldots, d+1$.

Remark. 1. Condition (i) ensures that the functionals A_k on $w(t+1)$ really are predicted by means of a (k-1)-th order lag from the past, i.e. there exists no $f \in \text{sp}\, A_k$ such that $f\,\hat{w}(t+1) = (r(\sigma^*)w)(t)$ with $d(r) < k-1$.

2. The essential condition is (iii), i.e. that the functionals A_k are orthogonal for different k. It could be reasonable to require more conditions, e.g. that $\text{sp}(A_k - \sigma^* R^{(k)}(\sigma^*)) \perp \text{sp}(\sigma^*(A_{k-1} - \sigma^* R^{(k-1)}(\sigma^*)))$, which expresses the idea that functionals which already have been predicted to be zero should not be used for further prediction. Conditions like these ones are currently under investigation.

3. COMPLEXITY AND ERROR

For given prediction model (A,R) in standard prediction form we define $m := q-p = \text{corank}(A)$ and for $k = 1, \ldots, T$ $c_k := \#$ (rows of R of degree $k-1$), $c_o := \#$ (rows of R equal to o).

Definition. The complexity of a prediction model (A,R) is $c(A,R) := (p, c_o, c_1, \ldots, c_T)$.

The complexity has the interpretation that models are more complex the less is predictable, i.e. the smaller is p, and that they are more complex the longer the part of the past, used for prediction. This is related to the number of inputs (m) and the number of state variables ($\sum_{t=o}^{T} tc_t$), c.f. Willems [11].

To define the error of a model (A,R) with respect to observations $w = (w(1), \ldots, w(T))$, assume (A,R) is in standard prediction form. Let $R^{(k)}$, $k = 1, \ldots, d+1$ and A_k, $k = 0,1, \ldots, d+1$ be as defined in section 2. For $k = 1, \ldots, d+1$, $f \in \text{sp } A_k$, $f = a^T A_k$, define for $t = k, \ldots, T$ $f\hat{w}(t+1) := a^T \{(R^{(k)}(\sigma^*)w)(t)\}$. Then for $k = o, \ldots, T$ define the errors $\varepsilon_k := (\varepsilon_k^1, \ldots, \varepsilon_k^{c_k})$ as follows

$$\varepsilon_o^i(A,R) := \inf_{\substack{L \subset \text{sp}A_o \\ \dim L \geqq c_o - i + 1}} \; \sup_{f \in L} \; \frac{||fw||}{||f||}, \quad i = 1, \ldots, c_o,$$

where $||fw||^2 := \frac{1}{T} \sum_{t=1}^{T} (fw(t))^2$ and $||f||^2 := \sum_{j=1}^{q} f_j^2$,

$$\varepsilon_k^i(A,R) := \inf_{\substack{L \subset \text{sp}A_k \\ \dim L \geqq c_k - i + 1}} \; \sup_{f \in L} \; \frac{||fw - f\hat{w}||}{||fw||_k}, \quad i = 1, \ldots, c_k,$$

where $||fw - f\hat{w}||^2 := \frac{1}{T-k} \sum_{t=k}^{T-1} (fw(t+1) - f\hat{w}(t+1))^2$ and $||fw||_k^2 := \frac{1}{T-k} \sum_{t=k}^{T-1} (fw(t+1))^2$. For any k for which $c_k = o$ we define $\varepsilon_k := o$.

These error measures have the following interpretation.

The error ε_o^1 measures the worst static law claimed by (A,R), as (A,R) predicts $A_o\hat{w}(t+1) = o$, so $f\hat{w}(t+1) = o$ for $f \in sp\, A_o$.

For $k = 1, \ldots, d+1$, ε_k^1 measures the worst prediction which is made by a(k-1)-th order lag on past observations. It is predicted that $A_k\hat{w}(t+1) = (R^{(k)}(\sigma^*)w)(t)$, and it is assumed that linear combinations of the components of $A_k\hat{w}(t+1)$, say $a^T A_k\hat{w}(t+1)$ with $a \in \mathbb{R}^{c_k}$, are predicted by $a^T\{(R^{(k)}(\sigma^*)w)(t)\}$. Let $f := a^T A_k$, then the mean error of this prediction is $\|fw-f\hat{w}\|$, and the relative error of prediction is $\|fw-f\hat{w}\| / \|fw\|$.

The errors ε_k^i, $i \geq 2$, have analogous interpretations. They measure the least attainable maximal error if one considers predicting w(t+1) by means of a (k-1)-th order lag on past observations and one restricts the functionals on w(t+1) to be predicted to a (c_k-i+1)-dimensional subspace of sp A_k, the space of all functionals on w(t+1) predicted by a (k-1)-th order lag. It can be shown that the procedure of section 5 also is optimal if instead of ε_k^i one considers $\hat{\varepsilon}_k^i$, which is defined as follows: $\hat{\varepsilon}_k^1 = \varepsilon_k^1$, if $\hat{\varepsilon}_k^1 = \|f_1w-f_1\hat{w}\| / \|f_1w\|$ then $\hat{\varepsilon}_k^2 := \sup \{\|fw-f\hat{w}\| / \|fw\| ; f \in spA_k \cap \{sp\, f_1\}^\perp\} = \|f_2w-f_2\hat{w}\| / \|f_2w\|$, $\hat{\varepsilon}_k^3 := \sup \{\|fw-f\hat{w}\| / \|fw\| ; f \in sp\, A_k \cap \{sp\,(f_1,f_2)\}^\perp\}$, and so forth.

The condition that (A,R) should be in standard prediction form is a reflection of the following intuitive requirement. Suppose for example that one accepts $f\hat{w}(t+1) = (r(\sigma^*)w)(t)$ because $\|fw-f\hat{w}\| / \|fw\|$ is small. Then for $\delta \in \mathbb{R}^{1xq}$, $\|\delta\|$ small there exists $\delta f \in \mathbb{R}^{1xq}$, $\|\delta f\|$ small such that $\|(f+\delta f)w - \{(r(\sigma^*) + \delta\sigma^{*k})w\}\| / \|(f+\delta f)w\|$ is small and one would accept $(f+\delta f)\hat{w}(t+1) = \{(r(\sigma^*) + \delta\sigma^{*k})w\}(t)$.
Assuming linearity of prediction, one hence would accept $\|\delta\|^{-1}\cdot\delta f\, \hat{w}(t+1) = \|\delta\|^{-1}\cdot(\delta\sigma^{*k}w)(t)$, although this need not be at all reasonable. In general, a k-th order prediction should not be accepted if its (good) quality is mainly due to the fact that the claimed prediction is close to being implied by predictions of lower order.

In section 5 we will describe a procedure to solve the following sequential optimization problem.

$\hat{\varepsilon}^1 = \varepsilon^1$, if $\hat{\varepsilon}^1 = \varepsilon(\ell_1^1, \ell_2^1)$ then $\hat{\varepsilon}^2(\mathcal{L}) := \sup\{\varepsilon(\ell_1,\ell_2)\ ;\ (\ell_1,\ell_2) \in \mathcal{L}^\perp$, $\ell_2 \in \mathcal{L}^* \cap \{sp\ell_2^1\}^\perp\} = \varepsilon(\ell_1^2, \ell_2^2)$, $\hat{\varepsilon}^3(\mathcal{L}) := \sup\{\varepsilon(\ell_1,\ell_2)\ ;\ (\ell_1,\ell_2) \in \mathcal{L}^\perp$, $\ell_2 \in \mathcal{L}^* \cap \{sp(\ell_2^1, \ell_2^2)\}^\perp\}$, and so forth.

We define a partial ordering as follows. Let $\varepsilon = (\varepsilon^1, \ldots, \varepsilon^d)$ and $\tilde{\varepsilon} = (\tilde{\varepsilon}^1, \ldots, \tilde{\varepsilon}^d)$, then $\{\varepsilon \leqq \tilde{\varepsilon}\} :\Leftrightarrow \{\varepsilon^i \leqq \tilde{\varepsilon}^i\ \forall i = 1, \ldots, d\}$.

Consider the following static prediction problem.

<u>Problem SP</u>. Given (a_i, b_i), $i = 1, \ldots, N$, and a maximal tolerated error $\varepsilon^* \geqq 0$, determine $\mathcal{L}$ such that $c(\mathcal{L}) = \min\{c(\tilde{\mathcal{L}}) : \varepsilon^1(\tilde{\mathcal{L}}) \leqq \varepsilon^*\}$ and such that $\varepsilon(\mathcal{L}) = \min\{\varepsilon(\tilde{\mathcal{L}}) : c(\tilde{\mathcal{L}}) \leqq c(\mathcal{L})\}$.

This problem always has a unique solution. Note that it is not a priori trivial that there exists a solution of minimal dimension which minimizes ε, as $\leqq$ is a partial ordering.

In this section we will describe the solution of SP assuming $N \geqq \max\{n_1, n_2\}$ and for generic data (a_i, b_i), $i = 1, \ldots, N$. Define the covariance matrix S by

$$S := \begin{bmatrix} S_{aa} & S_{ab} \\ S_{ba} & S_{bb} \end{bmatrix} := \frac{1}{N} \sum_{i=1}^{N} \begin{bmatrix} a_i \\ b_i \end{bmatrix} \cdot \begin{bmatrix} a_i \\ b_i \end{bmatrix}^T$$

and suppose $S_{aa} > 0$, $S_{bb} > 0$. Note that, as $N \geqq \max\{n_1, n_2\}$, this generically holds true. Let $S_{aa}^{-\frac{1}{2}} S_{ab} S_{bb}^{-\frac{1}{2}}$ have a singular value decomposition $U \Lambda V^T$, with $U \in \mathbb{R}^{n_1 \times n_1}$ and $V \in \mathbb{R}^{n_2 \times n_2}$ orthogonal, $\Lambda = \begin{bmatrix} D & 0 \\ 0 & 0 \end{bmatrix} \in \mathbb{R}^{n_1 \times n_2}$ with $D = \operatorname{diag}(\lambda_1, \ldots, \lambda_r)$ where $\lambda_1 \geqq \ldots \geqq \lambda_r > 0$, $r = \operatorname{rank} S_{ab}$. Denote the rows of $U^T S_{aa}^{-\frac{1}{2}}$ by ℓ_1^i, $i = 1, \ldots, n_1$, and those of $V^T S_{bb}^{-\frac{1}{2}}$ by ℓ_2^i, $i = 1, \ldots, n_2$. Then the following holds true.

<u>Lemma</u>. Let k be such that $1 - \lambda_k^2 \leqq \varepsilon^* < 1 - \lambda_{k+1}^2$.

Then $\mathcal{L} := \{(a,b)\ ;\ \ell_2^i b = \lambda_i \ell_1^i a,\ i = 1, \ldots, k\}$ solves SP uniquely.

formulate.

Suppose one wants to predict n_2 variables $b \in \mathbb{R}^{n_2}$ on the basis of n_1 other variables $a \in \mathbb{R}^{n_1}$ by means of a linear model $\mathcal{L} \subset \mathbb{R}^{n_1+n_2}$. The model will predict that, given a, b will belong to the set $\mathcal{L}_a := \{b \in \mathbb{R}^{n_2};\ (a,b) \in \mathcal{L}\}$.

Assume one has observed $(a_i, b_i) \in \mathbb{R}^{n_1} \times \mathbb{R}^{n_2}$ for $i = 1,\ldots,N$ and wants to construct $\mathcal{L}$ on the basis of these observations. We will now define the complexity of $\mathcal{L}$ and the error of $\mathcal{L}$, given the observations.

Note that due to linearity $\mathcal{L}_a = b + \mathcal{L}_o$ for any $b \in \mathcal{L}_a$.
Hence a natural measure of complexity is $c(\mathcal{L}) := \dim \mathcal{L}_o$.

Let $\mathcal{L}^{\perp} := \{(\ell_1,\ell_2) \in \mathbb{R}^{1\mathrm{x}n_1} \times \mathbb{R}^{1\mathrm{x}n_2}\ ;\ \ell_1 a + \ell_2 b = 0\ \ \forall (a,b) \in \mathcal{L}\}$.
Define the relative mean error of predicting $\ell_2 b$ by means of $-\ell_1 a$ by

$$\varepsilon(\ell_1,\ell_2) := \frac{\{\frac{1}{N}\sum_{i=1}^{N}(\ell_2 b_i + \ell_1 a_i)^2\}^{\frac{1}{2}}}{\{\frac{1}{N}\sum_{i=1}^{N}(\ell_1 b_i)^2\}^{\frac{1}{2}}}$$

For $\mathcal{L} \subset \mathbb{R}^{n_1+n_2}$ let $\mathcal{L}^* := \{\ell_2; \exists \ell_1\ (\ell_1,\ell_2) \in \mathcal{L}^{\perp}\}$ denote the space of functionals on b for which $\mathcal{L}$ makes predictions. It can be shown that $\mathcal{L}^* = \mathcal{L}_o^{\perp}$. Let $d := \dim \mathcal{L}^* = n_2 - c(\mathcal{L})$, then $\varepsilon(\mathcal{L}) := (\varepsilon^1(\mathcal{L}),\ldots,\varepsilon^d(\mathcal{L}))$ is defined by

$$\varepsilon^i(\mathcal{L}) := \inf_{\substack{\mathcal{L}_i^* \subset \mathcal{L}^* \\ \dim \mathcal{L}_i^* \geq d-i+1}} \ \sup_{\substack{(\ell_1,\ell_2) \in \mathcal{L}^{\perp} \\ \ell_2 \in \mathcal{L}_i^*}} \varepsilon(\ell_1,\ell_2) \quad , i = 1,\ldots,d.$$

This error has the following interpretation. By $\varepsilon^1(\mathcal{L})$ we measure the worst prediction made by $\mathcal{L}$. By $\varepsilon^i(\mathcal{L})$ we measure the worst prediction in a (d-i+1)-dimensional subspace of the space of functionals on b for which $\mathcal{L}$ makes predictions, if this (d-i+1)-dimensional subspace is chosen optimally. The space $\mathcal{L}$ determined by the lemma also can be shown to be optimal if instead of ε one considers $\hat{\varepsilon}$ which is defined as

Suppose ε_k^* are a priori chosen maximal tolerated error levels, i.e. one requires $\varepsilon_k^1 \leqq \varepsilon_k^*$, $k = 0,1,\ldots,T$.

Under these conditions and given the observations $w = (w(1),\ldots,w(T))$, the procedure determines a model (A,R) in standard prediction form which is optimal in the following sense.

First maximize c_o under the conditions that $\varepsilon_o^1 \leqq \varepsilon_o^*$. Among the optimal solutions, minimize ε_o under the partial ordering $\{\varepsilon_o \leqq \tilde{\varepsilon}_o\} :\Leftrightarrow \{\varepsilon_o^i \leqq \tilde{\varepsilon}_o^i \; \forall i = 1,\ldots,c_o\}$. It will be seen that indeed there exists a minimal element with respect to this partial ordering, and that the corresponding model is unique.

Given the solution for static laws, maximize c_1 under the condition $\varepsilon_1^1 \leqq \varepsilon_1^*$ and such that the newly determined laws together with the static ones are in standard prediction form, i.e. the linear combinations of $w(t+1)$ predicted by means of $w(t)$ should be orthogonal to the identified static laws. Among the optimal solutions, minimize ε_1 with the ordering $\{\varepsilon_1 \leqq \tilde{\varepsilon}_1\} : \Leftrightarrow \{\varepsilon_1^i \leqq \tilde{\varepsilon}_1^i \; \forall i = 1,\ldots,c_1\}$.

In general, when one has identified predictive laws $A_i\hat{w}(t+1) = (R^{(i)}(\sigma^*)w)(t)$ for $i = 1,\ldots,k-1$, then subsequently maximize c_k under the condition $\varepsilon_k^1 \leqq \varepsilon_k^*$ and such that the linear combinations of $\hat{w}(t+1)$ which are newly predicted are orthogonal to the rows of A_i , $i = 0,\ldots,k-1$. Among the optimal solutions, minimize ε_k with $\{\varepsilon_k \leqq \tilde{\varepsilon}_k\} :\Leftrightarrow \{\varepsilon_k^i \leqq \tilde{\varepsilon}_k^i \; \forall i = 1,\ldots,c_k\}$.

Remark. A more natural problem statement would be the following. Order the complexity according to e.g. $\{c \leqq \tilde{c}\} : \Leftrightarrow \{c = \tilde{c}$ or $p > \tilde{p}$ or $\exists k$ such that $c_i = \tilde{c}_i \; \forall i < k$, $c_k > \tilde{c}_k\}$. Under the condition that $\varepsilon_k^1 \leqq \varepsilon_k^*$, $k = 0,\ldots,T$, minimize c with respect to this total ordering. This problem seems very hard to solve.

4. CANONICAL VARIABLES

In section 5 we will describe an algorithm for the procedure given at the end of section 3. The algorithm is based on canonical variables. In this section we introduce this concept in our deterministic framework and we will prove a lemma which forms the basis of the algorithm. The lemma describes the solution of a static prediction problem, which we now will

If S is interpreted as a covariance matrix of random variables (a,b), as is done in multivariate statistics, then the pairs $(\ell_1^i a, \ell_2^i b)$ are called canonical variables. Note that these linear combinations here have an interpretation in terms of deterministic prediction.

5. A PREDICTIVE MODELLING PROCEDURE

We now will give a procedure, based on the lemma of section 4, which solves the dynamic prediction problem formulated at the end of section 3.

Problem DP.

Given $w = (w(1),\ldots,w(T))$ and maximal tolerated errors ε_k^*, $k = 0,\ldots,T$, determine a prediction model (A,R) such that

(0) c_o is maximal under the condition $\varepsilon_o^1 \leq \varepsilon_o^*$, and among these ε_o is minimal; let the optimal solution be $A_o\hat{w}(t+1) = 0$;

(1) c_1 is maximal under the condition $\varepsilon_1^1 \leq \varepsilon_1^*$ and $spA_1 \perp spA_o$, and among these ε_1 is minimal ; let the optimal solution be $A_1\hat{w}(t+1) = R^{(1)}w(t)$;

(k) given solutions $A_i\hat{w}(t+1) = (R^{(i)}(\sigma^*)w)(t)$, $i = 1,\ldots,k-1$, c_k is maximal under the condition $\varepsilon_k^1 \leq \varepsilon_k^*$ and $spA_k \perp spA_i$, $i = 0,\ldots,k-1$, and among these ε_k is minimal; let optimal solution be $A_k\hat{w}(t+1) = (R^{(k)}(\sigma^*)w)(t)$.

The procedure works as follows. By $col(a_1,\ldots,a_n)$ we denote the matrix with row i equal to a_i, $i = 1,\ldots,n$.

Step (0). (i) Calculate $\Pi_o := \frac{1}{T}\sum_{t=1}^{T} w(t)w(t)^T$.

(ii) Determine a singular value decomposition (SVD) $\Pi_o = U_o\Lambda_oU_o^T$, with $\Lambda_o = \text{diag}\,(\lambda_1^{(o)},\ldots,\lambda_q^{(o)})$, $\lambda_1^{(o)} \geq \ldots \geq \lambda_q^{(o)} \geq 0$. Let $\lambda_{d_o}^{(o)} > \varepsilon_o^{*2} \geq \lambda_{d_o+1}^{(o)}$, $U_o =: (u_1^{(o)},\ldots,u_q^{(o)})$,

$$P_o := I - \sum_{j=d_o+1}^{q} u_j^{(o)}u_j^{(o)T}.$$

(iii) Define $A_o := col(u_{d_o+1}^{(o)T},\ldots,u_q^{(o)T})$.

Step (1). (i) Calculate $\Pi_1 := \frac{1}{T-1} \sum_{t=1}^{T-1} \begin{bmatrix} w(t) \\ w(t+1) \end{bmatrix} \cdot \begin{bmatrix} w(t) \\ w(t+1) \end{bmatrix}^T$.

(ii) Let $\tilde{P}_1 := \begin{bmatrix} I_q & 0 \\ 0 & P_o \end{bmatrix}$, $\tilde{\Pi}_1 := \tilde{P}_1 \Pi_1 \tilde{P}_1 =: \begin{bmatrix} \Pi_1^- & \Pi_1^{-+} \\ \Pi_1^{+-} & \Pi_1^+ \end{bmatrix}$

with $\Pi_1^-, \Pi_1^+, \Pi_1^{+-T} = \Pi_1^{-+}$ all in $\mathbb{R}^{qxq}$.

Let $(\Pi_1^-)^{-\frac{1}{2}} \Pi_1^{-+} (\Pi_1^+)^{-\frac{1}{2}}$ have SVD $U_1 \Lambda_1 V_1^T$,

$\Lambda_1 = \mathrm{diag}(\lambda_1^{(1)}, \ldots, \lambda_{d_o}^{(1)}, 0, \ldots, 0)$, $\lambda_1^{(1)} \geq \ldots \geq \lambda_{d_o}^{(1)} \geq 0$ and

suppose $1 - \{\lambda_{c_1}^{(1)}\}^2 \leq \varepsilon_1^{*2} < 1 - \{\lambda_{c_1+1}^{(1)}\}^2$.

(iii) Let $V_1 = (v_1^{(1)}, \ldots, v_q^{(1)})$, $U_1^T (\Pi_1^-)^{-\frac{1}{2}} =: \mathrm{col}(\beta_1^{(1)}, \ldots, \beta_q^{(1)})$,

$V_1^T (\Pi_1^+)^{-\frac{1}{2}} =: \mathrm{col}(\alpha_1^{(1)}, \ldots, \alpha_q^{(1)})$, $P_1 := (I - \sum_{j=1}^{c_1} v_j^{(1)} v_j^{(1)T}) . P_o$

$d_1 := d_o - c_1$.

(iv) Define $A_1 := \mathrm{col}(\alpha_1^{(1)}, \ldots, \alpha_{c_1}^{(1)})$, $R^{(1)} :=$

$\mathrm{col}(\lambda_1^{(1)} \beta_1^{(1)}, \ldots, \lambda_{c_1}^{(1)} \beta_{c_1}^{(1)})$.

Step (k). (i) Calculate $\Pi_k := \frac{1}{T-k} \sum_{t=k}^{T-1} \begin{bmatrix} w(t-k+1) \\ \vdots \\ w(t) \\ w(t+1) \end{bmatrix} \cdot \begin{bmatrix} w(t-k+1) \\ \vdots \\ w(t) \\ w(t+1) \end{bmatrix}^T$.

(ii) Let $\tilde{P}_k := \begin{bmatrix} I_{kq} & 0 \\ 0 & P_{k-1} \end{bmatrix}$, $\tilde{\Pi}_k := \tilde{P}_k \Pi_k \tilde{P}_k =: \begin{bmatrix} \Pi_k^- & \Pi_k^{-+} \\ \Pi_k^{+-} & \Pi_k^+ \end{bmatrix}$

with $\Pi_k^- \in \mathbb{R}^{kqxkq}$, $\Pi_k^+ \in \mathbb{R}^{qxq}$, $\Pi_k^{+-T} = \Pi_k^{-+} \in \mathbb{R}^{kqxq}$.

Let $(\Pi_k^-)^{-\frac{1}{2}} \Pi_k^{-+} (\Pi_k^+)^{-\frac{1}{2}}$ have SVD $U_k \Lambda_k V_k^T$,

$\Lambda_k = \mathrm{diag}(\lambda_1^{(k)}, \ldots, \lambda_{d_{k-1}}^{(k)}, 0, \ldots, 0)$, $\lambda_1^{(k)} \geq \ldots \geq \lambda_{d_{k-1}}^{(k)} \geq 0$ and

suppose $1 - \{\lambda_{c_k}^{(k)}\}^2 \leq \varepsilon_k^{*2} \quad 1 - \{\lambda_{c_k+1}^{(k)}\}^2$.

(iii) Let $V_k = (v_1^{(k)},\dots,v_q^{(k)})$, $U_k^T(\Pi_k^-)^{-\frac{1}{2}} =: \mathrm{col}(\beta_1^{(k)},\dots,\beta_{kq}^{(k)})$,

$\beta_i^{(k)} =: (\beta_{i,k-1}^{(k)},\dots,\beta_{i,o}^{(k)})$, $\beta_{i,j}^{(k)} \in \mathbb{R}^{1\times q}$ $\quad i = 1,\dots,kq$,

$j = 0,\dots,k-1$, $V_k^T(\Pi_k^+)^{-\frac{1}{2}} =: \mathrm{col}(\alpha_1^{(k)},\dots,\alpha_q^{(k)})$,

$$P_k := (I - \sum_{j=1}^{c_k} v_j^{(k)} v_j^{(k)T}) . P_{k-1},\ d_k := d_{k-1} - c_k .$$

(iv) Define $A_k := \mathrm{col}(\alpha_1^{(k)},\dots,\alpha_{c_k}^{(k)})$, $R^{(k)} := \sum_{j=o}^{k-1} R_j^{(k)} \sigma^{*j}$

where $R_j^{(k)} := \mathrm{col}(\lambda_1^{(k)}\beta_{1,j}^{(k)},\dots,\lambda_{c_k}^{(k)}\beta_{c_k,j}^{(k)})$, $j = 0,\dots,k-1$.

Let A_k, $k \geq 0$ and $R^{(k)}$, $k \geq 1$ be as defined in the procedure. Let $A := \mathrm{col}(A_o,\dots,A_T)$, $R := \mathrm{col}(0,R^{(1)},\dots,R^{(T)})$. Then the following theorem is a direct consequence of the lemma in section 4.

<u>THEOREM</u>. The prediction model (A,R) as defined by the above procedure uniquely solves DP.

It is easy to see that $c(A,R) = (\sum_{t=o}^{T} c_t, c_o, c_1,\dots,c_T)$ where $c_o := q - d_o$ and c_t as defined in the procedure, $t \geq 1$. Moreover,

$\varepsilon_o^i(A,R) = \sqrt{\lambda_{d_o+i}^{(o)}}$, $\varepsilon_k^i(A,R) = \sqrt{1-\{\lambda_{c_k-i+1}^{(k)}\}^2}$.

<u>Remark</u>. 1. For step (k) (iv), note that $\alpha_i^{(k)}\hat{w}(t+1) = \lambda_i^{(k)}\beta_i^{(k)}\mathrm{col}(w(t-k+1),\dots,w(t))$ is equivalent to $\alpha_i^{(k)}\hat{w}(t+1) = e_i^T\{(R^{(k)}(\sigma^*)w)(t)\}$.

2. Generally one does not want to accept laws of too high order, e.g. one may require $\varepsilon_k^* < 0$ $\forall k \geq k_o$, in which case the procedure stops after step k_o-1.

3. Generically (A,R) is in standard prediction form.

6. REMARKS

The procedure of section 5 gives the solution to a special kind of optimal dynamic prediction for a deterministic time series. An application is to construct the model (A,R) on the basis of T observations and subsequently to predict $A\hat{w}(T+1) := (R(\sigma^*)w)(T)$.

Topics of current research on prediction of deterministic time series include the investigation of different orderings on the complexity other error measures, additional conditions on "allowable" prediction models and recursive identification.

REFERENCES

[1] H. Akaike, A new look at statistical model identification, IEEE AC-19, pp 716-723, 1974.

[2] K.J. Åström, Maximum likelihood and prediction error methods, Automatica 16, pp 551-574, 1980.

[3] E.J. Hannan and B.G. Quinn, The determination of the order of an autoregression, Journal of the Royal Statistical Society B 41, pp 190-195, 1979.

[4] T. Kailath, (ed.), Linear least-squares estimation, Benchmark papers in electrical engineering and computer science vol. 17, Dowden, Hutchinson and Ross, 1977.

[5] L. Ljung, A non-probalistis framework for signal spectra, Proceedings of the 24th Conference on Decision and Control, Fort Lauderdale, Florida, pp 1056-1060, 1985.

[6] G.S. Maddala, Econometrics, McGraw-Hill, New York, 1976.

[7] J. Makhoul, Linear prediction: a tutorial review, Proceedings of the IEEE 63, pp 561-580, 1975.

[8] J. Rissanen, Stochastic complexity and predictive modelling, The annals of statistics 14, 1986. To appear.

[9] R. Shibata, Asymptotically efficient selection of the order of the model for estimating parameters of a linear process, The annals of statistics 8, pp 146-164, 1980.

[10] H. Theil, Principles of econometrics, Wiley, New York, 1971.

[11] J.C. Willems, From time series to linear system. Part III: Approximate modelling, Automatica. To appear.

Some New Results on and Applications of an Algorithm of Agashe.

B. Hanzon
Department of Mathematics and Informatics, 2.317
Delft University of Technology
P.O. Box 356
2600 AJ Delft

1. Introduction

Identifying whether a system is (asymptotically) stable is important for modelling the system and for its robustness and sensitivity properties. In the case of linear systems, to check asymptotic stability, one has to check whether a certain polynomial has all its roots in the open left-half plane (l.h.p.).

In [1] Agashe presented a new algorithm (with a nice proof) for the determination of the number of right-half plane (r.h.p.) roots of any real or complex polynomial. A description of the Agashe algorithm in terms of the Euclidean algorithm is given in section 2. In section 3 it will be shown that from the <u>same</u> algorithm one can find not only the number of r.h.p.-roots but also the number of l.h.p.-roots and the number of purely imaginary roots.

In section 4.1 the results of section 3 are used to formulate a sufficient condition for <u>stability</u> of a matrix A, which involves only inspection of some intermediate results of Agashe's algorithm. In case this sufficient condition fails one can use a <u>necessary and sufficient</u> condition for stability of A that is presented next, in section 4.2. It involves a few computations with the polynomials that occur during the Agashe algorithm and substitution of the matrix A in a polynomial.

In section 5 the Lyapunov equation $MA + A^{*}M = -C$ is treated. By inspection of the Agashe algorithm one can check whether or not this linear equation in the unknown matrix M is nonsingular. If so, using polynomials produced by the Agashe algorithm one can compute the solution of the Lyapunov equation <u>without matrix inversion</u>, applying an idea of Kalman [6].

NATO ASI Series, Vol. F34
Modelling, Robustness and Sensitivity Reduction in Control Systems. Edited by R. F. Curtain

In section 6 an application to a feedback problem is presented. From Agashe's results it follows that certain changes in a polynomial will not change the (so-called) inertia (i.e. the distribution of its roots over left-half plane, imaginary axis and right-half plane). Using this it will be shown that not every complex reachable system can be stabilized by real feedback if the state dimension is larger than one. A simple three-dimensional counterexample is given.

2. A description of the Agashe algorithm in terms of the Euclidean algorithm.

Let $p_1 : s \mapsto p_1(s)$, $p_2 : s \mapsto p_2(s)$ denote two polynomials with real or complex coefficients. The well-known algorithm of Euclid (cf. e.g. [8]) to determine the g.c.d. of p_1 and p_2 can be described as follows. Let $p_1, p_2, \ldots, p_l;\ q_1, q_2, \ldots, q_l$ be the (uniquely determined) polynomials for which

$$\left\{\begin{array}{llll} p_1(s) & = q_1(s)p_2(s) & + p_3(s), & \deg(p_3) < \deg(p_2) \\ p_2(s) & = q_2(s)p_3(s) & + p_4(s), & \deg(p_4) < \deg(p_3) \\ & \vdots & & \\ p_{l-2}(s) & = q_{l-2}(s)p_{l-1}(s) & + p_l(s), & \deg(p_l) < \deg(p_{l-1}) \\ p_{l-1}(s) & = q_{l-1}(s)p_l(s). & & \end{array}\right. \tag{2.1}$$

Then p_l is the g.c.d. of p_1 and p_2. This algorithm is applied several times (in general) in the Agashe algorithm and each time that it is applied will be called a round of the Agashe algorithm. For each round the pair of polynomials to which the Euclidean algorithm is applied have to be specified. Except for the first round, this pair is formed by the last polynomial ('p_l') of the previous round together with its derivative ('$\frac{dp_l}{ds}$'). To explain the choice of the pair of polynomials in the first round, some definitions are needed.

2.2. Definition. A polynomial p (with real or complex coefficients) is called quasi-real (quasi-imaginary) if its values on the imaginary axis are real (purely imaginary). I.e. for all $\omega \in \mathbb{R}$, $p(i\omega) \in \mathbb{R}$ (for all

$\omega \in \mathbb{R}$, $ip(i\omega) \in \mathbb{R}$).

For each complex number s, the number $-\bar{s}$ will be called the <u>paraconjugate</u> of s. Note that paraconjugation corresponds to reflection with respect to the imaginary axis and that only purely imaginary numbers are equal to their paraconjugate.

2.3. <u>Notation</u>. For each polynomial p let the polynomial $\hat{p}$ be defined by

$$\hat{p}(s) = \overline{p(-\bar{s})} \tag{2.4}$$

and the polynomials p_r and p_i by

$$p_r = (p + \hat{p})/2; \quad p_i = (p - \hat{p})/2 \tag{2.5}$$

Then $p = p_r + p_i$, p_r is quasi-real and p_i is quasi-imaginary. This separation of the polynomial p into a quasi-real part p_r and a quasi-imaginary part p_i is clearly unique.

For a given polynomial p, the Agashe algorithm starts its first round with the two polynomials p_r and p_i. It is not difficult to show that therefore each of the polynomials $p_1, p_2, \ldots, p_l$ and $q_1, q_2, \ldots, q_{l-1}$ in (2.1) is quasi-real or quasi-imaginary. If p_l is quasi-real then its derivative is quasi-imagninary and vice versa. Therefore all polynomials that occur in the Agashe algorithm are quasi-real or quasi-imaginary.

The following terminology and notation will be used here

2.6. <u>Definition</u>.

(a) The <u>inertia</u> of a polynomial p is the triple (π, ν, δ), where
$\pi = \pi(p)$: = the number of zeroes of p, multiplicities <u>included</u>, which have <u>positive</u> real part,
$\nu = \nu(p)$: = the number of zeroes of p, multiplicities <u>included</u>, which have <u>negative</u> real part,
$\delta = \delta(p)$: = the number of purely imaginary zerces of p, multiplicities included.

(b) The inertia of a matrix A is the inertia of its characteristic polynomial.

Clearly $\pi(p) + \nu(p) + \delta(p) = \deg(p)$, cf. also [9] and [10]. In [1] Agashe presents a formula for the difference $\pi(p_i + p_{i+1}) - \pi(p_{i+1} + p_{i+2})$ $i \in 1,\ldots,l-1$, where $p_1,p_2,\ldots,p_l$ are as in (2.1) and $p_{l+1} = 0$ by convention and $p = p_1 + p_2$ is a decomposition of p into a quasi-real and a quasi-imaginary polynomial. For the details of this formula we refer to [1]. Making use of this one can for a given polynomial p, calculate $\pi(p) - \pi(p_l)$, where p_l is as in (2.1), so p_l is the g.c.d. of the quasi-real and the quasi-imaginary parts of p. In [1] it is also shown that if p is quasi-real (quasi-imaginary), then its derivative p' is quasi-imaginary (quasi-real) and $\pi(p) = \pi(p + p')$ The Agashe algorithm applied to a polynomial p ($\not\equiv 0$) consists of several rounds of the Euclidean algorithm. The last round ends with a nonzero polynomial of degree zero, i.e. a constant $\neq 0$. In each round the decrease in the number of r.h.p.-roots can be calculated and addition of those decrements gives the total number of r.h.p.-roots (with multiplicities) of the polynomial p.

In the following it will be shown that the Agashe algorithm gives (even) more information.

3. Determination of the number of purely imaginary roots from the Agashe algorithm.

3.1. Definition.

(i) A polynomial g will be called symmetric with respect to the imaginary axis if for some $c \in \mathbb{C} \setminus \{0\}$,

$$g(s) = c\,\hat{g}(s) \tag{3.2}$$

(if g is monic then $c = \pm 1$).

(ii) Let p be a polynomial. Let g be the highest degree polynomial factor of p that is symmetric with respect to the imaginary axis. Then g will be called the symmetric part of p with respect to the imaginary axis. (g is determined up to a constant nonzero factor). The polynomial $h = p/g$ is called the non-symmetric part of p, with respect to the imaginary axis.

It is not difficult to show that the g.c.d. of p and $\hat{p}$ is the symmetric part of p w.r.t. the imaginary axis. Consider the following.

3.3. Lemma. Let $p = p_r + p_i$, p_r quasi-real, p_i quasi-imaginary. Then the g.c.d. of p and $\hat{p}$ is equal (up to a constant nonzero factor) to the g.c.d. of p_r and p_i.

Proof. From (2.5) it follows that

$$p = p_r + p_i \quad \text{and} \quad \hat{p} = p_r - p_i \tag{3.4}$$

From this one can see that if k is a common factor of p_r and p_i then it is also a common factor of p and $\hat{p}$. Therefore the g.c.d. of p_r and p_i is a divisor of the g.c.d. of p and $\hat{p}$. Using (2.5) one shows in the same way that the g.c.d. of p and $\hat{p}$ is a divisor of the g.c.d. of p_i and p_r. It then follows that both g.c.d.'s must be equal up to a nonzero constant factor,

q.e.d.

3.4. Corollary. Let g be the polynomial that results after the first round of the Agashe algorithm applied to the polynomial p. Then g is the symmetric part of p with respect to the imaginary axis.

Proof. The polynomial g is equal to p_1 in (2.1) if p_1 and p_2 are taken equal to the quasi-real part p_r and quasi-imaginary part p_i of p. Therefore g is the g.c.d. of p_r and p_i and so, according to the previous lemma equal to the g.c.d. of p and $\hat{p}$, which is the symmetric part of p with respect to the imaginary axis,

q.e.d.

Let h = p/g. Then h is the non-symmetric part of p w.r.t. the imaginary axis. The decrement in r.h.p.-roots that is calculated in the first round of the Agashe algorithm is equal to $\pi(p) - \pi(g) = \pi(h)$, and the total decrement in r.h.p.-roots in the second and later rounds of the Agashe algorithm is $\pi(g)$. Because h is the nonsymmetric part of p w.r.t. the imaginary axis, h has no purely imaginary roots! So $\nu(h) = \deg(h) - \pi(h)$. And because g is symmetric with respect to the imaginary axis one has $\pi(g) = \nu(g)$, so $\delta(g) = \deg(g) - 2\pi(g)$. Combining this, one has:

3.5. Lemma. Let g be the polynomial that results after the first round of the Agashe algorithm applied to the polynomial p. Then

$$\delta(p) = \deg(g) - 2\pi(g) \tag{3.6}$$

and

$$\nu(p) = \deg(p) - \pi(p) - \deg(g) + 2\pi(g) \tag{3.7}$$

Because deg(g), $\pi(p)$ and $\pi(g)$ follow directly from the Agashe algorithm, and deg(p) is of course given, this lemma allows us to calculate the inertia $(\pi(p), \nu(p), \delta(p))$ directly from the Agashe algorithm.

3.6. Examples. Let us consider example 1 from [1], in which

$$p(s) = (1 + i)(s + 1 - i)(s + 2i)(s + 3)(s - 1 + 2i) =$$
$$= (1 + i)s^4 + 6is^3 + (-13 + 11i)s^2 + (-20 + 2i)s + (-24 - 12i) \tag{3.7}$$

So

$$p_r(s) = s^4 + 6i\,s^3 - 13s^2 + 2i\,s - 24 \text{ and}$$
$$p_i(s) = i\,s^4 \qquad + 11\,i\,s^2 - 20s - 12i \tag{3.8}$$

In [1] the algorithm is carried out for this case. One obtains $\pi(p) = 1$ (warning: in [1] the symbol m is used instead of π) and

$$g(s) = -204i\,s + 408 = -204i\,(s + 2i) \tag{3.9}$$

From this it follows that $\deg(g) = 1$ and $\pi(g) = 0$. So the number of purely imaginary roots of p is equal to $\delta(p) = \deg(g) - 2\pi(g) = 1$ and the number of l.h.p.-roots of p is equal to $\deg(p) - \pi(p) - \delta(p) = 4 - 1 - 1 = 2$. Using (3.7) one can verify that this is correct: the l.h.p.-roots are $(-1 + i)$ and -3, the purely imaginary root is $-2i$ (of course the same as the purely imaginary root of g!) and the r.h.p.-root is $1 - 2i$.

4. Checking the stability of a matrix.

The following definition will be used.

4.1 Definition.

(i) A square matrix A is called stable if the matrix function e^{At}, $t \geq 0$ is bounded (i.e. it is bounded componentwise)

(ii) A square matrix A is called asymptotically stable if for $t \to \infty$, e^{At} converges to zero (i.e. it converges to zero componentwise).

It is well-known that a (square) matrix A is asymptotically stable if its eigenvalues all lie in the (open) l.h.p. I.e. if p is the characteristic polynomial of A, then $\deg(p) = \nu(p)$. It is also well-known that a matrix A is stable if its eigenvalues all lie in the closure of the l.h.p., i.e. $\pi(p) = 0$ and the (algebraic) multiplicity of each of the purely imaginary eigenvalues is equal to their geometric multiplicity (i.e. the dimension of the corresponding eigenspace or equivalently, the maximal number of linearly independent corresponding eigenvectors). This can also be expressed as follows: all eigenvalues must lie in the closure of the left-half plane and all purely imaginary eigenvalues must correspond to a diagonal Jordan block in the Jordan form of the matrix A.

It is clear that asymptotic stability of A can be verified by applying the Agashe algorithm to the characteristic polynomial of A. To verify stability of A is somewhat more complicated. This is treated in sections 4.1 and 4.2 below.

4.1. A sufficient condition for stability of a matrix in terms of the Agashe algorithm.

From what is stated above it is clear that if the characteristic polynomial p of the (square) matrix A has all its roots in the closure of the left-half plane, i.e. $\deg(p) = \nu(p) + \delta(p)$, and if all the purely imaginary roots of p have multiplicity 1, then A is a stable matrix. Let us call this condition (C). This condition can easily be checked from the Agashe algorithm. First notice that $\pi(p) = 0$ implies that the symmetric part g of p, that results after the first round of the Agashe algorithm, has only purely imaginary roots, i.e. $\deg(g) = \delta(g)$. Furthermore $h = p/g$ has no purely imaginary roots as we have seen in section 3, so $p = hg$ satisfies condition (C) iff

(a) $\pi(p) = 0$

(b) all roots of g have multiplicity 1.

Now the following well known (and simple) lemma can be used.

4.1.1. Lemma. Let p be a polynomial and p' its derivative. Then the g.c.d. of p and p' is 1 (up to a nonzero constant factor) iff all roots of p have multiplicity one.

As described in section 2 in the second round of the Agashe algorithm the Euclidean algorithm is applied to g (the polynomial that results from the first round) and its derivative g'. It follows from the lemma that condition (b) above is satisfied iff the second round of the Agashe algorithm ends with a polynomial of degree zero, i.e. a nonzero constant. Therefore one has the following

4.1.2. Lemma. A polynomial p satisfies condition (C) iff

(i) the Agashe algorithm ends in at most two rounds (resulting in a polynomial of degree zero, i.e. a nonzero constant) and

(ii) the Agashe algorithm gives $\pi(p) = 0$.

4.1.3. Corollary. The matrix A is stable if the Agashe algorithm applied to the characteristic polynomial p of A ends in at most two rounds and gives $\pi(p) = 0$.

4.1.4. Remark. From the previous it follows that the matrix A is asymptotically stable iff the Agashe algorithm applied to the characteristic polynomial p of A ends in one round and gives $\pi(p) = 0$. (Then A is a fortiori stable).

4.2. A necessary and sufficient condition for the stability of a matrix.

The following well-known (and simple) result will be needed (this is a generalization of 4.1.1.)

4.2.1. Lemma. Let g be any polynomial, g' its derivative and k be the g.c.d. of g and g'. Then g/k is a polynomial of which all the roots have multiplicity one and each root of g is a root of the polynomial g/k.

Now, as before, let g be the polynomial that results after the first round of the Agashe algorithm applied to the characteristic polynomial p of A. Let $h = p/g$. Let k be the polynomial that results after the second round of the Agashe algorithm. Then k is the g.c.d. of g and its derivative g'. Furthermore let the polynomial q be given by

$$q = p/k \quad (= h.\ (g/k)) \tag{4.2.2}$$

Suppose g has only purely imaginary roots (if not, A is not stable). Then q has the same roots as p and with the same multiplicities as in p except for the purely imaginary roots which all have multiplicity one in q.

The necessary and sufficient condition for stability can now be stated.

4.2.3. Proposition. Let A,p,g,h,k be as described above. The matrix A is stable iff the following two conditions hold simultaneously.

(i) In the first two rounds the Agashe algorithm does not find r.h.p.-roots, i.e. $\pi(p) - \pi(k) = 0$

(ii) $q(A) = 0$. (4.24)

Proof. Note that (i) is equivalent to $\pi(q) = 0$ which is equivalent to $\pi(p) = 0$, because q has the same roots as p (be it with different multiplicities). It now suffices to show that under the condition that $\pi(p) = 0$, the condition $q(A) = 0$ is equal to the stability of A. By a change of basis A can be brought into Jordan form, and it is not difficult to see that without loss of generality it can be assumed that A is in Jordan form,

$$A = \begin{bmatrix} J_{\lambda_1} & 0 & \cdots & 0 \\ 0 & J_{\lambda_2} & \cdots & \vdots \\ \vdots & & \ddots & 0 \\ 0 & \cdots & 0 & J_{\lambda_m} \end{bmatrix} \tag{4.25}$$

where J_{λ_i}, $i \in \{1,2,\dots,m\}$ is the Jordan block that corresponds to the eigenvalue λ_i of A. Because $q = h.(g/k)$ and g has only purely imaginary roots here, the multiplicity of each root of p that is not purely imaginary has the same multiplicity as a root of q. Therefore, according to the Cayley-Hamilton theory, $q(J_{\lambda_i}) = 0$ if λ_i is not purely imaginary. (This holds both if A is stable and if A is unstable, as long as $\pi(p) = 0$ holds). So it remains to consider the Jordanblocks J_{λ_i} which correspond to a purely imaginary eigenvalue λ_i. From lemma 4.2.1. it follows that λ_i is a root of q with multiplicity one. Therefore $q(J_{\lambda_i}) = 0$ if and only if $J_{\lambda_i} = \lambda_i I$, i.e. J_{λ_i} is diagonal. So $q(J_{\lambda_i}) = 0$ for all $i = 1,2,\dots,m$ if and only if the Jordan blocks corresponding to purely imaginary roots λ_i are all diagonal. From the characterization of a stable matrix given at the beginning of this section 4 it follows that, under the assumption $\pi(p) = 0$, $q(J_{\lambda_i}) = 0$ for all $i = 1,2,\dots, m$ iff A is stable.

So A is stable iff $q(A) = 0$ and $\pi(p) = 0$,

q.e.d.

4.2.6. Remark. This proposition implies corollary 4.1.3, because if the Agashe algorithm ends in two rounds then $k = 1$ and then $q = p$ and

$q(A) = p(A) = 0$ follows from the Cayley-Hamilton theorem. So stability of A follows.

4.2.7. Example. Let $A = \begin{bmatrix} 0 & 1 \\ 1 & 2i \end{bmatrix}$. Then $p(s) = (s - i)^2 = p_r(s)$, quasi-real and

$p'(s) = 2s - 2i$, quasi-imaginary. It follows easily that $k(s) = 2(s - i)$.
So $q(s) = p(s)/k(s) = \frac{1}{2}(s - i)$. So $q(A) = \frac{1}{2}(A - iI) \neq 0$.
So A is not stable.

4.2.8. Example. $A = \begin{bmatrix} 0 & -1 & 0 & 0 \\ 1 & 0 & 0 & 0 \\ 0 & 0 & 0 & -1 \\ 0 & 0 & 1 & 0 \end{bmatrix}$, then

$p(s) = (s^2 + 1)^2 = p_r$; $k(s) = s^2 + 1$, so

$q(s) = p(s)/k(s) = s^2 + 1$. So

$q(A) = A^2 + I = -I + I = 0$. It follows that A is stable.

5. Application of the Agashe algorithm to a Lyapunov equation.

Consider the Lyapunov equation

$$MA + A^{*}M + C = 0 \tag{5.1}$$

for a given $n \times n$ matrix A, some arbitrary $n \times n$ matrix C and an unknown $n \times n$ matrix M. The equation (5.1) is a set of linear equations in the unknown components of the matrix M. It therefore has a solution for all possible C if and only if it is a nonsingular set of linear equations. In section 5.1. it will be shown how one can conclude from the Agashe algorithm (for the characteristic polynomial p of A) whether the set of linear equations is nonsingular or not. Then in section 5.2. it will be shown how one can find a solution in the nonsingular case, using polynomials that can be calculated recursively from polynomials that occur in the Agashe algorithm and <u>without</u> matrix inversion, using an idea of Kalman.

5.1. Checking the nonsingularity of a Lyapunov equation.

Consider the linear mapping

$$\mathcal{A} : B \mapsto BA + A^{*}B. \tag{5.1.1}$$

Let $\lambda_1,\dots,\lambda_m$ denote the eigenvalues of A. It is well-known that A is a nonsingular linear mapping iff

$$-\overline{\lambda_i} \neq \lambda_j , \quad i = 1,\dots,m, \quad j = 1,\dots,m \tag{5.1.2}$$

(cf. e.g. [2] app. F, or [3]). This means that the symmetric part of the characteristic polynomial p of A, with respect to the imaginary axis, must be a nonzero constant. Therefore from corollary 3.4 one has

5.1.3. Proposition. The linear mapping A is nonsingular iff the Agashe algorithm applied to the characteristic polynomial p of A consists of exactly one round.

5.2. Solution of a nonsingular Lyapunov equation.

In case A is nonsingular the Agashe algorithm consists of one round, i.e. it is a single application of the Euclidean algorithm (2.1) with $p_1 = p_r,\ p_2 = p_i,\ p = p_r + p_i,\ p_r$ quasi-real, p_i quasi imaginary. In this case the Lyapunov equation can be solved using the polynomials $q_1,q_2,\dots,q_{l-1}$ from (2.1). To start with define the polynomials $r_1,r_2,\dots,r_l$ and $t_1,t_2,\dots,t_l$ as follows

5.2.1. Definition. Let $r_1(s) = 1,\ r_2(s) = 0,\ t_1(s) = 0,\ r_2(s) = 1$ and define, recursively

$$\begin{aligned} r_k &= r_{k-2} - q_{k-2}r_{k-1} \quad \text{and} \\ t_k &= t_{k-2} - q_{k-2}t_{k-1} \end{aligned} \tag{5.2.2}$$

for all $k = 3,4,\dots,l$.

5.2.3. Remark. It follows easily that

$$\begin{bmatrix} r_{k-1} & t_{k-1} \\ r_k & t_k \end{bmatrix} = \begin{bmatrix} 0 & 1 \\ 1 & -q_{k-2} \end{bmatrix}\begin{bmatrix} 0 & 1 \\ 1 & -q_{k-3} \end{bmatrix} \cdots \begin{bmatrix} 0 & 1 \\ 1 & -q_1 \end{bmatrix}$$

for $k \geq 3$ and

$$\begin{bmatrix} r_1 & t_1 \\ r_2 & t_2 \end{bmatrix} = I_2.$$

5.2.4. Proposition.

$$p_k = r_k p_r + t_k p_i, \quad k = 1,2,3,4,\dots l.$$

Proof. If $k = 1$ then the equation states that $p_1 = p_r$ and if $k = 2$ then the equation states that $p_2 = p_i$, which is true by assumption. Now suppose the equation is true for all $k \le k_0$, $2 \le k_0 < l$. From (2.1) it follows that

$$p_{k_0+1} = p_{k_0-1} - q_{k_0-1} p_{k_0}$$

(let $p_{l+1} \equiv 0$ by convention).

It follows that

$$p_{k_0+1} = (r_{k_0-1} p_r + t_{k_0-1} p_i) - q_{k_0-1}(r_{k_0} p_r + t_{k_0} p_i) =$$

$$= (r_{k_0-1} - q_{k_0-1} r_{k_0}) p_r + (t_{k_0-1} - q_{k_0-1} t_{k_0}) p_i =$$

$$= r_{k_0+1} p_r + t_{k_0+1} p_i,$$

q.e.d.

5.2.5. Corollary. If A is nonsingular then

$$r_1 p_r + t_1 p_i = c, \quad c \text{ is a nonzero constant.}$$

Proof. If A is nonsingular then p_l, the symmetric part of p, is a nonzero constant. Therefore the previous porposition gives the result,

q.e.d.

5.2.6. Definition. If A is nonsingular, let

$$r = r_l/c, \quad t = t_l/c, \quad c \text{ as in the previous corollary.}$$

It follows that

$$r\, p_r + t\, p_i = 1 \tag{5.2.7}$$

By substitution of (2.5) it follows that

$$\frac{1}{2}(r + t)\, p + \frac{1}{2}(r - t)\hat{p} = 1 \tag{5.2.8}$$

For the following, note that $\hat{p}$ is the characteristic polynomial of $-A^*$, up to a nonzero factor. Therefore, according to the Cayley-Hamilton theorem,

$$\hat{p}(-A^*) = 0 \tag{5.2.9}$$

Now consider the following (socalled) Bezout form

$$\frac{p(x)\hat{p}(y) - p(y)\hat{p}(x)}{x - y} .$$

It is well-known (and not difficult to show) that this is a polynomial in x and y. Define the following polynomial in x and y

$$\psi(x,y) = -\frac{1}{4}\{r(y) + t(y)\}\{r(x) - t(x)\}\{\frac{p(x)\hat{p}(y) - p(y)\hat{p}(x)}{x - y}\}$$

5.2.10. Proposition. $\psi(x,y).(x-y) - 1$ is a multiple of $p(x)$ plus a multiple of $\hat{p}(y)$. In other words, $\psi(x,y)(x-y)$ is equal to 1 modulo $p(x)$ and $\hat{p}(y)$.

Proof. $\psi(x,y)(x-y) = \frac{1}{2}\{r(y) + t(y)\}\ p(y)\ \frac{1}{2}\{r(x) - t(x)\}\ \hat{p}(x)$ + a multiple of $p(x)$ + a multiple of $\hat{p}(y)$ = 1.1 + a multiple of $p(x)$ + a multiple of $\hat{p}(y)$, according to (5.2.8),

q.e.d.

Substitution of (3.4) gives the following formula for $\psi(x,y)$:

$$\psi(x,y) = \frac{1}{2}\{r(y) + t(y)\}\{r(x) - t(x)\}\ \frac{\{p_r(x)p_i(y) - p_r(y)p_i(x)\}}{x - y} \qquad (5.2.11)$$

Now let ξ,η be the linear mappings of matrices

$$\xi: M \mapsto MA \quad \text{and}$$
$$\eta: M \mapsto -A^* M \qquad (5.2.12)$$

Then $A = \xi - \eta$ and one has the following result, due to Kalman (cf. e.g. [6]).

5.2.13. Theorem. Let A be nonsingular.

Then $M = \psi(\xi,\eta)(-C)$ is the (unique) solution of the Lyapunov equation (5.1)

Proof. The Lyapunov equation (5.1) can be written as

$$(\xi - \eta)(M) = (-C) \qquad (5.2.14)$$

Premultiplication with $\psi(\xi,\eta)$ gives

$$\psi(\xi,\eta)(\xi - \eta)(M) = \psi(\xi,\eta)(-C) \qquad (5.2.15)$$

From the Cayley-Hamilton theorem (cf. also (5.2.9)) it follows that

$$p(\xi) = 0 \quad \text{and} \quad \hat{p}(\eta) = 0 \qquad (5.2.16)$$

Therefore proposition (5.2.10) tells us that

$$\psi(\xi,\eta)(\xi - \eta) = \text{identity} \tag{5.2.17}$$

and so (5.2.15) gives the desired result

$$M = \psi(\xi,\eta)(-C)$$

q.e.d.

Summarizing this, during the Agashe algorithm one can recursively compute the r_k and t_k, giving r_l and t_l at the end, from which r and t are found. Form the polynomial $\psi(x,y)$ in x and y as in (5.2.11). Then compute $M = \psi(\xi,\eta)(-C)$, the solution of the Lyapunov equation. Notice that A does not have to be stable.

5.2.18. Example. Let $A = \begin{bmatrix} 0 & 1 \\ 2 & 3 \end{bmatrix}$, then

$p(x) = x^2 - 3x - 2$; $p_r(x) = x^2 - 2$ and $p_i(x) = -3x$.

Let $p_1 = p_r$ and $p_2 = p_i$, then

$$p_1(x) = (-\frac{1}{3}x)\, p_2(x) + (-2)\;,\quad \text{so } q_1(x) = -\frac{1}{3}x\;,\; p_3(x) = -2,$$

$$p_2(x) = (\frac{3}{2}\, x)\, p_3(x)\;,\quad \text{so } q_2(x) = \frac{3}{2}\, x.$$

So $l = 3$ (compare (2.1)) and $p_l(x) = -2$.

According to (5.2.1),

$$r_3(x) = r_1(x) - q_1(x)r_2(x) = 1 \quad \text{and}$$

$$t_3(x) = t_1(x) - q_1(x)t_2(x) = -q_1(x) = \frac{1}{3}x\,.$$

Using 5.2.6. one finds

$$r(x) = -\frac{1}{2} \quad \text{and}$$

$$t(x) = -\frac{1}{6}x\,.$$

Substituting r,t,p_r,p_i in (5.2.11) one finds

$$\psi(x,y) = \frac{1}{2}(-\frac{1}{2} - \frac{1}{6}\, y)(-\frac{1}{2} + \frac{1}{6}x)\,\{\frac{(x^2 - 2)(-3y) - (y^2 - 2)(-3x)}{x - y}\} =$$

$$= \frac{1}{24}\,(-18 + 6x - 6y - 7xy + 3x^2y - 3xy^2 + x^2y^2)\,.$$

Now, for example, let $C = -I$ and let us compute

$$M = \psi(\xi,\eta)(-C) = \psi(\xi,\eta)(I) =$$

$$= \frac{1}{24}\left\{-18\, I_2 + 6\begin{pmatrix}0 & 1\\ 2 & 3\end{pmatrix} + 6\begin{pmatrix}0 & 2\\ 1 & 3\end{pmatrix} + 7\begin{pmatrix}0 & 2\\ 1 & 3\end{pmatrix}\begin{pmatrix}0 & 1\\ 2 & 3\end{pmatrix} + \right.$$

$$\left. - 3\left(\begin{pmatrix}0 & 2\\ 1 & 3\end{pmatrix}\begin{pmatrix}0 & 1\\ 2 & 3\end{pmatrix}\right)^2 - 3\begin{pmatrix}0 & 2\\ 1 & 3\end{pmatrix}^2\begin{pmatrix}0 & 1\\ 2 & 3\end{pmatrix} + \begin{pmatrix}0 & 2\\ 1 & 3\end{pmatrix}^2\begin{pmatrix}0 & 1\\ 2 & 3\end{pmatrix}^2\right\} =$$

$$= \frac{1}{12}\begin{pmatrix}-11 & 3\\ 3 & 1\end{pmatrix}.$$

One can easily check that this matrix M is the solution of the Lyapunov equation (5.1) for the given choices of A and C, i.e.

$$\frac{1}{12}\begin{pmatrix}-11 & 3\\ 3 & 1\end{pmatrix}\begin{pmatrix}0 & 1\\ 2 & 3\end{pmatrix} + \begin{pmatrix}0 & 2\\ 1 & 3\end{pmatrix}\frac{1}{12}\begin{pmatrix}-11 & 3\\ 3 & 1\end{pmatrix} = I_2.$$

6. Application of Agashe's results to a feedback problem.

The following result will be shown and then applied.

6.1. Lemma. Let p_r be a quasi-real polynomial and p_i a quasi-imaginary polynomial. Then the inertia of

$$p_{\lambda,\mu} = \lambda p_r + \mu p_i$$

is the same for all $\lambda > 0$, $\mu > 0$.

Proof. We make the following three observations.

(A). Consider the Euclidean algorithm (2.1). If p_1 and p_2 are multiplied by positive constants then all the polynomials $p_1, p_2, \ldots, p_l$ and $q_1, q_2, \ldots, q_{l-1}$ are all multiplied by positive constants.

(B). If a polynomial is multiplied by a positive constant its derivative is multiplied by the same positive constant.

(C). The rule for the calculation of the inertia is invariant under multiplication with positive constants of all the polynomials involved, because it only involves the degrees of the various polynomials and the signs of some of their coefficients.

From the structure of the Agashe algorithm as described in section 2 and from (A), (B) and (C) it follows that $p_{\lambda,\mu}$ has the same inertia for all $\lambda > 0$, $\mu > 0$,

q.e.d.

Using continuity of the roots of a polynomial as a 'function' of the coefficients (cf. e.g. [7], p.4) the previous lemma implies

6.2. Lemma. Let p be a polynomial and let $p = p_r + p_i$, p_r quasi-real, p_i quasi-imaginary. Then

$$\pi(p_r) = \nu(p_r),\ \pi(p_i) = \nu(p_i) \tag{6.3}$$

and

$$\pi(p) \geq \max(\pi(p_r),\ \pi(p_i)) \tag{6.4}$$

$$\nu(p) \geq \max(\pi(p_r),\ \pi(p_i)) \tag{6.5}$$

and so $\delta(p) \leq \deg(p) - 2\max(\pi(p_r),\ \pi(p_i))$ (6.6)

Proof. Let p_r be quasi-real, then on the imaginary axis one has (with $\omega \in \mathbb{R}$)

$$\hat{p}_r(i\omega) = \overline{p_r(-\overline{(i\omega)})} = \overline{\overline{p_r(i\omega)}} = p_r(i\omega).$$

It follows that $p_r = \hat{p}_r$ and so p_r is symmetric with respect to the imaginary axis. If p_i is quasi-imaginary, ip_i is quasi-real and therefore p_i is also symmetric with respect to the imaginary axis. Therefore (6.3) holds.

From the previous lemma one has $\pi(p) = \pi(\lambda p_r + \mu p_i)$ for all $\lambda,\mu > 0$. Because the roots of a polynomial are continuous with respect to the coefficients (cf. [7] ,p.4), one has (taking $\mu = 1$)

$$\pi(p) = \lim_{\lambda \downarrow 0} \pi(\lambda p_r + p_i) \geq \pi(p_i).$$

Similarly one finds

$$\pi(p) = \lim_{\mu \downarrow 0} \pi(p_r + \mu p_i) \geq \pi(p_r) \text{ and so (6.4) follows.}$$

Similarly one shows

$$\nu(p) \geq \max\ (\nu(p_r),\ \nu(p_i)).$$

Together with (6.3) one finds (6.5).

Finally (6.4) and (6.5) imply (6.6), q.e.d.

6.7. Corollary. If p is the characteristic polynomial of a stable matrix A and $p = p_r + p_i$, p_r quasi-real, p_i quasi-imaginary, then all the roots of p_r and p_i are purely imaginary.

Proof. If $\pi(p) = 0$ then $\pi(p_r) = \pi(p_i) = 0$ and so $\nu(p_r) = \nu(p_i) = 0$. So $\delta(p_r) = \deg(p_r)$ and $\delta(p_i) = \deg(p_i)$,

q.e.d.

6.8 Example.

$$A = \begin{pmatrix} 0 & 1 & 0 \\ 0 & 0 & 1 \\ 1 & a_1 & -1 \end{pmatrix}, \quad a_1 \in \mathbb{R}, \text{ has characteristic polynomial}$$

$p(s) = s^3 + s^2 - a_1 s - 1$ and so $p_r(s) = s^2 - 1$, $p_i(s) = s^3 - a_1 s$.
Because the roots ± 1 of p_r do not lie on the imaginary axis the matrix A is unstable for each real value of a_1.

The corollary can be applied to the following feedback problem. Consider the system

$$\begin{pmatrix} \dot{x}_1 \\ \dot{x}_2 \end{pmatrix} = \begin{pmatrix} F & -H \\ H & F \end{pmatrix} \begin{pmatrix} x_1 \\ x_2 \end{pmatrix} + \begin{pmatrix} b & 0 \\ 0 & b \end{pmatrix} \begin{pmatrix} u_1 \\ u_2 \end{pmatrix} \tag{6.9}$$

$x_1, x_2, b \in \mathbb{R}^n$, $u_1, u_2 \in \mathbb{R}$, F, H are n x n real matrices. Consider also a state feedback of the form

$$\begin{pmatrix} u_1 \\ u_2 \end{pmatrix} = \begin{pmatrix} k & 0 \\ 0 & k \end{pmatrix} \begin{pmatrix} x_1 \\ x_2 \end{pmatrix} \tag{6.10}$$

where k is an nx1 real row-vector. (This can be interpreted as a decentralized control problem, in which the control u_1 depends only on x_1 and u_2 only on x_2 and in which the same feedback law is applied to both x_1 and x_2 (cf. [6]). Let $\left[\begin{pmatrix} F & -H \\ H & F \end{pmatrix}, \begin{pmatrix} b \\ 0 \end{pmatrix} \right]$ be reachable.

The question arises whether or not such a system is always stabilizable with a feedback of the form presented. The answer to this question is negative, which can be seen as follows.

Let $A = F + iH \in \mathbb{C}^{n\times n}$,
$x = x_1 + ix_2 \in \mathbb{C}^n$ and
$u = u_1 + iu_2 \in \mathbb{C}$.

Then (6.9) can be written as

$$\dot{x} = Ax + bu \tag{6.11}$$

and (6.10) can be written as

$$u = kx \tag{6.12}$$

The combination of (6.11) and (6.12) leads to the autonomous system

$$\dot{x} = (A + bk)x \tag{6.13}$$

In this new terminology the question posed above can be replaced by the question: does there exist an $A \in \mathbb{C}^{n\times n}$, $b \in \mathbb{R}^n$, (A,b) reachable such that for all $k \in \mathbb{R}^n$ the matrix $A + bk$ is unstable? The answer to this question will be shown to be yes, which implies that the answer to the question above is no. Because of corollary (6.7) it suffices to find a pair (A,b) $A \in \mathbb{C}^{n\times n}$, $b \in \mathbb{R}^n$, (A,b) reachable such that the choice of $k \in \mathbb{R}^n$ only affects the quasi-real (quasi-imaginary) part of the characteristic polynomial of $A + bk$, while the quasi-imaginary (quasi-real) part has roots outside the imaginary axis. Such a pair (A,b) can be constructed as follows if $n \geq 2$: Let

$$\bar{A} = \begin{bmatrix} 0 & 1 & 0 & \cdots & \cdots & 0 \\ \vdots & 0 & 1 & \ddots & & \vdots \\ \vdots & & \ddots & \ddots & \ddots & 0 \\ 0 & \cdots & \cdots & 0 & 0 & 1 \\ 0 & \cdots & \cdots & 0 & 1 & 0 \end{bmatrix}, \quad P = \mathrm{diag}(\ldots -i,1,-i,1) \in \mathbb{C}^{n\times n},$$

$$b = \begin{bmatrix} 0 \\ 0 \\ \vdots \\ 0 \\ 1 \end{bmatrix} \quad \text{and} \quad A = P^{-1}\bar{A}\,P.$$

Of course $A + bk$ is asymptotically stable iff $P(A + bk)P^{-1}$ is asymptotically stable. Let $k = (k_0,k_1,\ldots,k_{n-1}) \in \mathbb{R}^n$ and consider

$$P(A + bk)P^{-1} = \bar{A} + (Pb)(kP^{-1}) =$$

$$\begin{bmatrix} 0 & 1 & 0 & \cdots & \cdots & \cdots & 0 \\ \vdots & \ddots & \ddots & \ddots & & & \vdots \\ \vdots & & & \ddots & \ddots & 1 & 0 \\ 0 & \cdots & \cdots & \cdots & \cdots & 0 & 1 \\ \cdots & ik_{n-4} & k_{n-3} & & 1+ik_{n-2} & & k_{n-1} \end{bmatrix}$$

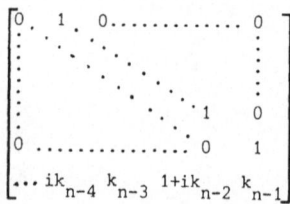

which has characteristic polynomial

$$p(s) = s^n - k_{n-1}s^{n-1} - (1 + ik_{n-2})s^{n-2} - k_{n-3}s^{n-3} - ik_{n-4}s^{n-4} \dots$$

If n is odd $p_i(s) = s^n - s^{n-2}$ and if n is even $p_r(s) = s^n - s^{n-2}$. In both cases it follows that p(s) has r.h.p. roots ($\pi(p) \geq 1$) for all possible choices of $k \in \mathbb{R}^n$. Notice that (A,b) is reachable because $(PAP^{-1},Pb) = (\bar{A},b)$ is reachable.

The conclusion is that for all $n \geq 2$ it is not always possible to stabilize the system (6.9) with a feedback of the form (6.10). Equivalently a complex system of the form (6.11) with $b \in \mathbb{R}^n$ can not always be stabilized by a real linear feedback (6.12) if $n \geq 2$. Of course, if n = 1 then one does have stabilizability, because then A + bk is a scalar with real part Re(A) + bk and $b \neq 0$, and $k \in \mathbb{R}$ can be chosen such that this real part is negative.

References

[1] Agashe, S.D. A new general Routh-like algorithm to determine the number of RHP roots of a real or complex polynomial. IEEE Trans. Automat. Contr., vol. AC-30, pp. 406-409, Apr. 1985.

[2] Chen, C.T. Linear system theory and design. Holt, Rinehart and Winston, 1984.

[3] Gantmacher, F.R. The theory of matrices, vol. 1,2, New York, Chelsea, 1959.

[4] Hahn, W. The stability of motion. Berlin, Springer Verlag, 1967.

[5] Hazewinkel, M. and C.F. Martin. On decentralization, symmetry and special structure in linear systems. Proc. 22-nd IEEE Conference on Decision and Control, San Antonio, 1983, p. 1405.

[6] Kalman, R.E., J. Coffy, P. Nicholson. Méthodes algebriques modernes appliquées à la théorie des systèmes linéaires. Fontainebleau, Centre d'Automatique, 1969.

[7] Marden, M. The Geometry of the zeros of a polynomial in the complex plane. Providence, R.I.: American Mathematical Society, 1949.

[8] Van der Waerden, B.L. Algebra I, II. Springer Verlag, Berlin 1971.

[9] Glover, K. All optimal Hankel-norm approximations of linear multivariable systems and their L^∞-error bounds. Int. J. Control, vol. 39, no. 6, pp. 1115-1193, 1984.

[10] Ostrowski, A. and Schneider, H. Some theorems on the inertia of general matrices. J. Math. Anal. Appl. 4, pp. 72-84, 1962.

An Application of H^∞-Design and Some Computational Improvements

I.Postlethwaite, D.W.Gu, S.D.O'Young and M.S.Tombs
Department of Engineering Science
University of Oxford
Parks Road
Oxford OX1 3PJ
U.K.

Abstract

In this paper, we present the results of a case study in which a recently developed H^∞- design package is used to design a full authority flight control system for a high performance helicopter. Simulation, using the full nonlinear model of the helicopter, indicates that performance and robustness requirements can easily be satisfied using the H^∞-approach.

We also give some new results on the solution to the algebraic Riccati equation. These lead to a numerically stable algorithm for reducing the size of $\bar{R}$ in the model-matching problem

$$\min_{\bar{Q}\epsilon H^\infty} \| \bar{R} - \bar{Q} \|_\infty$$

which is the central computational step in the H^∞-design procedure. For the 1-block problem, which arises when $\bar{Q}$ has neither zero rows nor zero columns, we obtain explicit formulae for a minimal realization of $\bar{R}$.

1. Introduction

The H^∞- design procedure and its computations, (Doyle 1984, Chu 1985, Francis and Doyle 1985), are now well understood by control theoreticians, and successful applications are beginning to be reported, (Safonov and Chiang 1986, Postlethwaite et al 1986b, 1987). An H^∞-design package has recently been developed at Oxford University and in the first part of this paper we present the results of its use in the design of a full authority flight control system for a high performance helicopter. The basic linear model of the helicopter excluding rotor dynamics has 8 states and 4 inputs,

NATO ASI Series, Vol. F34
Modelling, Robustness and Sensitivity Reduction in Control Systems. Edited by R. F. Curtain

is unstable and non-minimum phase. The rotor dynamics introduce a further 6 states, but we propose to treat these as uncertain and to leave them out of our nominal plant model. Simulation of the design, using the full nonlinear model of the helicopter with rotor dynamics, indicates that performance and robustness requirements are easily satistied.

In the second part of the paper we consider computational improvements. The central step in the state-space based approach to H^∞ design is the model-matching problem:

$$\min_{\bar{Q}\varepsilon H^\infty} || \bar{R} - \bar{Q} ||_\infty$$

where $\bar{R}$ is unstable and depends on the solutions to two algebraic Riccati equations. In this paper, we give some new results on the solution to the algebraic Riccati equation which can then be used to reduce the size of $\bar{R}$ and as a consequence the size of the optimal controller. The reduction of $\bar{R}$ is carried out using orthogonal transformations and so is numerically reliable. Furthermore, for the 1-block model-matching problem, we obtain explicit formulae for a minimal realization of $\bar{R}$. Limebeer and Hung (1986) also derived a minimal realization of $\bar{R}$ en route to their theoretical bound on the McMillan degree of the optimal controller. Their approach, however, although adequate for the theoretical purposes of their paper, does not lead to a numerically reliable algorithm. In addition, we believe the realizations of our paper, which retain the basic features of the original system data, will be more useful in computing the optimal controller of smallest McMillan degree.

The layout of the paper is as follows. In the next section, we give a brief description of the important steps in the H^∞-design procedure. Then in Section 3 we describe the design for the helicopter control problem. In Section 4, we present the new results on the solution to the algebraic Riccati equation which are then used to reduce the size of $\bar{R}$ in the model-matching problem. Conclusions are given in Section 5.

2. H^∞-Design Procedure

In this section, we give a brief summary of the H^∞-design procedure and introduce related formulae (Doyle 1984, Chu 1985, Francis and Doyle 1985). The conceptual H^∞-design procedure consists of four basic operations.

First, the design problem is formulated as a "standard problem" depicted in Figure 1, where the objective is to design a controller $\bar{K}$ such that the closed-loop system formed by $\bar{P}$ and $\bar{K}$ is internally stable, and the H^{∞}-norm of the transfer function $\bar{T}$ between the inputs v and errors e is minimized.

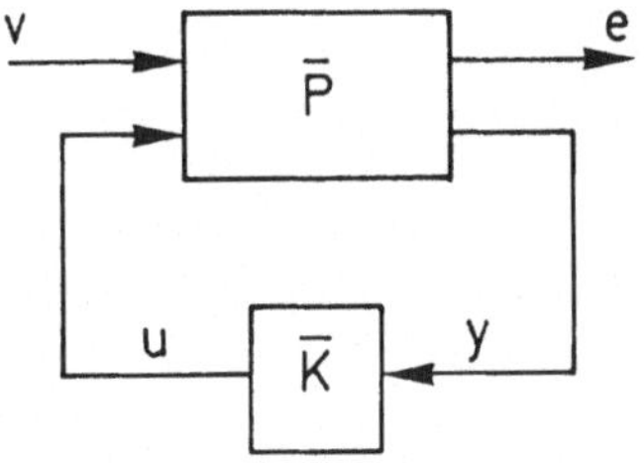

Figure 1. Standard Configuration

Secondly, the standard problem is transformed into an equivalent model-matching problem shown in Figure 2, where $\bar{R}$ is a fixed unstable transfer function derived from $\bar{P}$, and $\bar{Q}$ is a free parameter constrained to be stable and often structured to have identically zero columns and/or rows.

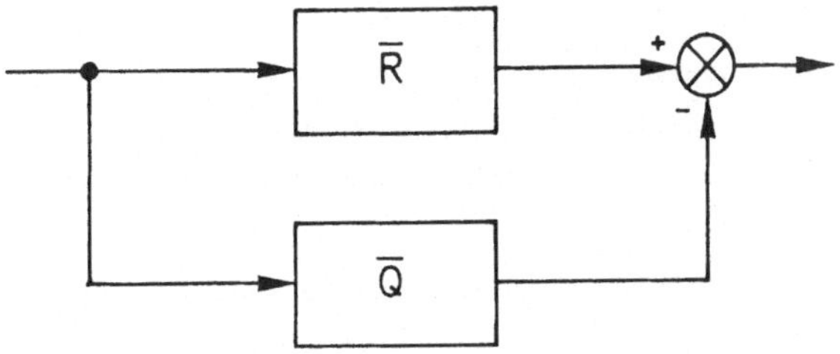

Figure 2. Model-Matching Problem

The third operation is to solve the model-matching problem $\min_{\bar{Q}\in\Pi^{\infty}} \|\bar{R} - \bar{Q}\|_{\infty}$. That is, to find a stable $\bar{Q}$ such that the supremum over all frequencies of the modelling error $\bar{\sigma}[(\bar{R}-\bar{Q})(j\omega)]$ is minimized, where $\bar{\sigma}[\cdot]$ denotes maximum singular value.

The final operation is to synthesize a (sub)optimal controller $\bar{K}$ from $\bar{Q}$ the (sub)optimal solution to the model-matching problem.

Let the state-space model of $\bar{P}$ in the first operation be

$$\bar{P} = \left(\begin{array}{c|cc} A & B_1 & B_2 \\ \hline C_1 & D_{11} & D_{12} \\ C_2 & D_{21} & D_{22} \end{array}\right) \tag{2.1}$$

where, A: nxn;

B_j: $n x m_j$;
C_i: $l_i x n$;
D_{ij}: $l_i x m_j$; $i,j = 1,2$, are real matrices and

$$\left(\begin{array}{c|c} A & B \\ \hline C & D \end{array}\right) := D + C(sI-A)^{-1}B \ .$$

Also, we can write

$$\bar{P} = \begin{pmatrix} \bar{P}_{11} & \bar{P}_{12} \\ \bar{P}_{21} & \bar{P}_{22} \end{pmatrix} ,$$

where, $$\bar{P}_{ij} = \left(\begin{array}{c|c} A & B_j \\ \hline C_i & D_{ij} \end{array}\right)$$

is an $l_i x m_j$ rational transfer matrix. The controller $\bar{K}$ is thus an $m_2 x l_2$ rational transfer function matrix. The transfer function $\bar{T}$ from v to e can be derived as

$$\bar{T} = F_l(\bar{P},\bar{K}),$$

where, $F_l(\bar{P},\bar{K})$ is called a linear fractional transformation (LFT) of $\bar{P}$ and $\bar{K}$ and is defined as

$$F_l(\bar{P},\bar{K}) = \bar{P}_{11} + \bar{P}_{12}\bar{K}(I-\bar{P}_{22}\bar{K})^{-1}\bar{P}_{21} \ . \tag{2.2}$$

For an algebraic Riccati equation

$$F^TX + XF - XWX + Q = 0 \ ,$$

we denote the stabilizing solution X (defined in Sec.4) as

$$X = \mathbf{Ric} \begin{pmatrix} F & -W \\ -Q & -F^T \end{pmatrix} \ .$$

The pseudo-inverse of a matrix X will be denoted as X^+ . Also, for a non-square matrix X comprising a set of orthonormal column (row) vectors, we call the matrix X_D the complementary orthogonal part of X if $[X \ X_D]$ ($\begin{pmatrix} X \\ X_D \end{pmatrix}$) is an orthogonal matrix.

We return now to the formulation of the model-matching problem from the standard configuration in (2.1). The $\bar{R}$ in the second operation is of the form $\bar{R} = \bar{T}_2{}^*\bar{T}_1\bar{T}_3{}^*$, where the $\bar{T}_i$ are given from an argument in (Chu 1985), by

$$\bar{T} = \bar{T}_1 - \bar{T}_2\bar{Q}\bar{T}_3 \ , \tag{2.3}$$

where,

$$\bar{T}_1 = \left(\begin{array}{cc|c} A+B_2F & -HC_2 & -HD_{21} \\ 0 & A+HC_2 & B_1+HD_{21} \\ \hline C_1+D_{12}F & C_1 & D_{11} \end{array}\right) , \tag{2.4}$$

$$\bar{T}_2 = \left(\begin{array}{c|cc} A+B_2F & B_2R_D^{-1/2} & -X^+C_1^TD_{co} \\ \hline C_1+D_{12}F & D_{12}R_D^{-1/2} & D_{co} \end{array}\right) , \tag{2.5}$$

$$\bar{T}_3 = \left(\begin{array}{c|c} A+HC_2 & B_1+HD_{21} \\ \hline R_D^{-1/2}C_2 & R_D^{-1/2}D_{21} \\ -\tilde{D}_{co}B_1^TY^+ & \tilde{D}_{co} \end{array}\right) , \tag{2.6}$$

and

$$F = -R_D^{-1}(D_{12}^T C_1 + B_2^T X) , \tag{2.7}$$

$$R_D = D_{12}^T D_{12} ,$$

$$X = \mathbf{Ric}\begin{pmatrix} A-B_2R_D^{-1}D_{12}^TC_1 & -B_2R_D^{-1}B_2^T \\ -C_1^TD_{co}D_{co}^TC_1 & -(A-B_2R_D^{-1}D_{12}^TC_1)^T \end{pmatrix} , \tag{2.8}$$

D_{co} is the complementary orthogonal part of $D_{12}R_D^{-1/2}$;

also,

$$H = -(B_1D_{21}^T + YC_2^T)\tilde{R}_D^{-1} , \tag{2.9}$$

$$\tilde{R}_D = D_{21}D_{21}^T ,$$

$$Y = \mathbf{Ric}\begin{pmatrix} (A-B_1D_{21}^T\tilde{R}_D^{-1}C_2)^T & -C_2^T\tilde{R}_D^{-1}C_2 \\ -B_1\tilde{D}_{co}^T\tilde{D}_{co}B_1^T & -(A-B_1D_{21}^T\tilde{R}_D^{-1}C_2) \end{pmatrix} , \tag{2.10}$$

$\tilde{D}_{co}$ is the complementary orthogonal part of $\tilde{R}_D^{-1/2}D_{21}$.

Note that since $\bar{T}_2$ and $\bar{T}_3$ are both square and inner the formula $\bar{R}=\bar{T}_2^*\bar{T}_1\bar{T}_3^*$ is easily derived from

$$||\bar{T}||_\infty = ||\bar{T}_2^*\bar{T}\bar{T}_3^*||_\infty = ||\ \bar{T}_2^*\bar{T}_1\bar{T}_3^* - \bar{Q}\ ||_\infty \ .$$

When $l_1=m_2$ and $l_2=m_1$, there are no D_{co} and $\tilde{D}_{co}$. Hence, $\bar{Q}$ has no identically zero columns and rows. This is the simplest model-matching problem and is usually referred to as the 1-block problem. When $l_1=m_2$ ($l_2=m_1$) and $l_2<m_1$ ($l_1>m_2$), $\bar{Q}$ has identically zero columns (rows) and these cases are called 2-block problems. When $l_1>m_2$ and $l_2<m_1$, $\bar{Q}$ has both identically zero columns and rows and this is called a 4-block problem. The problems which arise from other relationships between l_i and m_j (i,j=1,2) cannot yet be solved efficiently. Fortunately, these problems are usually considered impractical (in the sense of being overdetermined).

The fourth operation involves the recovery from the (sub)optimal $\bar{Q}_{opt}$ of the controller $\bar{K}_{opt}$ which is given by the formula

$$\bar{K} = F_1(\bar{J},\bar{Q}),$$

where,

$$\bar{J} = \left(\begin{array}{c|cc} A+B_2F+HC_2+HD_{22}F & -H & B_2+HD_{22} \\ \hline F & 0 & I \\ -(C_2+D_{22}F) & I & -D_{22} \end{array}\right). \tag{2.11}$$

The controller $\bar{K}$ is an interconnection of $\bar{J}$ and $\bar{Q}$ and so should have a state dimension equal to the sum of the states of $\bar{J}$ and $\bar{Q}$ together. However, a straightforward application of the LFT formula (2.2), using state-space models, will result in a controller of higher dimension. Not wishing to use a minimal realization routine which might be sensitive and which would obscure the structure of $\bar{J}$ and $\bar{Q}$ from $\bar{K}$, we suggest using the realization given in (Postlethwaite et al 1986) for computing the LFT. For a reliable minimal realization routine see (Tombs and Postlethwaite, 1986).

3. Helicopter Design

A comprehensive nonlinear model of a high performance helicopter has been provided by the Royal Aircraft Establishment (Bedford) for control system design and simulation studies. The aim is to design a full authority flight control system. The major deficiencies of rotary wing aircraft are nearly always associated with excessive cross-axis couplings and inadequate dynamic stability. The requirements, then, on any high performance helicopter is to stabilize the aircraft and decouple the controlled inputs, thus reducing workload, but still allowing fast response to pilot demands.

We will consider the helicopter in a high speed flight condition of 80 knots, straight and level at a height of 100 ft. The basic linear model of the helicopter has 8 states and 4 inputs (collective pitch, longitudinal cyclic pitch, lateral cyclic pitch and tail rotor collective pitch), and is unstable and non-minimum phase. If we include the rotor dynamics a 14 state model is obtained, but we propose to treat the rotor dynamics as uncertain (which they are) and leave them out of our nominal plant description.

The performance objectives are to minimize cross-coupling, to achieve improved stability margins, and to meet specifications on time response characteristics required for good handling quality.

Robustness is a primary issue in the design because of model uncertainty (especially that due to the omission of high frequency rotor dynamics), sensor noise and wind gusts. The problem is also complicated by rate and amplitude constraints on the actuators.

3.1 Prolem Formulation and Weighting Function Selection

Let the helicopter transfer function be $\bar{G}$ and define the sensitivity function $\bar{S}$ as $\bar{S} := (I+\bar{G}\bar{K})^{-1}$. We propose to find a stabilizing controller $\bar{K}$ that minimizes

$$\left\| \begin{matrix} W_1\bar{S}W_i \\ W_2\bar{K}\bar{S}W_i \end{matrix} \right\|_\infty \qquad (3.1)$$

where the diagonal weights W_1, W_2, W_i are chosen to meet the design objectives. The problem is illustrated in Figure 3. By defining $r = v$ and $e = (e_1\ e_2)^T$ we see how the problem can easily be reformulated into the standard form of Figure 1.

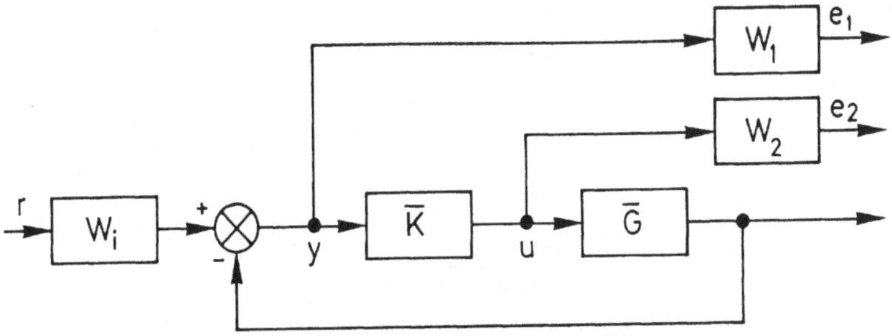

Figure 3. Helicopter Design Problem

The following 6 outputs are selected for control: flight path angle (γ), roll rate (p), pitch rate (q), lateral velocity (v), pitch attitude (θ), and bank angle (ϕ). The outputs, roll rate and pitch rate, are not to be controlled directly but are included to improve control. The reference input vector r will have two zero elements corresponding to the two rates.

Selection of W_1

The Bode magnitude plots of the weighting function W_1 are shown in Figure 4. First-order low-pass filters are used on γ, v, θ and ϕ to ensure that these outputs can be controlled accurately (d.c. sensitivity of 0.025) with good disturbance rejection up to about 1 rad/s. With only 4 plant inputs, no attempt is made to control directly the extra rate outputs p and q at low frequencies, but second-order band-pass filters are used on each of these variables to reject disturbances and cross-coupling effects in the frequency range 1 to 5 rad/s. Note that the low-pass filters on γ and v are given a finite attenuation which has the affect of reducing overshoot in these channels.

Selection of W_2

The same first-order high-pass filter is used on each of the plant inputs and is shown in Bode diagram form in Figure 5. A cut-off frequency of about 10 rad/s is used to limit the system bandwidth robustness, and also to limit the magnitudes of the poles of the controller. A low frequency gain of -30 dB was used so that $W_1\bar{S}W_i$ clearly dominates the cost function at low frequencies.

Selection of W_i

The diagonal weighting function W_i is chosen to be a constant matrix with a weight of 0.1 on each of the rates and 1 on each of the other output demands. The reduced weighting on the rates (which are not directly controlled) is so that some disturbance rejection is obtained on these outputs without them significantly affecting the cost. That is, the primary aim of W_i is to force good tracking in γ, v, θ and ϕ.

3.2 Results

The standard problem has a $\bar{P}$ of 20 states and corresponds to a 2-block optimization. The singular values of the optimum cost function are shown in Figure 6. As expected the maximum singular value is flat and approximately 1.11. The optimal controller had 59 states but this was easily reduced to 18, using the minimal realization routine of Tombs and Postlethwaite (1986), without any significant change in the cost.

A typical step response for the linear system is shown in Figure 7 with the corresponding actuator signals in Figure 8. Note the coordinated use of all four plant inputs to produce a step change in flight path angle. The response was not significantly different on the full nonlinear simulation with rotor dynamics indicating the robustness of the design.

For a more detailed investigation of the helicopter control problem see Tombs (1987).

To give an idea of the performance of the H^∞ design package, the complete helicopter design problem with the above weights took approximately 3.5 minutes of CPU time on a VAX 11/780.

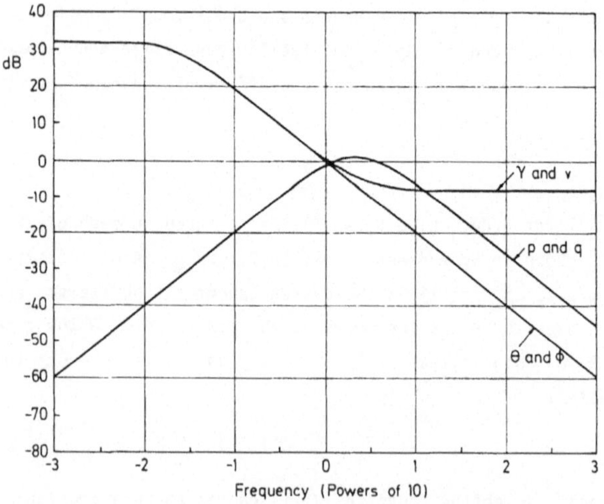

Figure 4. Bode magnitude plots for the diagonal weight W_1

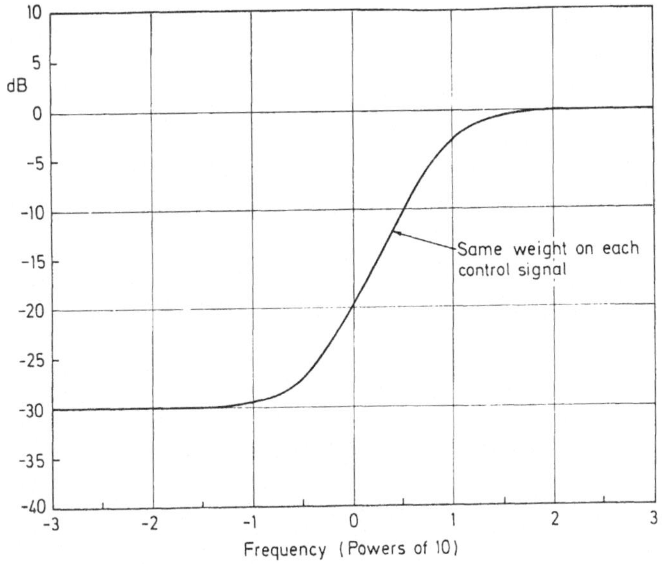

Figure 5. Bode magnitude plots for the diagonal weight W_2

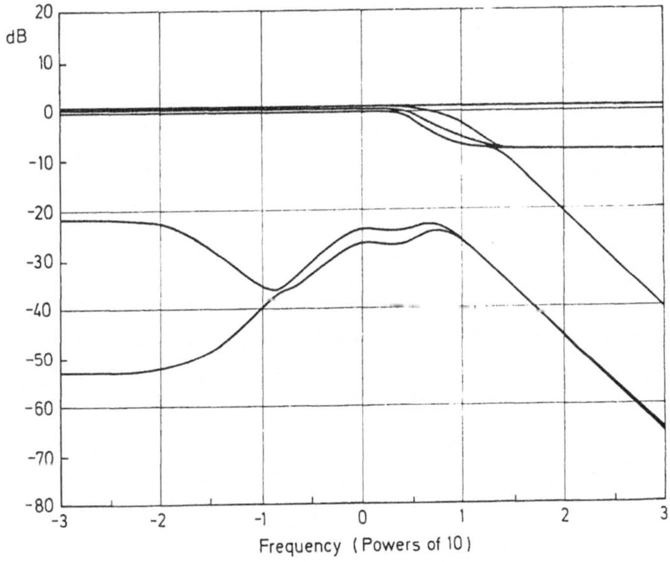

Figure 6. Singular values of the optimum cost function

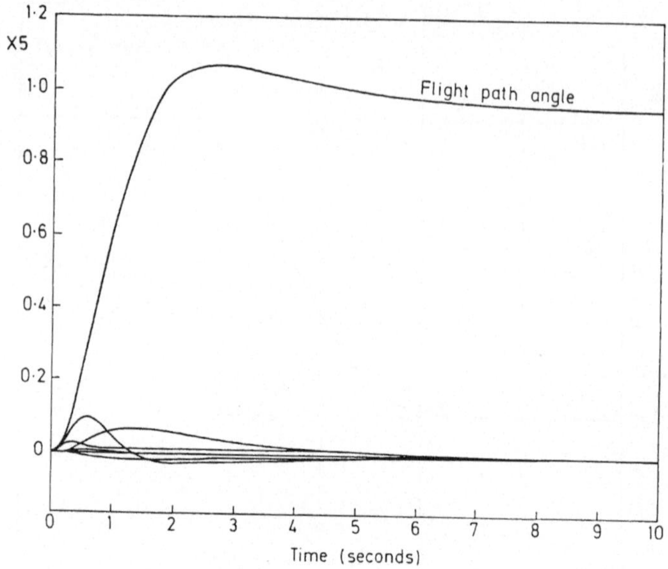

Figure 7. Output response to a 5° step demand in flight path angle.

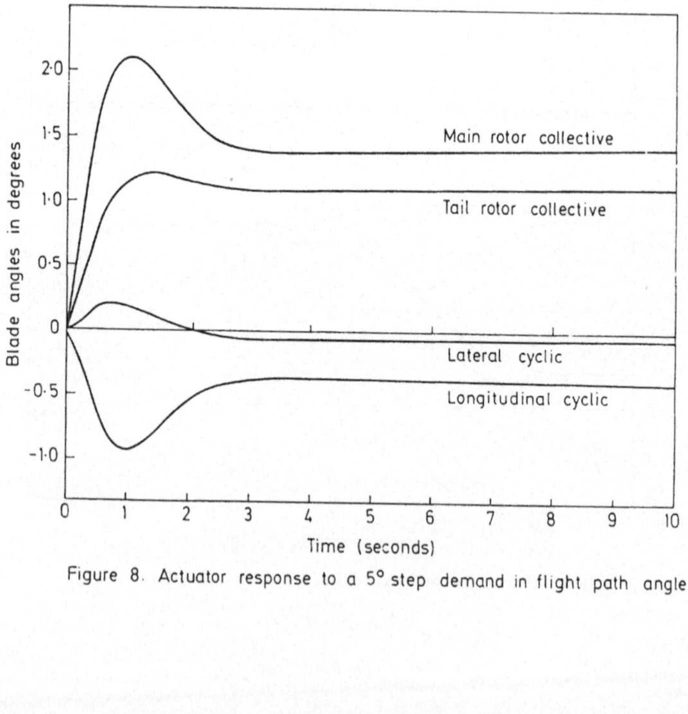

Figure 8. Actuator response to a 5° step demand in flight path angle

4. Computational Improvements

A major difficulty in the state-space based computations is the rapidly increasing problem size which results in controllers of artificially high McMillan degree. Many of the controller states, however, are often uncontrollable and/or unobservable. This is because $\bar{R}$ in the model-matching problem will (without special attention) be non-minimal. The recovery of the optimal $\bar{K}$ from $\bar{Q}$ using the LFT is also a source of extra states.

It is crucially inmportant therefore to have available a reliable minimal realization routine. In the Oxford H^{∞}-design package we use the algorithm of Tombs and Postlethwaite (1986) which finds a truncated balanced realization of a stable non-minimal state-space system. The algorithm can be used either as a form of minimal realization or as a form of model reduction; it appears to be very reliable and accurate.

In the remainder of this section we show how it is possible to reduce the size of the computations leading up to the calculation of $\bar{R}$ in the model-matching problem. $\bar{R}$ is correspondingly of smaller state dimension and is in fact minimal in the 1 block case. The procedure stems from new results on the solution to the algebraic Riccati equation which are given first.

4.1 The Algebraic Riccati Equation

From formulae (2.4)--(2.6) we can see that the algebraic Riccati equation (ARE) plays an important role in the H^{∞}-design procedure. In this section, we will investigate the rank of the solution of ARE. The results will then be used in our size reduction procedure for the model-matching problem given in Section 4.2.

For convenience we rewrite the algebraic Riccati equation as follows,

$$A^TX + XA - XBB^TX + Q^TQ = 0 \quad . \tag{4.1}$$

Several results have already been established about ARE (4.1), (Barnett, 1971):

1) If $Q^TQ > 0$, then the real symmetric solution $X > 0$;
2) If $[A, Q]$ is observable, then $X > 0$;
3) If $[A, Q]$ is only detectable, then X is not necessarily positive definite;
4) The stabilizability of $[A, B]$ and the condition that the Hamiltonian matrix

$\begin{pmatrix} A & -BB^T \\ -Q^TQ & -A^T \end{pmatrix}$ has no eigenvalues on the $j\omega$-axis are necessary and sufficient for the existence of a stabilizing solution X of (4.1); where, 'stabilizing solution' means that the symmetric solution X also makes $(A-BB^TX)$ stable, (Doyle 1984);

5) If a stabilizing solution exists, it is unique, (Doyle 1984).

Throughout this section, we assume that ARE(4.1) satisfies some conditions (as 4) above) such that it has a stabilizing solution.

The following two definitions will be found useful:

Definition 1: For a pair of matrices [A, C], the observability index (**o.i.**) is the number of states in A which can be observed via C.

Definition 2: The unobservable modes (**uno.m.**) of [A, C] are those eigenvalues in the unobservable block A_{11} of A when [A, C] is in observable canonical form [$\begin{pmatrix} A_{11} & A_{12} \\ 0 & A_{22} \end{pmatrix}$, $(0 \quad C_2)$] .

The main theorem of this section can now be stated.

Theorem 1: The rank of the stabilizing solution X of ARE(4.1) equals the sum of the **o.i.** of [A, Q] and the number of unstable eigenvalues in the **uno.m.** of [A, Q]. Also, there exists an orthogonal matrix U such that

U^TXU has the form : $\begin{pmatrix} 0 & 0 \\ 0 & \tilde{X}_2 \end{pmatrix}$;

U^TAU has the form : $\begin{pmatrix} \tilde{A}_1 & \tilde{A}_2 \\ 0 & \tilde{A}_3 \end{pmatrix}$;

QU has the form : $[\ 0 \quad \tilde{Q}_2\ ]$.

Proof: (Postlethwaite et al 1986a)

Obviously, we also have the corollary:

Corollary : The rank of the stabilizing solution of ARE(4.1) equals the sum of the number of unstable eigenvalues of A and the number of stable eigenvalues of A which can be observed via Q.

In the special case of ARE with zero constant term, that is

$$A^TX + XA - XBB^TX = 0 \tag{4.2}$$

we have the theorem:

Theorem 2: The rank of the stabilizing solution X of (4.2) equals the number of unstable eigenvalues of A. There also exists an orthogonal transformation which transforms A into a Schur form with its stable and unstable eigenvalues separated into the lower and upper block diagonals respectively and transforms X into a form as in Theorem 1.

Proof: Since in this case the Q in ARE(4.1) is zero, it means the A is completely unobservable via Q. The observable index (o.i.) of [A Q] is thus zero. Applying Theorem 1 directly here yields the conclusion of this theorem. The existence of the transformation that puts A and X into the required forms is clear from the proof of Theorem 1.

Q.E.D.

We notice here that if an orthogonal transformation U is used to turn A into a Schur form and the eigenvalues of A are separated into the stable part A_- (in the 1-1 block) and unstable part A_+ (in the 2-2 block), then the solution X of (4.2) possesses the special form

$$X = U \begin{pmatrix} 0 & 0 \\ 0 & \tilde{X}_{22} \end{pmatrix} U^T ,$$

where, $\tilde{X}_{22}$ satisfies

$$\tilde{A}_+^T \tilde{X}_{22} + \tilde{X}_{22} A_+ - \tilde{X}_{22} \tilde{B}_2 \tilde{B}_2^T \tilde{X}_{22} = 0 , \tag{4.3}$$

and $\tilde{B}_2$ is defined by the appropriately partitioned matrix

$$U^T B = \begin{pmatrix} B_1 \\ \tilde{B}_2 \end{pmatrix} .$$

Also, it is easy to show that $\tilde{X}_{22}^{-1}$ is the solution of the Lyapunov equation

$$(-A_+)Y + Y(-A_+)^T = -\tilde{B}_2 \tilde{B}_2^T \quad . \tag{4.4}$$

Both (4.3) and (4.4) are smaller in dimension than (4.2), and therefore more useful in practical computations.

Some engineering interpretations of Theorems 1 and 2 might be helpful. Consider the linear quadratic Gaussian regulator problem of finding a state feedback such that the closed-loop system is stable, and the cost function

$$J = \int_0^\infty (x^T Q^T Q x + u^T R^T R u) \, dt$$

is minimized, where x and u satisfy

$$\dot{x} = Ax + Bu \quad .$$

The weights $Q^T Q$ and $R^T R$ are chosen to be positive semi-definite and positive definite, respectively.

If (A, B) is stabilizable, it is well-known that the minimum cost J_{opt} is obtained by a constant state feedback matrix $K = -(R^TR)^{-1}B^TP$, where P is the solution to the algebraic Riccati equation

$$A^TP + PA - PB(R^TR)^{-1}B^TP + Q^TQ = 0 ,$$

and the minimum cost $J_{opt} = x^T(0)Px(0)$, where x(0) is the initial condition of the state vector at time zero.

In order to minimize J and stabilize the feedback system, Theorem 1 tells us that the subspace of the state-space chosen to be fed back, via P in $K = -(R^TR)^{-1}B^TP$, must take into account the state trajectories within the subspace which is observable, and hence penalized, through Q, plus the additional unstable states which are not observable by Q but must be stabilized. Theorem 2 applies when no penalty is put on the states in the cost function. The subspace fed back via P takes into account only unstable dynamics. In addition, if the original system is also stable, no feedback is necessary, and hence $J_{opt} = 0$.

This interpretation is most apparent in the special case when A is diagonal and hence the states are decoupled.

4.2 Size Reduction in the Model-Matching Problem

(a) 1-Block Case

In this case, Theorem 2 can be applied and a minimal realization of $\bar{R}$ derived. Without loss of generality, we can assume that D_{12} and D_{21} in (2.1) are identity matrices by absorbing them in the controller $\bar{K}$ as shown in (Limebeer and Hung 1986). Formulae (2.4)--(2.10) can then be simplified to :

$$\bar{T}_1 = \left(\begin{array}{cc|c} A-B_2C_1-B_2B_2^TX & B_1C_2+YC_2^TC_2 & B_1+YC_2^T \\ 0 & A-B_1C_2-YC_2^TC_2 & -YC_2^T \\ \hline -B_2^TX & C_1 & D_{11} \end{array}\right) , \quad (4.5)$$

$$\bar{T}_2 = \left(\begin{array}{c|c} A-B_2C_1-B_2B_2^TX & B_2 \\ \hline -B_2^TX & I \end{array}\right) , \quad (4.6)$$

$$\bar{T}_3 = \left(\begin{array}{c|c} A-B_1C_2-YC_2^TC_2 & -YC_2^T \\ \hline C_2 & I \end{array}\right) , \quad (4.7)$$

where,

$$X = \textbf{Ric}\begin{pmatrix} A-B_2C_1 & -B_2B_2^T \\ 0 & -(A-B_2C_1)^T \end{pmatrix} , \quad (4.8)$$

$$Y = \textbf{Ric}\begin{pmatrix} (A-B_1C_2)^T & -C_2^TC_2 \\ 0 & -(A-B_1C_2) \end{pmatrix} . \quad (4.9)$$

The following computation steps, which actually follow the proofs of Theorems 2 and 1 (Postlethwaite et al 1986a), can then be carried out.

Step 1: Choose orthogonal matrix $U = (U_1\ U_2)$ such that

$$U^T(A-B_2C_1)U = \begin{pmatrix} W_- & W_{12} \\ 0 & W_+ \end{pmatrix},$$

where, W_- is stable and W_+ completely unstable.

Solve the following ARE:

$$W_+^T\hat{X} + \hat{X}W_+ - \hat{X}U_2^TB_2B_2^TU_2\hat{X} = 0 .$$

Let the stabilizing, positive definite solution be $\tilde{X}_{22}$;

Step 2: Choose orthogonal matrix $V = (V_1\ V_2)$ such that

$$V^T(A-B_1C_2)^TV = \begin{pmatrix} Z_- & Z_{12} \\ 0 & Z_+ \end{pmatrix} .$$

Solve the following ARE and let the stabilizing, positive definite solution be $\tilde{Y}_{22}$:

$$Z_+^T\hat{Y} + \hat{Y}Z_+ - \hat{Y}V_2^TC_2^TC_2V_2\hat{Y} = 0 \qquad ;$$

Step 3: Choose orthogonal matrix $S = (S_1\ S_2)$ such that

$$S^T\tilde{X}_{22}^{1/2}W_+\tilde{X}_{22}^{-1/2}\ S = \begin{pmatrix} S_1^T\tilde{X}_{22}^{1/2}W_+\tilde{X}_{22}^{-1/2}S_1 & 0 \\ S_2^T\tilde{X}_{22}^{1/2}W_+\tilde{X}_{22}^{-1/2}S_1 & S_2^T\tilde{X}_{22}^{1/2}W_+\tilde{X}_{22}^{-1/2}S_2 \end{pmatrix},$$

and

$$S^T\tilde{X}_{22}^{1/2}U_2^T(B_2D_{11}-B_1) = \begin{pmatrix} 0 \\ S_2^T\tilde{X}_{22}^{1/2}U_2^T(B_2D_{11}-B_1) \end{pmatrix} ;$$

Step 4: Choose orthogonal matrix $R = (R_1\ R_2)$ such that

$$R^T\tilde{Y}_{22}^{-1/2}Z_+^T\tilde{Y}_{22}^{1/2}R = \begin{pmatrix} R_1^T\tilde{Y}_{22}^{-1/2}Z_+^T\tilde{Y}_{22}^{1/2}R_1 & R_1^T\tilde{Y}_{22}^{-1/2}Z_+^T\tilde{Y}_{22}^{1/2}R_2 \\ 0 & R_2^T\tilde{Y}_{22}^{-1/2}Z_+^T\tilde{Y}_{22}^{1/2}R_2 \end{pmatrix},$$

and

$$(D_{11}C_2-C_1-B_2^TU_2\tilde{X}_{22}^{1/2}S_1S_1^T\tilde{X}_{22}^{1/2}U_2^T)V_2\tilde{Y}_{22}^{1/2}R$$
$$= (\ 0 \quad (D_{11}C_2-C_1-B_2^TU_2\tilde{X}_{22}^{1/2}S_1S_1^T\tilde{X}_{22}^{1/2}U_2^T)V_2\tilde{Y}_{22}^{1/2}R_2\ ) .$$

Now we can apply these orthogonal matrices to $\bar{T}_1$, $\bar{T}_2$ and $\bar{T}_3$.

By state similarity transformations SSTs $[U^T, U]$ and $[\tilde{X}_{22}^{-1/2}, \tilde{X}_{22}^{1/2}]$,

$$\bar{T}_2 = \left(\begin{array}{c|c} -W_+^T & \tilde{X}_{22}U_2^TB_2 \\ \hline -B_2^TU_2 & I \end{array}\right)$$
$$= \left(\begin{array}{c|c} -\tilde{X}_{22}^{-1/2}W_+^T\tilde{X}_{22}^{1/2} & \tilde{X}_{22}^{1/2}U_2^TB_2 \\ \hline -B_2^TU_2\tilde{X}_{22}^{1/2} & I \end{array}\right) .$$

Similarly,

$$\bar{T}_3 = \left(\begin{array}{c|c} -Z_+ & -V_2^TC_2^T \\ \hline C_2V_2\tilde{Y}_{22} & I \end{array}\right)$$
$$= \left(\begin{array}{c|c} -\tilde{Y}_{22}^{1/2}Z_+\tilde{Y}_{22}^{-1/2} & -\tilde{Y}_{22}^{1/2}V_2^TC_2^T \\ \hline C_2V_2\tilde{Y}_{22}^{1/2} & I \end{array}\right) ;$$

also, by SST [diag(U^T,V^T), diag(U,V)],

$$\bar{T}_1 = \left(\begin{array}{cc|c} -W_+^T & \tilde{X}_{22}U_2^T(B_1+YC_2^T)C_2V_2\tilde{Y}_{22} & \tilde{X}_{22}U_2^T(B_1+YC_2^T) \\ 0 & -Z_+ & -V_2^TC_2^T \\ \hline -B_2^TU_2 & C_1V_2\tilde{Y}_{22} & D_{11} \end{array}\right)$$

$$= \left(\begin{array}{cc|c} -\tilde{X}_{22}^{-1/2}W_+^T\tilde{X}_{22}^{1/2} & \tilde{X}_{22}^{1/2}U_2^T(B_1+YC_2^T)C_2V_2\tilde{Y}_{22}^{1/2} & \tilde{X}_{22}^{1/2}U_2^T(B_1+YC_2^T) \\ 0 & -\tilde{Y}_{22}^{1/2}Z_+\tilde{Y}_{22}^{-1/2} & -\tilde{Y}_{22}^{1/2}V_2^TC_2^T \\ \hline -B_2^TU_2\tilde{X}_{22}^{1/2} & C_1V_2\tilde{Y}_{22}^{1/2} & D_{11} \end{array}\right).$$

Then, notice that

i) $\tilde{X}_{22}^{1/2}W_+\tilde{X}_{22}^{-1/2} + \tilde{X}_{22}^{-1/2}W_+^T\tilde{X}_{22}^{1/2} - \tilde{X}_{22}^{1/2}U_2^TB_2B_2^TU_2\tilde{X}_{22}^{1/2} = 0$; (4.10)

ii) $U_2^TB_2C_1V_2 - U_2^TB_1C_2V_2$

$$= U_2^T\{(A-B_1C_2) - (A-B_2C_1)\}V_2$$

$$= U_2^T\{V\begin{pmatrix} Z_-^T & 0 \\ Z_{12}^T & Z_+^T \end{pmatrix}V^T - U\begin{pmatrix} W_- & W_{12} \\ 0 & W_+ \end{pmatrix}U^T\}V_2$$

$$= U_2^TV_2Z_+^T - W_+U_2^TV_2 \quad ; \qquad (4.11)$$

iii) $YC_2^TC_2V_2 = VV^TYVV^TC_2^TC_2V_2 = V_2\tilde{Y}_{22}V_2^TC_2^TC_2V_2$; (4.12)

iv) $Z_+^T - \tilde{Y}_{22}V_2^TC_2^TC_2V_2 = -\tilde{Y}_{22}Z_+\tilde{Y}_{22}^{-1}$. (4.13)

By routine calculation and applying SSTs

$$[\begin{pmatrix} I & -I & 0 \\ 0 & I & 0 \\ 0 & 0 & I \end{pmatrix}, \begin{pmatrix} I & I & 0 \\ 0 & I & 0 \\ 0 & 0 & I \end{pmatrix}],\ [\begin{pmatrix} I & -\tilde{X}_{22}^{1/2}U_2^TV_2\tilde{Y}_{22}^{1/2} \\ 0 & I \end{pmatrix}, \begin{pmatrix} I & \tilde{X}_{22}^{1/2}U_2^TV_2\tilde{Y}_{22}^{1/2} \\ 0 & I \end{pmatrix}] \text{ and } [S^T,S]$$

$$\bar{T}_2^*\bar{T}_1 = \left(\begin{array}{cc|c} S_2^T\tilde{X}_{22}^{1/2}W_+\tilde{X}_{22}^{-1/2}S_2 & 0 & S_2^T\tilde{X}_{22}^{1/2}U_2^T(B_2D_{11}-B_1) \\ 0 & -\tilde{Y}_{22}^{1/2}Z_+\tilde{Y}_{22}^{-1/2} & -\tilde{Y}_{22}^{1/2}V_2^TC_2^T \\ \hline B_2^TU_2\tilde{X}_{22}^{1/2}S_2 & (C_1+B_2^TU_2\tilde{X}_{22}U_2^T)V_2\tilde{Y}_{22}^{1/2} & D_{11} \end{array}\right)$$

$$= \bar{T}_1\bar{T}_r \ ,$$

where,

$$\bar{T}_1 = \left(\begin{array}{c|c} S_2^T\tilde{X}_{22}^{1/2}W_+\tilde{X}_{22}^{-1/2}S_2 & S_2^T\tilde{X}_{22}^{1/2}U_2^TB_2 \\ \hline B_2^TU_2\tilde{X}_{22}^{1/2}S_2 & I \end{array}\right) ,$$

$$\bar{T}_r = \left(\begin{array}{cc|c} -S_2^T\tilde{X}_{22}^{-1/2}W_+^T\tilde{X}_{22}^{1/2}S_2 & S_2^T\tilde{X}_{22}^{1/2}U_2^T(B_2C_1+B_2B_2^TU_2\tilde{X}_{22}U_2^T)V_2\tilde{Y}_{22}^{1/2} & S_2^T\tilde{X}_{22}^{1/2}U_2^TB_1 \\ 0 & -\tilde{Y}_{22}^{1/2}Z_+\tilde{Y}_{22}^{-1/2} & -\tilde{Y}_{22}^{1/2}V_2C_2 \\ \hline -B_2^TU_2\tilde{X}_{22}^{1/2}S_2 & (C_1+B_2^TU_2\tilde{X}_{22}U_2^T)V_2\tilde{Y}_{22}^{1/2} & D_{11} \end{array}\right).$$

Similarly, we can calculate

$$\bar{T}_r\bar{T}_3^{\star} = \left(\begin{array}{cc|c} -S_2^T\tilde{X}_{22}^{-1/2}W_+^T\tilde{X}_{22}^{1/2}S_2 & 0 & S_2^T\tilde{X}_{22}^{1/2}U_2^T(B_1+V_2\tilde{Y}_{22}V_2^TC_2^T) \\ 0 & R_2^T\tilde{Y}_{22}^{-1/2}Z_+^T\tilde{Y}_{22}^{1/2}R_2 & R_2^T\tilde{Y}_{22}^{1/2}V_2^TC_2^T \\ \hline -B_2^TU_2\tilde{X}_{22}^{1/2}S_2 & (D_{11}C_2-C_1-B_2^TU_2\tilde{X}_{22}^{1/2}S_1S_1^T\tilde{X}_{22}^{1/2}U_2^T)V_2\tilde{Y}_{22}^{1/2}R_2 & D_{11} \end{array}\right)$$

Thus, the transfer function matrix R equals

$$\begin{aligned} \bar{R} &= \bar{T}_2^{\star}\bar{T}_1\bar{T}_3^{\star} \\ &= \bar{T}_1(\bar{T}_r\bar{T}_3^{\star}) \\ &= \left(\begin{array}{cc|c} S_2^T\tilde{X}_{22}^{1/2}W_+\tilde{X}_{22}^{-1/2}S_2 & S_2^T\tilde{X}_{22}^{1/2}U_2^TB_2(D_{11}C_2-C_1-B_2^TU_2\tilde{X}_{22}^{1/2}S_1S_1^T\tilde{X}_{22}^{1/2}U_2^T)V_2\tilde{Y}_{22}^{1/2}R_2 & S_2^T\tilde{X}_{22}^{1/2}U_2^T(B_2D_{11}-B_1-V_2\tilde{Y}_{22}V_2^TC_2^T) \\ 0 & R_2^T\tilde{Y}_{22}^{-1/2}Z_+\tilde{Y}_{22}^{1/2}R_2 & R_2^T\tilde{Y}_{22}^{1/2}V_2^TC_2^T \\ \hline B_2^TU_2\tilde{X}_{22}^{1/2}S_2 & (D_{11}C_2-C_1-B_2^TU_2\tilde{X}_{22}^{1/2}S_1S_1^T\tilde{X}_{22}^{1/2}U_2^T)V_2\tilde{Y}_{22}^{1/2}R_2 & D_{11} \end{array}\right) \end{aligned} \tag{4.14}$$

In fact, (4.14) is a minimal realization of $\bar{R}$.

Theorem 3: In the 1-block case, the transfer function matrix $\bar{R}$ in the model-matching problem has a minimal realization given by (4.14).

Proof: (Postlethwaite et al 1986a)

(b) 2-Block and 4-Block Cases

By Theorem 1, in these more general cases, we can choose orthogonal matrices U and V and operate on the AREs (2.8) and (2.10) such that the solutions and also the coefficient matrices of the AREs possess special forms. Since a state similarity transformation (SST) does not change the input/output mapping of a system, we can apply SST on $\bar{T}_2$ (in (2.5)) with $[U^T,U]$, on $\bar{T}_3$ (in (2.6)) with $[V^T,V]$ and on $\bar{T}_1$ (in (2.4)) with $[\mathrm{diag}(U^T,V^T), \mathrm{diag}(U,V)]$. In this way, we can get rid of redundant parts in these transfer function matrices resulting in an $\bar{R}$ of smaller state dimension although not necessarily minimal.

5. Conclusions

The helicopter control problem demonstrated the power of H^{∞}-design in satisfying robustness and performance requirements. Robustness was achieved by appropriately weighting $\bar{K}\bar{S}$ in the cost function to control the system bandwidth, and performance was achieved by weighting $\bar{S}$. New results on the rank of the solution to the algebraic Riccati equation led to an efficient algorithm for reducing the size of $\bar{R}$ in the model-matching problem. In the 1-block problem, the method was taken further to obtain explicit formulae for a minimal realization of $\bar{R}$. Compared with Limebeer and Hung (1986) we believe our approach to be more direct and numerically more reliable being based on orthogonal transformations.

6. Acknowledgements

The authors are grateful to the U.K. Science and Engineering Research Council for financial support and the Royal Aircraft Establishment (Bedford) for providing the helicopter data.

References

Barnett S (1971) Matrices in control theory. VRN London

Chu CC (1985) H^{∞} - optimization and robust multivariable control. Ph.D. thesis, Department of Electrical Engineering, University of Minnesota

Doyle JC (1984) Advances in multivariable control. Lecture notes presented at ONR/Honeywell Workshop, Minneapolis

Francis BA, Doyle JC (1985) Linear control theory with an H^{∞} - optimality criterion. Systems control group report, No.8501, University of Toronto

Glover K (1984) All optimal Hankel-norm approximations of linear multivariable systems and their L^{∞} - error bounds. Int. J. Control, 39

Limebeer DJN, Hung YS (1986) An analysis of the pole-zero cancellations in H^{∞} - optimal control. Internal report, Dept of Electrical Engineering, Imperial College, London

Postlethwaite I, Gu DW, O'Young SD (1986a) Some Computational Results on Size Reduction in H^{∞} - Design. OUEL Report No 1648/86, University of Oxford

Postlethwaite I, Gu DW, O'Young SD, Tombs MS (1986b) Industrial control system design using H^{∞} - optimization. The 25th CDC conference, Athens, Greece

Postlethwaite I, O'Young SD, Gu DW, Tombs MS, Yue A (1987) H^{∞} - control system design: a critical assessment based on industrial applications. The IFAC Congress, Munich, FRG

Safonov MG, Chiang RY (1986) CACSD Using the State-Space L^{∞} -Theory -- A Design Example. Proc. IEEE Conf. on CACSD, Washington

Tombs MS, Postlethwaite I (1986) Truncated Balanced Realization of a Stable Non-Minimal State-Space System. To appear in Int. J. Control

Tombs MS (1987) Robust Control System Design with Application to High Performance Helicopters. D.Phil. Thesis, University of Oxford

Hankel and Toeplitz operators in linear quadratic and $\mathbf{H}^\infty$- designs[1]

Edmond A. Jonckheere and Jyh-Ching Juang

Department of Electrical Engineering-Systems
University of Southern California
Los Angeles, CA 90089-0781

The structures of the Hankel and Tthe oeplitz operators of the linear quadratic optimization and the $\mathbf{H}^\infty$- design are studied. The relations between these two design methodologies can be identified via this common operator theoretic interpretation. In particular, the computational burden of the $\mathbf{H}^\infty$- synthesis can be alleviated by the linear quadratic approximation. The alternative algorithms, convergence properties, and error bounds of the linear quadratic approximation are discussed. The similarities and differences between the linear quadratic approximation and the ϵ-iteration are presented. Finally, simulation results are provided.

1 Introduction

This paper addresses the problem of the common Hankel and Toeplitz operator structure of the linear quadratic optimization and the $\mathbf{H}^\infty$-design. Emphasis will be put on the utilization of linear quadratic techniques to compute the $\mathbf{H}^\infty$-optimal performance. The Hankel and Toeplitz operator structure of the linear quadratic design has been exploited by Jonckheere and Silverman[1], [2], [3]. On the other hand, the Hankel and Toeplitz operator interpretation of the $\mathbf{H}^\infty$-design has been worked out by Jonckheere and Verma[4] and Jonckheere and Juang[5]. This paper will state the recent results along this direction.

A mixed sensitivity $\mathbf{H}^\infty$-design problem is formulated as follows[4], [6], [7], [8], [9]. Referring to the system in Figure 1, the rational $p * m$ plant under consideration can, in general, be represented in coprime factorization form[10]:

$$G = N_r D_r^{-1} = D_l^{-1} N_l \tag{1}$$

where $N_r \in \mathbf{H}^\infty_{p*m}$ and $D_r \in \mathbf{H}^\infty_{m*m}$ ($N_l \in \mathbf{H}^\infty_{p*m}$ and $D_l \in \mathbf{H}^\infty_{p*p}$) are right (left) coprime and $U_r \in \mathbf{H}^\infty_{m*p}$, $V_r \in \mathbf{H}^\infty_{m*m}$, $U_l \in \mathbf{H}^\infty_{m*p}$, and $V_l \in \mathbf{H}^\infty_{p*p}$ satisfy the Bezout equation:

$$\begin{bmatrix} V_r & U_r \\ -N_l & D_l \end{bmatrix} \begin{bmatrix} D_r & -U_l \\ N_r & V_l \end{bmatrix} = I \tag{2}$$

A feedback controller K that stabilizes the system can then be parameterized as

$$\begin{aligned} K &= (V_r - Z_r N_l)^{-1}(U_r + Z_r D_l) \\ &= (U_l + D_r Z_l)(V_l - N_r Z_l)^{-1} \end{aligned} \tag{3}$$

[1]This reasearch was supported by the Joint Service Electronics Program (JSEP F-49620-85-C-0071)

NATO ASI Series, Vol. F34
Modelling, Robustness and Sensitivity Reduction in Control Systems. Edited by R. F. Curtain

for some $Z_r, Z_l \in \mathbf{H}^\infty_{m*p}$, and provided $(V_r - Z_r N_l)^{-1}$, $(V_l - N_r Z_l)^{-1}$ exist. Using this parameterization, the sensitivity function (at the plant output node) S_{out} and the complementary sensitivity function T_{out} can be written as

$$S_{out} = I - N_r(U_r + Z_r D_l)$$

and

$$T_{out} = N_r(U_r + Z_r D_l)$$

A typical mixed sensitivity $\mathbf{H}^\infty$-design is to minimize the following objective function

$$J = \left\| \begin{bmatrix} W_1 S_{out} W_3 \\ W_2 T_{out} W_3 \end{bmatrix} \right\|_\infty \tag{4}$$

over all stabilizing controllers. The weighting functions W_1, W_2, and W_3 are the parameters that can be adjusted to reflect physical disturbances and desired performance.

Substituting Eqn.(4) into S_{out} and T_{out} yields

$$J = \|[A_1 + B_1 Z_r C_1]\|_\infty$$

where

$$A_1 = \begin{bmatrix} W_1(I - N_r U_r)W_3 \\ W_2 N_r U_r W_3 \end{bmatrix} \quad B_1 = \begin{bmatrix} -W_1 N_r \\ W_2 N_r \end{bmatrix}$$

and

$$C_1 = D_l W_3.$$

The dimension of A_1, B_1, and C_1 is $2p * p$, $2p * m$, and $p * p$, respectively.

A property of the $\mathbf{L}^\infty$- norm is that multiplication by unitary operator preserves the norm. Recall that a function is inner if it is in $\mathbf{H}^\infty$ and isometric a.e., while a function is outer if it is in $\mathbf{H}^\infty$ and its range is dense. We can inner-outer factorize C_1 as

$$C_1 = C_{1o} C_{1i}$$

where C_{1o} is outer and C_{1i} is inner. Similarly, we can have an outer-inner factorization for B_1

$$B_1 = B_{1i} B_{1o}$$

where $B_{1i} \in \mathbf{H}^\infty_{2p*m}$ and 'inner' in the sense that $B_{1i}^* B_{1i} = I^2$ and $B_{1o} \in \mathbf{H}^\infty_{m*m}$ and outer. Let $B_{1i}^\perp$ be the orthonormal complement of B_{1i}, *i.e.*, the matrix that makes $\begin{bmatrix} B_{1i} & B_{1i}^\perp \end{bmatrix}$ square and inner.

After some manipulations, the objective function J can be rewritten as

$$J = \left\| \begin{bmatrix} A_2 + Y_1 \\ A_3 \end{bmatrix} \right\|_\infty$$

where

$$A_2 = B_{1i}^* A_1 C_{1i}^*, \; Y_1 = B_{1o} Z_r C_{1o}, \text{ and } A_3 = B_{1i}^{\perp *} A_1 C_{1i}^*.$$

[2] Assume $2p \geq m$

Note that $A_2 \in \mathbf{L}^{\infty}_{m*p}$ and $A_3 \in \mathbf{L}^{\infty}_{(2p-m)*p}$ and $Y_1 \in \mathbf{H}^{\infty}_{m*p}$. By partial fraction decomposition, we can decompose A_2 as

$$A_2 = A_4 + H$$

where $A_4 \in \mathbf{H}^{\infty}_{m*p}$ and $H \in \mathbf{H}^{\infty\perp}_{m*p}$. Define $Y = -(A_4 + Y_1)$; we then have

$$J = \left\| \begin{bmatrix} H - Y \\ A_3 \end{bmatrix} \right\|_{\infty}$$

We can further simplify J as

$$J = \left\| \begin{bmatrix} H - Y \\ T \end{bmatrix} \right\|_{\infty}$$

where T is the outer function satisfying $T^*T = A_3^*A_3$.

After the above manipulations, the design problem boils down to the evaluation of the achievable performance ϵ_o and the computation of the optimal Y of the following problem:

$$\epsilon_o = \inf_{Y \in \mathbf{H}^{\infty}_{m*p}} \left\| \begin{bmatrix} H - Y \\ T \end{bmatrix} \right\|_{\infty} \tag{5}$$

In fact, many $\mathbf{H}^{\infty}$- design problems can be reduced to the above form; for example, the problem of probing the performance at several nodes against one pertubation, the problem of quantifying the performance at one node against several uncertainties, or the problem of dealing with different numbers of inputs and outputs. The bottleneck in this type of $\mathbf{H}^{\infty}$-synthesis is the computation of the achievable performance ϵ_o. Theoretically[6] [9], ϵ_o equals the spectral radius of the Hankel+Toeplitz operator $\mathbf{H}_H^*\mathbf{H}_H + \mathbf{T}_T^*\mathbf{T}_T$; however it is not easy to compute practically. Finite dimensional approximation[5], [9] and an iteration method, called ϵ-iteration[6], [11], [12], have been proposed. The linear quadratic approximation, based upon the common Hankel and Toeplitz operator structure of the linear quadratic design and the $\mathbf{H}^{\infty}$-design, provides a new avenue to investigate this problem.

This paper is organized as follows. The Hankel and Toeplitz operators of the linear quadratic and the $\mathbf{H}^{\infty}$-designs are identified in Section 2. The algorithms for evaluating the $\mathbf{H}^{\infty}$-achievable performance using the linear quadratic mapping are then proposed in Section 3. In Section 4, the relations between the linear quadratic approximation and the ϵ-iteration are discussed. The simulation results are provided in Section 5. Finally, some conclusions are drawn.

2 Hankel and Toeplitz operators

In this section, we will first identify the Hankel and Toeplitz operators that are associated with the reverse time linear quadratic optimization problem. The spectral properties will also be discussed. The operators that are associated with the $\mathbf{H}^{\infty}$-design, which turn out to be Hankel and Toeplitz operators, are then investigated. This common operator structure makes it possible to consider the two different optimization problems under the same framework. In particular, we can map the $\mathbf{H}^\infty$-data to the corresponding linear quadratic data, and evaluate the $\mathbf{H}^\infty$-optimal performance based on the well established linear quadratic control theory. This will be the topic of the next section.

A few historical remarks are in order here. The role and use of the Toeplitz and Hankel operators and their spectra in the linear quadratic problem have been identified by Jonckheeere and Silverman in a series of papers [1,2,3]. The common Toeplitz and Hankel operator structure shared by the $\mathbf{H}^\infty$ and the linear quadratic problems was brought to light in Jonckheere and Verma[4] and Jonckheere and Juang[5].

Other relations between the linear quadratic optimization problem and the $\mathbf{H}^\infty$-design have been investigated by Doyle[8], by Kwakernaak[13], and by Grimble[14]. Doyle viewed these as two kinds of norm approximation problems; namely, linear quadratic design or $\mathbf{L}^2$design is a norm projection problem, whereas $\mathbf{L}^\infty$-design is a norm dilation problem. Kwakernaak exploited the all-pass solution structure of the two problems. Grimble attempted to use Kwakernaak's result to solve the SISO $\mathbf{H}^\infty$-optimization problem. The Hankel and Toeplitz operator interpretation specific to our approach is, however, quite different; we do not consider the same system under different norm measures or objective functions; rather we study different systems under different norm measures with the same achievable performance. This approach is believed to be an important linkage between the celebrated linear quadratic control theory and the emerging $\mathbf{H}^\infty$-design.

2.1 Hankel and Toeplitz operators of linear quadratic optimization problem

Consider the reverse time continuous linear finite dimensional dynamic system:

$$\begin{aligned} \dot{x}(t) &= Ax(t) + Bu(t), \quad -\infty < t \leq 0 \\ x(0) &= \eta \end{aligned} \tag{6}$$

where (A,B) is controllable and the system is asymptotically stable. We also assume that $u \in \mathbf{L}_r^2(-\infty,0)$ and $x \in \mathbf{L}_n^2(-\infty,0)$. The design objective is to minimize the performance index $J(\eta, u(t))$ defined as

$$J(\eta, u(t)) \triangleq \int_{-\infty}^{0} \begin{bmatrix} x^T(t) & u^T(t) \end{bmatrix} \begin{bmatrix} Q & S \\ S^T & R \end{bmatrix} \begin{bmatrix} x(t) \\ u(t) \end{bmatrix} dt \tag{7}$$

where the weighting matrix $\begin{bmatrix} Q & S \\ S^T & R \end{bmatrix}$ is time invariant and symmetric, but not necessarily positive semidefinite.

From Eqn.(6), the state $x(t)$ admits the functional representation

$$x(t) = \int_{-\infty}^{t} e^{A(t-\tau)} Bu(\tau) d\tau \tag{8}$$

The cost functional $J(\eta, u(t))$ can be expanded as

$$\begin{aligned} J(\eta, u(t)) &= \int_{-\infty}^{0} x^T(t)Qx(t)dt + \int_{-\infty}^{0} x^T Su(t)dt \\ &\quad + \int_{-\infty}^{0} u^T(t)S^T x(t)dt + \int_{-\infty}^{0} u^T Ru(t)dt \end{aligned}$$

Define

$$1(t-\tau) = \begin{cases} 1 & \text{if } t > \tau \\ 0 & \text{otherwise} \end{cases}$$

The term $\int_{-\infty}^{0} u^T(t)S^T x(t)dt$ can be rewriten as

$$\begin{aligned} &\int_{-\infty}^{0} u^T(t)S^T x(t)dt \\ &= \int_{-\infty}^{0}\int_{-\infty}^{t} u^T(t)S^T e^{A(t-\tau)}u(\tau)d\tau dt \\ &= \int_{-\infty}^{0}\int_{-\infty}^{0} u^T(t)S^T e^{A(t-\tau)}Bu(\tau)1(t-\tau)d\tau dt \end{aligned}$$

Similarly, we have

$$\int_{-\infty}^{0} x^T(t)Su(t)dt = \int_{-\infty}^{0}\int_{-\infty}^{0} u^T(t)B^T e^{-A^T(t-\tau)}Su(\tau)1(\tau-t)d\tau dt$$

Substituting Eqn.(8) in $\int_{-\infty}^{0} x^T(s)Qx(s)ds$ yields

$$\begin{aligned} &\int_{-\infty}^{0} x^T(s)Qx(s)ds \\ &= \int_{-\infty}^{0}[\int_{-\infty}^{s} u^T(t)B^T e^{A^T(s-t)}\,dt]Q[\int_{-\infty}^{s} e^{A(s-\tau)}Bu(\tau)d\tau]ds \\ &= \int_{-\infty}^{0}\int_{-\infty}^{0}\int_{-\infty}^{0} u^T(t)B^T e^{A^T(s-t)}\,Qe^{A(s-\tau)}Bu(\tau)1(s-t)1(s-\tau)d\tau dt ds \\ &= \int_{-\infty}^{0}\int_{-\infty}^{0} u^T(t)B^T[\int_{\max(t,\tau)}^{0} e^{A^T(s-t)}Qe^{A(s-\tau)}\,ds]Bu(\tau)d\tau dt \\ &= \int_{-\infty}^{0}\int_{-\infty}^{0} u^T(t)B^T[e^{-A^T t}(-Y + e^{A^T t}Ye^{At})e^{-A\tau}1(t-\tau) \\ &\quad + e^{-A^T\tau}(-Y + e^{A^T\tau}Qe^{A\tau})e^{-At}1(\tau-t)]Bu(\tau)d\tau dt \end{aligned}$$

where Y solves

$$A^T Y + YA + Q = 0 \tag{9}$$

We therefore have

$$J(\eta, u(t)) = \int_{-\infty}^{0}\int_{-\infty}^{0} u^T(t)k_{LQ}(t,\tau)u(\tau)d\tau dt \tag{10}$$

where the kernel $k_{LQ}(t,\tau)$ is defined as

$$\begin{aligned} &k_{LQ}(t,\tau) \\ &= (B^T Y + S^T)e^{A(t-\tau)}B1(t-\tau) + B^T e^{-A^T(t-\tau)}(S+YB)1(\tau-t) \\ &\quad + R\delta(t-\tau) - B^T e^{-A^T t}Ye^{-A\tau}B \end{aligned} \tag{11}$$

Since $u \in \mathbf{L}^2_r(-\infty, 0)$, we can write

$$J(\eta, u(t)) = \left\langle\ u,\ \ \mathbf{K}u\ \right\rangle$$

where the inner product is defined in the usual way and $\mathbf{K}$: $\mathbf{L}^2_r(-\infty,0) \rightarrow \mathbf{L}^2_r(-\infty,0)$ is defined as

$$\mathbf{K} : u(t) \mapsto (\mathbf{K}u)(t) = \int_{-\infty}^{0} k_{LQ}(t,\tau)u(\tau)d\tau$$

Thus, the reverse time linear quadratic optimization problem is equivalent to the problem of minimizing the quadratic form induced by the self-adjoint operator $\mathbf{K}$ and defined over the past controls.

The operator $\mathbf{K}$ can further be decomposed as

$$\mathbf{K} = \mathbf{T} + \mathbf{H} \tag{12}$$

where $\mathbf{T} : \mathbf{L}^2_r(-\infty,0) \rightarrow \mathbf{L}^2_r(-\infty,0), v(t) = (\mathbf{T}u)(t) = \int_{-\infty}^{0}[R\delta(t-\tau) + (B^TY + S^T)e^{A(t-\tau)}B1(t-\tau) + B^Te^{-A^T(t-\tau)}(S+YB)1(\tau-t)]u(\tau)d\tau$ and $\mathbf{H}$ is the operator having kernel $-B^Te^{-A^Tt}Ye^{-A\tau}B$. Clearly, $\mathbf{T}$ is a Wiener-Hopf operator; it is self-adjoint and bounded because the system is stable. On the other hand, $\mathbf{H}$ can be factored as

$$\mathbf{H} = -\mathbf{C}^*Y\mathbf{C}$$

where $\mathbf{C}$ is the reachability operator $\mathbf{C} : \mathbf{L}^2_r(-\infty,0) \rightarrow \mathbf{R}_n$ defined as

$$\eta = \mathbf{C}u(t) = \int_{-\infty}^{0} e^{-At}Bu(t)dt$$

$\mathbf{H}$ is compact since Y is of finite dimension and $\mathbf{C}$ is bounded (it is in fact compact). Recall that Y solves Eqn.(9); if we let[3] $Q = C^TC$ then Y is the observability Grammian of the system with the artificial output $y(t) = Cx(t)$. The operator $\mathbf{H}$ can then be represented as follows[4]:

$$\mathbf{H} = -\mathcal{H}^*\mathcal{H}$$

where $\mathcal{H} : \mathbf{L}^2_r(-\infty,0) \rightarrow \mathbf{L}^2_p(-\infty,0)$ is

$$(\mathcal{H}u)(t) = \int_{-\infty}^{0} Ce^{-A(t+\tau)}Bu(\tau)d\tau$$

Clearly, $\mathcal{H}$ is a Hankel operator. The Wiener-Hopf operator $\mathbf{T}$ is isomorphic to a Toeplitz operator. This occurs because the matrix representation of a Wiener-Hopf operator under the Laguerre functions basis of $\mathbf{H}^2$ has constant terms along the diagonals and is hence a Toeplitz matrix[15]. This is why we call $\mathbf{H}+\ \mathbf{T}$ the Hankel plus Toeplitz operator associated with the linear quadratic problem, although $\mathbf{H}$ is not, strictly speaking, a Hankel operator.

[3]Here we assume that Q is positive definite; if Q is negative semidefinite then we can have $Q = -C^TC$. If Q is not sign definite, we can factorize it in a nonsymmetric way only. In the linear quadratic problem we are going to consider, we do have the property that Q is sign definite.

[4]Assume C is of rank $p \leq n$

It will be shown that the spectral radius of the operator $\mathbf{K} = \mathbf{T} + \mathbf{H}$ plays a central role in bridging the $\mathbf{H}^\infty$ and linear quadratic designs. Let us, first of all, investigate the spectrum of the operators $\mathbf{T}$ and $\mathbf{H}$. The spectrum of $\mathbf{H}$, $\sigma(\mathbf{H})$, is a finite set of real eigenvalues with finite multiplicities because $\mathbf{H}$ is compact (of finite range) and self-adjoint. Let $\phi(jw)$ be the symbol of the Wiener-Hopf operator $\mathbf{T}$, *i.e.*, the Fourier transform of its kernel

$$\begin{aligned} &\phi(jw) \\ = &\int_{-\infty}^{\infty} [R\delta(t) + (B^TY + S^T)e^{At}B1(t) + B^Te^{-A^Tt}(S + YB)1(-t)]e^{-jwt}dt \\ = &R + (S^T + B^TY)(jwI - A)^{-1}B - B^T(jwI + A^T)^{-1}(S + YB) \end{aligned}$$

We state the following two lemmas of Jonckheere[3] without proof.

Lemma 2.1

$$\sigma_{ess}(\mathbf{T}) = \overline{\cup_{i=1}^{r}\{\lambda_i(\phi(jw)) : w \in R\}}$$

and

Lemma 2.2

$$\sigma(\mathbf{T}) = \sigma_{ess}(\mathbf{T}) \cup \Lambda$$

where Λ is a finite set of real eigenvalues with finite multiplicities, located outside $\sigma_{ess}(\mathbf{T})$ and between the connected components of $\sigma_{ess}(\mathbf{T})$.

One can then regard $\mathbf{K}$ as a perturbed Wiener-Hopf operator; its spectrum , $\sigma(\mathbf{K})$, is a combination of an essential spectrum and some extra eigenvalues. Since the compact perturbation $\mathbf{H}$ does not change the essential spectrum, we have

$$\sigma_{ess}(\mathbf{K}) = \sigma_{ess}(\mathbf{T})$$

On the other hand, the extra eigenvalues, in particular the extreme one, are the major objects of our concern in this paper. This type of spectral phenomena,*i.e.*, an essential spectrum in addition to a finite number of eigenvalues, have occurred in quantum mechanics in the problem of computing the distribution of energy levels of systems of particles[16], among other things. A typical approach to estimate the eigenvalue is to apply analytic perturbation theory[17]. More precisely, we allude to the Kato-Rellich theorem, the Neumann series expansion of the resolvent, the Rayleigh-Schrödinger series expansion of the eigenelement, and so on. However, this approach is not appropriate here, for the 'perturbation' we have might be very large.

Finally, note that the optimal control u^* and the optimal cost functional $J^*(\eta)$ can be characterized as follows:

$$u^* = \mathbf{T}^{-1}\mathbf{C}^*(\mathbf{C}\mathbf{T}^{-1}\mathbf{C}^*)^{-1}\eta \tag{13}$$

and

$$J^*(\eta) = -\eta^T P \eta \tag{14}$$

where

$$P = Y - (\mathbf{C}\mathbf{T}^{-1}\mathbf{C}^*)^{-1} \tag{15}$$

This follows from a classical variational argument applied to the operator interpretation of the reverse time linear quadratic problem.

Having understood the structure, in particular the spectral interpretation, of the reverse time linear quadratic optimization problem, we turn our attention to the linear operator interpretation and the spectral theory of the $\mathbf{H}^\infty$-design.

2.2 Hankel and Toeplitz operators associated with the $\mathbf{H}^\infty$-design problem

The mixed sensitivity $\mathbf{H}^\infty$ design problem reduces to the following:

$$\epsilon_o = \inf_{Y \in \mathbf{H}^\infty_{m*p}} \left\| \begin{bmatrix} H - Y \\ T \end{bmatrix} \right\|_\infty$$

Once we have ϵ_o, the optimal Y can be computed and the optimal controller can be derived by back substitution. In this subsection, we will show that the operator associated with the above problem is indeed a Hankel+Toeplitz operator. Recall that in the $\mathbf{H}^\infty$-design problem of the first kind (sensitivity minimization), we can reduce the underlying problem to the $T = 0$ particular case of the above general problem, which then becomes a Hankel approximation problem. In other words, the achievable performance is the norm of the Hankel operator associated with H. The problem of evaluating the achievable performance ϵ_o and deriving the optimal solution of the mixed sensitivity $\mathbf{H}^\infty$-design problem is a generalization of the Hankel approximation problem. In fact, it is a Hankel+Toeplitz approximation problem. The achievable performance ϵ_o is the norm of a Hankel+Toeplitz operator. Since this result is crucial, we state it as a theorem.

Theorem 2.1

$$\begin{aligned} \epsilon_o &= \left\| \begin{bmatrix} \mathbf{H}_H \\ \mathbf{T}_T \end{bmatrix} \right\|_2 \\ &= \sqrt{\lambda_{max}(\mathbf{H}_H^* \mathbf{H}_H + \mathbf{T}_{T^*T})} \end{aligned}$$

where $\mathbf{H}_H$ and $\mathbf{T}_T$ are the Hankel and Toeplitz operators associated with H and T, respectively.

Proof:

This can be proved by Sz. Nagy-Foias lifting theorem[18] (see Feintuch and Francis [9]) or by the broadband matching of Helton[19] (see Verma and Jonckheere[6]) , among other proofs.

□

We can then focus our attention on the Hankel operator $\mathbf{H}_H$ and the Toeplitz operator $\mathbf{T}_T$.

Assume that $T(s)$ and $H(s)$ have the following minimal realizations:

$$\begin{aligned} T(s) &= D_T + C_T(sI - A_T)^{-1}B_T \in \mathbf{H}^{\infty}_{p*p} \\ H(s) &= C_H(sI - A_H)^{-1}B_H \in \mathbf{H}^{\infty\,\perp}_{m*p} \end{aligned}$$

Let $\mathbf{T}_T$ be the Toeplitz operator associated with $T(s)$; then $\mathbf{T}_T : \mathbf{L}^2_p(0,\infty) \to \mathbf{L}^2_p(0,\infty)$

$$\begin{aligned} (\mathbf{T}_T u)(t) &= D_T u(t) + \int_0^{\infty} C_T e^{A_T(t-\tau)} B_T 1(t-\tau) u(\tau) d\tau \\ &= \int_0^{\infty} [D_T \delta(t-\tau) + C_T e^{A_T(t-\tau)} B_T 1(t-\tau)] u(\tau) d\tau \end{aligned}$$

The adjoint operator $\mathbf{T}^*_T$ of $\mathbf{T}_T$, or the Toeplitz operator with symbol $T^T(-jw)$ is

$$(\mathbf{T}^*_T u)(t) = \int_0^{\infty} [D^T_T \delta(t-\tau) + B^T_T e^{-A^T_T(t-\tau)} C^T_T 1(\tau - t)] u(\tau) d\tau \tag{16}$$

The operator T_{T^*T} associated with the symbol $T^T(-jw)T(jw)$ is also Toeplitz and $T_{T^*T} = T^*_T T_T$[5]. Decompose $T^T(-s)T(s)$ as

$$T^T(-s)T(s) = F^T(-s) + F(s) + F_0$$

with $F(s) \in \mathbf{H}^{\infty}_{p*p}$ and strictly proper, $F^T(-s) \in \mathbf{H}^{\infty\,\perp}_{p*p}$, and F_0 is a constant and symmetric matrix. The state space realization of $F(s)$ can be easily derived

$$F(s) = C_F(sI - A_T)^{-1}B_T$$

where

$$C_F = B^T_T V + D^T_T C_T \tag{17}$$

and V satisfies

$$V A_T + A^T_T V + C^T_T C_T = 0 \tag{18}$$

Also, $F_0 = D^T_T D_T$. The Toeplitz operator $\mathbf{T}_{T^*T} : \mathbf{L}^2_p(0,\infty) \to \mathbf{L}^2_p(0,\infty)$ can be represented as follows:

$$v(t) = (\mathbf{T}_{T^*T} u)(t) = F_0 u(t) + \int_0^t f(t-\tau) u(\tau) d\tau + \int_t^{\infty} f^T(\ t + \tau) u(\tau) d\tau$$

where $f(t)$ is the impulse response of $F(s)$, which is

$$f(t) = C_F e^{A_T t} B_T \qquad t \geq 0$$

Hence,

$$(\mathbf{T}_{T^*T} u)(t) = F_0 u(t) + \begin{cases} \int_0^t C_F e^{A_T(t-\tau)} B_T u(\tau) d\tau & t > \tau \\ \int_t^{\infty} B^T_T e^{-A^T_T(t-\tau)} C^T_F u(\tau) d\tau & t < \tau \end{cases}$$

[5] A well known result: If $f \in \mathbf{L}^{\infty}$, $g \in \mathbf{H}^{\infty}$, then $\mathbf{T}_f \mathbf{T}_g = \mathbf{T}_{fg}$.

In other words, the kernel of the Toeplitz operator $\mathbf{T}_{T^*T}$ is

$$F_0\delta(t-\tau)+C_Fe^{A_T(t-\tau)}B_T1(t-\tau)+B_T^Te^{-A_T^T(t-\tau)}C_F^T1(\tau-t) \tag{19}$$

The Hankel operator induced by $H(s)$ is $\mathbf{H}_H : \mathbf{L}_p^2(0,\infty) \rightarrow \mathbf{L}_m^2(0,\infty)$

$$(\mathbf{H}_Hu)(t)=\int_0^\infty h(-t-\tau)u(\tau)d\tau$$

Since $H(s)=C_H(sI-A_H)^{-1}B_H \in \mathbf{H}_{m*p}^{\infty\perp}$, the impulse response is anticausal:

$$h(t)=\begin{cases} 0 & \text{for } t>0 \\ C_He^{A_Ht}B_H & \text{for } t\leq 0. \end{cases}$$

Thus,

$$(\mathbf{H}_Hu)(t)=\int_0^\infty C_He^{-A_H(t+\tau)}B_Hu(\tau)d\tau$$

The adjoint operator of $\mathbf{H}_H$, $\mathbf{H}_H^*$, is

$$(\mathbf{H}_H^*u)(t)=\int_0^\infty B_H^Te^{-A_H^T(t+\tau)}C_H^Tu(\tau)d\tau$$

Therefore,

$$\begin{aligned} &(\mathbf{H}_H^*\mathbf{H}_Hu)(t) \\ =\ &\int_0^\infty B_H^Te^{-A_H^Tt}[\int_0^\infty e^{-A_H^Ts}C_H^TC_He^{-A_Hs}ds]e^{-A_H\tau}B_Hu(\tau)d\tau \\ =\ &\int_0^\infty B_H^Te^{-A_H^Tt}Y_He^{-A_H\tau}B_Hu(\tau)d\tau \end{aligned}$$

where

$$Y_H=\int_0^\infty e^{-A_H^Ts}C_H^TC_He^{-A_Hs}ds$$

or Y_H satisfies the Lyapunov equation

$$A_H^TY_H+Y_HA_H-C_H^TC_H=0 \tag{20}$$

The kernel of the Hankel-like operator $\mathbf{H}_H^*\mathbf{H}_H$ is

$$B_H^Te^{-A_H^Tt}Y_He^{-A_H\tau}B_H \tag{21}$$

We can then conclude that the kernel of the Hankel+Toeplitz operator associated with the $\mathbf{H}^\infty$design problem is

$$\begin{aligned} &F_0\delta(t-\tau)+C_Fe^{A_T(t-\tau)}B_T1(t-\tau)+B_T^Te^{-A_T^T(t-\tau)}C_F^T1(\tau-t) \\ +\ &B_H^Te^{-A_H^Tt}Y_He^{-A_H\tau}B_H \end{aligned} \tag{22}$$

Like its linear quadratic counterpart, it appears that the spectrum of this operator is a combination of an essential spectrum and some extra eigenvalues. The specific feature of this spectrum is that the extra eigenvalues are located on the real axis to the right of the essential spectrum because the operator $\mathbf{H}_H^*\mathbf{H}_H$ and $\mathbf{T}_{T^*T}$ are self-adjoint and positive (semi)definite. Moreover, the square root of the spectral radius of $\mathbf{H}_H^*\mathbf{H}_H+$ $\mathbf{T}_{T^*T}$ is the achiveable performance as was stated in Theorem 2.1. Some more spectral interpretation can be found in Jonckheere and Verma[4].

2.3 Linear quadratic mapping

Theorem 2.2 *The kernel of the Hankel plus Toeplitz operator* $\mathbf{H}_H^*\mathbf{H}_H + \mathbf{T}_{T^*T}$ *of the* $\mathbf{H}^\infty$ *design equals the kernel* $\mathbf{H} + \mathbf{T}$ *of the linear quadratic problem if*

$$A = \begin{bmatrix} -A_H & 0 \\ 0 & A_T^T \end{bmatrix} \quad B = \begin{bmatrix} B_H \\ C_F^T \end{bmatrix}$$

$$Q = \begin{bmatrix} -C_H^T C_H & 0 \\ 0 & 0 \end{bmatrix} \quad S = \begin{bmatrix} Y_H B_H \\ B_T \end{bmatrix}$$

and

$$R = D_T^T D_T$$

Proof:

By comparing the kernel of the Hankel + Toeplitz operator of the linear quadratic optimization problem, Eqn.(11), and the kernel of the Hankel + Toeplitz operator of the $\mathbf{H}^\infty$- design, Eqn.(22), this theorem is easy to verify. In particular, note that $Y = \begin{bmatrix} -Y_H & 0 \\ 0 & 0 \end{bmatrix}$.

□

This theorem bridges the two kinds of design in the following sense: Given an $\mathbf{H}^\infty$-design data, its achievable performance ϵ_o can be computed through this mapping using linear quadratic technique.

Observe that the mapped linear quadratic system may not be controllable although $H(s)$ and $T(s)$ are minimal.

Proposition 2.1 *The Hankel and Toeplitz operators of a not completely controllable linear quadratic problem remain unchanged after discarding the uncontrollable modes.*

Proof:

First we will prove that the Hankel and Toeplitz operators are invariant under similarity transformation. A similarity transformation T transforms the system quintuple (A, B, Q, R, S) to $(\hat{A}, \hat{B}, \hat{Q}, \hat{R}, \hat{S}) = (TAT^{-1}, TB, T^{-T}QT^{-1}, R, T^{-T}S)$ and the matrix Y that solves the Lyapunov equation becomes $\hat{Y} = T^{-T}YT^{-1}$. Hence the new Hankel kernel

$$\begin{aligned} & -\hat{B}^T e^{-\hat{A}^T t}\hat{Y} e^{-\hat{A}\tau}\hat{B} \\ = \; & -B^T T^T T^{-T} e^{-A^T t} T^T T^{-T} Y T^{-1} T e^{-A\tau} T^{-1} T B \\ = \; & -B^T e^{-A^T t} Y e^{-A\tau} B \end{aligned}$$

and, similarly, the new Toeplitz kernel are the same as the original kernels.

Now, suppose that the nonsingular matrix T transforms the system quintuple to the canonical form: $(\hat{A}, \hat{B}, \hat{Q}, \hat{R}, \hat{S}) = (\begin{bmatrix} A_c & A_x \\ 0 & A_u \end{bmatrix}, \begin{bmatrix} B_c \\ 0 \end{bmatrix}, \begin{bmatrix} Q_{11} & Q_{12} \\ Q_{21} & Q_{22} \end{bmatrix}, R, \begin{bmatrix} S_1 \\ S_2 \end{bmatrix}$

) with (A_c, B_c) controllable. Also, $\hat{Y} = \begin{bmatrix} Y_{11} & Y_{12} \\ Y_{12} & Y_{22} \end{bmatrix}$, where Y_{11} satisfies the reduced order (controllable part only) Lyapunov equation. The Hankel kernel becomes

$$\begin{aligned} & -\hat{B}^T e^{-\hat{A}^T t} \hat{Y} e^{-\hat{A}\tau} \hat{B} \\ = & -\begin{bmatrix} B_c^T & 0 \end{bmatrix} \begin{bmatrix} e^{-A_c^T t} & 0 \\ * & e^{-A_u^T t} \end{bmatrix} \begin{bmatrix} Y_{11} & Y_{12} \\ Y_{21} & Y_{22} \end{bmatrix} \begin{bmatrix} e^{-A_c \tau} & * \\ 0 & e^{-A_u \tau} \end{bmatrix} \begin{bmatrix} B_c \\ 0 \end{bmatrix} \\ = & -B_c^T e^{-A_c^T t} Y_{11} e^{-A_c \tau} B_c \end{aligned}$$

The Toeplitz kernel is

$$\begin{aligned} & R\delta(t-\tau) \\ & +(B^T Y + S^T) e^{A(t-\tau)} B 1(t-\tau) + B^T e^{-A^T(t-\tau)} (S + YB) 1(\tau - t) \\ = & R\delta(t-\tau) \\ & + \left(\begin{bmatrix} B_c^T & 0 \end{bmatrix} \begin{bmatrix} Y_{11} & Y_{12} \\ Y_{21} & Y_{22} \end{bmatrix} + \begin{bmatrix} S_1^T & S_2^T \end{bmatrix} \right) \\ & \qquad \begin{bmatrix} e^{A_c(t-\tau)} & * \\ 0 & e^{A_u(t-\tau)} \end{bmatrix} \begin{bmatrix} B_c \\ 0 \end{bmatrix} 1(t-\tau) \\ & + \begin{bmatrix} B_c^T & 0 \end{bmatrix} \begin{bmatrix} e^{-A_c^T(t-\tau)} & 0 \\ * & e^{-A_u^T(t-\tau)} \end{bmatrix} \\ & \qquad \left(\begin{bmatrix} S_1 \\ S_2 \end{bmatrix} + \begin{bmatrix} Y_{11} & Y_{12} \\ Y_{21} & Y_{22} \end{bmatrix} \begin{bmatrix} B_c \\ 0 \end{bmatrix} \right) 1(\tau - t) \\ = & R\delta(t-\tau) \\ & +(B_c^T Y_{11} + S_1^T) e^{A_c(t-\tau)} B_c 1(t-\tau) + B_c^T e^{-A_c^T(t-\tau)} (S_1 + Y_{11} B_c) 1(\tau - t) \end{aligned}$$

□

Thus, if the mapped linear quadratic system is uncontrollable, we can always apply the similarity transformation that yields the Kalman decomposition and discard the uncontrollable modes.

After establishing the relations between the $\mathbf{H}^\infty$-design and the linear quadratic optimization problem on the basis of this Hankel and Toeplitz operator interpretation, we will, in the next section, apply this mapping to devise algorithms to efficiently compute the achievable performance.

3 Algorithms for evaluating the achievable performance

In this section we will investigate the algorithms for evaluating the achievable performance ϵ_o of the $\mathbf{H}^\infty$-design via linear quadratic mapping. The convergence properties as well as some alternative algorithms to speed up the computation will be proposed.

3.1 Linear quadratic approximation

Recall that the spectrum of the Hankel+Toeplitz operator associated with the $\mathbf{H}^{\infty}$-design is an essential spectrum with some eigenvalues to the right of it, the whole spectrum being on the real axis. The achievable performance ϵ_o is the square root of the maximal eigenvalue or, more precisely, the spectral radius. According to Theorem 2.2, we can map the $\mathbf{H}^{\infty}$-design data, $H(s)$ and $T(s)$, to its linear quadratic counterpart, $\hat{A}, \hat{B}, \hat{Q}, \hat{R}$, and $\hat{S}$. This yields

$$\begin{aligned} &\epsilon_o^2 \\ &= \lambda_{max}(\mathbf{H}_H^*\mathbf{H}_H + \mathbf{T}_{T^*T}) \\ &= \sup_{u(t)} \frac{\int_{-\infty}^{0} \begin{bmatrix} x^T(t) & u^T(t) \end{bmatrix} \begin{bmatrix} \hat{Q} & \hat{S} \\ \hat{S}^T & \hat{R} \end{bmatrix} \begin{bmatrix} x(t) \\ u(t) \end{bmatrix} dt}{\int_{-\infty}^{0} u^T(t)u(t)dt} \\ &\text{subject to } \dot{x}(t) = \hat{A}x(t) + \hat{B}u(t) \end{aligned} \tag{23}$$

To convert the supremum problem into the more conventional infimum problem of the linear quadratic theory. Note that

$$\lambda_{max}(\mathbf{H}_H^*\mathbf{H}_H + \mathbf{T}_{T^*T}) = \gamma - \lambda_{min}(\gamma I - \mathbf{H}_H^*\mathbf{H}_H - \mathbf{T}_{T^*T}).$$

for any constant γ. Since $\gamma I - \mathbf{H}_H^*\mathbf{H}_H - \mathbf{T}_{T^*T}$ is also a Toeplitz+Hankel operator, we can map the kernel of $\gamma I - \mathbf{H}_H^*\mathbf{H}_H - \mathbf{T}_{T^*T}$ of the $\mathbf{H}^{\infty}$-design data to the kernel of $\mathbf{H} + \mathbf{T}$ of the linear quadratic data. In this section we will adopt this mapping; the approach using the mapping of Section 2 can be found in Jonckheere and Juang[5].

Let $T(s)$ and $H(s)$ admits the (minimal) state space representations:

$$\begin{aligned} T(s) &= D_T + C_T(sI - A_T)^{-1}B_T \\ H(s) &= C_H(sI - A_H)^{-1}B_H \end{aligned}$$

The adjoint linear quadratic system admit the following data:

$$\begin{aligned} A &= \begin{bmatrix} -A_H & 0 \\ 0 & A_T^T \end{bmatrix} B = \begin{bmatrix} B_H \\ C_F^T \end{bmatrix} \\ Q &= \begin{bmatrix} C_H^T C_H & 0 \\ 0 & 0 \end{bmatrix} S = \begin{bmatrix} -Y_H B_H \\ -B_T \end{bmatrix} \\ R &= \gamma I - D_T^T D_T. \end{aligned} \tag{24}$$

where C_F and Y_H are defined in Eqn.(17) and Eqn.(20), respectively. The matrix Y that solves the Lyapunov equation (Eqn.(9)) is

$$Y = \begin{bmatrix} Y_H & 0 \\ 0 & 0 \end{bmatrix} \tag{25}$$

The pair (A, B) is assumed to be controllable; if not, we can use Proposition 2.1 of the previous section to derive a controllable subsystem to work with.

We then have

$$\epsilon_o^2 = \gamma - \inf_{u(t)} \frac{\int_{-\infty}^{0} \begin{bmatrix} x^T(t) & u^T(t) \end{bmatrix} \begin{bmatrix} Q & S \\ S^T & R \end{bmatrix} \begin{bmatrix} x(t) \\ u(t) \end{bmatrix} dt}{\int_{-\infty}^{0} u^T(t)u(t)\,dt} \tag{26}$$

subject to the state equation

$$\dot{x}(t) = Ax(t) + Bu(t) \tag{27}$$

Let

$$\begin{aligned} \mu &\triangleq \gamma - \epsilon_o^2 \\ &= \inf_{u(t)} \frac{\int_{-\infty}^{0} \begin{bmatrix} x^T(t) & u^T(t) \end{bmatrix} \begin{bmatrix} Q & S \\ S^T & R \end{bmatrix} \begin{bmatrix} x(t) \\ u(t) \end{bmatrix} dt}{\int_{-\infty}^{0} u^T(t)u(t)\,dt} \end{aligned}$$

Instead of computing μ, we will compute the related quantity λ_1 defined by the variational problem :

$$\lambda_1 \triangleq \inf_{\eta} \frac{\inf_{u(t)} \int_{-\infty}^{0} \begin{bmatrix} x^T(t) & u^T(t) \end{bmatrix} \begin{bmatrix} Q & S \\ S^T & R \end{bmatrix} \begin{bmatrix} x(t) \\ u(t) \end{bmatrix} dt}{\int_{-\infty}^{0} u^T(t)u(t)\,dt} \tag{28}$$

where the infimum of the numerator is subject to the constraints:

$$\dot{x}(t) = Ax(t) + Bu(t) \text{ and } x(0) = \eta$$

and the denominator is evaluated for the optimal control $u^*(t)$ that achieves the infimum of the numerator. The essence of the linear quadratic approximation is to approximate μ using λ_1 or, more exactly, using $\beta \triangleq \gamma\text{-}\lambda_1$ to approximate $\epsilon_o^2 = \gamma\text{-}\mu$.

The above infimum problem (Eqn.(28)) is the standard linear quadratic optimization problem; λ_1 can then be easily derived. Let P be the matrix that satisfies

$$\begin{aligned} &-\eta^T P \eta \\ &= \inf_{u(t)} \int_{-\infty}^{0} \begin{bmatrix} x^T(t) & u^T(t) \end{bmatrix} \begin{bmatrix} Q & S \\ S^T & R \end{bmatrix} \begin{bmatrix} x(t) \\ u(t) \end{bmatrix} dt \end{aligned}$$

subject to the constraints:

$$\dot{x}(t) = Ax(t) + Bu(t) \text{ and } x(0) = \eta$$

Then, P is the symmetric, antistabilizing, negative extremal solution of the algebraic Riccati equation:

$$A^T P + PA + Q - (PB + S)R^{-1}(PB + S)^T = 0 \tag{29}$$

The symmetric, antistablizing solution P of the algebraic Riccati equation does not necessarily exist. If it exists (see Proposition 3.1 below), then the optimal control $u^*(t)$ is of feedback type:

$$u^*(t) = Kx^*(t) = -R^{-1}(S^T + B^T P)x^*(t) \tag{30}$$

and the optimal state $x^*(t)$ is

$$x^*(t) = e^{(A+BK)t}\,\eta$$

The denominator term of Eqn.(28) becomes

$$\int_{-\infty}^{0} u^{*T}(t)u^*(t)dt = \eta^T Z\eta$$

where Z is the positive definite solution of the Lyapunov equation:

$$(A + BK)^T Z + Z(A + BK) - K^T K = 0 \tag{31}$$

The optimal λ_1 of Eqn.(28) is

$$\begin{aligned} &\lambda_1 \\ = \; &\inf_{\eta} \frac{-\eta^T P\eta}{\eta^T Z\eta} \\ = \; &\inf_{\varsigma} \frac{-\varsigma^T L^{-T} PL^{-1}\varsigma}{\varsigma^T\varsigma} \\ = \; &\lambda_{min}(-L^{-T} PL^{-1}) \end{aligned} \tag{32}$$

where L factorizes Z as follows (Cholesky factorization):

$$Z = L^T L \tag{33}$$

As mentioned above, we will use $\beta(\gamma)$ to approximate ϵ_o^2. The methodology is as follows:

1. select an a priori estimate γ,
2. compute the a posterori estimate $\beta(\gamma)$,
3. update new estimate γ based on the correction $\gamma - \beta(\gamma)$ and go on iterating.

To justify this methodology, we have to address the following issues :

- How to characterize ϵ_o^2; namely, what is the relation between γ, $\beta(\gamma)$, and ϵ_o^2 ?
- How to update the new γ to ensure that the algorithm is convergent and furthermore has fast convergence rate ?
- Are the estimates γ and $\beta(\gamma)$ close to ϵ_o^2 or could they be arbitrarily far?

- Is this procedure better than other existing methods ?

The relations between ϵ_o^2, λ_1, and γ will be discussed here below. Based on these relations, we can devise algorithms to approximate ϵ_o. First let's establish the condition for the existence of the symmetric, antistablizing solution of the algebraic Riccati equation.

Proposition 3.1 *There exists a symmetric, antistablizing solution P of the algebraic Riccati equation* (Eqn.(29)) *if and only if (A, B) is controllable and* $\gamma \geq \|T\|_\infty^2$.

Proof: The controllability requirement is obvious. A well known fact is that existence of a symmetric solution of the algebraic Riccati equation is equivalent to its associated Popov function $\varphi(jw)$ being positive semidefinite for all real w (except for some system poles on the jw axis) and vice versa[20]. The Popov function $\varphi(jw)$ associated with the algebraic Riccati equation (Eqn.(29)) is

$$\begin{aligned}\varphi(jw) &= R + S^T(jwI - A)^{-1}B + B^T(-jwI - A^T)^{-1}S \\ &\quad + B^T(-jwI - A^T)^{-1}Q(jwI - A)^{-1}B\end{aligned}$$

Substituting the data of the $\mathbf{H}^\infty$design (Eqn.(24)) into the above equation gives

$$\begin{aligned}\varphi(jw) &= R - B_T^T(jwI - A_T^T)^{-1}C_F^T - C_F(-jwI - A_T)^{-1}B_T \\ &\quad - B_H^T Y_H(jwI + A_H)^{-1}B_H - B_H^T(-jwI - A_H^T)^{-1}Y_H B_H \\ &\quad + B_H^T(jwI - A_H^T)^{-1}C_H^T C_H(-jwI - A_H)^{-1}B_H \\ &= \gamma I - D_T^T D_T - B_T^T(jwI - A_T^T)^{-1}C_F^T - C_F(-jwI - A_T)^{-1}B_T \\ &\quad \text{since } Y_H \text{ solves Eqn.(20).} \\ &= \gamma I - \phi^*(jw)\phi(jw)\end{aligned}$$

where $\phi(jw)$ is the frequency response of the transfer function matrix $T(s)$. The assumption $\gamma \geq \|T\|_\infty^2$ implies that the Popov function $\gamma I - \phi^*(jw)\phi(jw)$ is positive semidefinite and, hence, there exists a symmetric solution. The converse can be argued similarly.

□

Thus, the admissible region of γ in our approximation is $[\|T\|_\infty^2, \infty)$. We will assume that (A,B) is controllable and γ greater than $\|T\|_\infty^2$ throughout this section without further explanation.

Define

$$\begin{aligned}J(\eta) &\triangleq \inf_{u(t)} \int_{-\infty}^{0} \begin{bmatrix} x^T(t) & u^T(t) \end{bmatrix} \begin{bmatrix} Q & S \\ S^T & R \end{bmatrix} \begin{bmatrix} x(t) \\ u(t) \end{bmatrix} dt \\ &\quad \text{subject to } \dot{x}(t) = Ax(t) + Bu(t) \text{ and } x(0) = \eta \\ &= \inf_{u} \left\langle u, \ (\gamma I - \mathbf{H} - \mathbf{T})u \right\rangle \\ &\quad \text{subject to } \dot{x}(t) = Ax(t) + Bu(t) \text{ and } x(0) = \eta\end{aligned}$$

It can be proved that $J(\eta) = 0$ for some $\eta \neq 0$ if and only if $\gamma I - \mathbf{H} - \mathbf{T}$ has an eigenvalue at 0; see Jonckheere and Silverman[1,3].

Lemma 3.1 $\beta \triangleq \gamma - \lambda_1 \leq \epsilon_o^2$ *for all* $\gamma \geq \|T\|_\infty^2$.

Proof:
This is a trivial consequence of the fact $\mu \leq \lambda_1$. □

Theorem 3.1

$\gamma > \epsilon_o^2$ *if and only if* $\lambda_1 > 0$.
$\gamma < \epsilon_o^2$ *if and only if* $\lambda_1 < 0$.
$\gamma = \epsilon_o^2$ *if and only if* $\lambda_1 = 0$.

Proof:
Recall that ϵ_o is the square root of the maximal eigenvalue of $\mathbf{H}+\mathbf{T}$. Now, $\gamma > \epsilon_o^2$ implies that $\gamma I - \mathbf{H} - \mathbf{T}$ is (strictly) positive definite. Thus $J(\eta) > 0$ and $\lambda_1 > 0$. On the other hand, if $\lambda_1 > 0$, then $\gamma I - \mathbf{H} - \mathbf{T}$ is positive definite and $\gamma > \epsilon_o^2$. This proves the first statement. The third statement is proved as follows:

$$\gamma = \epsilon_o^2 \Longleftrightarrow J(\eta) = 0 \Longleftrightarrow \lambda_1 = 0.$$

Finally, if $\gamma < \epsilon_o^2$, then $\gamma I - \mathbf{H} - \mathbf{T}$ has at least one negative eigenvalue. Hence $J(\eta) < 0$ for some $\eta \neq 0$, and $\lambda_1 < 0$. The converse can then be easily proved by contradiction. □

The above lemma and theorem are crucial in the linear quadratic approximation. In particular, Theorem 3.1 says that ϵ_o^2 is the unique fixed point of the iteration. Besides, if we select γ between $\|T\|_\infty^2$ and ϵ_o^2, then the computed $\beta(\gamma)$ stays between γ and ϵ_o^2. Having these properties at hand, we can devise an algorithm to compute the achievable performance ϵ_o as follows:

Algorithm 1:
Given $H(s) \in \mathbf{H}_\perp^\infty$ and $T(s) \in \mathbf{H}^\infty$ and an a priori estimate γ_o,

1. Map the $\mathbf{H}^\infty$design data to the linear quadratic data via Eqn.(24).
2. $\beta \leftarrow \gamma_o; \gamma \leftarrow 0$ // initialization //
3. **while** β not sufficiently close to γ **do**
4. $\gamma \leftarrow \beta$
5. $R \leftarrow \gamma I - D_T^T D_T$
6. $P \leftarrow$ antistablizing solution of Eqn.(29)
7. $Z \leftarrow$ solution of the Lyapunov equation, Eqn.(31)
8. $\lambda_1 \leftarrow$ minimal eigenvalue, Eqn.(32)
9. $\beta \leftarrow \gamma - \lambda_1$.

10. if $\beta \leq \|T\|_\infty^2$, then $\beta \leftarrow \|T\|_\infty^2 + \varepsilon$, where ε is a small number.

11. endif

12. repeat

The a priori estimate γ_o is assumed to be greater than $\|T\|_\infty^2$ to ensure existence of P, the solution of the algebraic Riccati equation. Note also that if the a posteriori estimate β is less than $\|T\|_\infty^2$, we shouldn't assign the value β to the new estimate γ. That's why Step 10 is included. Simulations have shown that this algorithm works pretty well; in general, 2 or 3 iterations lead to an estimate with error ($|\sqrt{\gamma} - \sqrt{\beta}|$) less than 10^{-4}; see Jonckheere and Juang[5] for more details.

3.2 Nonlinear programming problem

The problem of finding γ such that $\beta(\gamma) = \gamma$ can be stated as the problem of finding the root of $\lambda_1(\gamma)$=0. This in turn can be regarded as a nonlinear programming problem. Extensive research has been pursued on the root finding problem; the algorithms, their convergence properties and related issues have been studied.

Let η_o be the eigenvector satisfying the general eigenvalue problem

$$\lambda_1(\gamma)Z(\gamma)\eta_o(\gamma) + P(\gamma)\eta_o(\gamma) = 0$$

with $\lambda_1(\gamma)$ being the infimum eigenvalue of Eqn.(32).

Take the derivative of the above equation with respect to γ and multiply by $\eta_o{}^T$ from the left,

$$\eta_o{}^T[\lambda_1'(\gamma)Z(\gamma) + \lambda_1(\gamma)Z'(\gamma) + P'(\gamma)]\eta_o = 0 \tag{34}$$

Differentiating the algebraic Riccati equation (Eqn.(29)) with respect to γ, we have

$$(A + BK(\gamma))^T P'(\gamma) + P'(\gamma)(A + BK(\gamma)) + K^T(\gamma)K(\gamma) = 0 \tag{35}$$

Comparing Eqn.(35) and Eqn.(31) yields

$$P'(\gamma) = -Z(\gamma)$$

because the Lyapunov equation allows for only one solution as the consequence of $(A + BK)$ being unstable[21].

By Eqn.(34),

$$\eta_o{}^T[(-1 + \lambda_1'(\gamma))Z(\gamma) + \lambda_1(\gamma)Z'(\gamma)]\eta_o = 0$$

Indeed, we have

$$\lambda_1'(\gamma) = 1 - \lambda_1(\gamma)\frac{\eta_o{}^T Z'(\gamma)\eta_o}{\eta_o{}^T Z(\gamma)\eta_o} \tag{36}$$

To see what $Z'(\gamma)$ is, we differentiate Eqn.(29) twice with respect to γ. This leads to

$$[A + BK]^T Z'(\gamma) + Z'(\gamma)[A + BK] + 2[K^T - ZB]R^{-1}[K - B^T Z] = 0 \tag{37}$$

Since $(A + BK)$ is unstable, $Z'(\gamma)$ is negative definite. Hence, $\frac{\eta_o{}^T Z'(\gamma)\eta_o}{\eta_o{}^T Z(\gamma)\eta_o}$ is always negative. Thus as long as $\frac{\eta_o{}^T Z'(\gamma)\eta_o}{\eta_o{}^T Z(\gamma)\eta_o}$ is bounded, the derivative $\lambda_1{}'(\gamma) > 1$ when $\lambda_1 > 0$ or $\epsilon^2 > \epsilon_o^2$, $\lambda_1{}'(\gamma) < 1$ when $\lambda_1 < 0$ or $\epsilon^2 < \epsilon_o^2$, and $\lambda_1{}'(\gamma) = 1$ when $\lambda_1 = 0$ or $\epsilon^2 > \epsilon_o^2$. Also,

$$\beta'(\gamma) = 1 - \lambda_1{}'(\gamma) = \lambda_1(\gamma)\frac{\eta_o{}^T Z'(\gamma)\eta_o}{\eta_o{}^T Z(\gamma)\eta_o}$$

This is to say that the a posteriori estimate $\beta(\gamma)$ is monotonously decreasing when $\gamma \geq \epsilon_o^2$ and is monotonously increasing when $\epsilon_o^2 \geq \gamma \geq \|T\|_\infty^2$. Furthermore, we can evaluate the derivative $\lambda_1{}'(\gamma)$ (or equivalently, $\beta'(\gamma)$) by solving the Lyapunov equation, Eqn.(37).

Theorem 3.2 *Assume $\|T\|_\infty^2 < \gamma \leq M < \infty$ and $\epsilon_o^2 > \|T\|_\infty^2$, where M is some real number. Then the following definitions are equivalent characterizations of ϵ_o^2.*

1. *ϵ_o^2 is the fixed point of the iteration, i.e., $\beta(\epsilon_o^2) = \epsilon_o^2$.*
2. *ϵ_o^2 is the supremum of the function $\beta(\gamma)$, i.e., $\beta(\epsilon_o^2) > \beta(\gamma)$ for all $\gamma \in (\|T\|_\infty^2, \infty)$, $\gamma \neq \epsilon_o^2$.*
3. *ϵ_o^2 is the unique solution of $\beta'(\gamma) = 0$.*

Proof:

The monotonously increasing property of $\beta(\gamma)$ as well as $\beta(\gamma) \leq \epsilon_o^2$ ensures that the sequence γ_n generated by the fixed point iteration is convergent and converges to ϵ_o^2.

Lemma 3.1 shows that $\beta(\gamma) \leq \epsilon_o^2$. In fact we have $\beta(\epsilon_o^2) = \epsilon_o^2$ and it is unique. This proves the second statement.

Finally, $\beta(\epsilon_o^2) = \epsilon_o^2 \iff \lambda_1 = 0 \iff \lambda_1{}' = 1 \iff \beta'(\gamma) = 0$.

□

In Algorithm 1, fixed point iteration is used; its convergence rate is linear. Modifications, such as interpolation method, quasi-Newton method, inverse interpolation method *et al* to speed up the convergence rate are available; see for example [22]. Listed below are some alternatives that can be applied to the iterative computation of the achievable performance. This listing is of course far from exhaustive.

bisection method:

rule: $\gamma_{n+1} = \frac{1}{2}(\gamma_n + \gamma_{n-1})$

convergence rate: linear, order of convergence $p = 1$

assume ϵ_o^2 lies between γ_n and γ_{n-1}.

parabolic interpolation:

rule: $\gamma_{n+1} = \gamma_n - \frac{2\lambda_1(\gamma_n)sgn(w)}{|w| + \sqrt{w^2 - 4\beta(\gamma_n)\beta(\gamma_{n-2})}}$

where $w = \beta(\gamma_{n-1}) + (\gamma_n - \gamma_{n-1})\beta(\gamma_{n-1})$

convergence rate: $p = 1.84$

secant method:

rule: $\gamma_{n+1} = \gamma_n - \frac{(\gamma_n - \gamma_{n-1})\lambda_1(\gamma_n)}{\lambda_1(\gamma_n) - \lambda_1(\gamma_{n-1})}$

convergence rate: $p = 1.618$

Newton-Raphson's method:

rule: $\gamma_{n+1} = \gamma_n - \frac{\lambda_1(\gamma_n)}{\lambda_1'(\gamma_n)}$

convergence rate: quadratic, $p = 2$

The algorithm that makes use of Newton-Raphson's method is stated below. The algorithms using other iteration schemes can be modified accordingly.

Algorithm 2

Given $H(s) \in \mathbf{H}^\infty_\perp$ and $T(s) \in \mathbf{H}^\infty$ and an a priori estimate γ_o,

1. Map the $\mathbf{H}^\infty$design data to the linear quadratic data via Eqn.(24)
2. $\gamma \leftarrow 0; \lambda_1 \leftarrow -\gamma_o; \lambda_1' \leftarrow 1$ // initialization //
3. **while** λ_1 not sufficiently close to 0 **do**
4. $\quad \gamma \leftarrow \gamma - \frac{\lambda_1}{\lambda_1'}$
5. $\quad R \leftarrow \gamma I - D_T^T D_T$
6. $\quad P \leftarrow$ antistablizing solution of Eqn.(29)
7. $\quad Z \leftarrow$ solution of the Lyapunov equation, Eqn.(31)
8. $\quad \lambda_1, \eta_o \leftarrow$ minimal eigenvalue and eigenvector, Eqn.(32)
9. $\quad Z' \leftarrow$ solution of the Lyapunov equation, Eqn.(37)
10. $\quad \lambda_1' \leftarrow \lambda_1 \frac{\eta_o^T Z' \eta_o}{\eta_o^T Z \eta_o}$.
11. **repeat**
12. $\beta \leftarrow \gamma - \lambda_1$.

Note that although fast convergence rate is a desired property, it is never free. One extra Lyapunov equation has to be solved in order to have quadratic convergence rate in computing ϵ_o^2.

There are two features of $\lambda_1(\gamma)$ that are relevant from the programming point of view. One is $\beta(\gamma)=\gamma$ or $\lambda_1(\gamma)=0$; the other is $\lambda_1'(\gamma)=1$. Using the former feature to characterize ϵ_o is equivalent to a root finding routine as Algorithm 2, while Algorithm 1 uses the latter feature. We can combine them and propose a hybrid algorithm. At each point $(\gamma_n, \lambda_1(\gamma_n))$ we can construct two line segments. One with slope $\lambda_1'(\gamma_n)$, *i.e.*,

$$\text{line } 1 : \lambda_1(\gamma) - \lambda_1(\gamma_n) = \lambda_1'(\gamma_n)(\gamma - \gamma_n)$$

the other is of slope 1,

$$\text{line } 2 : \lambda_1(\gamma) - \lambda_1(\gamma_n) = \gamma - \gamma_n.$$

The intersection of line 1 with the line $\lambda_1(\gamma)$=0 ,*i.e.*, $\gamma^\alpha = \gamma_n - \frac{\lambda_1(\gamma_n)}{\lambda_1'(\gamma_n)}$, is to the right of ϵ_o^2; while the intersection of line 2 with $\lambda_1(\gamma)$=0 , $\gamma^\beta = \gamma_n - \lambda_1(\gamma_n)$, is to the left of ϵ_o^2 on the real line. We can then, for example, pick the point $\gamma = \frac{1}{2}(\gamma^\alpha + \gamma^\beta)$ as the new estimate.

Algorithm 3

Given $H(s) \in \mathbf{H}_\perp^\infty$ and $T(s) \in \mathbf{H}^\infty$ and an a priori estimate γ_o,

1. Map the $\mathbf{H}^\infty$design data to the linear quadratic data via Eqn.(24)
2. $\gamma \leftarrow 0; \lambda_1 \leftarrow -\gamma_o; \lambda_1' \leftarrow 1$ // initialization //
3. **while** λ_1 not sufficiently close to 0 **do**
4. **if** $\lambda_1' \leq 0$ **then** $\gamma \leftarrow \gamma - \lambda_1$
5. **else** $\gamma \leftarrow \gamma - \frac{\lambda_1(1+\lambda_1')}{2\lambda_1'}$
6. **endif**
7. $R \leftarrow \gamma I - D_T^T D_T$
8. $P \leftarrow$ antistablizing solution of Eqn.(29)
9. $Z \leftarrow$ solution of the Lyapunov equation, Eqn.(31)
10. $\lambda_1,\eta_o \leftarrow$ infimum eigenvalue and eigenvector, Eqn.(32)
11. $Z' \leftarrow$ solution of the Lyapunov equation, Eqn.(37)
12. $\lambda_1' \leftarrow \lambda_1 \frac{\eta_o^T Z' \eta_o}{\eta_o^T Z \eta_o}$.
13. **repeat**
14. $\beta \leftarrow \gamma\text{-}\lambda_1$

Bisection technique is used at Step 5. This can be replaced by any other search strategy.

Still another way to view this problem is to consider Eqn.(36). Define the function f as the map from the a priori error, $\gamma - \epsilon_o^2$, to the a posteriori error, $\epsilon_o^2 - \beta(\gamma)$; then by Lemma 3.1, $f(x) \geq 0, \forall x \in [-(\epsilon_o^2 - \|T\|_\infty^2), +\infty)$. Obviously, $f(x)|_{x=0} = 0$. Also, Theorem 3.1 indicates that $x + f(x) \leq 0$ for $x \leq 0$.

Let $\alpha(x) = -\frac{\eta_o{}^T Z'(x)\eta_o}{\eta_o{}^T Z(x)\eta_o}$. Clearly $\alpha(x) \geq 0$. We then have

$$f'(x) - \alpha(x)f(x) = \alpha(x)x$$

Together with the 'boundary condition' $f(0) = 0$, $f(x)$ can be solved as

$$f(x) = \int_0^x \alpha(v)ve^{\int_v^x \alpha(u)du}dv$$

Suppose that $0 \leq \underline{\alpha} \leq \alpha(u) \leq \hat{\alpha}$ for u under consideration, then

$$\text{for } x > 0, \quad [\frac{1}{\underline{\alpha}}e^{\underline{\alpha}x} - (x + \frac{1}{\underline{\alpha}})] \leq f(x) \leq [\frac{1}{\hat{\alpha}}e^{\hat{\alpha}x} - (x + \frac{1}{\hat{\alpha}})]$$

and

$$\text{for } x < 0, \quad [\frac{\underline{\alpha}}{\hat{\alpha}^2}e^{\hat{\alpha}x} - \underline{\alpha}(\frac{x}{\hat{\alpha}} + \frac{1}{\hat{\alpha}^2})] \leq f(x) \leq [\frac{\hat{\alpha}}{\underline{\alpha}^2}e^{\underline{\alpha}x} - \hat{\alpha}(\frac{x}{\underline{\alpha}} + \frac{1}{\underline{\alpha}^2})]$$

This indicates that, given the derivative, the iteration is quadratically convergent. Moreover, if the span of $\alpha(x)$ is small, *i.e.*, $\hat{\alpha}$ close to $\underline{\alpha}$, then a posteriori error $f(x)$ can be more accurately estimated. On the other hand, if $\alpha(x)$ is small , the error $f(x)$ is small.

The above discussion has also provided a posteriori bounds ontheestimate; namely, given any γ, one lower bound on ϵ_o^2 is $\beta(\gamma)$ and an upper bound is $\frac{\beta(\gamma)-\gamma\beta'(\gamma)}{1-\beta'(\gamma)}$. The a priori (lower) bounds on the estimate is derived in the following theorem.

Theorem 3.3

For $\|T\|_\infty^2 < \gamma < \epsilon_o^2$, *we have* $\epsilon_o^2 > \beta(\gamma) > \|T\|_\infty^2$.

For $\epsilon_o^2 < \gamma < \infty$, *we have* $\epsilon_o^2 > \beta(\gamma) > \left\| \begin{bmatrix} \mathbf{T}_T\Pi \\ \mathbf{H}_H \end{bmatrix} \right\|_2^2$.

where Π *is the projection onto the reachability subapce of* (A, B).

Proof : The first statement is a consequence of Theorem 3.1. The monotonously decreasing property of $\beta(\gamma)$ over the range $\gamma > \epsilon_o^2$ allows us to bound $\beta(\gamma)$ from below by $\beta(\gamma)|_{\gamma\to\infty}$.

$$\begin{aligned} & \lambda_1(\gamma) \\ = & \inf_\eta \frac{-\eta^T P\eta}{\eta^T Z\eta} \\ = & \inf_\eta \frac{\eta^T[-Y + (\mathbf{C}\mathbf{T}^{-1}\mathbf{C}^*)^{-1}]\eta}{\eta^T[(\mathbf{C}\mathbf{T}^{-1}\mathbf{C}^*)^{-1}(\mathbf{C}\mathbf{T}^{-2}\mathbf{C}^*)(\mathbf{C}\mathbf{T}^{-1}\mathbf{C}^*)^{-1}]\eta} \end{aligned}$$

where $\mathbf{T} = \gamma I - \mathbf{T}_T^*\mathbf{T}_T$ and $\mathbf{H} = -\mathbf{H}_H^*\mathbf{H}_H$. As $\gamma \to \infty$, $\mathbf{T}^{-1} \to \gamma^{-1}I + \gamma^{-2}\mathbf{T}_T^*\mathbf{T}_T$. Thus,

$$\begin{aligned} & -Y + (\mathbf{C}\mathbf{T}^{-1}\mathbf{C}^*)^{-1} \\ \to & -Y + [\mathbf{C}(\gamma^{-1}I + \gamma^{-2}\mathbf{T}_T^*\mathbf{T}_T)\mathbf{C}^*]^{-1} \\ \to & \gamma(\mathbf{C}\mathbf{C}^*)^{-1} - Y - (\mathbf{C}\mathbf{C}^*)^{-1}\,\mathbf{C}\mathbf{T}_T^*\mathbf{T}_T\mathbf{C}^*(\mathbf{C}\mathbf{C}^*)^{-1} \end{aligned}$$

On the other hands,

$$
\begin{aligned}
& (\mathbf{C}\mathbf{T}^{-1}\mathbf{C}^*)^{-1}(\mathbf{C}\mathbf{T}^{-2}\mathbf{C}^*)(\mathbf{C}\mathbf{T}^{-1}\mathbf{C}^*)^{-1} \\
\to\ & [\mathbf{C}(\gamma^{-1}I+\gamma^{-2}\mathbf{T}_T^*\mathbf{T}_T)\mathbf{C}^*]^{-1}\gamma^{-2}(\mathbf{C}\mathbf{C}^*)^{-1}\,[\mathbf{C}(\gamma^{-1}I+\gamma^{-2}\mathbf{T}_T^*\mathbf{T}_T)\mathbf{C}^*]^{-1} \\
\to\ & (\mathbf{C}\mathbf{C}^*)^{-1}
\end{aligned}
$$

Therefore,

$$
\begin{aligned}
& \beta(\gamma) \\
=\ & \gamma-\lambda_1(\gamma) \\
\geq\ & \gamma-\inf_{\eta}\frac{\eta^T[\gamma(\mathbf{C}\mathbf{C}^*)^{-1}\ -Y-(\mathbf{C}\mathbf{C}^*)^{-1}\,\mathbf{C}\mathbf{T}_T^*\mathbf{T}_T\mathbf{C}^*(\mathbf{C}\mathbf{C}^*)^{-1}\,]\eta}{\eta^T(\mathbf{C}\mathbf{C}^*)^{-1}\,\eta} \\
& \qquad \text{as } \gamma\to\infty \\
=\ & \gamma-\lambda_{min}(\gamma I-\mathbf{C}^*Y\mathbf{C}-\Pi\mathbf{T}_T^*\mathbf{T}_T\Pi) \\
=\ & \lambda_{max}(\mathbf{C}^*Y\mathbf{C}+\Pi\mathbf{T}_T^*\mathbf{T}_T\Pi) \\
=\ & \left\|\begin{bmatrix}\mathbf{T}_T\Pi\\ \mathbf{H}_H\end{bmatrix}\right\|_2^2
\end{aligned}
$$

where $\Pi \triangleq \mathbf{C}^*\ (\mathbf{C}\mathbf{C}^*)^{-1}\ \mathbf{C}$.

□

Under some extreme circumstances, the operator $\mathbf{H}_H^*\mathbf{H}_H+\ \mathbf{T}_T^*\mathbf{T}_T$ can be modelled as a perturbed Toeplitz operator where the perturbation $\mathbf{H}_H^*\mathbf{H}_H$ is small, and hence $\left\|\begin{bmatrix}\mathbf{T}_T\\ \mathbf{H}_H\end{bmatrix}\right\|_2 \simeq \|T\|_\infty$. Under the opposite extreme circumstance, the operator $\mathbf{H}_H^*\mathbf{H}_H+$ $\mathbf{T}_T^*\mathbf{T}_T$ can be regarded as a perturbed Hankel operator where the contribution of $\mathbf{T}_T^*\mathbf{T}_T$ is small , in which case $\left\|\begin{bmatrix}\mathbf{T}_T\\ \mathbf{H}_H\end{bmatrix}\right\|_2 \simeq \left\|\begin{bmatrix}\mathbf{T}_T\Pi\\ \mathbf{H}_H\end{bmatrix}\right\|_2$. Thus, the bounds are not conservative. Note also that the asymptotic lower bound $\left\|\begin{bmatrix}\mathbf{T}_T\Pi\\ \mathbf{H}_H\end{bmatrix}\right\|_2$ can be evaluated by solving one Lyapunov equation.

4 Relations between the linear quadratic approximation and the ϵ-iteration

We have shown how to map the $\mathbf{H}^\infty$-design data to the linear quadratic optimization data and utilize the well established linear quadratic theory to find a good estimate of ϵ_o , the $\mathbf{H}^\infty$-optimal performance. Also, we have investigated the convergence properties and reformulated the problem as a nonlinear programming problem which can be easily tackled. In this section, we will reformulate the equations relevant to the computation of the a posteriori estimate β. It turns out that the computations can be reduced substantially. Moreover, the relations between the linear quadratic approximation and the ϵ-iteration become self-evident from this viewpoint. The ϵ-iteration can then be regarded as a special technique of the more general linear quadratic approximation.

4.1 A partitioning of the linear quadratic approximation

Assume that $(A,B) = (\begin{bmatrix} -A_H & 0 \\ 0 & A_T^T \end{bmatrix}, \begin{bmatrix} B_H \\ C_F^T \end{bmatrix})$ is controllable and the a priori estimate $\gamma > \|T\|_\infty^2$. In view of Eqn.(15), the solution P of the algebraic Riccati equation, Eqn.(29), has been shown to be $P = Y - W^{-1}$ where $W \triangleq \mathbf{C}\mathbf{T}^{-1}\mathbf{C}^*$. $\mathbf{T}^{-1}$ is well defined since $\gamma > \|T\|_\infty^2$ and W is positive definite if, in addition, (A, B) is controllable. Substitute these into Eqn.(29) and after some manipulations, we get

$$WA^T + AW + (B + WS_1)R^{-1}(B + WS_1)^T = 0 \tag{38}$$

where $S_1 = \begin{bmatrix} 0 \\ B_T \end{bmatrix}$.

Partitioning W as $W = \begin{bmatrix} W_1 & W_2 \\ W_2^T & W_3 \end{bmatrix}$ with $W_1 = W_1^T, W_3 = W_3^T$ and using Eqn.(24), we can decouple Eqn.(38) and rewrite it as the following set of equations:

$$W_3A_T + A_T^TW_3 + (C_F^T + W_3B_T)R^{-1}(C_F + B_T^TW_3) = 0 \tag{39}$$

$$W_2[A_T + B_TR^{-1}(C_F + B_T^TW_3)] - A_HW_2 + B_HR^{-1}(C_F + B_T^TW_3) = 0 \tag{40}$$

$$-W_1A_H^T - A_HW_1 + (B_H + W_2B_T)R^{-1}(B_H^T + B_T^TW_2^T) = 0 \tag{41}$$

Hence, P can be computed by solving a set of equations of smaller dimension. Notice that in this mapping, the positive definite solution W_3 in Eqn.(39) is selected in order to have an antistabilizing P. W_3 being positive definite implies that 'the closed-loop system matrix' $A_T + B_TR^{-1}(C_F + B_T^TW_3)$ is stable which further ensures that the Sylvester equation (Eqn.(40)) is well posed since $-A_H$ is also stable. The condition $(\gamma \leq \|T\|_\infty^2)$ is still the necessary and sufficient condition for the algebraic Riccati equation (Eqn.(39)) to have a symmetric, stablizing solution. The computations required for solving this set of equations is less than that required for solving Eqn.(29). It has been known[23] that it requires approximately $75n^3$ operations to solve an algebraic Riccati equation of order n. To solve a Sylvester equation $(AX + XB = C)$ with $A \in \mathbf{R}^{n*n}$ and $B \in \mathbf{R}^{m*m}$, one may need $\frac{5}{3}m^3 + 10n^3 + 5m^2n + \frac{5}{2}mn^2$ operations[24]. Let the degree of $H(s)$ and the degree of the $T(s)$ be δ_h and δ_t, respectively. The operations needed to solve Eqn.(29) are approximately $75(\delta_h + \delta_t)^3$. In contrast to this, the operations needed to solve P based on the partitioning are

Riccati equation	Eqn.(39)	$75{\delta_t}^3$
Sylvester equation	Eqn.(40)	$\frac{5}{3}{\delta_t}^3 + 10{\delta_h}^3 + 5{\delta_t}^2\delta_h + \frac{5}{2}\delta_t{\delta_h}^2$
Lyapunov equation	Eqn.(41)	$(15 + \frac{25}{6}){\delta_t}^3$
Total		$\frac{230}{3}{\delta_t}^3 + 5{\delta_t}^2\delta_h + \frac{5}{2}\delta_t{\delta_h}^2 + \frac{175}{6}{\delta_h}^3$

In the case where the degree of $H(s)$ equals the degree of $T(s)$, the amount of operations needed is about one third of the former one, a substantial saving.

The optimal control u^* can be derived using Eqn.(30). The partitioning for solving P or W can also be used to save on the operations needed for solving Z and Z' of the Lyapunov equations (Eqn.(31) and Eqn.(37)). Let $WZW = \begin{bmatrix} Z_a & Z_b \\ Z_b^T & Z_c \end{bmatrix}$ in which Z solves Eqn.(31).

After some manipulations, we have

$$A_{Tc}^T Z_c + Z_c A_{Tc} + B_{Tc} R^{-2} B_{Tc}^T = 0 \tag{42}$$

$$-A_H Z_b + Z_b A_{Tc} + B_{Hc} R^{-1} B_T^T Z_c + B_{Hc} R^{-2} B_{Tc}^T = 0 \tag{43}$$

$$-A_H Z_a - Z_a A_H^T + B_{Hc} R^{-1} B_T^T Z_b^T + Z_b B_T R^{-1} B_{Hc}^T + B_{Hc} R^{-2} B_{Hc}^T = 0 \tag{44}$$

where

$$\begin{aligned} A_{Tc} &= A_T + B_T R^{-1}(C_F + B_T^T W_3) \\ B_{Tc} &= C_F^T + W_3 B_T \\ B_{Hc} &= B_H + W_2 B_T \end{aligned}$$

The matrix Z' of Eqn.(37) can also be solved via this partitioning because the matrix $A + BK$ is block upper triangular. Only $\frac{3}{8}$ of the effort is needed for solving the Lyapunov equation in comparison with the former approach. All the programming schemes discussed above such as Newton-Raphson's method, interpolation method, and so on can be applied directly. The algorithms stated above can be modified to take advantage of this computational feature.

4.2 ϵ-iteration

The usual way to evaluate the achievable performance ϵ_o^2 is the ϵ-iteration[11]. The procedure of ϵ-iteration is as follows:

Algorithm 4

Given $H(s) \in \mathbf{H}_\perp^\infty$ and $T(s) \in \mathbf{H}^\infty$ and a priori estimate γ_o,

1. $\gamma \leftarrow \gamma_o; \sigma_\Gamma \leftarrow 0$ // initialization //
2. **while** $\sigma_\Gamma$not sufficiently close to 1 **do**
3. Spectral factorization yielding M ; $M^*M = \gamma I - T^*T$
4. Unstable and stable projections of HM^{-1}, H_u and H_s ; $HM^{-1} = H_u + H_s, H_u$ unstable and H_s stable.
5. Computation of the maximal Hankel singular value σ_Γ of H_u^*
6. **repeat**

The updating rule of γ in the ϵ-iteration is according to the lemma due to Doyle[7] and Francis[25].

Lemma 4.1

$\gamma > \epsilon_o^2$ *if and only if* $\sigma_\Gamma < 1$.
$\gamma < \epsilon_o^2$ *if and only if* $\sigma_\Gamma > 1$.
$\gamma = \epsilon_o^2$ *if and only if* $\sigma_\Gamma = 1$.

This iteration has been investigated by Chu and Doyle[11], Chang and Pearson[12], and more recently Chang *et al*[26]. We'd like to work out this iteration more explicitly.

The transfer function matrix $M(s)$ that satisfies

$$M^T(-s)M(s) = \gamma I - T^T(-s)T(s)$$

can be parameterized as

$$M(s) = R^{\frac{1}{2}} - R^{-\frac{1}{2}}(B^T W_c + C_F)(sI - A_T)^{-1}B_T$$

where W_c satisfies

$$A_T^T W_c + W_c A_T + (C_F^T + W_c B_T)R^{-1}(C_F + B_T^T W_c) = 0 \tag{45}$$

We then evaluate $H(s)M^{-1}(s)$ and its unstable projection $H_u(s)$. It turns out that

$$H_u(s) = C_H(sI - A_H)^{-1}(B_H + W_b B_T)R^{-\frac{1}{2}}$$

where W_b solves

$$-A_H W_b + W_b[A_T + B_T R^{-1}(C_F + B_T^T W_c)] + B_H R^{-1}(C_F + B_T^T W_c) = 0 \tag{46}$$

The complex conjugate transpose of $H_u(s)$ is

$$H_u^T(-s) = -R^{-\frac{1}{2}}(B_H^T + B_T^T W_b^T)(sI + A_H^T)^{-1}C_H^T$$

The controllability Grammian W_o of $H_u^*(s)$ satisfies

$$-A_H^T W_o - W_o A_H + C_H^T C_H = 0 \tag{47}$$

while the observability Grammian W_a satisfies

$$-A_H W_a - W_a A_H^T + (B_H + W_b B_T)R^{-1}(B_H + W_b B_T)^T = 0 \tag{48}$$

The Hankel singular value σ_Γ is

$$\sigma_\Gamma = \sqrt{\lambda_{max}(W_o W_a)}$$

Comparing the equations used in ϵ-iteration, Eqns.(45, 46, 48, 47) , with that used in the linear quadratic approximation , Eqns.(39, 40, 41, 20), yields

$$\begin{aligned} W_c &= W_3 \\ W_b &= W_2 \\ W_a &= W_1 \end{aligned}$$

and

$$W_o = Y_H \tag{49}$$

The ϵ-iteration and the linear quadratic approximation involve the same operations, although they are motivated independently.

Theorem 4.1

$\sigma_\Gamma > 1$ if and only if $\lambda_1 < 0$.
$\sigma_\Gamma = 1$ if and only if $\lambda_1 = 0$.
$\sigma_\Gamma < 1$ if and only if $\lambda_1 > 0$.

Proof:

This theorem is self-evident and indeed can be proved using Theorem 3.1 and Lemma 4.1. We will prove this in a purely algebraic way. Since W is positive definite,

$$\begin{aligned} \lambda_1(\gamma) &= \inf_\eta \frac{-\eta^T P \eta}{\eta^T Z \eta} \\ &= \inf_\eta \frac{-\eta^T W P W \eta}{\eta^T W Z W \eta} \end{aligned}$$

Also,

$$\begin{aligned} -WPW &= W - WYW \\ &= \begin{bmatrix} W_1 - W_1 Y_H W_1 & W_2 - W_1 Y_H W_2 \\ W_2^T - W_2^T Y_H W_1 & W_3 - W_2^T Y_H W_2 \end{bmatrix} \end{aligned}$$

Define the inertia of a matrix M as $I(M) = (I_+(M), I_=(M), I_-(M))$ where $I_+(M)$ is the number of eigenvalues of M wih positive real part, $I_=(M)$ is the number of zero real part eigenvalues of M, and $I_-(M)$ stands for the number of eigenvalues with negative real part. Clearly,

$I_-(W - WYW) = 0, I_=(W - WYW) = 0$ if and only if $\lambda_1 > 0$.
$I_-(W - WYW) = 0, I_=(W - WYW) \neq 0$ if and only if $\lambda_1 = 0$.
$I_-(-WPW) = I_-(W - WYW) \neq 0$ if and only if $\lambda_1 < 0$.

Since W_1 is nonsingular, it follows that

$I_-(W_1 - W_1 Y_H W_1) = 0, I_=(W_1 - W_1 Y_H W_1) = 0$ if and only if $\sigma_\Gamma < 1$.
$I_-(W_1 - W_1 Y_H W_1) = 0, I_=(W_1 - W_1 Y_H W_1) \neq 0$ if and only if $\sigma_\Gamma = 1$.
$I_-(W_1 - W_1 Y_H W_1) \neq 0$ if and only if $\sigma_\Gamma > 1$.

Finally, $I(W - WYW) = I(W_1 - W_1Y_HW_1) + I(W^{\perp})$ where $W^{\perp}$ is the Schur complement[27].

$$\begin{aligned} & W^{\perp} \\ = \ & (W_3 - W_2^TY_HW_2) - (W_2^T - W_2^TY_HW_1)(W_1 - W_1Y_HW_1)^{-1}(W_2 - W_1Y_HW_2) \\ = \ & W_3 - W_2^TW_1^{-1}W_2 \end{aligned}$$

$W^{\perp}$ is positive definite since W is positive definite. Hence,

$$I_{=}(W_1 - W_1Y_HW_1) = I_{=}(W - WYW)$$

and

$$I_{-}(W_1 - W_1Y_HW_1) = I_{-}(W - WYW)$$

The theorem is proved.

□

It is the submatrix $W_1 - W_1Y_HW_1$ that is crucial in determining the condition $\beta(\gamma)=\gamma$. We can, when evaluating the minimal eigenvalue λ_1, drop the complementary parts. Indeed,

$$\begin{aligned} & \lambda_1 \\ = \ & \inf_{\eta} \frac{\eta^T(W - WYW)\eta}{\eta^TWZW\eta} \\ = \ & \inf_{\eta} \frac{\eta^T \begin{bmatrix} W_1 - W_1Y_HW_1 & W_2 - W_1Y_HW_2 \\ W_2^T - W_2^TY_HW_1 & W_3 - W_2^TY_HW_2 \end{bmatrix} \eta}{\eta^T \begin{bmatrix} Z_a & Z_b \\ Z_b^T & Z_c \end{bmatrix} \eta} \\ = \ & \inf_{\eta} \frac{\eta^T \begin{bmatrix} W_1 - W_1Y_HW_1 & 0 \\ 0 & W_3 - W_2^TW_1^{-1}W_2 \end{bmatrix} \eta}{\eta^T \begin{bmatrix} Z_a & Z_aW_1^{-1}W_2 + Z_b \\ W_2^TW_1^{-1}Z_a + Z_b^T & Z_{22} \end{bmatrix} \eta} \\ \leq \ & \inf_{\varsigma} \frac{\varsigma^T[W_1 - W_1Y_HW_1]\varsigma}{\varsigma^TZ_a\varsigma} \\ \triangleq \ & \lambda_2 \end{aligned}$$

where $Z_{22} = Z_c + W_2^TW_1^{-1}Z_b + Z_b^TW_1^{-1}W_2 + W_2^TW_1^{-1}Z_aW_1^{-1}W_2$.

λ_2 can be used as a replica of λ_1 in the iteration for they are of the same sign — leading to the same iteration with small difference in correction. The computational load for evaluating λ_2 is about $\frac{1}{8}$ of that of λ_1 since the computation of the generalized eigenvalue of a symmetric matrix is about $7n^3$ where n is the dimension of the matrix[28].

The relations between the ϵ-iteration and the linear quadratic approximation are clear from the proof of Theorem 4.1. ϵ-iteration tests the measure of singularity of the matrix $W_1 - W_1 Y_H W_1$ and offers an indication of the direction of next try; the linear quadratic approximation evaluates the eigenvalue and generates an improved estimate. The error in the judgement step of ϵ-iteration; namely, $\sigma_\Gamma - 1$, is not the of same scale as the probing estimate γ. The Hankel singular value σ_Γ being very close to 1 does not mean that $|\gamma - \epsilon_o^2|$ is very small (it only implies that $\frac{|\gamma-\epsilon_o^2|}{\epsilon_o^2}$ is small); that is to say, the termination condition in the ϵ-iteration is vague. Suitable normalization, therefore, must be included. In constrast, a priori estimate γ and a posteriori estimate $\beta(\gamma)$ in the linear quadratic approximation are of the same scale. Furthermore, the linear quadratic approximation has a lower-bound guarantee (Theorem 3.3); ϵ-iteration has to evaluate the bounds beforehand to limit the searching domain.

4.3 Remarks

In the case that the assumption $(A, B) = (\begin{bmatrix} -A_H & 0 \\ 0 & A_T^T \end{bmatrix}, \begin{bmatrix} B_H \\ C_F^T \end{bmatrix})$ being controllable does not hold[6], the computations of β can be reduced further. Indeed, we then only have to solve an algebraic Riccati equation of order δ_t, a Sylvester equation of order $(\delta_l - \delta_t)/\delta_t$[7], and a Lyapunov equation of order $(\delta_l - \delta_t)$ to derive the solution of the original Riccati equation, where δ_l is the order of the controllable part of (A, B). When $\delta_t = \delta_h = \delta_l$[8], the computations needed to compute β are an algebraic Riccati equation of order δ_t, a Lyapunov equation of order δ_t, and an eigenvalue evaluation routine; these in totals are less than that required of ϵ-iteration. For more details, see Juang[29].

5 Simulation results

The first example is taken from [26]. The objective is to minimize the $\mathbf{H}^\infty$-norm of

$$\left\| \begin{bmatrix} W_1 S_{out} \\ W_2 T_{out} \end{bmatrix} \right\|_\infty$$

over all stablizing controllers. The plant $G(s)$ and the weighting functions $W_1(s)$ and $W_2(s)$ are given as follows:

$$G(s) = \frac{2}{s-1}, \quad W_1(s) = \frac{s+10}{100(s+0.1)}, \quad W_2(s) = 0.1s + 1$$

After some manipulations, this is converted to a problem like Eqn.(5) with

$$H(s) = \frac{2.3061}{s-1} - \frac{0.0559}{s-0.1414}$$

[6] Still, we assume $H(s)$ and $T(s)$ are minimal

[7] A Sylvester equation $AX + XB + C = 0$ with $A \in \mathbf{R}_{(\delta_l - \delta_t)*(\delta_l - \delta_t)}$ and $B \in \mathbf{R}_{\delta_t * \delta_t}$.

[8] Surprisingly enough, this case happens often in the mixed $\mathbf{H}^\infty$-design problem.

and

$$T(s) = 0.01 + \frac{0.0986}{s + 0.1414}$$

The linear quadratic mapping yields the following quintuple:

$$A = \begin{bmatrix} -1.0000 & 0.0000 \\ 0.0000 & -0.1414 \end{bmatrix} \quad B = \begin{bmatrix} 2.3061 \\ 0.5585 \end{bmatrix}$$

$$Q = \begin{bmatrix} 1.0000 & -0.1000 \\ -0.1000 & 0.0100 \end{bmatrix} \quad S = \begin{bmatrix} -1.1041 \\ 0.1190 \end{bmatrix}$$

and

$$R = 0.0001$$

Note that the state associated with $T(s)$ has been cancelled; the computations can then be reduced. This cancellation phenomenon happens often in the $\mathbf{H}^\infty$-design since $H(s)$ and $T(s)$ are derived out of the same inner factors. The computational burden can then be cut down substantially. As for the accuracy, Chang *el al* picked the a priori estimate $\gamma_o = (1.100436643272142)^2$ and gave the lower and upper bounds after the second iteration

$$1.100437963936794 < \epsilon_o < 1.100437963947313$$

Using the same starting point the linear quadratic approximation gives

$$\beta(\gamma_o) = (1.10043796394726817)^2$$

in one try. This result can be regarded as the actual solution (up to the machine epsilon), which can only be reached with more than 3 iterations using Chang's iteration (and more than 54 iterations using bisectional ϵ-iteration according to [26]). The superiority of the linear quadratic approximation is evident.

The linear quadratic approximation has also been used in the $\mathbf{H}^\infty$-design of the TRW flexible truss structure to speed up the computations. The structure under consideration has 4 inputs, 4 outputs, and 20 vibrating modes. The $\mathbf{H}^\infty$-design objective is

$$\inf_{\text{stablizing controller}} \left\| \begin{bmatrix} W_1 S_{out} \\ W_2 T_{out} \end{bmatrix} \right\|_\infty$$

where S_{out} and T_{out} are the sensitivity function and the complementary sensitivity function at the plant output nodes, respectively. The weighting functions W_1 and W_2 are selected to reflect the requirements on the disturbance rejection and the robustness. The procedure for synthesizing the controller follows mainly [8] and [30]. A computer aided design package is developed and added to the existing CTRL-C™[31] controller design tool; see [32] for the descriptions of the $\mathbf{H}^\infty$-design package and see the report [33] for the details of the $\mathbf{H}^\infty$-design of the TRW structure. A linear quadratic Gaussian design with frequency dependent weighting has also been presented in [34]. For the underlying $\mathbf{H}^\infty$-design problem, after parameterizations, factorizations, and some manipulations as outlined in Section 1, the problem becomes that of Eqn.(5). The matrix $H(s)$ contains 14 states, while $T(s)$ contains 10 states. The linear quadratic approximation is then applied. The a priori estimate is selected to be $\gamma = 100$ on purpose. The evaluated $\beta(\gamma)$ and the accuracy of the three algorithms proposed in Section 3 are tabulated here below.

Algorithm	no. of iteration	a posteriori estimate	accuracy
1	1	0.0946916692747	$4.5593 * 10^{-3}$
	2	0.0955608800611	$9.9884 * 10^{-7}$
	3	0.0955610709608	$4.6846 * 10^{-14}$
2	1	0.0946916692747	$4.5593 * 10^{-3}$
	2	0.0955610062238	$3.3872 * 10^{-7}$
	3	0.0955610700946	$4.5324 * 10^{-9}$
	4	0.0955610709606	$6.2628 * 10^{-13}$
	5	0.0955610709608	$1.1223 * 10^{-16}$
3	1	0.0946916692747	$4.5593 * 10^{-3}$
	2	0.0955609516635	$6.2419 * 10^{-7}$
	3	0.0955610701557	$4.2124 * 10^{-9}$
	4	0.0955610709608	$1.3508 * 10^{-13}$
	5	0.0955610709608	$4.4893 * 10^{-17}$

The accuracy and convergence properties of all three algorithms are satisfactory. The total $\mathbf{H}^\infty$ synthesis procedure can then be speeded up. Algorithm 1 converges fastest in this example although theoretically it is the lowest.

6 Conclusion

The Hankel and Toeplitz operators structure of the reverse time linear quadratic optimization and the $\mathbf{H}^\infty$-design have been studied. This common structure is a powerful tool in bridging two seemingly unrelated design methodologies. Several versions of the linear quadratic approximation have been proposed to compute the $\mathbf{H}^\infty$-achievable performance efficiently. The detailed operations that are involved in the linear quadratic approximation as well as the ϵ-iteration are worked out; they turn out to involve the same operations. In some sense, we can then regard the ϵ-iteration as a special technique of the more general linear quadratic approximation. However, the major advantage of the linear quadratic approximation is that it provides a good estimate after one step! Numerical examples have shown that the linear quadratic approximation is efficient in alleviating the computational burden of the $\mathbf{H}^\infty$-synthesis.

References

[1] E. A. Jonckheere and L. M. Silverman. Spectral theory of the linear quadratic optimal control problem: discrete time case. *IEEE Trans. on Circuits and Systems,*

CAS-25:, 1978.

[2] E. A. Jonckheere and L. M. Silverman. Spectral theory of the linear quadratic optimal control problem: A new algorithm for spectral computations. *IEEE Trans. on Automatic Control*, AC-25:, 1980.

[3] E. A. Jonckheere and L. M. Silverman. The linear quadratic optimal control problem – the operator theoretic viewpoint. *Operator theory: Advances and Applications*, 12:, 1984.

[4] E. A. Jonckheere and M. Verma. A spectral characterization of $\mathbf{H}^{\infty}$–optimal feedback performance – the multivariable case. *Systems and Control Letters*, 1986.

[5] E. A. Jonckheere and J. C. Juang. Toeplitz+Hankel structure in $\mathbf{H}^{\infty}$-design and fast computation of achievable performance. In *Proceedings of American Control Conference*, Seattle WA., June 1986.

[6] M. Verma and E. A. Jonckheere. $\mathbf{L}^{\infty}$-compensation with mixed sensitivity as a broadband matching problem. *Systems and Control Letters*, 4:, 1984.

[7] J. C. Doyle. Synthesis of robust controller and filters. In *Proceedings of IEEE Control and Decision Conference*, page , IEEE, IEEE, San Antonio, TX., December 1983.

[8] J. C. Doyle. *Advances in multivariable control.* Lecture Notes, ONR/Honeywell Workshop, October 1984.

[9] A. Feintuch and B. A. Francis. Uniformly optimal control of linear systems. *Automatica*, 21(5):, 1985.

[10] C. A. Desoer, R. W. Lin, J. Murray, and R. Saeks. Feedback system design: the fractional representation approach t to analysis and synthesis. *IEEE Trans. on Automatic Control*, AC-25(6):, June 1980.

[11] C. C. Chu and J. C. Doyle. The general distance problem in $\mathbf{H}^{\infty}$-synthesis. In *Proceedings of IEEE Control and Decision Conference*, Fort Lauderdale, FL., December 1985.

[12] B. C. Chang and J. B. Pearson. Iterative computation of minimal $\mathbf{H}^{\infty}$-norm. In *Proceedings of IEEE Control and Decision Conference*, page , Fort Lauderale, FL., 1985.

[13] H. Kwakernaak. Minimax frequency domain performance and robustness optimization of linear feedback systems. *IEEE Trans. on Automatic Control*, AC-30(10):, 1985.

[14] M. J. Grimble. Optimal $\mathbf{H}^{\infty}$–robustness and the relationship to LQG design problems. *International Journal of Control*, 43(2):, 1986.

[15] A. Devinatz. On Wiener-Hopf operators. In *Proc. Conf. Functional Analysis*, page , Irvine, CA., 1966.

[16] M. Reed and B. Simon. *Analysis of operators.* Academic press, 1978.

[17] T. Kato. *Pertubation theory of linear operators.* Springer-Verlag, 1976.

[18] B. Sz.-Nagy and C. Foias. *Harmonic analysis of operators on Hilbert space.* American Elsevier, N.Y., 1970.

[19] J. W. Helton. Broadbanding: gain equalization directly from data. *IEEE Trans. on Circuits and Systems*, CAS-28(12):, 1981.

[20] J. C. Willems. Least squares stationary optimal control and the algebraic Riccati equation. *IEEE Trans. on Automatic Control*, AC-16(6):, December 1971.

[21] F. R. Gantmacher. *Theory of Matrices.* Chelsea Publishing Company, 1960.

[22] T. R. F. Nonweiler. *Computational mathematics.* Ellis Horwood Limited, 1984.

[23] A. Laub. A Schur method for solving algebraic Riccati equations. *IEEE Trans. on Automatic Control*, AC-24(6):, 1979.

[24] G. H. Golub, S. Nash, and C. Van Loan. A Hessenberg-Schur method for the problem $AX + XB = C$. *IEEE Trans. on Automatic Control*, AC-24(6):, December 1979.

[25] B. A. Francis. Optimal disturbance attenuation with control weighting. In A. Bagchi and H. Th. Jongen, editors, *Systems and optimization, Lecture Notes in Control and Information Science*, Springer-Verlag, 1984.

[26] B. C. Chang, S. S. Banda, and T. E. McQuade. Fast iterative algorithm for $\mathbf{H}^\infty$-optimization problems. 1986. preprint.

[27] E. V. Haynsworth. Determination of the inertia of a partitioned hermitian matrix. *Linear Algebra and its application*, 1:, 1968.

[28] G. E. Golub and C. F. Van Loan. *Matrix computations.* The Johns Hopkins university press, 1983.

[29] J. C. Juang. $\mathbf{H}^\infty$*-design: its structure, computations, and applications.* PhD thesis, University of Southern California, 1986.

[30] M. G. Safonov, E. A. Jonckheere, M. Verma, and D. Limbeer. Synthesis of positive real multivariable feedback systems. *International Journal of Control*, 1986.

[31] *CTRL-C manual.* Systems and Control Technology, 1983.

[32] J. C. Juang and E. A. Jonckheere. A package for $\mathbf{H}^{\infty}$-design. 1986. Internal report. Department of Electrical Engineering-Systems, University of Southern California.

[33] J. C. Juang and E. A. Jonckheere. $\mathbf{H}^{\infty}$-design of the TRW truss structure. 1986. Technical report. Submitted for publication.

[34] Ph. C. Opdenacker, E. A. Jonckheere, and M. G. Safonov. Reduced order compensator design for an experimental large flexible structure. In *Proceedings of IEEE Control and Decision Conference*, Fort Lauderdale, FL., December 1985.

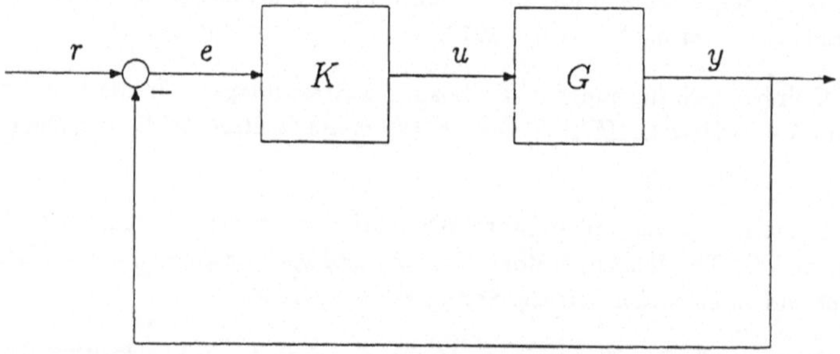

Figure 1. The system under considerations.

H_∞- OPTIMAL CONTROL, LQG POLYNOMIAL SYSTEMS TECHNIQUES AND NUMERICAL SOLUTION PROCEDURES

by

M. Saeki*, M.J. Grimble**, E. Kornegoor***, and M.A. Johnson**

*Institute of Information Sciences and Electronics,
University of Tsukuba,
Niiharigun,
Ibaraki, 305,
JAPAN.

**Industrial Control Unit,
University of Strathclyde
Glasgow G1 1XW
SCOTLAND, UK.

***Department of Applied Mathematics,
Twente University of Technology,
7500 AE, Enschede,
THE NETHERLANDS.

Notation

$\mathbb{R}$	~ set of real numbers
$\mathbb{R}^n$	~ n-tuple of real numbers
$\mathbb{R}^{nxm}$	~ nxm - tuple of real numbers
$\mathbb{P}(.)$	~ set of polynomials with real coefficients
$\mathbb{R}(.)$	~ set of ratios a(.)/b(.) where $a,b \in \mathbb{P}(.)$
Toep(A)	~ Toeplitz matrix formed from $A \in \mathbb{P}(.)$

$$\mathrm{Toep}(A) \triangleq \begin{bmatrix} a_o & 0 & \cdots & \cdots & 0 \\ a_1 & a_o & \ddots & & \vdots \\ \vdots & \ddots & \ddots & a_o & 0 \\ a_n & a_{n-1} & \cdots & a_1 & a_o \\ 0 & a_n & & \vdots & a_1 \\ \vdots & \vdots & & \vdots & \vdots \\ 0 & \cdots & \cdots & 0 & a_n \end{bmatrix} \in \mathbb{R}^{\alpha \times \beta}$$

where $A = a_o + a_1 z^{-1} + \ldots + a_n z^{-n} \in \mathbb{P}(z^{-1})$ and dimensions α, β depend on the order of the polynomial multiplication being represented.

NATO ASI Series, Vol. F34
Modelling, Robustness and Sensitivity Reduction in Control Systems. Edited by R. F. Curtain

1. Introduction

The introductory material first relates the general background of the robustness problem. This is followed by more specific material giving the context for the LQG and polynomial systems approach to the problem of H_∞ optimal control design.

1.1 General background

Uncertainty in the models of a system and its disturbance inputs is usually mitigated by using a feedback policy. Indeed Horowitz (1963,[1]) strongly makes the point that the true importance of feedback lies in the ability to achieve a desired performance despite model uncertainty. It is argued that the actual design and cost of feedback should be closely related to the extent of the uncertainty and the narrowness of performance tolerances.

In the closed loop design problem for multivariable systems it is generally required that:

(i) The closed loop system is **asymptotically stable.**

(ii) Asymptotic tracking results and is independent of any disturbances appearing in the system to produce **good regulation.**

(iii) The multivariable closed loop system has **low interaction.**

(iv) A fast response and **good transient behaviour** is obtained.

In addition there is often the unstated requirement that characteristics (i) to (iv) are achieved when the system parameters are perturbed from their nominal values. This desirable property is termed **robustness.** A system has **stability robustness** if it remains stable under a limited range of parameter variations. Similarly, if the quality of system regulation does not deteriorate over a limited range of system or disturbance parameter variations the system is **disturbance robust.**

Uncertainties in the system description can arise from:

(i) **Modelling errors.** Imprecise knowledge of the physical mechanisms, poor (or limited) parameter identification and simplifying assumptions often leads to incorrect models, or model parameters.

(ii) **Time-varying parameters.** Most systems suffer from wear of mechanical parts, degradation of electrical parts, changes in the quality of feedstock materials, environmental changes and other extraneous influences causing the time variation of system parameters.

(iii) **Nonlinear behaviour.** The system description is often given in terms of equations linearised about an operating point. If the operating point changes the form and/or the parameters of the linearised equations may change.

There are two other design requirements related to the robustness concepts which are often specified:

(i) **Property of high integrity:** The closed loop system must remain stable under sensor and/or actuator failure conditions.

(ii) **Good closed loop reliability:** The closed loop system must retain the property to track or regulate under sensor and/or actuator failure conditions.

The important problem of controller design involves achieving a satisfactory trade off between a subset of the above design requirements. Recent robustness research is a direct attempt to formulate these desirable properties more precisely and construct methods to meet the design requirements optimally.

1.2 Optimal methods for robust system design

There are two generally accepted basic approaches to the design of optimal linear systems and to robustness improvement. The LQ (linear quadratic) state feedback approach has certain inherent guaranteed properties (Safonov and Athans 1977[2]). Unfortunately the LQG (linear quadratic Gaussian) output feedback controller does not have equivalent properties (Doyle 1978[3]), except under certain limiting conditions (Doyle and Stein 1979[4]; Moore and Blight 1981[5]), or in cases where state and state estimate feedback can be employed (Grimble, 1984[6]). However, the sensitivity and robustness characteristics of LQG controllers can be modified and improved by an appropriate choice of cost weighting functions (Safonov, Laub and Hartman, 1981[7]; Grimble, 1983[8]).

The second basic design approach is still relatively new and has not reached maturity but it does offer potential advantages in some practical problems where least squares criteria are not appropriate. This method was introduced by Zames (1981[9]) in a seminal paper which demonstrated the utility of minimising a H_∞-norm (viz, maximum value of a frequency response), rather than the usual L_2-quadratic norm. Zames noted that for some problems good spectrum information on the disturbances is not available and it is then desirable to limit the maximum value of the disturbance frequency response, rather than to use an LQG design which assumes good disturbance models. Zames also showed that the controllers obtained were similar in structure to those found by classical design techniques.

In its initial form the approach had difficulties due to the abstract nature of the mathematical theory and because of the simplicity of problem definition. For example, a weighted sensitivity function was minimised and no other performance requirements were considered. This led to the result that the shape of the optimal sensitivity function was exactly inversely proportional to the shape of the weighting function. Thus, as Freudenberg and Looze (1983[10]) noted, the design work in choosing the weighting function was the same as that in classical control design and it was of little practical advantage to use the optimisation procedure. The controller also had the effect that it cancelled all stable plant poles and zeros which may not be desirable if these are lightly damped.

The H_∞ optimization theory was advanced substantially by Zames and Francis (1981[11]) and several other researchers using a diverse range of abstract mathematical techniques. Kwakernaak (1982[12]; 1983[13];1984[14]; 1985[15]) has employed a more conventional polynomial systems method to solve the problem, and has generalised the cost function to include both sensitivity and complementary weighting terms. A key lemma introduced by Kwakernaak in this work relates the H_∞ and ℓ_2 optimization problems. This lemma is employed in the following analysis.

The solution of the LQG problems for systems with dynamic cost weighting functions using the polynomial system descriptions was presented by Grimble (1986[16]). If the H_∞ optimal robustness control problem is embedded within a conventional LQG problem, the solution procedures of Grimble, (1986[16]) may be used to readily obtain the equations which prescribe the controller (Grimble, 1986[17]). Saeki used this philosophy to solve the problem formulated by Kwakernaak (1982[12]). Following this route Saeki (1986[18]) was able to show the exact equivalence between the solution of the resulting Diophantine equations and the controller equations previously obtained by Kwakernaak (1982[12]). Numerical routines to solve the set of controller equations were proposed in [18] and [19]. In this paper the techniques of [17], [18], and [19] are applied to the dual criterion performance index proposed by Grimble (1985[20]).

2. H_∞-optimal control, LQG and polynomial system techniques

In this section, the relationship between the dual criterion performance index and a simple robustness cost function comprising only sensitivity and complementary sensitivity costings is considered. This is followed by a section containing solution theorems which use the auxiliary lemma of Kwakernaak to embed the H_∞-optimal control problem in a LQG - framework. This route previously pursued by Grimble (1986[17]) is extended here for systems with dynamic weightings which have simple unit circle poles. A second feature of the analysis of this paper is that the dual criterion is condensed algebraically to permit the application of the numerical techniques of Saeki (1986[18]), and Saeki and Kornegoor (1986[19]).

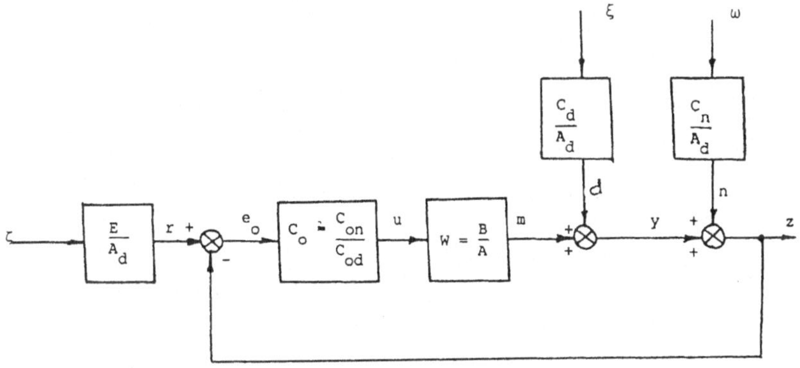

Fig. 1: System description

2.1 System description and cost functions

Consider the single-input, single output discrete time system as shown in Fig. 1. The following system relationships obtain:

Plant

$$m(t) = W(z^{-1})u(t) = \frac{A(z^{-1})}{B(z^{-1})}u(t), \quad W \in \mathbb{R}(z^{-1}); A, B \in \mathbb{P}(z^{-1}) \qquad (1)$$

Controller

$$u(t) = C_o(z^{-1})e_o(t) = \frac{C_{on}(z^{-1})}{C_{od}(z^{-1})} e_o(t), \qquad (2)$$

with

$$C_o \varepsilon\ \mathbb{R}(z^{-1});\ C_{on}, C_{od} \varepsilon \mathbb{P}(z^{-1})$$

Reference

$$r(t) = \frac{E(z^{-1})}{A_d(z^{-1})} \zeta(t) \qquad E, A_d\ \varepsilon\ \mathbb{P}\ (z^{-1}) \qquad (3)$$

Input disturbance

$$d(t) = \frac{C_d(z^{-1})}{A_d(z^{-1})} \xi(t) \qquad C_d,\ A_d\ \varepsilon\ \mathbb{P}\ (z^{-1}) \qquad (4)$$

Output disturbance

$$n(t) = \frac{C_n(z^{-1})}{A_d(z^{-1})} \omega(t) \quad C_n,\ A_d\ \varepsilon\ \mathbb{P}\ (z^{-1}) \qquad (5)$$

where the plant (1) is assumed to be free of unstable hidden modes, and $A_d\ \varepsilon\ \mathbb{P}\ (z^{-1})$ is strictly Hurwitz. The driving signals for the reference and input and output disturbance processes, namely $\{\zeta(t)\}$, $\{\xi(t)\}$ and $\{\omega(t)\}$ are assumed (without loss of generality) to be white noise processes with zero mean and unit covariances.

The following system signals are of interest

Plant output	$y(t) = m(t) + d(t)$	(6)
Measured output	$z(t) = y(t) + n(t)$	(7)
Controller input	$e_o(t) = r(t) - z(t)$	(8)
Tracking error	$e(t) = r(t) - y(t)$	(9)

Define the sensitivity function $S\ \varepsilon\ \mathbb{R}\ (z^{-1})$ as:

$$S(z^{-1}) \triangleq (1 + WC_o)^{-1}\ \varepsilon\ \mathbb{R}\ (s) \qquad (10)$$

and the complementary sensitivity function $T\ \varepsilon\ \mathbb{R}(z^{-1})$ as:

$$T(z^{-1}) \triangleq (1 + WC_o)^{-1} WC_o\ \varepsilon\ \mathbb{R}(z^{-1}) \qquad (11)$$

$$= 1 - S = WU \qquad (12)$$

where

$$U(z^{-1}) \triangleq C_o S\ \varepsilon\ \mathbb{R}\ (z^{-1}) \qquad (13)$$

From the system of Fig. 1, the following output equation (using the definitions of sensitivity and complementary sensitivity functions) results:

$$y = (1 + WC_o)^{-1} d + (1 + WC_o)^{-1} WC_o(r-n) \qquad (14)$$

$$y = Sd + T(r-n) \qquad (15)$$

The tracking performance arises from the requirement for the system output y(t) to follow the reference signal r(t) so as to minimise some measure of the tracking error given by (10). Robustness requirements are met by selecting a controller $C_o(z^{-1})$ to produce closed loop stability and to minimise some measure of the sensitivity function S, and complementary function T given by equations (12) and (13). The dual criterion of Grimble (1985[20]) permits the composite costing of performance and robustness properties.

<u>Lemma 1</u> **Dual cost criterion**

Let Φ_{ee}, $\Phi_{e_oe_o}$, and Φ_{uu} be the covariances of signals e, e_o and u for the system of Fig. 1. The dual criterion defined by:

$$J_\infty^D \triangleq \sup_{|z|=1} \{Q_c\Phi_{ee} + P_c\Phi_{e_oe_o} + R_c\Phi_{uu}\} \tag{16}$$

$$= \sup_{|z|=1} \{Q_D\Phi_{ee} + P_DSS^* + R_DTT^*\} \tag{17}$$

where Q_c, P_c, R_c, Q_D, P_D, R_D ε $\mathbb{R}(z^{-1})$. The relationship between the weights is given by:

$$Q_D = Q_c \tag{18}$$

$$P_D = P_c\Phi_{cc} \tag{19}$$

$$R_D = R_c(WW^*)^{-1}\Phi_{cc} \tag{20}$$

and $$\Phi_{cc} = \Phi_{rr} + \Phi_{nn} + \Phi_{dd} \tag{21}$$

•

<u>Proof</u> From the system equations, e_o and u follow as:

$$e_o = S(r-n-d) \tag{22}$$

and $$u = U(r-n-d) = W^{-1}T(r-n-d) \tag{23}$$

The covariances $\Phi_{e_oe_o}$ and Φ_{nn} may be written:

$$\Phi_{e_oe_o} = SS^*\Phi_{cc} \tag{24}$$

and $$\Phi_{uu} = (WW^*)^{-1}\Phi_{cc}TT^* \tag{25}$$

where $$\Phi_{cc} = \Phi_{rr} + \Phi_{nn} + \Phi_{dd}. \tag{26}$$

Equation (17) can be obtained from (16) by direct substitution of (24) and (25) and vice versa. •

<u>Lemma 2</u> **Robustness cost criterion**

Given the system configuration of Fig. 1 then:

$$J_\infty^D = \sup_{|z|=1} \{\tilde{Q}\Phi_{ee} + \tilde{P}SS^* + \tilde{R}TT^*\} \tag{27}$$

$$= \sup_{|z|=1} \{PSS^* + RTT^*\} \tag{28}$$

where $\tilde{Q}, \tilde{P}, \tilde{R}, P, C \in \mathcal{R}(z^{-1})$ and the relationship between the weights is given by:

$$P = \tilde{P} + \tilde{Q}\Phi_o \tag{29}$$

$$R = \tilde{R} + \tilde{Q}\Phi_{nn} \tag{30}$$

where $\Phi_o \triangleq \Phi_{rr} + \Phi_{dd}$. (31)

Proof Error signal $e(t)$ may be written:

$$e = S(r-d) + Tn \tag{32}$$

hence $\Phi_{ee} = SS^*\Phi_o + TT^*\Phi_{nn}$ (33)

where $\Phi_o = \Phi_{rr} + \Phi_{dd}$. Write therefore

$$P = \tilde{P} + \tilde{Q}\Phi_o$$

and $R = \tilde{R} + \tilde{Q}\Phi_{nn}$

hence direct substitution in (28) yields (27). Similarly starting from (27) use (33) to obtain (28) with definitions (29) and (30) for weights P and R. •

These lemmas usefully clarify the relationship between performance costings and robustness measures. If the problem is formulated using the Dual criterion (16) then Lemma 1 indicates the cost weights which accrue to the sensitivity and complementary sensitivity terms via P_D and R_D (namely equations (19) and (20)). Note that there is a direct link between $\Phi_{e_o e_o}$ and SS^* via the tunable weighting P_c, and similarly between Φ_{nn} and TT^* via the tunable weighting R_c.

Lemma 2 shows the relationship between a cost function comprising Φ_{ee}, SS^* and TT^* to be dependent on only SS^* and TT^*. The weights of the cost function (27) may be combined to give the condensed robustness cost function (28) or the weights of (28) parameterised to incorporate a performance costing (viz on Φ_{ee}).

The problem of solving the optimization associated with cost function (28) was pursued by Saeki and Kornegoor (1986[18]; 1986[19]). In the sequel, the application of the numerical techniques to the formulation arising from the Dual criterion approach is pursued.

2.2 Problem formulation

Consider the following H_∞ optimization problem:

$$\min_{\text{w.r.t.}C_o} J_\infty^D = \sup_{|z|=1} \{Q_c\Phi_{ee} + P_c\Phi_{e_o e_o} + R_c\Phi_{uu}\}$$

where C_o is selected to ensure closed loop stability and the cost weighting functions have the following form:

$$Q_c(z^{-1}) = \frac{\tilde{B}_q^* \tilde{B}_q}{A_w^* A_w} \; \varepsilon \; \mathbb{R}(z^{-1}) \tag{34}$$

$$P_c(z^{-1}) = \frac{\tilde{B}_p^* \tilde{B}_p}{A_w^* A_w} \; \varepsilon \; \mathbb{R}(z^{-1}) \tag{35}$$

$$R_c(z^{-1}) = \frac{\tilde{B}_r^* \tilde{B}_r}{A_w^* A_w} \; \varepsilon \; \mathbb{R}(z^{-1}) \tag{36}$$

where $\tilde{B}_q$, $\tilde{B}_p$, $\tilde{B}_r$, $A_w \; \varepsilon \; \mathbb{P}(z^{-1})$ and A_w is strictly Hurwitz. From Lemma 1:

$$J_\infty^D = \sup_{|z|=1} \{\tilde{Q}_D \Phi_{ee} + \tilde{P}_D SS^* + \tilde{R}_D TT^*\} \tag{37}$$

where (equations (18), (19) and (20)):

$$\tilde{Q}_D = Q_c \tag{38}$$

$$\tilde{P}_D = P_c \Phi_{cc} \tag{39}$$

$$\tilde{R}_D = R_c (WW^*)^{-1} \Phi_{cc} \tag{40}$$

where Φ_{cc} is given by (26).

From Lemma 2, the cost function (37) becomes:

$$J_\infty^D = \sup_{|z|=1} \{P_D SS^* + R_D TT^*\} \tag{41}$$

where $P_D = \tilde{P}_D + \tilde{Q}_D \Phi_o$ (42)

$$R_D = \tilde{R}_D + \tilde{Q}_D \Phi_{nn} \tag{43}$$

with Φ_o given by (31).

combining equation (38) to (40) and (42) to (43) yields:

$$P_D = P_c \Phi_{cc} + Q_c \Phi_o \tag{44}$$

$$R_D = R_c (WW^*)^{-1} \Phi_{cc} + Q_c \Phi_{nn} \tag{45}$$

Using transfer relationships (3), (4) and (5) identify:

$$\Phi_{rr} = \frac{EE^*}{A_d A_d^*} \tag{46}$$

$$\Phi_{dd} = \frac{C_d C_d^*}{A_d A_d^*} \tag{47}$$

$$\Phi_{dd} = \frac{C_n C_n^*}{A_d A_d^*} \tag{48}$$

so that using (34) to (36) and the above equations (44) to (48) yields:

$$P_D \triangleq \frac{B_q^* B_q}{A_q^* A_q} \quad \text{(subscript q is used to be consistent with later sections)} \tag{49}$$

where $A_q = A_w A_d$ (50)

$$B_q^* B_q = \tilde{B}_p^* \tilde{B}_p (EE^* + C_d C_d^* + C_n C_n^*) + \tilde{B}_q \tilde{B}_q (EE^* + C_d C_d^*) \tag{51}$$

and $$R_D \triangleq \frac{B_r^* B_r}{A_r^* A_r} \tag{52}$$

where $A_r = A_w A_d B^+ B^o (B^-)^* z^{-deg(B^-)}$ (53)

$$B_r^* B_r = \tilde{B}_r^* \tilde{B}_r AA^* (EE^* + C_d C_d^* + C_n C_n^*) + BB^* \tilde{B}_q^* \tilde{B}_q^* C_n C_n^* \tag{54}$$

and $B = B^+ B^o B^-$ (55)

where $B^+ \in \mathbb{P}(z^{-1})$ has minimum phase zeros, $B^o \in \mathbb{P}(z^{-1})$ has unit circle zeros and $B^- \in \mathbb{P}(z^{-1})$ has non-minimum phase zeros.

In this case, A_q and A_r contain the common factor $A_w A_d$, and A_r can be Hurwitz due to the presence of B^o. The H_∞ optimal control problem for cost (44) with A_q and A_r being strictly Hurwitz was solved in [18].

To summarise, the formulation of a dual criterion is algebraically reduced to the condensed cost function:

$$J_\infty^D = \sup_{|z|=1} \{P_D SS^* + R_D TT^*\}$$

where P_D has the form specified by equations (49) to (51), and R_D has the form specified by equations (52) to (54). The solution of this optimization problem is considered next.

2.3 Solution theorems

The approach utilized is to follow Grimble ([17], 1986) and embed the H_∞-optimal control problem within an LQG-optimal control problem for which the solution theorems using polynomial techniques have already been investigated (Grimble, 1986[16], 1985[20]).

In the framework of the system shown in Fig. 1, set n = d = 0 to obtain the reduced system of Fig. 2. Note the re-definition of the transfer relating ξ and r. For the reduced system description the following relations pertain:

$$\Phi_{cc} = \Phi_{rr} = \frac{B_\sigma B_\sigma^*}{A_\sigma A_\sigma^*} \triangleq \Sigma \in \mathbb{R}(z^{-1}) \tag{56}$$

with B_σ, $A_\sigma \in \mathbb{P}(z^{-1})$ strictly Hurwitz.
From equation (25):

$$\Phi_{uu} = (WW^*)^{-1}\Phi_{cc}TT^*$$
$$= (WW^*)^{-1}\Sigma TT^* \tag{57}$$

From equation (33):

$$\Phi_{ee} = SS^*\Phi_o + TT^*\Phi_{nn}$$

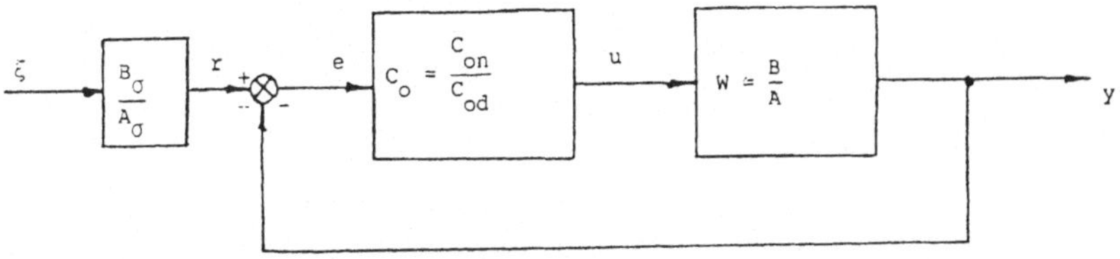

Fig. 2

however $n = 0$ thus $\Phi_{nn} = 0$ and $\Phi_o = \Phi_{rr} + \Phi_{dd} = \Phi_{rr} = \Sigma$

$$= SS^*\Sigma \tag{58}$$

Consider an LQG cost function of the form:

$$J = \frac{1}{2\pi j}\oint_{|z|=1} \{Q\Phi_{ee} + RWW^*\Phi_{uu}\}\frac{dz}{z} \tag{59}$$

which using equations (57) and (58) becomes:

$$J = \frac{1}{2\pi j}\oint_{|z|=1} \{QSS^* + RTT^*\}\,\Sigma\,\frac{dz}{z} \tag{60}$$

where $\Sigma \in \mathbb{R}(z^{-1})$ is given by (56).

The solution theorem for the optimal control problem based on the LQG cost function (59) follows:

Theorem 1 **LQG optimal control solution**

Assume that A_q and A_r are Hurwitz and the zeros of A_q and A_r are simple on the unit circle. The spectral factor $D_e \in \mathbb{P}(z^{-1})$ is defined by:

$$D_e^* D_e = A_r^* A_r B_q^* B_q + A_q^* A_q B_r^* B_r \tag{61}$$

where D_e is assumed to be strictly Hurwitz. Introduce the polynomial equations with respect to G_1, H_1, $F_1 \in P(z^{-1})$:

$$D_e^* B^- z^{-n_2} G_1 + F_1 A_q A_\sigma = B_q^* B_q A^{-*} B^{-*} A_r^* B_\sigma z^{-n_3} \tag{62}$$

$$D_e^* A^- z^{-n_2} H_1 - F_1 A_r A_\sigma = B_r^* B_r A^{-*} B^{-*} A_q^* B_\sigma z^{-n_3} \tag{63}$$

If $n_2 = \deg(D_e)$ and $n_3 = \deg(D_e A^- B^-)$, equations (62) and (63) can be solved for the unique minimal degree solution with respect to F_1 (a solution, F_1 which satisfies $\deg(F_1) < n_3$ always exist and is the minimal degree solution). Hence, the LQG optimal controller which minimizes (59) is given by:

$$C_o = (B^o B^+ A_q H_1)^{-1}(A_o A^+ A_r G_1) \tag{64}$$

and the minimum value of the cost function is given by:

$$J_{min} = \frac{1}{2\pi j} \oint_{|z|=1} \left\{ \frac{F_1^* F_1}{D_e^* D_e A^{-*} A^- B^{-*} B^-} + \frac{B_q^* B_q B_r^* B_r}{D_e^* D_e} \Sigma \right\} \frac{dz}{z} \tag{65}$$

•

Proof (Saeki, 1986[18])

This theorem is based on previous results of Grimble (1986[17]) extending the results to weight denominator polynomials which are Hurwitz rather than strictly Hurwitz. In particular, emphasis is given to reducing the degrees of coefficient and unknown solution polynomials using factorizations for the system polynomials $A, B \in \mathcal{P}(z^{-1})$.

The link between the LQG and H_∞-optimal control problems is given via the lemma introduced by Kwakernaak and employed in [17] and [18].

Lemma 3 **The auxiliary lemma of Kwakernaak**

Consider the auxiliary problem of minimizing

$$J = \frac{1}{2\pi j} \oint_{|z|=1} X(z^{-1})\Sigma(z^{-1}) \frac{dz}{z} \tag{66}$$

Suppose that for some real rational $\Sigma = \Sigma^* > 0$ on $|z|=1$ is minimized by a function $X = X^*$ for which $X = \lambda^2$ where λ is a real constant, then this function also minimizes $\sup_{|z|=1} X(z^{-1})$. •

The equivalence between the cost functions (59) and (60) and the structure of (60) when compared to (66) implies $X(z^{-1}) = QSS^* + RTT^* \in \mathcal{R}(z^{-1})$. Lemma 3 shows that if the LQG optimal controller which minimises (66) (equivalently (59)) also satisfies $X(z^{-1}) = \lambda^2 \in \mathcal{R}^+$, then the controller is the H_∞ optimal controller which minimises

$$J_\infty = \sup_{|z|=1} X(z^{-1}) = \sup_{|z|=1} (QSS^* + RTT^*) \tag{67}$$

and the optimal value is given by:

$$J_\infty^o = \lambda^2 \tag{68}$$

The result is translated into the following conditions for the H_∞ optimal controller.

Theorem 2 **H_∞ optimal controller**

Assume that $D_e^*D_e\lambda^2 - B_q^*B_qB_r^*B_r$ is positive on $|z|=1$ and that there exists a solution $(G_3, H_3, F_1^-, A_\sigma, \lambda)$ which satisfies:

$$D_e^*D_eB^- z^{-n_2}G_3 + F_1^-A_q\tilde{A}_\sigma = B_q^*B_qA_r^*F_{1s}^- z^{-n_2} \tag{69}$$

$$D_e^*D_eA^- z^{-n_2}H_3 + F_1^-A_r\tilde{A}_\sigma = B_r^*B_rA_q^*F_{1s}^- z^{-n_2} \tag{70}$$

$$\tilde{A}_\sigma^*\tilde{A}_\sigma = D_e^*D_e\lambda^2 - B_q^*B_qB_r^*B_r \tag{71}$$

where $\deg(F_1^-) < n_3$, F_1^- has all its zeros outside the unit circle or is constant, $\tilde{A}_\sigma$ is strictly Hurwitz, $F_{1s}^- = (F_1^-)^* z^{-n_4}$ and $n_4 = \deg(F_1^-)$. Then the optimal controller which minimizes (67) is given by

$$C_o = (B^oB^+A_qH_3)^{-1}(A^oA^+A_rG_3) \tag{72}$$

and the minimum value of J_∞ is λ^2. •

Proof Saeki, 1986[18]. •

Corollary 1 **Equivalent controller solution equations**

Equations (69), (70) and (71) are equivalent to

$$A_rB^-G_3 + A_qA^-H_3 - F_{1s}^- = 0 \tag{73}$$

$$B_r^*B_rA_q^*B^- z^{-n_2}G_3 - B_q^*B_qA_r^*A^- z^{-n_2}H_3 + F_1^-\tilde{A}_\sigma = 0 \tag{74}$$

$$\tilde{A}_\sigma^*\tilde{A}_\sigma = D_e^*D_e\lambda^2 - B_q^*B_qB_r^*B_r \tag{75}$$

Proof: •

These equations, when the weighting elements are strictly Hurwitz, follow directly from the results in Grimble [17] and may be obtained using Theorem 2 for the Hurwitz case as follows:

(i) Equation (73) follows from $A_r \times (69) + A_q \times (70)$

(ii) Equation (74) follows from $A_q^*B_r^*B_r \times (69) - A_r^*B_q^*B_q(70)$

(iii) Equivalence follows from: (a) any solution of (69) to (71) satisfies (73) to (75) and (b) any solution of (73) to (75) for which $\lambda^2 < (\lambda^o)^2$ also satisfies (69) to (71) and contradicts the optimality of $(\lambda^o)^2$.

Corollary 1 corresponds to the equations obtained by Kwakernaak (1985[15]) for the continuous time case by a different polynomial route. Applying Theorem 2, and Corollary 1, the solution for the H_∞ optimal control problem for the dual performance index case obtains:

<u>Theorem 3</u> **H_∞-optimal controller for dual performance index**

Assume that all the zeros of B^o are simple. Let D_e be defined as in Theorem 1. $n_5 \triangleq \deg(D_e) - \deg(A_w A_d)$ and $n_6 \triangleq \deg(D_e A^-) - \deg(A_w A_d)$. Assume that $D_e^* D_e \lambda^2 - B_q^* B_q B_r^* B_r$ is positive on $|z|=1$ and that there exists a solution $(G_4,\ H_4,\ F_2^-,\ \tilde{A}_\sigma, \lambda)$ which satisfies:

$$BG_4 + A^- H_4 - F_{2s}^- = 0 \tag{76}$$

$$B_r B_r^* z^{-n_5} G_4 - B_q B_q^* B^* A^- z^{-n_5} H_4 + F_2^- \tilde{A}_\sigma = 0 \tag{77}$$

$$\tilde{A}_\sigma \tilde{A}_\sigma^* = D_e D_e^* \lambda^2 - B_q^* B_q B_r^* B_r \tag{78}$$

where $\deg(F_2^-) < n_6$, F_2^- has all its zeros outside the unit circle or is constant, $\tilde{A}_\sigma$ is strictly Hurwitz, $F_{2s}^- = (F_2^-)^* z^{-n_7}$ and $n_6 > n_7 \triangleq \deg(F_2^-)$. Then the optimal controller which minimizes (16) is given by:

$$C_o = H_4^{-1}(A^o A^+ G_4) \tag{79}$$

<u>Proof</u> Since A_r and A_q contain a common factor $A_w A_d$ in (50) and (53), F_{1s}^- includes a factor $A_w A_d$ from (73) and so F_1^- includes a factor $A_w^* A_d^* z^{-\deg(A_w A_d)}$. Since A_r includes $(B^-)^*$ in (53), F_1^- includes a factor B^- from (74) because $\tilde{A}_\sigma$ is strictly Hurwitz and does not include a factor whose zeros are outside the unit circle. From this and (73), H_3 includes a factor$(B^-)^*$. Thus introduce new variables $G_3 \triangleq G_4$, $H_3 = H_4(B^-)^* z^{-\deg(B^-)}$ and $F_1^- = F_2^- B^- A_w^* A_d^* z^{-\deg(A_w A_d)}$. Then the equations (73) and (74) are simplified to (76) and (77), respectively.

3. <u>Numerical algorithms</u>

The solution theorems utilized the LQG route to obtain polynomial equations delineating the H_∞-optimal controller. In particular Corollary 1, and Theorem 3 show that the simultaneous solution of two linear equations and a spectral factorization are required. The problem is essentially nonlinear in the cost parameter λ^2, and two numerical techniques for a solution are presented in this section along with a numerical example.

3.1 Problem reformulation

Recall from Theorem 3, that it is required to solve:

$$BG_4 + A^- H_4 - F_{2s}^- = 0 \tag{76}$$

$$B_r B_r^* z^{-n_5} G_4 - B_q B_q^* B^* A^- z^{-n_5} H_4 + F_2^- \tilde{A}_\sigma = 0 \tag{77}$$

$$\tilde{A}_\sigma \tilde{A}_\sigma^* = D_e D_e^* \lambda^2 - B_q^* B_q B_r^* B_r \tag{78}$$

$$D_e D_e^* = A_r^* A_r B_q B_q^* + A_q^* A_q B_r^* B_r \tag{61}$$

with respect to unknown polynomials G_4, H_4, and F_2^- for the maximum value of $\lambda \in \mathbb{R}^+$. Denote $G_4, H_4, F_2^- \in \mathbb{P}(z^{-1})$ as:

$$F_2^- = f_o + f_1 z^{-1} + \ldots + f_{n_f-1} z^{-(n_{f-1})} \tag{80}$$

$$G_4 = g_o + g_1 z^{-1} + \ldots + g_{n_g-1} z^{-(n_{g-1})} \tag{81}$$

$$H_4 = h_o + h_1 z^{-1} + \ldots + h_{n_h-1} z^{-(n_{h-1})} \tag{82}$$

where

$$\begin{aligned} n_f &= \deg(D_e A^-) - \deg(A_w A_d) \\ n_g &= \deg(A^-) + \max(\deg(A_q), \deg(B_q)) \\ n_h &= \max(\deg(A_r), \deg(B_r)) \end{aligned} \tag{86}$$

The degrees of F_2^- is given by Theorem 3. The degrees of G_4 and H_4 are determined by the degrees of F_2^- and the degrees of the solutions G_4 and H_4 of (76) and (74). From (80):

$$F_{2s}^- = f_{n_f-1} + f_{n_f-2} z^{-1} + \ldots + f_1 z^{-(n_f-2)} + f_o z^{-(n_f-1)} \tag{84}$$

By expanding the equations in powers of z^{-1} and equating the coefficients of like powers, the following linear algebraic equations obtain from equations (76) and (77) respectively:

$$\Gamma_{11} g + \Gamma_{12} h + \Gamma_{13} f = 0 \tag{85}$$

$$\Gamma_{21} g + \Gamma_{22} h + \Gamma_{23} f = 0 \tag{86}$$

where $g \in \mathbb{R}^{n_g}$, $h \in \mathbb{R}^{n_h}$, and $f \in \mathbb{R}^{n_f}$ are vectors of the polynomial coefficients in equations (80) to (82). The matrices Γ_{ij} are given by:

$\Gamma_{11} = \text{Toep}(B) \in \mathbb{R}^{n_f x n_g}$, $\Gamma_{12} = \text{Toep}(A^-) \in \mathbb{R}^{n_f x n_h}$,

$\Gamma_{21} = \text{Toep}(B_r B_r^* z^{-n_5}) \in \mathbb{R}^{m x n_g}$, $\Gamma_{22} = \text{Toep}(B_q^* B_q B^* A^- z^{-n_5}) \in \mathbb{R}^{m x n_h}$

$\Gamma_{23} = \text{Toep}(\tilde{A}_r) \in \mathbb{R}^{m x n_f}$ where $m \triangleq \deg(D_e) + n_f$, and

$\Gamma_{13} \in \mathbb{R}^{n_f x n_f}$ is a matrix whose (i, n_f-i+1)th element is -1 for i=1,..,n_f and the other elements are zero. The numerical problem is thus the solution of equations (85), and (86) simultaneous with (78) and (61) for the minimum value of λ^2, and $\bar{F}_{2s}$ strictly Hurwitz.

3.2 A singular value/condition number method

The first method proposed utilizes singular values and, an indication of rank deficiency using the matrix condition number. Using the property that $\Gamma_{13}^2 = I_{n_f}$, $f \in \mathbb{R}^{n_f}$ may be eliminated from (85) to yield:

$$\tilde{\Gamma}\begin{bmatrix} g \\ h \end{bmatrix} = [\Gamma_{21} - \Gamma_{23}\Gamma_{13}\Gamma_{11} \vdots \Gamma_{22} - \Gamma_{23}\Gamma_{13}\Gamma_{12}]\begin{bmatrix} g \\ h \end{bmatrix} = 0 \tag{87}$$

Since $\tilde{\Gamma}$ is a $(\deg(D_e) + n_f) \times n_f$ matrix, the homogeneous equation (87) has a nonzero solution if and only if

$$\text{rank } \tilde{\Gamma} < n_f \tag{88}$$

The rank can be obtained by the singular value decomposition of $\tilde{\Gamma}$. Let the singular values of a real mxn matrix with m > n be denoted by:

$$\theta_1 > \theta_2 > \ldots > \theta_{n_f} > 0$$

The rank deficiency is given by the number of zero singular values.

The optimality of a solution can be considered as the solvability of (87) with a nonzero solution. The reciprocal of the condition number of $\tilde{\Gamma}$, namely:

$$E = \theta_{n_f}/\theta_1$$

can be adopted as the optimality index. When the index E is zero, (88) is satisfied. Index E can be calculated for a given λ by factorizing (78) and calculating the singular value of $\tilde{\Gamma}$. If E is zero at several values of λ, the largest absolute value of λ gives the optimal λ. These results are used to construct the algorithm:

Algorithm 1

Step 1 Calculate the singular values of $\tilde{\Gamma}$ for given values of λ and obtain the index E .

Step 2 Draw the graph of (E,λ). The largest absolute value of λ for which E is zero gives the optimal λ. The solution g and h of (87) gives the optimal solution. •

The graph (E,λ) usefully reveals the global characteristic of (87). The drawback of this algorithm is that the spectral factorization and singular value decomposition necessary to compute E for each λ, makes the algorithm computationally demanding.

3.3 A generalised eigenvalue method

Algorithm 1 is basically a heuristic trial and error technique which is readily improved using a scheme based on first order expansions with respect to λ. Incorporate the dependence on λ, of (78) via the notation $A_\sigma(z^{-1},\lambda)$ then for $\lambda \to \infty$:

$$A_\sigma(z^{-1},\lambda) \cong \lambda D_e(z^{-1}) \tag{89}$$

for which (87) can be approximated as:

$$(\Gamma_a - \lambda\Gamma_b)x = o \tag{90}$$

where $\Gamma_a \triangleq [\Gamma_{21}\ \Gamma_{22}]$, $\Gamma_b \triangleq \theta_o\Gamma_{13}[\Gamma_{11}\ \Gamma_{12}]$, $\theta_o = \text{Toep}(D_e) \in \mathbb{R}^{m x n_f}$ and $x = (g^T,h^T)^T$; Γ_a is a $mx(n_g + n_h)$ matrix; Γ_b is the product of these matrices with dimensions $m \times n_f$, $n_f \times n_f$, $n_f \times (n_g + n_h)$, respectively. Since $n_f < \min(m, n_g+n_h)$, rank $(\Gamma_b) < n_f$. If deg $(D_e) > 0$, $\Gamma_a - \lambda\Gamma_b$ is a rectangular matrix. In this case the approximate equation is usually overdetermined and there exists no nonzero solution x. Therefore, instead of (90), consider the generalized eigenvalue problem:

$$\Gamma_a^T\Gamma_a x = \lambda(\Gamma_a^T\Gamma_b)x \tag{91}$$

Since rank $(\Gamma_b) < n_f$, this generalized eigenvalue problem has at least $n_g + n_h - n_f$ eigenvalues at $\lambda = \infty$. Numerical examples show that the largest absolute value of the finite values of λ gives the good approximation for the optimal λ and the corresponding eigenvector gives a good approximation of the optimal solution vector.

A similar results can be obtained for the approximation of $\lambda_o(z^{-1},\lambda)$ at a finite value of $\lambda = \lambda_a$:

$$A_\sigma(z^{-1},\lambda) \cong A_\sigma(z^{-1},\lambda_a) + (\lambda-\lambda_a)\left.\frac{dA_\sigma}{d\lambda}\right|_{\lambda=\lambda_a} \tag{92}$$

where $\tilde{A}_\sigma(z^{-1},\lambda_a)$ is the spectral factor of $D_e^*D_e\lambda_a^2 - B_q^*B_qB_r^*B_r$. Polynomial $dA_\sigma/d\lambda$ at $\lambda = \lambda_a$ is given as the solution of a polynomial equation:

$$\left(\frac{dA_\sigma}{d\lambda}\right)^* A_\sigma(z^{-1},\lambda_a) + A_\sigma^*(z^{-1},\lambda_a)\left(\frac{dA_\sigma}{d\lambda}\right) = 2\,\lambda_a D_e^* D_e \tag{93}$$

where $dA_\sigma/d\lambda$, $A_\sigma(z^{-1},\lambda_a)$, $D_e \in \mathbb{P}(z^{-1})$. Denote these polynomials as:

$$A_\sigma(z^{-1},\lambda_a) = a_o + a_1z^{-1} + \ldots + a_{n_2}z^{-n_2} \tag{94}$$

$$\frac{d}{d\lambda} A_\sigma = v_o + v_1 z^{-1} + \ldots + v_{n_2} z^{-n_2} \tag{95}$$

$$D_e^* D_e = d_{n_2} z^{n_2} + \ldots + d_1 z + d_o + d_1 z^{-1} + \ldots + d_{n_2} z^{-n_2} \tag{96}$$

Then the equation (93) is equivalent to the linear algebraic equation:

$$\Sigma_a v = 2\lambda_o d \tag{97}$$

where $v = (v_o, v_1, \ldots, v_{n_2})^T$, $d = (d_o, d_1, \ldots, d_{n_2})^T$

$\Sigma_a = \Sigma_1 + \Sigma_2$, and

$$\Sigma_1 = \begin{pmatrix} 0 & \cdots & 0 & a_o \\ \vdots & & \cdot^{\cdot^{\cdot}} & \vdots \\ 0 & \cdot^{\cdot^{\cdot}} & & \vdots \\ a_o & a_1 & \cdots & a_{n_2} \end{pmatrix} \qquad \Sigma_2 = \begin{pmatrix} a_{n_2} & 0 & \cdots & 0 \\ \vdots & \ddots & & \vdots \\ \vdots & & \ddots & 0 \\ a_o & \cdots & \cdots & a_{n_2} \end{pmatrix}$$

The coefficients of $dA_\sigma/d\lambda$ at $\lambda = \lambda_a$ are given by the solution of (97). Since $A_\sigma(z^{-1},\lambda_a)$ is strictly Hurwitz, $A_\sigma(z^{-1},\lambda_a)$ and $A_\sigma^*(z^{-1},\lambda_a)$ are coprime and

so the equation (46) has a unique solution $\frac{dA_\sigma}{d\lambda}$ with deg $(\frac{dA_\sigma}{d\lambda}) = n_2$. This means that the algebraic equation (97) always has a unique solution.

Depending upon the approximation (92), the matrix Γ_{23} is approximated as

$$\Gamma_{23} \cong \theta_1 + (\lambda - \lambda_a)\,\theta_2 \tag{99}$$

where θ_1 = Toep $(A_\sigma(z^{-1},\lambda_a)) \in \mathbb{R}^{m x n_h}$ and θ_2 = Toep $(dA_\sigma/d\lambda|_{\lambda=\lambda_a}) \in \mathbb{R}^{m x n_h}$. By the substitution of (99) into (87), the approximation of (87) becomes:

$$(\Gamma_c - \lambda\Gamma_d)\, x = 0 \tag{100}$$

where $\Gamma_c = [\Gamma_{21} \;\; \Gamma_{22}] - (\theta_1 - \lambda_a\theta_2)\Gamma_{13}[\Gamma_{11} \;\; \Gamma_{12}]$ and $\Gamma_d = \theta_2\Gamma_{13}[\Gamma_{11} \;\; \Gamma_{12}]$. A better approximation of the solution of (87) may be obtained by solving the generalized eigenvalue problem:

$$\Gamma_c^T\Gamma_c x = \lambda\Gamma_c^T\Gamma_d x \tag{101}$$

Numerical examples show that the largest absolute value of λ except for $\lambda = \infty$ gives a good approximation of the optimal λ and the corresponding eigenvector gives a good approximation of the optimal solution.

These are summarized as the algorithm:

Algorithm 2

Step 1 Set k = 0 and set $\lambda = \lambda_o$.

Step 2 Obtain $A_\sigma(\lambda_k)$ by the spectral factorization of

$$D_e^* D_e \lambda_k^2 - B_q^* B_q B_r^* B_r .$$

Step 3 Obtain $dA_\sigma/d\lambda$ at $\lambda = \lambda_k$ by solving (93)

Step 4 Solve (101) and obtain λ. Update $k:=k+1$ and $\lambda_k := \lambda$.

If $|\lambda_k - \lambda_{k-1}|$ is sufficiently small, the optimal λ is obtained. Otherwise go to Step 2.

Algorithm 2 needs only the value of λ_o as the starting point. Numerical examples show that Algorithm 2 converges for a wide interval of λ_o which includes $\lambda_o = \infty$ in most cases and that the eigenvalue of (91) usually gives a good approximation of the optimal value of λ. Therefore the starting point λ_o can be adequately given by solving (91).

3.4 An example for the numerical procedures

Consider the case:

$A = 1 + 2z^{-1} + 3z^{-2}$, $B = 16z^{-1}$

$B_q = 1 + 0.75z^{-1}$, $B_p = 1$, $B_r = 2-2z^{-1}$, $E = 1$

$C_d = 1.5 + 0.8z^{-1}$, $C_n = 1$, $A_d = 1.05z^{-1}$, $A_w = 1+0.5z^{-1}$

The eigenvalues of (91) are given by Table 1.

Table 1 Eigenvalues of equation (91)

1	-0.271554D+18	0.000000D+00
2	0.280059D+18	0.000000D+00
3	0.987582D+01	0.000000D+00
4	-0.494904D+00	0.000000D+00
5	0.227309D+00	0.000000D+00
6	0.252699D-01	0.000000D+00
7	0.330348D-03	0.958163D-03
8	0.330348D-03	-0.958163D-03

The largest two eigenvalues correspond to $\lambda = \infty$, which results from the singularity of Γ_6 matrix. The third largest eigenvalue gives a good approximation of optimal λ. Following algorithm 1, the graph (E,λ) is drawn in Fig. 3.

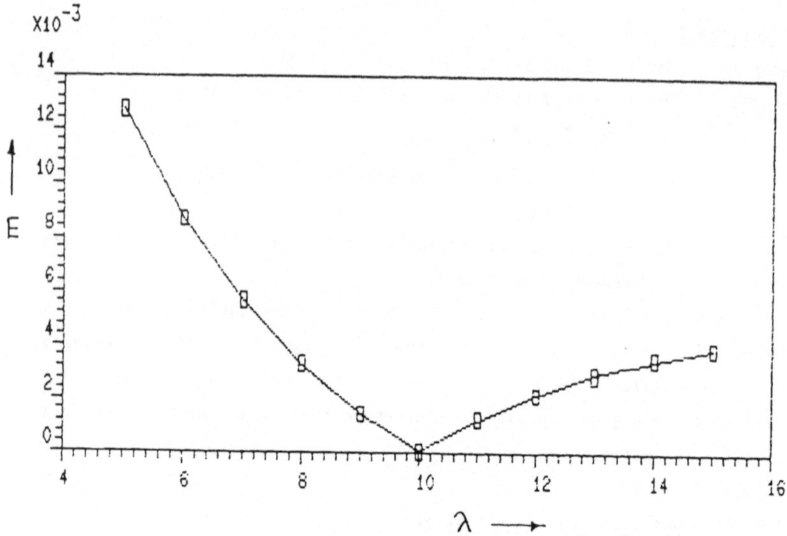

Fig. 3

The sequence of λ_k obtained by Algorithm 2 are given in Table 2 where the starting values of λ is chosen as $\lambda_o = 10$.

Table 2 The sequence of λ_k

λ_o	10
λ_1	9.974 141 406 94322
λ_2	9.974 142 091 82644
λ_3	9.974 142 091 82640
λ_4	9.974 142 091 82640

The optimal controller is given by

$$C_o = \frac{-0.05079 - 0.1273z^{-1} + 0.04343z^{-2} + 0.04794z^{-3}}{1 + 0.002053z^{-1} - 0.2495z^{-2} - 0.0004535z^{-3}}$$

4. Conclusions

The paper presented several new developments of the polynomial system methods for solving an H_∞-optimal control problem via an LQG framework. As in previous formulations this was achieved using the auxiliary lemma first introduced by Kwakernaak. The solution theorems first presented by Grimble (1985[17]) were extended for a larger class of cost function weights. The algebraic manipulation of the dual criterion to a condensed cost function was also a feature of the analysis presented.

The second part of the paper comprised the development of new numerical algorithms to solve the polynomial equations (two Diophantine equations and one spectral factorization) specifying the controller. Two algorithms were proposed and the details of a numerical example presented.

Future work will be directed at further clarifications of the utility of the various cost functions, since it is clear that the full flexibility of the dual criterion cost function has yet to be pursued in depth. The numerical algorithms proposed are a first attempt at devising polynomial solution procedures. The success of the methods is encouraging but further analysis (convergence properties, for example) and algorithmic development is required and is a subject of current research.

5. Acknowledgements

M. Saeki would like to acknowledge the support of the Ministry of Education, Japanese Government for a sabbatical year spent in the Industrial Control Unit at the University of Strathclyde where the research contribution to this paper was performed. M.J. Grimble and M.A. Johnson gratefully acknowledge the support of the Science and Engineering Research Council Control and Instrumentation Subcommittee for funds to pursue research in the area of H_∞-optimal control and LQG system design. E. Kornegoor would like to thank the Industrial Control Unit, University of Strathclyde and Twente University of Technlogy for the provision of funds to support a sojourn at the Industrial Control Unit where the research contribution to this paper was performed.

6. References

1. Horowitz, I., 'Synthesis of Feedback Systems', Academic Press, New York, 1963.
2. Safonov M.G., and Athans, M., 'Gain and phase margin for multiloop LQG regulators', IEEE Trans. on Automatic Control, Vol. AC-22, No. 2, pp. 173-179, 1977.
3. Doyle, J.G., 'Guaranteed margins for LQG regulators', IEEE Trans. Automatic Control, Vol. AC-23, No. 4, pp. 756-757, 1978.
4. Doyle J.C., and Stein, G., 'Robustness with observers', IEEE Trans. Automatic Control, Vol. AC-24, No. 4, pp. 607-611, August 1979.

5. Moore, J.B. and Blight, J.D., 'Performance and robustness trade-offs in LQG regulator design', IEEE CDC Conference, San Diego, USA, pp. 1191-1200, 1981.

6. Grimble, M.J., 'Robustness of combined state and state-estimate feedback control schemes', IEEE Trans. Automatic Control, Vol. AC-29, No. 7, pp. 667-669, July 1984.

7. Safonov, M.G., Laub, A.J., and Hartmann, G.L., 'Feedback properties of multivariable systems : The role and use of the return difference matrix', IEEE Trans. Automatic Control, Vol. AC-26, pp. 47-65, 1981.

8. Grimble, M.J., 'Robust LQG design of discrete systems using a dual criterion', IEEE CDC Conference, San Antonio, Texas, 1983.

9. Zames, G., 1981, 'Feedback and optimal sensitivity : Model reference transformations, multiplicative seminorms, and approximate inverses', IEEE Trans. Automatic Control Vol. AC-26, No. 2, pp. 301-320, 1981.

10. Freudenberg J.S. and Looze, D.P., 'An analysis of H_∞-optimization design methods', Research Report, Co-ordinated Science Laboratory University of Illinois, 1983.

11. Zames, G., and Francis, B.A., 'A new approach to classical frequency methods : Feedback and minimax sensitivity', IEEE CDC Conference, San Diego, California, 1981.

12. Kwakernaak, H., 'Optimal robustness of linear feedback systems', Research Report, Mem. Nr.395, Dept. of Appl. Maths, Twente University of Tech. Enschede, The Netherlands, July, 1982.

13. Kwakernaak, H., 'Robustness optimization of linear feedback systems', 22nd IEEE CDC Conference, San Antonio, Texas, Dec. 1983.

14. Kwakernaak, H., 'Minimax frequency domain optimization of multivariable linear feedback systems', IFAC World Congress, Budapest, Hungary, 1984.

15. Kwakernaak, H., 'Minimax frequency domain performance and robustness optimization of linear feedback systems', IEEE Trans. Automatic Control, AC-30, No. 10, pp. 994-1004, 1985.

16. Grimble, M.J., 'Controller for LQG self-tuning applications with coloured measurement noise and dynamic costing', Proc. IEE Pt. D, Vol. 133, No. 1, Jan. 1986.

17. Grimble, M.J., 'Optimal H_∞-robustness and the relationship to LQG design problems', Int. J. of Control, Vol. 43, No. 2, pp. 351-372, January 1986.

18. Saeki, M., 'Polynomial approach to H_∞-optimal control problem for a discrete time system'. Report ICU/103/1986, Industrial Control Unit, University of Strathclyde, Glasgow, Scotland, UK, (Submitted for Publication).

19. Saeki, M., and Kornegoor, E., ' Numerical algorithm for solving a polynomial equation in an H_∞-optimization problems; Report ICU/121 /1986, Industrial Control Unit, University of Strathclyde, Glasgow, Scotland, UK.

20. Grimble, M.J., 'LQG design of discrete systems using a dual criterion', IEE Proc. Vol. 132, No. 2, pp. 612-68, 1985.

OPTIMAL H^{∞}- INTERPOLATION: A NEW APPROACH

George Zames * Allen Tannenbaum † Cyprian Foias ‡

* Department of Electrical Engineering, McGill University, 3480 University St., Montréal, P.Q., Canada H3A 2A7

† Department of Electrical Engineering, University of Minnesota, Minneapolis, Minn. 55455

‡ Department of Mathematics, Indiana University, Bloomington, Indiana, U.S.A. 47405

Abstract

Explicitly computable solutions to the problem of L^{∞}- sensitivity minimization for (a possibly infinite-dimensional plant represented by) an inner function $M \in H^{\infty}$, subject to a rational weighting $W \in H^{\infty}$, are obtained. This is equivalent to the problem of best approximation of $M^*W \in L^{\infty}$ by $Q \in H^{\infty}$ (more generally, $Q \in H^{\infty}_{[\kappa]}$).

The main new idea involves the representation of the Hankel operator Γ of M^*W as a finite-rank perturbation of the multiplication operator $\mathbf{M^*W}$. The perturbation takes the form of a "Complementary Hankel Operator" determined by W. This idea is exploited to obtain explicit formulas for: (a) All discrete eigenvalues and eigenvectors of $\Gamma^*\Gamma$; (b) all S-numbers of $\Gamma^*\Gamma$; (c) the optimal $H^{\infty}_{[\kappa]}$ approximations; (d) the essential spectrum of $\Gamma^*\Gamma$.

The formulas obtained are surprisingly simple when the order of W is small, even for infinite-dimensional M, and therefore appear to be particularly well suited to control-sensitivity problems. The case of 1st-order W is worked out in detail.

NATO ASI Series, Vol. F34
Modelling, Robustness and Sensitivity Reduction in Control Systems. Edited by R. F. Curtain

1. INTRODUCTION

A large class of optimal "generalized" interpolation problems are subsumed by the minimization problem

$$\inf_{Q \epsilon H^\infty} \|M^*W - Q\|_\infty \triangleq \tilde{\rho} \tag{1.1}$$

M, W, $\epsilon\ H^\infty$, M inner and nonconstant. For many L^∞-functions the problem of best approximation by an H^∞-function can be reduced to (1.1). In particular this is true of the Control Theory problem of minimizing the sensitivity of a plant M subject to a weighting W ([10], [11], [13], [4], [17]). The solution of (1.1) is completely determined by a "Hankel operator" Γ depending on M^*W. For rational W and completely general $M \epsilon H^\infty$, we will obtain *explicitly computable* formulas or descriptions of (a) the discrete eigenvalues and eigenvectors of $\Gamma^*\Gamma$; (b) The essential spectrum of $\Gamma^*\Gamma$; (c) the optimal interpolant solving (1.1); (d) the s-numbers of $\Gamma^*\Gamma$, and (e) solutions to the more general problem of Adamjan et al [1],

$$\inf_{Q \epsilon H^\infty_{[\kappa]}} \|M^*W - Q\|_\infty \triangleq \tilde{\rho}_\kappa \tag{1.2}$$

where $H^\infty_{[\kappa]} = H^\infty + \{$ rational functions in H^∞_- with $\leq \kappa$ poles$\}$. The basic idea of the paper is that if W has finite degree then Γ can be viewed as a finite-dimensional perturbation of the multiplication operator $\mathbf{M^*W}$, even if $\mathbf{M}$ is infinite dimensional.

This paper is a continuation of the work described in [5-9]. The original objective of [5] was to find explicitly computable solutions to H^∞-minimization problems involving delays, where infinite dimensionality of M posed difficulties. However, it gradually became apparent that what has evolved, from [5] to the present, amounts to a new approach to H^∞-interpolation, which is advantageous when the weighting W has lower degree than the "plant" inner-function M (as it usually has in Control) whether M is infinite dimensional or not.

The motivating idea for [5] was a simple example of Sarason's [16] involving a Volterra Operator. D.S. Flamm [2,3], starting with the same idea, independently got results closely related to those of [5] in the course of his Ph.D project under the supervision of S.J. Mitter. Mitter and Flamm have since obtained additional results, including some on uniqueness in delay problems [14].

1.1 Notation

$\mathbb{R}$ denotes the reals.

$\mathbb{C}$, $\mathbb{C}_+$, $\mathbb{C}_-$, $\mathbb{C}_0$ are the complex plane, open right and left half planes, and imaginary axis respectively. The point ∞ is excluded from each. For any $x \in C$, $\bar{x}$ is the complex conjugate. The imaginary unit is j.

$\bar{\mathbb{C}}$, $\bar{\mathbb{C}}_+$, $\bar{\mathbb{C}}_-$, $\bar{\mathbb{C}}_0$ are the corresponding closed sets, which include the point ∞.

A *multiplication operator* $\mathbf{K}$ in L^2 satisfies

$$(\mathbf{K}u)(j\omega) = K(j\omega)u(j\omega), \qquad \forall\omega \tag{1.2a}$$

for all $u \in L^2$, where $K \in L^\infty$. In general, H^∞ functions will be represented by capitals, and multiplication operators generated by them will be represented by the corresponding boldface capitals, as in (1.2a)

The space $H^p(\mathbb{C}_+)$ will be denoted by H^p and viewed as a subspace of $L^p(-\infty,\infty)$. $H^p_- = \{x^* \in L^p(-\infty,\infty) : x \in H^p\}$.

$[x]_+$ ($[x]_-$) are projections of any $x \in L^2$ onto H^2 ($H^2_- := H^{2\perp} := L^2 \ominus H^2$).

$(\cdot)^*$ denotes the adjoint of an operator or the conjugation $x^*(s) = \overline{x(-\bar{s})}$ which takes functions $x \in H^2$ into $x^* \in H^2_-$.

1.2 Preliminaries

M^*W can be decomposed into two components, $[M^*W]_+ \in H^\infty$ and $[M^*W]_- \in H^\infty_-$. The solution of (1.1) is completely determined by the component in H^∞_-, which is unaffected by $Q \in H^\infty$. Equivalently, the solution is (1.2) determined by an operator Γ_{M^*W}, which is defined in general as follows: For any $K \in L^\infty, \Gamma_K$ denotes the *reversed Hankel Operator*(1), $H^2 \to H^2_-$, $\Gamma_K u = [Ku]_-$.

It is standard that the minimal norm $\tilde{\rho}$ satisfies

$$\tilde{\rho} = \|\Gamma_{M^*W}\| = \inf\{\|\mathbf{K}\|_\infty : K \in H^\infty,\ \Gamma_{M^*K} = \Gamma_{M^*W}\}.$$

The conventional approach to H^∞ interpolation exploits the finite dimensionality or compactness of Γ_{M^*W} to compute the largest eigenvalue $\rho_{\max}$ of $\Gamma^*_{M^*W}\Gamma_{M_*W}$ ($\rho_{\max}$ must equal $\tilde{\rho}$) and the associated eigenvector $\tilde{x} \in H^2$. By the Nehari-AAK-Sarason theories, the optimal interpolant is then $\rho_{\max}\tilde{x}^*/\tilde{x}$. However, we are interested in systems which may be noncompact, e.g., when delays are present, and will take a different tack to find $\rho_{\max}$ and x.

Let Γ denote Γ_{M^*W}. All eigenvectors of $\Gamma^*\Gamma$ must lie in the(2) subspace $H^2 \ominus MH^2$ of H^2. The multiplication operator $\mathbf{WM}^*$ restricted to $H^2 \ominus MH^2$ can be expressed as a sum of reversed Hankel operators,

$$\begin{aligned} \mathbf{WM}^*x &= [\mathbf{WM}^*x]_- + [\mathbf{WM}^*x]_+ \\ &= \Gamma_{WM^*}x + \Gamma^*_{W^*}(\mathbf{M}^*x), \quad x \in H^2 \ominus MH^2, \end{aligned} \tag{1.3}$$

Here $\Gamma^*_{W^*}$ is the adjoint of Γ_{W^*}. Note that for any $K \in L^\infty$, the adjoint $\Gamma^*_K : H^2_- \to H^2$ is the unique operator satisfying $< \Gamma_K x, y >_{H^2_-} =< x, \Gamma^*_K y >_{H^2}$ for all $x \in H^2, y \in H^2_-$. This adjoint exists and is given by $\Gamma^*_k y =< K^* y >_+$. Denote $\Gamma^*_{W^*}$ by Φ, to obtain,

$$\mathbf{WM}^*x = \Gamma x + \Phi(\mathbf{M}^*x). \tag{1.3'}$$

We will call Φ the complementary (reversed) Hankel operator of Γ. If W is rational the complementary Hankel Φ has finite rank even if the original Hankel Γ doesn't, and (1.3′) expresses Γ (after rearrangement) as a finite rank perturbation of the multiplication operator M^*W. This fact provides the basis for our approach.

Assumption: *From now on, the domain of Γ will be restricted to $H^2 \ominus MH^2$. (Elements of MH^2 are not eigenvectors of Γ) and the range to $M^*H^2 \ominus H^2 (= H^2_- \ominus M^*H^2_-)$.*

1.3 An Eigenvector Equation

Multiply (1.3) by $\mathbf{W}^*\mathbf{M}$ to get, for any $x \in H^2 \ominus MH^2$,

$$\begin{aligned} \mathbf{W}^*\mathbf{W}x &= \mathbf{W}^*\mathbf{M}[\mathbf{WM}^*x]_- + \mathbf{W}^*\mathbf{M}[\mathbf{WM}^*x]_+ \\ &= [\mathbf{W}^*\mathbf{M}[\mathbf{M}^*\mathbf{W}x]_-]_+ + [\mathbf{W}^*\mathbf{M}[\mathbf{M}^*\mathbf{W}x]_-]_- \\ &\quad + \mathbf{W}^*\mathbf{M}[\mathbf{WM}^*x]_+ \end{aligned}$$

It is not hard to show that (1) $\mathbf{M}[\mathbf{M}^*u]_-$ projects any $u \in H^2$ onto $H^2 \ominus MH^2$, and that (2) $[\mathbf{M}^*u]_-$ is in $M^*H^2 \ominus H^2$ which is the domain of Γ^*. Using these facts, we get

$$\mathbf{W}^*\mathbf{W}x = \Gamma^*\Gamma x + \Phi^*(\mathbf{M}\Gamma x) + \mathbf{W}^*\mathbf{M}\Phi(\mathbf{M}^*x) \tag{1.5}$$

After subtracting $\rho^2 x$ from both sides (ρ real), adding and subtracting $\rho\Phi^*(\mathbf{M}x^*)$ and rearranging, we get an equation

$$\begin{aligned} (\Gamma^*\Gamma - \rho^2\mathbf{I})x &= (\mathbf{W}^*\mathbf{W} - \rho^2\mathbf{I})x - \rho\Phi^*(\mathbf{M}x^*) - \mathbf{W}^*\mathbf{M}\Phi(\mathbf{M}^*x) \\ &\quad - \Phi^*[\mathbf{M}(\Gamma x - \rho x^*)], \quad x \in H^2 \ominus \mathbf{M}H^2 \end{aligned} \tag{1.6}$$

which relates the reciprocal resolvents $(\Gamma^*\Gamma - \rho^2\mathbf{I})$ and $(\mathbf{W}^*\mathbf{W} - \rho^2 I)$..

The solution of (1.6) for eigenvectors of $\Gamma^*\Gamma$ can be simplified for many purposes by the choice of a "symmetrized" basis.

It is shown in Appendix 0 that Γ has the property $(\Gamma y^*) = \Gamma^* y^*$, $y \in H^2 \ominus MH^2$, and that consequently it is always possible to span the eigenspace of $\Gamma^*\Gamma$ associated with an eigenvalue ρ^2 by eigenvectors x which generate conjugate (x, x^*) or anticonjugate $(x, -x^*)$ Schmidt pairs, i.e., $\Gamma x = \pm\rho x^*, \Gamma^* x^* = \pm\rho x, \rho \geq 0$. Eigenvectors x generating such pairs will be called *symmetrized*. The dimension of $\ker(\Gamma^*\Gamma - \rho^2 \mathbf{I})$ is the number of linearly independant such pairs. For purposes of computing eigenvalues/vectors we will restrict attention to symmetrized eigenvectors without loss of generality[3].

Returning now to (1.6), suppose for the moment that an eigenvalue ρ^2 of $\Gamma^*\Gamma$ exists in (1.6), and that $x \in H^2 \ominus MH^2$ is the associated eigenvector. The first and last terms of (1.6) vanish and we get an equation for the eigenvector x

$$\begin{aligned}(W^*W - \rho^2)x &= \rho\phi^*_{(M^*x)} + W^*M\phi_{(M^*x)} \\ &:= \Delta\end{aligned} \tag{1.7}$$

in which $\rho \in \mathbb{R}$, multiplication operators have been replaced by H^∞_- functions, and $\phi_{(M^*x)}$ denotes an H^∞ function depending on M^*x.

If W is rational, then Φ has finite rank, ϕ has the same poles as W, and ϕ is therefore a linear combination of known vectors. Moreover the zeros of $W^*(s)W(s) - \rho^2$ which lie in $\bar{\mathbb{C}}_+$, and which are finite in number must be zeros of the RHS of (1.7), provided we can show that $x(s)$ is analytic and $M(s)$ well defined at these zeros. We will show that these zeros provide a set of independent equations which can be solved for ρ and ϕ whenever $W^*W \not\equiv$ const.; x is then given by the formula

$$x = \frac{\rho\phi^* + W^*M\phi}{W^*W - \rho^2}, \quad \rho = \pm|\rho|. \tag{1.8}$$

More generally, if W (has a continuation which) is meromorphic on $\underline{\mathbb{C}}$ (but not necessarily rational) we have the following result.

1.4 A Criterion for the Existence of Eigenvalues

Lemma 0 If $W^*(s)W(s) \not\equiv \rho^2$ identically, then ρ^2 is an eigenvalue of $\Gamma^*\Gamma$ if and only if the expression

$$\frac{\rho\phi^*(s) + W^*(s)M(s)\phi(s)}{W^*(s)W(s) - \rho^2} \triangleq x(s) \tag{1.9}$$

defines a function $x \in H^2, x \neq 0$, for some $\phi \in [WH^2_-]_+, \rho = \pm|\rho|$. If $x \in H^2, x \neq 0$, then x is an eigenvector with eigenvalue ρ^2, which satisfies $\Gamma x = \rho x^*$.

□

2. RESULTS FOR RATIONAL WEIGHTINGS W

Henceforth, suppose except where noted that $W \in H^\infty$ is a rational ("weighting") function of order N. The $\mathbb{C}_-$ poles of W^*W are a subset of the poles of W consisting of N^0 poles, of which N_0 are distinct and denoted by $p_0, p_1, \ldots, p_{N_0}$, $(N_0 \leq N^0)$.

2.1 Determination of Discrete Eigenvalues and Eigenvectors

Under these assumptions, the RHS of (1.7) is finite dimensional. Indeed, the function $\phi(s)$ can be expressed as a linear combination of vectors as follows: ϕ lies in the subspace $[W(H^2_-)]_+$ which is spanned by the functions $\psi_{im}(s) \triangleq (s-p_i)^m$, $p_i \in \mathbb{C}$, $i = 1, 2, \ldots, N$, $m = 1, 2, \ldots, m_i$, $\Sigma_i m_i \triangleq N$. Order the ψ_{im} to form a sequence $\{\psi_k\}_{k=1}^N$. Then we get

$$\Delta(s) = \rho \sum_{k=1}^{N} \bar{c}_k \psi_k^*(s) + W^*(s) M(s) \sum_{k=1}^{N} c_k \psi_k(s) \tag{2.1}$$

Let $\Omega(\rho)$ denote the set of zeros of $W^*(s)W(s) - \rho^2$ in $\mathbb{C}$, each zero repeated according to multiplicity. Whenever $W^*W \not\equiv$ const., $W^*W - \rho^2$ has $2N^0$ zeros in $\bar{\mathbb{C}}$. Multiply (1.7) by $(1+s)$ (this step can be omitted if $\Omega(\rho)$ contains no zeros at $\infty \in \bar{\mathbb{C}}$) to get

$$(1+s)\Delta = (1+s)(W^*W - \rho^2)x = (1+s)[\rho\phi^* + W^*M\phi] \tag{2.2}$$

An eigenvalue of ρ^2 of $\Gamma^*\Gamma$ will be called *discrete* if ρ^2 is not a limit point of the spectrum of $\Gamma^*\Gamma$ and its multiplicity is finite. We will show in Prop. 3.2 that if ρ^2 is a discrete eigenvalue of $\Gamma^*\Gamma$, then all $\bar{\mathbb{C}}_+$ zeros of $W^*W - \rho^2$ are points of analyticity of $M(s)$, which implies that $M(s)$ is well defined at these zeros as is $\Delta(s)$. By evaluating (2.2) at these zeros we get $N^0 + N^1$ equations in the $2N$ coefficients $c_k, \bar{c}_k$, $k = 1, 2, \ldots, N$ of (2.1). After multiplying (2.1) by $M^*(s)$ and observing that $M^*x \in H^2_-$, we get an additional $N^0 - N^1$ equations at the $\mathbb{C}_-$ zeros.

Finally, consider the degenerate situation $N^0 < N$, in which poles of $W(s)$ coincide with zeroes of $W^*(s)$. Then ϕ and ϕ^* on the RHS of (2.2) have poles in $\mathbb{C}_-$ and $\mathbb{C}_+$ respectively which are not poles of $W^*(s)W(s)$, and are therefore absent from the LHS. There are $2(N - N^0)$ such *degenerate poles*. The negative Laurent coefficients of $\Delta(s)$ evaluated at these poles must vanish, giving $2(N - N^0)$ additional equations, for a total of $2N$. These equations can be written in the matrix form

$$A(\rho)d = 0$$

in which ρ will be restricted, $\rho > 0$.

Here $A(\rho)$ is a $2N^0 \times 2N^0$ *characteristic matrix* of $\Gamma^*\Gamma$, whose construction is described in greater detail in Appendix I; and d is the transpose of $[\bar{c},\ c]$, $c := [c_1, \ldots, c_N]$. It will be shown that if $W^*W - \rho^2 \not\equiv 0$ identically, then $A(\rho) \not\equiv 0$, and there is a $1:1$ correspondence between $\ker A(\rho)$ and the subspace of symmetrized eigenvectors associated with the constant ρ. Therefore $\det A(\rho)$ must vanish at that value ρ. The converse will also appear to be true[4].

In the following theorems, the case where $W^*W \equiv$ const. and $\deg W < \deg M < \infty$ will be called *totally degenerate*. In this case the weighting W is itself the optimal interpolant, by Nevanlinna-Pick Theory, and $\phi \equiv 0$.

Let $I(\rho)$ be the $2N \times 2N$ identity matrix I_{2N} for $\rho > 0$, and $I(\rho) = \mathrm{diag}\ [0_N, I_N]$ for $\rho = 0$.

Theorem 1[0] In the totally degenerate case, ρ^2 is a discrete eigenvalue of $\Gamma^*\Gamma$ whose eigenvectors are described below in Prop. 2.1.

Otherwise, ρ^2 is a discrete eigenvalue of $\Gamma^*\Gamma$ if and only if $A(\rho)$ is well defined[5], $\det A(\rho) = 0$, and $W^*W \not\equiv \rho^2$. If $\det A(\rho) = 0$, $W^*W \not\equiv \rho^2$, then the multiplicity of ρ^2 as an eigenvalue of $\Gamma^*\Gamma$ is $\dim\ker A(\rho)I(\rho) \leq N$, and the associated (symmetrized) eigenvectors x are given by

$$x(s) = \frac{\pm\rho\phi^*(s) + W^*(s)M(s)\phi(s)}{W^*(s)W(s) - \rho^2} \tag{2.3}$$

where $\phi(s) = \sum_{i=1}^{N} c_k\psi_k(s)$, $(\psi_k(s) = (s - p_i)^{-m})$, $c := [c_1, \ldots, c_{N0}]$, and $[\bar{c},\ c]^{Tr}$ is any vector in $\ker A(\rho)$.

□

Prop. 2.1 If $W^*W \equiv$ const. and $\deg W < \deg M$ then W has the factorization $W = \rho w_0/w_0^*$, where either $w_0 \in H^2$ is rational and has no zeros, or $W \equiv \rho$ and $w_0 \equiv 1$; ρ^2 is now an eigenvalue of multiplicity $\dim X$, where $X \triangleq w_0^{-1}(H^2 \ominus MH^2)$, and any element of X is an eigenvector of $\Gamma^*\Gamma$. In particular, if $\deg M = \infty$, then ρ^2 has infinite multiplicity.

The proof follows that of Theorem 1 in Section 3.

2.2 Optimal Interpolation in H^∞

Let σ_{ess}^0 be the set of essential singularities[6], of $M(s)$ ($\sigma_{ess}^0 \subset \bar{\mathbb{C}}_0$) and

$$\rho_{ess} = \sup\{|W(j\omega)| : \omega \in \sigma_{ess}^0\}$$

The interpolation problem (1.1) *attains a minimum* $\widetilde{W} \in H^\infty$ if $\tilde{\rho} \triangleq \inf_{Q \in H^\infty} \|WM^* - Q\|_\infty = \|\widetilde{W}M^*\|_\infty$.

An eigenvector x and associated eigenvalue ρ of $\Gamma^*\Gamma$ are *maximal* iff $\rho = \tilde{\rho}$.

Theorem 2[0] $\tilde{\rho} \geq \rho_{ess}$ always. In the totally degenerate case $\rho_{ess}^0 = 0$ and $\widetilde{W} = W$. Otherwise, $\rho > \rho_{ess}$ if and only if $\det A(\rho) = 0$ for some $|\rho| \in (\rho_{ess}, \|W\|)$ in which case

$$\tilde{\rho} = \max\{|\rho| : |\rho| \in (\rho_{ess}, \|W\|),\ \det A(\rho) = 0\} \tag{2.4a}$$

and problem (1.1) attains a unique minimum

$$\widetilde{W}(s) = \tilde{\rho}\left[\frac{\tilde{\rho}\phi(s) + W(s)M^*(s)\phi^*(s)}{\tilde{\rho}\phi^*(s) + W^*(s)M(s)\phi(s)}\right] \tag{2.4b}$$

where ϕ is specified as in Theorem 1.

□

An Analytic Characteristic function

$\det A(\rho)$ is analytic between the (finite number of) branching values of ρ (at which $W^*(s)W(s) - \rho^2$ has multiple roots), but may be discontinuous at the branching values. However, $\det A(\rho)$ may be shown to have the same zeros as a somewhat more complicated characteristic function $D^{-1}(\cdot)\det A_0(\cdot)$, which is defined as follows.

For simplicity suppose that there are no degenerate poles ($N^0 = N$), and that no pole of $M(s)$ is a pole of $W(s)$ or point of $\Omega(\rho)$. Let $D(\rho)$ denote the determinant of the $2N \times 2N$ alternant matrix $[s_i(\rho)^{j-1}]_{i,j=1,\ldots,2N}$ formed from the roots $s_i(\rho)$ of $W^*(s)W(s) - \rho^2 = 0$. Then $D(\rho)$ is the difference product $\Pi_{1\leq i<j\leq N}[s_j(\rho) - s_i(\rho)]$. Let $D^{-1}(\cdot)\det A_0(\cdot)$ denote the function equalling $D^{-1}(\rho)\det A(\rho)$ for ρ not a branching value, and $\lim_{\rho\to\rho_0} D^{-1}(\rho)\det A_0(\rho)$ for ρ_0 a branching value.

Prop. 2.2 $D^{-1}(\cdot)\det A_0(\cdot)$ is an analytic function of ρ for ρ in the interval (ρ_{ess}, ∞), or for ρ in some neighbourhood of a nonzero discrete eigenvalue of $\Gamma^*\Gamma$.

The proof is in Section 3. □

It follows from Prop. 2.2 that in any compact subinterval of (ρ_{ess}, ∞), the number of points ρ at which $\det A(\rho) = 0$ is finite.

2.3 Results Linking the Essential Spectra of $\Gamma^*\Gamma$ and $\mathbf{M}\Gamma$

The proof of Theorem 1 makes use of a result on the essential spectrum of $\Gamma^*\Gamma$ which is stated in Lemma 1 below.

Let $\mathbf{K}$ be any operator in a Hilbert space $\mathcal{H}$, whose spectrum is $\sigma(\mathbf{K})$. The *essential spectrum*[7] of $\mathbf{K}$ is $\sigma_{ess}(\mathbf{K}) \triangleq \{\lambda \in \mathbb{C} : \lambda$ is a limit point of $\sigma(\mathbf{K})$ or an eigenvalue of infinite multiplicity; or $\bar{\lambda}$ is a limit point of $\sigma(\mathbf{K}^*)$ or an eigenvalue of infinite multiplicity of $\sigma(\mathbf{K}^*)\}$. For any bounded, self adjoint $\mathbf{K}$, $\sigma(\mathbf{K})$ is real, and the discrete eigenvalues of $\mathbf{K}$ are precisely the points of $\sigma(\mathbf{K})\backslash\sigma_{ess}(\mathbf{K})$.

Lemma 1

$\sigma_{ess}(\Gamma^*\Gamma) = \{|\lambda|^2 : \lambda \in \mathbb{C},\ \lambda \in \sigma_{ess}(\mathbf{M}\Gamma)\}$ where $\sigma_{ess}(\mathbf{M}\Gamma) = W(\sigma^0_{ess})$.

The proof is in Sect. 3.

2.4 Optimal Interpolation in $H^\infty_{[\kappa]}$, S-numbers

The S-numbers of finite order κ of any operator $\mathbf{X}$ are its distances from the sets of operators of rank $\kappa = 0,\ 1,\ \ldots,$ i.e.,

$$S_\kappa \triangleq \inf\{\|\mathbf{X} - \mathbf{X}_\kappa\| : \mathbf{X}_\kappa \text{ has rank } \kappa\}.$$

Also, $S_\infty \triangleq \lim_{\kappa\to\infty} S_\kappa$. We have $\|\mathbf{X}\| = S_0 \geq S_1 \geq \ldots \geq S_\infty$.

For any self-adjoint $\mathbf{X}$ there is the result (see Gohberg-Krein [12]) that $S_\infty = \sup\{\rho : \rho \in \sigma_{ess}(\mathbf{X})\}$, and if $n \leq \infty$ is the number of eigenvalues $\lambda_0 \leq \lambda_1 \leq \ldots$ in $(S_\infty,\ \infty)$, each repeated according to multiplicity, then: if $n = \infty$,

$$S_\kappa = \lambda_{k+1}, \quad \kappa = 0,\ 1,\ \ldots,$$

and if $n < \infty$,

$$\begin{aligned} S_\kappa &= \lambda_{\kappa+1}, & \kappa &= 0,\ 1,\ \ldots,\ (n-1) \\ &= S_\infty & \kappa &= n,\ (n+1),\ \ldots \end{aligned}$$

The following theorem is a corollary to Theorem 1.

Theorem 3 the S-numbers $S_\kappa(\Gamma^*\Gamma)$ are:

a) $S_\infty(\Gamma^*\Gamma) := \sup\{|W\omega)| : j\omega$ is an essential singularity of $M(s)\} := \rho_{ess}$.

b) Except in the case $\kappa = 0$ and W degenerate, the S-numbers of finite order are the ordered points ρ_κ satisfying $\|W\|_\infty \geq \rho_0 \geq \rho_1 \geq \ldots \geq \rho_{ess}$ at which $\det A(\rho_\kappa) = 0$, each value ρ_κ repeated with multiplicity $\det A(\rho)I(\rho)$ and followed, if the total of such points is $n < \infty$, by the numbers $\ldots > S_n = S_{n+1} = \ldots$ where $S_\kappa = S_\infty$ for $\kappa \geq n$.

In the degenerate case $S_0 = \|W\|$ (see Prop. 2.1). □

The extremal approximation problem 1.2 attains a solution $\tilde{Q} \in H^\infty_{[\kappa]}$ if

$$\min_{Q\in H^\infty_{[\kappa]}} \|M^*W - Q\| = \|M^*W - Q_\kappa\| = \tilde{\rho}_\kappa, \qquad \kappa = 0, 1, \ldots \tag{2.5}$$

The following result is an immediate corollary to Theorem 3 and the theorem of Adamjan et al [1].

Theorem 4

(a) If $\|W\| = \rho_{ess}$, then $\widetilde{\rho}_\kappa = \rho_{ess}$ and $\widetilde{Q}_\kappa = 0$ for all $\kappa = 0, 1, \ldots$.

(b) If $\deg M = \infty$, $\|W\| > \rho_{ess}$, and $\det A(\rho_i) = 0$ on some (nonempty) set of points $\rho_\kappa \in (\rho_{ess}, \|W\|]$, $\kappa = 1, 2, \ldots$, then the extremal approximation problem (1.2) attains a unique solution if $\rho_{k+1} > \rho_{ess}$ (ρ_κ defined as in Theorem 3) whereupon the S-number $S_\kappa(\Gamma^*\Gamma)$ lies in $(\rho_{ess}, \|W\|)$. The solution is

$$\begin{aligned}\widetilde{\rho}_\kappa &= S_\kappa(\Gamma^*\Gamma)\\ \widetilde{Q}_\kappa &= M^*W - \rho_\kappa \widetilde{x}^*/\widetilde{x}\end{aligned} \tag{2.6}$$

where $\rho_\kappa = \widetilde{\rho}_\kappa$, and $\widetilde{x}_\kappa$ is any eigenvalue given by (2.3), ($\rho = \rho_\kappa$).

(c) If $\deg M < \infty$, (1.2) attains a solution given by (2.6), except in the case $\kappa = 0$, W degenerate, where $\widetilde{Q}_0 = 0$.

2.5 Existence of a Maximal Eigenvector Corollary to Theorem 2

If $|W(j\omega)| \not\equiv$ Const., then $\widetilde{\rho} > \rho_{ess}$, there is a maximal eigenvector, and $\widetilde{W}$ is unique under either of the following hypotheses:

(a) $M(s)$ has a zero, and for some $\omega_0 \in \bar{\mathbb{C}}_0$, $|W(j\omega_0)| > \rho_{ess} = \inf\{|W(j\omega)| : \omega \in \bar{\mathbb{C}}_0, \omega \in \sigma_{ess}^0\}$.

(b) $|W(j\omega)| > |W(\infty)|$ for $|\omega| < \infty$, and $M(s)$ has no essential singularities except at $s = \infty$ (i.e., except for a delay or Blaschke product whose zeros approach ∞).

2.6 Angle-Magnitude Formulas for First Order W

The complexity of the formulas in Theorems 1-4 depends primarily on $\deg W$. The general approach is therefore particularly suited to Control problems, where $\deg W$ is typically small whereas $\deg M$ may be large. As even first order W enable us to draw qualitative conclusions about many control problems, and the theory is exceptionally simple for them, it is worth pursuing in some detail. Although in the 1-st order case it is possible to obtain explicit formulas for the eigenvalues of $\Gamma^*\Gamma$ (as in the examples worked out in [5-9] using $\tan^{-1}$ functions), for many engineering applications it may be more fruitful to state the results in terms of angle-magnitude functions ("Bode Diagrams") of $M(j\omega)$ and $W(j\omega)$.

Consider the special case of all functions in $\widetilde{H}^\infty := \{x \in H^\infty : x(s) = \overline{x}(\overline{s})\}$. For problems (1.1, 1.2), there is no loss of generality in restricting first order W to the normalized form $W(s) = (\alpha + \beta s)(\alpha + s)^{-1}$, $\alpha > 0$, β real. Then $\phi(s) = c\Psi(s) = c(\alpha + s)^{-1}$. To exclude degeneracies, suppose that $|\beta| \neq 1$, so that $W^*(s)W(s) \not\equiv$ const.

Here $N = N^0 = 2$, and the root loci of $W^*(s)W(s) - \rho^2 = 0$ are symmetric around the origin and lie on two branches, $\Omega(\rho) := \{s_p, -s_p\}$. The points in $\Omega(p)$ at values of ρ at which the characteristic determinant vanishes will be called critical points and will be denoted by s_{ρ_i}.

By Theorem 1, the discrete eivenvalues of $\Gamma^*\Gamma$ are the squared magnitudes $|W(s_{\rho_i})|$ of W at those critical points which are not essential singularities (on $\overline{\mathbb{C}}_0$) of $M(s)$. the s_{ρ_i} are solutions to the characteristic equation, which here assumes the form $\rho\phi^*(s_\rho) + W^*(s_\rho)M(s_\rho)\phi(s_\rho) = 0$, $(\rho > 0)$, and can be expressed as

$$\frac{W^*(s_\rho)}{\rho}M(s_\rho) = -\phi^*(s_\rho)\phi^{-1}(s_\rho) = -\frac{\alpha+s}{\alpha-s}. \tag{2.7}$$

Case 1. $|\beta| < 1$. $|W(j\omega)|$ is monotonely decreasing and $\|W\| = 1$.

(a) For $\rho \in [|\beta|, 1]$ both branches of $\Omega(\rho)$ lie on the imaginary axis. The critical points are the "frequencies" ω_n at which (by (2.7)),

$$\arg M(j\omega_n) - \arg W(j\omega_n) = (2n-1)\pi + 2\tan^{-1}\omega_n/\alpha, \tag{2.8}$$
$$n = \pm 0, \pm 1, \ldots$$

i.e. *the discrete eigenvalues of* $\Gamma^*\Gamma$ *are the squared magnitudes* $|W(j\omega_n)|$ *of the weighting* W *at those* ω_n *which are not essential singularities of* $M(s)$*, and at which the angle function of the plant* $M(j\omega)$*, reduced by the angle function of the weighting* $W(j\omega)$*, intersets lines* π *apart.* ($\arg W$ is monotonely decreasing here, and $\arg M$ is always essentially so).

(b) For $\rho \in [0, |\beta|]$, both branches of $\Omega(\rho)$ lie on the segments $|\sigma| > |\beta|$ of the real axis. By (2.7), the critical points are those real σ, $(|\sigma| > \alpha)$, at which

$$|M(\sigma)| = |(\alpha+\sigma)/(\alpha-\sigma)|, \tag{2.9}$$

as $(W(\sigma)) = \rho$ on the critical points).

By Theorem 2, $\widetilde{\rho}$ is the maximal eigenvalue, which here is $\widetilde{\rho} = |W(j\omega_1)|$, where ω_1 is the smallest solution of (2.8) obtained for $n = 1$, provided that $M(s)$ has no essential singularities in $[0, \omega_1]$. The minimal H^∞ function satisfying (1.1) is then

$$\widetilde{W}(s) := \widetilde{\rho}\left(\frac{\widetilde{\rho}(\alpha-s) + (\alpha+\beta s)M(-s)}{\widetilde{\rho}(\alpha+s) + (\alpha-\beta s)M(s)}\right) \tag{2.10}$$

(2.10) holds for completely arbitrary $M \in \widetilde{H}^\infty$*, subject only to the restriction of essential singularities outside* $[0,\ \omega_1]$.

If $\deg M < \infty$, then all the eigenvalues evaluated at the critical points (2.8, 2.9) are S-numbers. If $\deg M = \infty$, then only those corresponding to critical points on the imaginary axis such that $|W(j\omega_n)| \geq \rho_{ess}$ are S-numbers. In control problems, the only essential singularities are isolated at $s = \infty$, and are produced by a Blaschke product (which is present in any accurate model) or a delay. in that case, $\rho_{ess} = |\beta|$, $\arg M(j\omega) \to -\infty$ as $\omega \to \infty$, and there is an infinity of S-numbers greater than S_∞.

Case 2. $|\beta| > 1$. $|W(j\omega)|$ is monotonely increasing, and $\|W\| = \beta$. The eigenvalues are given by the same formulas as in Case 1, except that now: the real critical points are confined to the interval $|\sigma| < \alpha$; $\rho_{ess} = W(j\omega^0)$, $\omega^0 := \sup\{\omega : j\omega$ is an essential singularity of $M(s)\}$; and $\widetilde{p}$ is determined by the imaginary critical point $j\omega_n$ corresponding to the *largest* finite n in (2.8). If ∞ is essentially singular, then $\widetilde{\rho} = \rho_{ess}$, and there is no *discrete* maximal eigenvector.

APPENDIX 0. SCHMIDT PAIR SYMMETRIZATION

Prop A0.1 For any $K \in L^\infty, y^* \in H^2_-$,

$$\begin{aligned} &(a) \quad \Gamma_K^* y^* = [\overline{K} y^*]_+ \\ &(b) \quad (\Gamma_K y)^* = \Gamma_{K^*} y^* \end{aligned} \tag{A0.1}$$

Proof

(a) For any $x \in H^2$ we have

$$\begin{aligned} <\Gamma_K x, y^*> &= <[\mathbf{K}x]_-, y^*> && \text{by definition of } \Gamma_K \\ &= <\mathbf{K}x, y^*> && \text{as } [y^*]_- = y^*) \\ &= \frac{1}{2\pi}\int_{-\infty}^{\infty} x(j\omega)K(j\omega)y(j\omega)d\omega \\ &= <x, \overline{K}y^*> \\ &= <x, [\overline{K}y^*]_+> && (\text{as } x = [x]_+) \end{aligned}$$

which implies (a).

(b) From (a) we get

$$(\Gamma_K^* y^*)(j\omega) = [\overline{K}y^*]_+(j\omega) \tag{A0.2}$$

From the definition of Γ_K we get

$$(\Gamma_K y)^*(j\omega) = (\overline{[Ky]_-})(j\omega) = [\overline{K}y^*]_+(j\omega) \tag{A0.3}$$

The last identity is obtained from the rule that for any $x \in L^2$, $\overline{[x]_-} = [\overline{x}]_+$, which is proved as follows: Let $\mathcal{F}$ denote the Fourier transformation in $L^2(-\infty,\infty)$, $\mathbf{J}$ the reversal operator $(\mathbf{J}x)(t) = x(-t)$, and $\overline{(\cdot)}$ complex conjugation. $\mathcal{F}$ has the property that $\overline{(\cdot)}\mathcal{F} = \mathcal{F}\overline{(\cdot)}J$ and $\mathcal{F}^{-1}\overline{(\cdot)} = \mathbf{J}\overline{(\cdot)}\mathcal{F}^{-1}$. Let $\Pi_\pm$ denote the projection operator $\mathcal{F}^{-1}[\cdot]_+\mathcal{F}$. Observe that $\mathbf{J}\Pi_+ = \Pi_-\mathbf{J}$, and $(\cdot)$ commutes with $\mathbf{J}$ and $\Pi_\pm$. Consequently

$$\begin{aligned}[\overline{x}]_- &= \overline{(\cdot)}\mathcal{F}\Pi_-\mathcal{F}^{-1}x = \mathcal{F}\overline{(\cdot)}\mathbf{J}\Pi_-\mathcal{F}^{-1}x \\ &= \mathcal{F}\Pi_+\mathbf{J}\overline{(\cdot)}\mathcal{F}^{-1}x = \mathcal{F}\Pi_+\mathcal{F}^{-1}\overline{(\cdot)}x \\ &= [\overline{x}]_+\end{aligned}$$

as claimed. From (A0.2) and (A0.3), (b) is true. □

Let $\sigma_{ev}(\mathrm{G})$ denote the set of eigenvalues of any linear operator $\mathbf{G}$. Consider the equation

$$\Gamma x - \rho x^* = 0, \qquad x \in H^2 \ominus MH^2 \tag{A0.4}$$

$\rho \in \mathbb{R}$, which is equivalent to $(\Gamma x)^* - \rho x = 0$. Although $[\Gamma(\cdot)]^*$ is not linear over the field $\mathbb{C}$, we shall employ the notation

$$\begin{aligned}\sigma_{ev}[\Gamma(\cdot)^*] &= \{\rho \in \mathbb{R} : \quad \text{(A0.4) has a solution } \mathrm{x} \neq 0\} \\ \ker[\Gamma(\cdot)^* - \rho I] &= \ \text{span}\ \{x : x \neq 0 \text{ solves (A0.4) }\}\end{aligned}$$

Lemma A0.1

$$\sigma_{ev}(\Gamma^*\Gamma) = \{\rho^2 : \rho \in \mathbb{R}, \quad \rho \in \sigma_{ev}[\Gamma(\cdot)]^*\} \tag{A0.5}$$

$$\ker[\Gamma^*\Gamma - \rho^2] = \ker[\Gamma(\cdot)^* - \rho I] \oplus \ker[\Gamma(\cdot)^* + \rho I] \tag{A0.6}$$

Proof. Let $K := H^2 \ominus MH^2$, $K_- := H^2_- \ominus M^*H^2_-$, and $\Gamma_e : K \oplus K_- \to K \oplus K_-$,

$$\Gamma_e : \ = \ \begin{bmatrix} & \Gamma^* \\ \Gamma & \end{bmatrix} \tag{A0.7}$$

Note that $\Gamma_e^2 = \text{diag}\ (\Gamma^*\Gamma, \Gamma\Gamma^*)$

The property (A0.1) ensures that $\sigma_{ev}(\Gamma^*\Gamma) = \sigma_{ev}(\Gamma\Gamma^*)$, because $[(\Gamma^*\Gamma - \rho^2 I)x]^* = (\Gamma\Gamma^* - \rho^2 I)x^*$. As Γ_e^2 is a direct sum of $\Gamma^*\Gamma$ and $\Gamma\Gamma^*$, $\sigma_{ev}(\Gamma_e^2)$ coincides with the preceeding two spectra. Since Γ_e is bounded and self adjoint, its eigenvalues are real and we have

$$\sigma_{ev}(\Gamma^*\Gamma) = \sigma_{ev}(\Gamma_e^2) = \{\rho^2 : \rho \in \sigma_{ev}(\Gamma_e)\} \tag{A0.8}$$

and for any $\rho \in \mathbb{R}$ such that $\rho^2 \in \sigma_{ev}(\Gamma^*\Gamma)$ we have

$$\ker(\Gamma_e^2 - \rho^2 I) = \ker(\Gamma_e - \rho I) \oplus \ker(\Gamma_e + \rho I)$$

Furthermore, since K reduces Γ_e, $\ker(\Gamma^*\Gamma - \rho^2 I) = Proj_K \ker(\Gamma_e^2 - \rho^2 I)$ and consequently

$$\ker(\Gamma^*\Gamma - \rho^2 I) = Proj_K[\ker(\Gamma_e - \rho I) \oplus \ker(\Gamma_e + \rho I)] \tag{A0.9}$$

The Lemma will follow from (A0.8) and (A0.9) if we show that

$$Proj_K \ker(\Gamma_e \pm \rho I) = \ker[\Gamma(\cdot)^* \pm \rho I] \tag{A0.10}$$

Consider the case $-\rho$, (the proof for $+\rho$ will be similar). Suppose $\{\varsigma \in Proj_K \ker(\Gamma_e - \rho I)$, and let $\eta^* := \Gamma\varsigma$. By (A0.1), (ς, η^*) and (η, ς^*) both belong to $\ker(\Gamma_e - \rho I)$, and consequently so do $x_1 := (\varsigma + \eta)$ and $x_2 := j(\varsigma - \eta)$. Since $\varsigma = x_1 - jx_2$, and x_1, x_2 are solutions of (A0.4), $\varsigma \in \ker[\Gamma(\cdot)^* - \rho I]$. Conversely, suppose $\varsigma \in \ker[\Gamma(\cdot)^* - \rho I]$. Therefore $\varsigma = \Sigma c_i x_i$, $c_i \in \mathbb{R}$, where x_i satisfies (A0.4). By (A0.1), $\Gamma^* x_i^* - \rho x_i = 0$, and therefore $y_i := \Sigma c_i (x_i, x_i^*)^{\mathrm{Tr}} \in \ker(\Gamma_e - \rho I)$, and $\varsigma = Proj_K y_i$. Therefore (A0.10) is true.

APPENDIX 1. CONSTRUCTION OF THE CHARACTERISTIC DETERMINANT

Recall that $\deg W = N; \deg W^*W = 2N^0 \leq 2N$. Let

$$\Omega(\rho) := \{s \in \bar{\mathbb{C}} : W^*(s)W(s) - \rho^2 = 0\}, \quad \rho \geq 0$$

be the set of zeros of $W^*(s)W(s) - \rho^2$, which depends on ρ. If $W^*W \equiv \rho^2$, then $\Omega(\rho) = \bar{\mathbb{C}}$. Otherwise, the points of $\Omega(\rho)$ are the $2N^0$ roots of an algebraic equation, which lie on branches ("root-loci"). The roots are distinct except at a finite number of *branching values* of $\rho \in [0, \infty)$. Inbetween branching values, the roots can be ordered $s_1(\rho), \ldots, s_{2N^0}(\rho)$ so that each (branch) $s_i(\rho)$ is analytic on the open interval bounded by consecutive branching values of ρ.

The roots occur in conjugate pairs, $[s_i(\rho), -\bar{s}_i(\rho)]$, and therefore an even number of them, say $2N'(\rho)$, lie on $\bar{\mathbb{C}}_0$. This symmetry also implies that each open (i.e., at its end-points) branch lies entirely in one of three regions $\mathbb{C}_+, \bar{\mathbb{C}}_0$, or $\mathbb{C}_-$.

At each branching value of ρ, there are $2N_0(\rho)$ $(N_0(\rho) < N^0(\rho))$ distinct *roots* of which $2N_1(\rho)$ $(N_1(\rho) \leq N'(\rho))$ lie on $\bar{\mathbb{C}}_0$.

It is possible to order the distinct roots of $\Omega(\rho)$, by reordering the branches if necessary, so that for each $\rho \geq 0$ the $\mathbb{C}_+$ roots are followed by the $\bar{\mathbb{C}}_0$ roots and then the $\mathbb{C}_-$ roots, i.e.,

$$\Omega(\rho) = \{s_1 \ldots s_{N_0 - N_1}; \underline{j\omega}_{N_0 - N_1 + 1}, \ldots, \underline{j\omega}_{N_0 + N_1}; s_{N_0 + N_1 + 1}, \ldots, s_{2N^0}\}$$

where $s_i = s_i(\rho)$, $N_i = N_i(\rho)$, and the last N_0 roots equal $\{-\bar{s}_i, \ldots, -\bar{s}_{N_0}\}$. Assume the roots have been so ordered. Under this ordering, $s_i(\rho)$ is continuous between branching values but may discontinuous at the branching values.

Equations at the zeros of $W^*(s)W(s) - \rho^2$

Various functions of ρ will be defined under the assumption that each ρ in their domain of definion satisfies:

Assumption: $M(s)$ is analytic in $\Omega(\rho) \cap \bar{\mathbb{C}}_+$

Under this assumption the RHS of (2.2) is well defined; otherwise, a point of $\Omega(\rho) \cap \bar{\mathbb{C}}_0$ is an essential singularity of $M(s)$, and neither $M(s)$ nor $\Delta(s)$ is well defined there. By evaluating (2.2) on the $\bar{\mathbb{C}}_0$ zeros of $\Omega(\rho)$ we get $N^0(\rho) + N^1(\rho)$ equations; after multiplying (2.2) by $M^*(s)$ and noting that $x \in H^2 \ominus MH^2$ implies that M^*x is analytic in $\mathbb{C}_-$, we get an additional $N^0(\rho) - N^1(\rho)$ equations at the $\mathbb{C}_-$ zeros, for a total of $2N^0$ equations in the $2N$ coefficients $c_k, \bar{c}_k$ of (2.1), as follows: At each finite zero $s_i \in \mathbb{C}$ we get m_i (:= multiplicity of s_i) equations,

$$M^1(s_i)\Delta(s_i) = 0, \ldots, \frac{d^{m_i - 1}}{ds_i^{m_i - 1}} M^1(s)\Delta(s) \Bigg|_{s=s_i} = 0 \tag{A1.1}$$

where $M^1(s_i) = 1$ for $s_i \in \mathbb{C}_+$, and $M^1(s_i) = M^*(s_i)$ for $s_i \in \mathbb{C}_-$; and at a zero at ∞ of multiplicity m^∞ we get m^∞ equations

$$(1+s)s^m \Delta(s) \Bigg|_{s \to \infty} = 0, \quad m = 0, 1, \ldots, m^\infty - 1 \tag{A1.2}$$

where $\Sigma_i m_i + m^\infty = N^0 + N^1$. (Observe that $\phi(s)$ in (2.2) is strictly proper, so the limit in (A1.2) exists). Arrange these $2N^0$ equations in order of the zeros $s_i(\rho)$.

Let us write (A1.1,-2) in matrix form. Let $F_i := [F_{mk}]$, $G_i = [G_{mk}]$, $m = 1, 2, \ldots, m_i$, $k = 1, 2, \ldots, N$, be the $m_i \times N^0$ matrices such that at any finite zero s_i,

$$\bar{F}^z_{mk} := d^{m-1}/ds^{m-1}(M^1(s)\psi_k^*(s)) \Bigg|_{s=s_i},$$

$$G^z_{mk} - d^{m-1}/ds^{m-1}(M^1(s)W^*(s)M(s)\psi_k(s)) \Bigg|_{s=s_i}$$

and if $s_{N_0+N_1} = \infty$,

$$\overline{F}^z_{mk} = (1+s)^{m-1} M^1(s)\psi^*_k(s)\Big|_{s\to\infty},$$

$$G^z_{mk} = (1+s)s^{m-1} M^1(s)W^*(s)M(s)\psi_k(s)\Big|_{s\to\infty}.$$

Then (A1.1,2) can be expressed as

$$\rho\overline{F}^z_i\overline{c} + G^z_i c = 0, \qquad i = 1,2,\dots,2N_0$$

where $c = [c_1,\ c_2,\ \dots,\ c_N]^{tr}$.

Equations at the Degenerate Poles

Those poles of $W(s)$ and $W^*(s)$ which are not poles of $W^*(s)W(s)$ are *degenerate.* $W(s)$ has $N - N^0$ degenerate poles of which N_d are distinct. Order the distinct degenerate poles in $\mathbb{C}_-$ after those in $\mathbb{C}_+$, i.e. $\{p_1,\dots,p_{N_d}; p_{N_d+1},\dots,p_{2N_d}\}$, where $\{p_{N_d+1},\dots,p_{2N_d}\} = \{-\overline{p}_1,\dots,-\overline{p}_{N_d}\}$.

At each degenerate pole $p_i \in \mathbb{C}_+ \cap \mathbb{C}_-$ of multiplicity m_i as a ple of $W(s)$, and multiplicity δ_i as a pole of $W^*(s)W(s)$, $(m_i > \delta_i \geq 0)$ we get δ_i equations

$$(s-p_i)^{m_i} M^1(\rho_i)\Delta(\rho_i) = 0,\dots, d^{\delta_i-1}/ds^{\delta_i-1}(s-p_i)^{m_i} \quad M^1(s)\Delta(s)\Big|_{s=p_i} = 0 \qquad (A1.4)$$

After expressing $\Delta(s)$ in terms of the basis $\psi^*_i(s), \psi_i(s),\ i = 1,\dots,N$, (A1.4) can be written in terms of $\delta_i \times N$ matrices $F^p(s_i), G^p(s_i)$,

$$\rho\overline{F}^p_i\overline{c} + G^p_i c = 0, \qquad i = 1,\dots,N_d \qquad (A1.5)$$

A Characteristic Matrix

We construct a matrix $A(\rho)$ out of (A1.3), (A1.5); the matrices at the $\mathbb{C}_+$ singularities (i.e. zeros in $\Omega(\rho)$ followed by degenerate poles), followed by the $\overline{\mathbb{C}}_0$ zeros in $\Omega(\rho)$, followed by the $\mathbb{C}_-$ singularities. Let

$$F^+(\rho) = \begin{bmatrix} F^z_1(\rho \\ \vdots \\ F^z_{N_0}(\rho) \\ \cdots \\ F^p_1 \\ \vdots \\ F^p_{N_d} \end{bmatrix}, F^0(\rho) = \begin{bmatrix} F^z_{N_0+1}(\rho) \\ \vdots \\ F^z_{N_0+N_1}(\rho) \end{bmatrix}, F^-(\rho) = \begin{bmatrix} F^z_{N_0+N_1+1}(\rho) \\ \vdots \\ F^z_{2N_0}(\rho) \\ \cdots \\ F^p_{N_d+1} \\ \vdots \\ F^p_{2N_d} \end{bmatrix} \qquad (A1.6)$$

(where the $\Omega(\rho)$ zero matrices depend on ρ, but the degenerate pole matrices do not). The matrices G^+, G^0, G^- are defined similarly by substituting G for F in (A1.6).

Then

$$A(\rho) := \begin{bmatrix} \rho\overline{F}^+(\rho) & G^+(\rho) \\ \rho\overline{F}^0(\rho) & G^0(\rho) \\ \rho\overline{F}^-(\rho) & G(\rho) \end{bmatrix}$$

is a $2N \times 2N$ matrix, and the equations [A1.1, 2, 4] can be expressed in the form

$$A(\rho)[\bar{c}, c]^{\mathrm{tr}} = 0$$

$\det A(\rho)$ will be called a *characteristic determinant.*

Equivalent Characteristic Functions

Two scalar (matrix) valued characteristic functions of ρ are *equivalent* if they have equal zeros (kernels for all ρ). The characteristic matrix

$$A_1(\rho) := \begin{bmatrix} \rho\overline{F}^+ & G^+ \\ \rho\overline{F}^0 & G^0 \\ \overline{G}^+ & \rho F^+ \end{bmatrix}$$

can be shown to be equivalent to $A(\rho)$ for $\rho > 0$, and $\det A_1(\rho)$ is equivalent to $\det A(\rho)$ for $\rho \geq 0$.

FOOTNOTES

(0) Proofs are in the enlarged report obtainable from the authors.

(1) This terminology is justified because the operator $\mathbf{J}\Gamma_K$, where $\mathbf{J} : H^2 \to H^2_-$ denotes the reversal $(\mathbf{J}x)(s) \triangleq x(-s)$, is unitarily equivalent to the usual Hankel operator in $\rho^2[0,\infty)$. Indeed, the usual Hankel can be expressed as $\mathcal{F}^{-1}\mathbf{V}^{-1}\mathbf{J}^{-1}\Gamma_K\mathbf{V}\mathcal{F}$, where $\mathcal{F}$ is the Fourier transformation from $\ell^2[0,\infty)$ to $H^2(\text{disk})$, and $\mathbf{V}$ is the isometry from $H^2(\text{disk})$ to $H^2(\mathbb{C}_+)$, $(\mathbf{V}x)(s) \triangleq (1+s)^{-1}K[(1-s)(1+s)^{-1}]$. (See Sarason [16]).

(2) Observe that $\ker\Gamma \supset \mathbf{M}H^2$, as $[\mathbf{WM}^*(\mathbf{MH}^2)]_- = [\mathbf{WH}^2]_- = 0$. Therefore $(\ker\Gamma)^\perp \subset H^2 \ominus MH^2$. (Indeed, Γ maps $H^2 \ominus MH^2$ into $M^*H^2 \ominus H^2$.)

(3) In the case of the restricted H^p-spaces with the complex conjugate symmetry $\{x \in H^p : x(\bar{s}) = \overline{x(s)}\}$, the operator $x \to (\Gamma x)^*$ is self-adjoint. Eigenvectors x generating conjugate Schmidt pairs span each eigenspace of $\Gamma^*\Gamma$, and anticonjugate pairs $(x, -x^*)$ can be discarded.

(4) $\det A(\rho)$ can be related to a function $D^{-1}(\rho)\det A(\rho)$ which is analytic in ρ whenever ρ^2 is in some neighbourhood of a discrete eigenvalue, and which has the same zeros as $\det A(\rho)$. See Prop. 2.2.

(5) $A(\rho)$ is well defined (see proof of Lemma 2) $\iff \rho^2 \neq |W(j\omega_0)|^2$ for any $j\omega_0 \in \bar{\mathbb{C}}_0$ which is an essential singularity of $M(s)$.

(6) i.e., points in $\bar{\mathbb{C}}_0$ at which $M(j\omega)$ can not be continued analytically from $\mathbb{C}_+$ to C_- by arcs crossing; i.e. limit points of zeros of $M(s)$, or points, $j\omega \in \bar{\mathbb{C}}_0$ at which $dM^1/d\omega \neq 0$, where dM^1 is the singular measure defining M (see Nikolskii [15]). Also, see Footnote(7).

(7) The notation σ^0_{ess} introduced in 2.1.1 reflects the well-known fact (Nikolsii [15]) that $\sigma^0_{ess} = S(\sigma_{ess}(\mathbf{T}))$ where $\mathbf{T}$ is Sarason's compressed shift operator and S the conformal map from the disk to $\bar{\mathbb{C}}_+$. ρ_{ess} is the radius of $W(\sigma^0_{ess})$. Also (See Appendix 2, Lemma A2.2), $W(\sigma^0_{ess}) = \sigma_{ess}(\mathbf{M}\Gamma)$, $\mathbf{M}\gamma$ being isomorphic to Sarason's $S^{-1}W\,S(\mathbf{T})$.

REFERENCES

[1] V.M. Adamjan, D.Z. Arov, and M.G. Krein, "Analytic properties of Schmidt pairs for a Hankel operator and the generalized Schur-Takagi problem", Math. USSR Sbornik, **15**: 31-73, 1971.

[2] D.S. Flamm, "H^∞-Optimal Sensitivity for Delay Systems", M.I.T. L.I.D.S. report, Nov. 1985.

[3] D.S. Flamm, "Control of Delay Systems for Minimax Sensitivity", M.I.T. Ph.D. Thesis and L.I.D.S. Report, June 1986.

[4] B.C. Chang and J.B. Pearson, Jr., "Optimal Disturbance Reduction in Linear Multivariable Systems", IEEE Trans. Automatic Control, Vol. 22, No. 10, October 1984, pp. 880-887.

[5] C. Foias, A. Tannenbaum, and G. Zames, "Weighted sensitivity minimization for delay systems", Proc. C. Dec. Contr., 1985, 244-249; Also IEEE Trans. on Automatic Control, **AC-31**: 763-766, August 1986.

[7] ------------------------, "On the H^∞-optimal sensitivity problem for systems with delays", SIAM Journal on Control and Optimization, to appear.

[8] ------------------------, "On decoupling the H^∞-optimal sensitivity problem for products of plants", Systems and Control Letters, **7**: 239-246, 1986.

[9] C. Foias and A. Tannenbaum, "On the Nehari problem for a certain class of L^∞-functions appearing in control theory", Technical Report, Department of Electrical Engineering, McGill University, February 1986. To appear in the Journal of Functional Analysis.

[10] B.A. Francis, "Notes on H^∞-optimal linear feedback systems", Lecture Notes, Linköping University, 1983.

[11] B.A. Francis and J. Doyle, "Linear control theory with an H^∞ optimality criterion", Systems Control Group report #8501, University of Toronto, October 1985.

[12] I.C. Gohberg and M.G. Krein, *Introduction to the Theory of Linear Nonselfadjoint Operators*", Am. Math. Soc., Translations, Vol. 18, AMS, Providence, 1965.

[13] J.W. Helton, "Non-Euclidean Functional Analysis and Electronics", AMS Bulletin, Vol.7, No.3, July 1982, 1-63.

[14] S.J. Mitter, Private Communications, Nov. 1985 and August 1986.

[15] N.K. Nikolskii, *Treatise on the Shift Operator*, Springer, 1980.

[16] D. Sarason, "Generalized interpolation in H^∞", Trans. AMS, **127**: 179-203, 1967.

[17] M.S. Verma and E. Jonkheere, "L^∞-Compensation with Mixed Sensitivity as a Broad Band Matching Problem", Systems and Control Letters, May 1984, pp. 125-131.

Some Min-Max Optimization Problems in Infinite-Dimensional Control Systems[1]

F. Fagnani
Scuola Normale Superiore
Pisa, Italy

D. Flamm[2]

S.K. Mitter[3]

Abstract

In this paper we discuss certain optimization problems which have their origin in infinite-dimensional control problems. The optimization problems are of the form:

$$\inf_{h \varepsilon H^\infty} \| \Phi - h \|_\infty, \text{ where } \Phi \varepsilon L^j, \text{ can}$$

be factored as $\Phi = \overline{\psi} W$, ψ an inner function and W a proper rational funtion. Problems of sensitivity minimization for plants with a rational function with a single delay in the input lead to an optimization problem of the above type. Existence of a solution to such problems is easy to prove but in general there is no unique solution. We discuss the question of uniqueness based on a criterion due to Adamjan, Arov and Krein and recent work of Sarason. For some special cases, we give a parametrization of all solutions when there is no uniqueness. These problems are equivalent ot extension problems for Hankel operators.

We discuss how the ideas and methods discussed in this paper can be used to solve problems of sensitivity minimization for certain distributed parameter systems.

[1] This research has been supported by the Air Force Office of Scientific Research under Grant AFOSR 85-0227, Army Research Office under Contract DAAG29-84-0005.

[2] Aerospace Corporation, Los Angeles, California, U.S.A.

[3] Department of Electrical Engineering and Computer Science and Laboratory for Information and Decision Systems, Massachusetts Institute of Technology Cambridge, MA 02139, U.S.A.

NATO ASI Series, Vol. F34
Modelling, Robustness and Sensitivity Reduction in Control Systems. Edited by R. F. Curtain

1. Problem Formulation

Consider the following feedback control system:

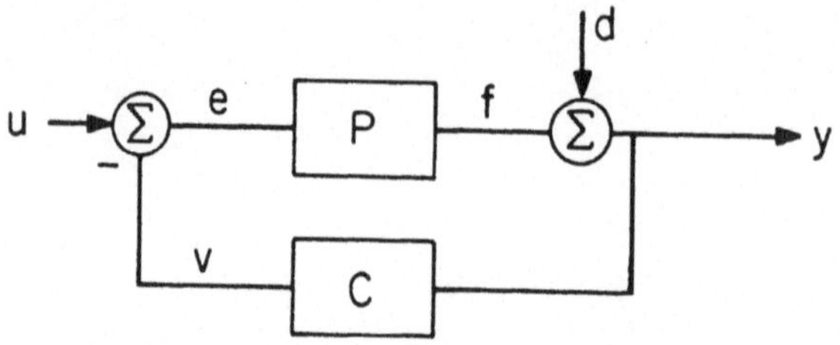

Here P(s) is a causal transfer function, the plant, and it is required to find a compensator, with causal transfer function C(s) which achieves internal stability and minimizes the weighted sensitivity function in the infinity norm:

$$||W(s)[1+P(s)C(s)]^{-1}||_\infty , \tag{1.1}$$

where W(s) is a proper, stable, outer rational function, normalized to be 1 at infinity. The plant P(s) is not assumed to be a rational function. In this paper we concentrate on the case

$$P(s) = e^{-S\Delta}P_0(s) \tag{1.2}$$

where $P_0(s)$ is a strictly proper rational function which is outer.

The choice of W as proper (and not strictly proper) makes the problem difficult, since in general there is a high degree of non-uniqueness in the solution.

2. Notation, Mathematical Formulation and Preliminaries

We follow the basic notation of Hoffman [1962] and Garnett [1981].

Let $D = \{z\varepsilon C \,|\, |z|<1\}$ and $\pi^+ = \{z\varepsilon C \,|\, \mathrm{Re}\ z>0\}$. By $H^p(D)$ or $H^p(\pi^+)$, $1\leq p\leq\infty$, we denote the Banach space of holomorphic functions in D or π^+ with the usual norms. Given a $g\varepsilon H^p(D)$, let

$$h(w) = (1+w)^{-2/p}g((w-1)(w+1)^{-1}). \tag{2.1}$$

Then $h\varepsilon H^p(\pi^+)$. The mapping $g \rightarrow h$ is an isometric isomorphism of $H^p(D)$ to

$H^p(\pi^+)$. We shall usually work with $H^p(\pi^+)$ and the argument will be omitted.

We assume the plant admits a co-prime factorization:

$$\begin{cases} P(s) = \varphi(s)\phi^{-1}(s), \quad \varphi,\phi \varepsilon H^{\infty} \\ \exists\ a,b,\varepsilon H^{\infty}, \text{ such that, } a\varphi+b\phi = 1. \end{cases} \tag{2.2}$$

By the Corona Theorem (cf. Garnett [1981], p. 324) this will be true if, there exists a $\delta>0$, such that

$$|\varphi(w)| + |\phi(w)| \geq \delta \quad \forall\ w\varepsilon\pi^{+}. \tag{2.3}$$

We also assume that P(s) is the transfer function corresponding to a causal, time-invariant, linear bounded operator

$$T:L_e^2(R) \rightarrow L_e^2(R), \tag{2.4}$$

where L_e^2 denotes the extended L^2-space (cf. for example Desoer and Vidyasagar [1975]. A characterization of such transfer functions can be given in terms of the Paley-Wiener-Schwartz theorem (c.f. Segal [1955], Theorem 3).

The Youla parametrization of compensators achieving stability is then valid for this more general class. We say that the causal feedback compensator C(s) renders the feedback system stable of

$$u,d\varepsilon L^2 \rightarrow e,v,f,y\varepsilon L^2. \tag{2.5}$$

Such a compensator will be termed admissible and (2.5) is true if and only if

$$(1+P(s)C(s))^{-1},\ P(s)(1+P(s)C(s))^{-1},\ C(s)(1+P(s)C(s))^{-1}\ \varepsilon\ H^{\infty}. \tag{2.6}$$

Note however internal asymptotic stability of the canonical state-space realizations corresponding to each of these transfer functions is not necessarily guaranteed in this general setting. Some form of exact controllability and observability conditions will have to be verified for this to be true. We then have

Proposition 2.1

C is an admissible controller $\leftrightarrow \exists$ $Z \varepsilon H^{\infty}$, $Z \neq b\varphi^{-1}$ such that

$$C = \frac{a+\phi Z}{b-\varphi Z} . \tag{2.7}$$

□

As is well known, the original sensitivity minimization problem does not admit a solution and it is necessary to consider a relaxed problem. In the standard way, we consider the relaxed problem

$$\operatorname*{Min}_{h \varepsilon H^{\infty}} ||W+\varphi h||_{\infty}, \tag{2.8}$$

where $W \varepsilon H^{\infty}$ and is outer and $\varphi \varepsilon H^{\infty}$ inner. (2.8) is equivalent to the problem of minimizing the distance

$$d(\bar{\varphi}W, H^{\infty}), \tag{2.9}$$

where $\bar{\varphi}$ denotes the conjugate of φ.

If W is proper and not strictly proper, $\bar{\varphi}W \varepsilon L^{\infty}$ and in general there is no unique solution. If W is strictly proper, $\bar{\varphi}W \varepsilon H^{\infty}+C$ ($H^{\infty}+C$ = set of functions $f+g$, $f \varepsilon H^{\infty}$, $g \varepsilon C$ (the space of continuous functions)), then there is a unique solution (c.f. Garnett [1981], Theorem 1.7). If φ has an essential singularity on the imaginary axis at the point μ then $\bar{\varphi}W \varepsilon H^{\infty}+C \leftrightarrow W(\mu) = 0$. In this case there is a unique solution to (2.8).

3. Generalized Interpolation and the Theory of Hankel Operators

The results of this section are contained in Sarason [1967, 1985] and Adamjan, Arov and Krein [1968].

Let H be a Hilbert space and K a closed subspace. Let $S \varepsilon \mathscr{L}(H)$, the space of bounded linear operators from H to H. Let

$$T = P_K S|_K . \tag{3.1}$$

T is the compression of S on K and S is the dilation of T on H. Now let

$H=H^2$ and $K=(\varphi H^2)^\perp$ where φ is an inner function. Let $W\varepsilon H^\infty$ and let M_W be the corresponding multiplication operator on H^2. Clearly $M_W \ \varepsilon \ \mathscr{L}(H^2)$ and $||M_W||_{\mathscr{L}(H^2)} = ||W||_\infty$.

Let $T = P_K M_W|_K$. W is termed the interpolating symbol of T. Consider the semigroup of translation $s \to e^{-s\Delta}$ and the compressions $S_\Delta = P_K M_{e^{-s\Delta}}|_K$. Then,

<u>Proposition 3.1</u>. $T\varepsilon\mathscr{L}(K)$, $TS_\Delta = S_\Delta T \ \forall\Delta>0 \to \exists \ Z\varepsilon H^\infty$ s.t. $T = P_K M_Z|_K$ and $||T||_{\mathscr{L}(K)} = ||Z||_\infty$. □

Z is the minimal interpolating symbol of T. Returning to the relaxed sensitivity minimization problem (2.8), if we take $K = (\varphi H^2)$, $T = P_K M_W|_K$, then

$$\min_{h\varepsilon H^\infty} ||W+\varphi h||_\infty = ||T||_{\mathscr{L}(K)}. \tag{3.2}$$

□

Let $f\varepsilon K$ and $f\neq 0$. f is called a maximal vector of T of

$$||Tf||=||T||.||f|| \tag{3.3}$$

<u>Proposition 3.2</u>. Let f be a maximal vector T. Then

$$\phi = \frac{Tf}{f}$$

is a minimal interpolating symbol of T, which is a constant times an inner function.

It is clear that T has a maximal vector iff T^*T has a maximal eigenvalue, and $\rho(T^*T) = ||T^*T|| = ||T||^2$ and hence $||T|| = (\rho(T^*T))^{1/2}$. Finally if T is compact then T has a maximal vector and T is compact $\leftrightarrow \bar{\varphi}W\varepsilon H^\infty+C$.

An operator $\mathscr{H}: H^2 \to (H^2)^\perp$ is a Hankel operator $\leftrightarrow \exists$ a symbol $\phi\varepsilon L^\infty$ such that $\mathscr{H}=P_- M_\phi|_{H^2}$, where P_- is the projection on $(H^2)^\perp$. If $\mathscr{H}$ is a Hankel operator with symbol ϕ, then ϕ' is a symbol of $\mathscr{H} \leftrightarrow \phi-\phi' \ \varepsilon \ H^\infty$. The operator $T = P_K M_W|_K$ defines a Hankel operator $\mathscr{H}= P_- M_{\bar{\varphi}} T$, with symbol $\bar{\varphi}W$. Clearly ϕ is a symbol of $\mathscr{H} \leftrightarrow \varphi\phi$ is a symbol of T.

Let $\mathscr{H}$ be a Hankel operator with $||\mathscr{H}|| = s$. Then on $H^2(D)$, we have the following criterion for the uniqueness of the minimal symbol of $\mathscr{H}$.

<u>Theorem 3.3</u>. H has a unique minimal symbol $\longleftrightarrow$

$$\lim_{\rho \downarrow s} \langle (\rho^2 I - \mathcal{H}^* \mathcal{H})^{-1} .1, 1 \rangle = \infty,$$

where 1 is the function identically 1 on $H^2(D)$. □

The function 1 transforms to $\frac{1}{s+1}$ on $H^2(\pi^+)$. In case there is no unique symbol, one can obtain a parametrization of all minimal symbols of by the following generalization of the Schur algorithm for interpolation. Let

$$\mathcal{H}_\varepsilon = (1-\varepsilon)\mathcal{H}, \; 0<\varepsilon<1. \quad \text{Let} \tag{3.4}$$

$$q_\varepsilon = \| (I - \mathcal{H}_\varepsilon^* \mathcal{H}_\varepsilon)^{-1/2} 1 \|_2^{-1} (I - \mathcal{H}_\varepsilon^* \mathcal{H}_\varepsilon)^{-1} 1.$$

$$r_\varepsilon = \overline{\mathcal{H}_\varepsilon q_\varepsilon}$$

The sequences $\{q_\varepsilon\}$ and $\{r_\varepsilon\}$ are uniformly bounded on compacts of D and hence there exists a sequence $\varepsilon_n \to 0$ such that q_ε r_ε converge uniformly on the compacts of D and hence to q, r in the space of holomorphic functions on D. Let

$$U = \begin{pmatrix} \bar{q} & \bar{r} \\ r & q \end{pmatrix}. \tag{3.5}$$

<u>Theorem 3.4</u>. ϕ is a minimal symbol for $\mathcal{H} \longleftrightarrow \exists$ $\varphi \varepsilon \mathcal{B}(H^\infty)$ (the unit ball of $H^\infty(D)$) such that

$$\phi = U\varphi = \frac{\bar{q}\varphi + \bar{r}}{r\varphi + q}$$

ϕ is unimochular $\longleftrightarrow \varphi$ is inner. □

On $H^2(\pi^+)$, we have to use q' and r' in the definition of U, where

$q = (s+1)q'$

$r = (s+1)r'$.

4. Sensitivity Minimization for Delay Systems.

This section follows Flamm [1985], Flamm and Mitter [1986], and Fagnani [1986]. See also Foias, Tannenbaum and Zames [1986].

In the first instance, we take the plant to be $P(s) = e^{-s\Delta}$, and the weighting function $W(s) = n(s)d^{-1}(s)$, with $n(s) = \prod_{i=1}^{n}(s+a_i)$ and $d(s) = \prod_{i=1}^{n}(s+b_i)$, $a_i, b_i > 0$. In this case $K = (e^{-s\Delta}H^2)^{\perp}$ and $T = P_K M_W|_K$. It is easy to show that

$$\mathscr{L}^{-1}(K) = L^2(0,\Delta) \subseteq L^2(0,\infty), \text{ where}$$

$\mathscr{L}$ is the Laplace Transform considered as a unitary operator between $L^2(0,\infty)$ and H^2 and $\mathscr{L}^{-1}$ denotes the inverse Laplace Transform. Defining

$$V = \mathscr{L}^{-1}T\mathscr{L}\Big|_{L^2(0,\Delta)}, \text{ we see that } V \in \mathscr{L}(L^2(0,\Delta))$$

and

$$V = I+S \text{ where } S \text{ is the Volterra operator} \tag{4.1}$$

$$(Sf)(t) = \int_0^t G(t-s)f(s)ds \quad \forall t \in [0,\Delta] \text{ and} \tag{4.2}$$

$$G(t) = \mathscr{L}^{-1}(W(s)^{-1}). \tag{4.3}$$

The kernel G(t) has the explicit representation

$$G(t) = \sum_{i=1}^{n} \alpha_i e^{-b_j t}, \text{ where} \tag{4.4}$$

$$\alpha_j = [n(-b_j)-d(-b_j)] \prod_{i\neq j}(b_i - b_j)^{-1}. \tag{4.5}$$

Since T and V are unitarily equivalent, in order to calculate the maximal eigenvalue of T^*T (if it exists) it is useful to obtain a state-space realization of V^*V.

Proposition 4.1. The operator V^*V is realized as the map $f \to z$, where

$$\begin{cases} \dot{x}_1 = -Dx_1 + \alpha' f; \quad x_1(0) = 0 \\ y = c'x_1 + f \end{cases} \tag{4.6}$$

$$\begin{cases} \dot{x}_2 = Dx_2 - \alpha' y; \quad x_2(\Delta) = 0 \\ z = c'x_2 + y, \end{cases}$$

where $D = \text{diag.}\ (b_1,\dots,b_n)$ $c = (1,1,\dots,1)'$ and $\alpha = (\alpha_1,\dots,\alpha_n)'$. □

Now,

$$V^*V = I + (S+S^*+S^*S)$$

$$= \text{Identity + Compact operator, and hence}$$

$$\sigma(V^*V) = 1+\sigma(S+S^*+S^*S).$$

Therefore, the spectrum of V^*V consists of a succession of eigenvalues with a possible point of accumulation at 1 and the point 1 whose spectral type is not known a priori.

From this we may arrive at the following criteria for the existence of maximal eigenvalues:

(i) If $U = S+S^*+S^*S$ is a non-negative operator, then V^*V has a maximal eigenvalue.

(ii) If $\exists\ M\geq 0$ such that $|W(i\omega)|>1$, $\forall\omega\geq M$, then V^*V has a maximal eigenvalue greater than 1.

Now,

$$|W(i\omega)| > 1 \leftrightarrow \prod_{i=1}^{n}(\omega^2+a_i^2) > \prod_{i=1}^{n}(\omega^2+b_i^2), \tag{4.7}$$

and we may conclude that

$$\sum_{i=1}^{n} a_i^2 > \sum_{i=1}^{n} b_i^2 \rightarrow V^*V \text{ has a maximal eigenvalue.} \tag{4.8}$$

Also,

$$\sum_{i=1}^{n} a_i \geq \sum_{i=1}^{n} b_i \rightarrow V^*V \text{ has a maximal eigenvalue.} \tag{4.9}$$

Let us denote by λ^2 the maximal eigenvalue and g the corresponding eigenvector. Then from the theory of Sarason it follows that the optimal sensitivity is given by

$$\chi = \frac{\mathcal{L}(Vg)}{\mathcal{L}(g)} . \tag{4.10}$$

Indeed, we have the following expression for the optimal sensitivity.

<u>Proposition 4.2.</u> There exist polynomials v,c of degree (n-1) such that $v\tilde{v} = \lambda^2 c\tilde{c}$ and

$$\chi(s) = \lambda^2 \frac{n(s)c(s) - e^{-s\Delta}\tilde{d}(s)v(s)}{\lambda^2 d(s)c(s) - e^{-s\Delta}\tilde{n}(s)v(s)} , \tag{4.11}$$

where for a polynomial P(s), $\tilde{P}(s) = (-1)^{\text{degree } P} P(-s)$.

<u>Proof:</u> From (4.6)

$$\hat{f}(s) = \mathscr{L}(f(s)) = \frac{c'(sI-D)^{-1}x_2(0)}{\lambda^2-(I-c'(sI-D)^{-1}b)(I+c'(sI+D)^{-1}b)}$$

$$= \frac{c(s)d(s)}{\lambda^2 d(s)\tilde{d}(s)-n(s)\tilde{n}(s)}$$

where c(s) is a polynomial of degree (n-1), and hence

$$f(t) = \sum_{i=1}^{n} (\alpha_i e^{\beta_i t} + \tilde{\alpha}_i e^{-\beta_i t}), \tag{4.12}$$

where β_i are the solutions of

$$\lambda^2 d\tilde{d}-n\tilde{n} = 0 \text{ and } \alpha_i,\ \tilde{\alpha}_i \tag{4.13}$$

are appropriate coefficients.

The eigenvalues of V^*V are obtained by restricting the functions given by (4.12) to the interval $[0,\Delta]$ and are of the form

$$F(s) = \mathscr{L}(f|_{[0,\Delta]}) = \frac{cd-e^{-s\Delta}b}{\lambda^2 d\tilde{d}-n\tilde{n}}, \tag{4.14}$$

where b is an appropriate polynomial. One can then compute

$$Vf = \sum_{i=1}^{n} (\alpha_i W(\beta_i)e^{\beta_i t} + \tilde{\alpha}_i \tilde{W}(\beta_i)e^{-\beta_i t}). \tag{4.15}$$

By computing, $V^*Vf = Vf + S^*Vf$, and imposing the condition $V^*Vf = \lambda^2 f$, we obtain

$$\sum_{i=1}^{n} \frac{\alpha_i W(\beta_i)e^{\beta_i}}{-b_j+\beta_i} + \frac{\tilde{\alpha}_i \tilde{W}(\beta_i)e^{-\beta_i}}{-b_j-\beta_i} = 0 \quad \forall j. \tag{4.16}$$

By computing, $\mathscr{L}((Vf))$, one obtains the expression for the optimal sensitivity

$$\chi(s) = \frac{n(s)c(s)-e^{-s\Delta}v(s)\tilde{d}(s)}{d(s)c(s)-e^{-s\Delta}b(s)} \,. \tag{4.17}$$

It remains to find the relation between the polynomials c,v and b. It can be shown

$$b(s) = \frac{1}{\lambda^2}\,\tilde{n}(s)v(s), \text{ and} \tag{4.18}$$

$$v(s)\tilde{v}(s) = \lambda^2 c(s)\tilde{c}(s), \tag{4.19}$$

which concludes the proof. □

4.1 The One-Pole One-Zero Case

In this case $W(s) = \frac{s+1}{s+\beta}$. We have the following theorem.

Theorem 4.3. If $\beta<1$, then there exists a maximal vector and the optimal sensitivity is given by

$$\chi(s) = \lambda\,\frac{n-\lambda e^{-s\Delta}\tilde{d}}{\lambda d-e^{-s\Delta}\tilde{n}} \,. \tag{4.20}$$

If $\beta>1$, no unique solution exists. All minimal symbols are given by

$$\chi_\phi(s) = \frac{n(s)-e^{-s\Delta}\tilde{d}(s)\phi(s)}{d(s)-e^{-s\Delta}\tilde{n}(s)\phi(s)}, \quad \phi \in \mathscr{B}(H^\infty). \tag{4.21}$$

χ_ϕ is inner $\longleftrightarrow \phi$ inner.

Proof. The first part of the theorem follows from Proposition 4.2. The second part of the theorem follows from Theorem 4.3 and 4.4. □

Remark 4.4. If we take $\phi=0$ in (4.14) then we recover $W(s) = \frac{n(s)}{d(s)}$ as a minimal symbol and this corresponds to applying open-loop control. By taking $\phi=1$, we obtain the minimal symbol

$$\chi(s) = \frac{n(s)-e^{-s\Delta}\tilde{d}(s)}{d(s)-e^{-s\Delta}\tilde{n}(s)},$$

which is inner.

4.2 The Case where $W(s) = \frac{(s+1)^2}{(s+\alpha)(s+\beta)}$, $\alpha, \beta>0$.

This case is far more complicated. If $\alpha+\beta\geq 2$, we obtain a unique solution. This follows by applying the criterion (4.9). Now

$$||W||_\infty = 1 \longleftrightarrow \alpha\beta \geq 1.$$

In this case, one can show that W is a minimal symbol. Now assuming $\alpha\neq 1$, $\beta\neq 1$, we obtain the following theorem which is the analog of Theorem 4.3

Theorem 4.4. $\alpha\beta\geq 1 \longrightarrow$ no unique solution exists. All minimal symbols are given by

$$\chi_\phi = \frac{n-e^{-s\Delta}\tilde{d}\varphi_\phi}{d-e^{-s\Delta}\tilde{n}\varphi_\phi}, \text{ where} \quad (4.22)$$

$$\chi_\phi = \frac{b(s+1) + \tilde{V}\phi}{V+b(s-1)\phi}, \ \phi \in \mathscr{B}(H^\infty), \text{ and} \quad (4.23)$$

V is a polynomial of degree 1 and b is a constant.

Proof. The proof of this theorem follows from a detailed application of Theorems 4.3 and 4.4. □

Remark 4.5. It should be noted that in the above φ does not necessarily lie in H^∞. □

Remark 4.6. One may conclude that in the "neighbourhood" of $\alpha+\beta=2$, a unique solution exists. However in the region $\{(\alpha,\beta)\,|\,\alpha+\beta>2,\ \alpha\beta<1\}$, it is not known whether the solution is unique or non-unique. It appears to be very difficult to calculate the norm of the operator T in this region.

5. Concluding Remarks.

A far more general theory can be obtained when we combine the ideas presented in this paper (and earlier in Flamm [1986], Flamm and Mitter [1985], [1986]) with the Scattering Theory ideas implicit in Adamjan, Arov and Krein [1968] and the theory of realizations of infinite-dimensional systems, as for example discussed in Fuhrmann [1981]. The basis for these ideas are the following:

We define:

$$K = H^2 \ominus \varphi H^2 \doteq (\varphi H^2)^{\perp} \quad \text{(as before)},$$

$$K^{\#} = (H^2)^{\perp} \ominus \bar{\varphi}(H^2)^{\perp} \doteq (\bar{\varphi}(H^2)^{\perp})^{\perp}.$$

We have the decomposition:

$$L^2 = \varphi H^2 \oplus K \oplus K^{\#} \oplus \bar{\varphi}(H^2)^{\perp}.$$

Define the Hankel operator with symbol $\bar{\varphi}W$

$$\mathcal{H}_{\bar{\varphi}W} : H^2 \rightarrow (H^2)^{\perp}.$$

Then since $W \varepsilon H^{\infty}$ is proper (but not strictly proper) and outer, $W^{-1} \varepsilon H^{\infty}$. Hence

$$\operatorname{Ker}(\mathcal{H}_{\bar{\varphi}W}) = \varphi H^2$$

Therefore we have the canonical factorization

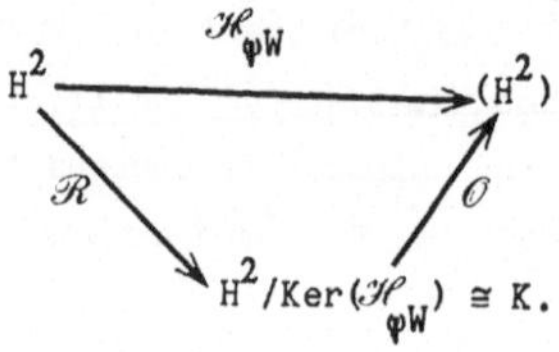

where $\mathcal{R}$ and $\mathcal{O}$ are the reachability and observability operators. Moreover $_{\bar{\varphi}W}$ with domain restricted to K has image $K^{\#}$. We also have $\mathcal{H}^{*}_{\bar{\varphi}W}: (H^2 \to H^2$, with symbol $\varphi\bar{W}$, with domain restricted to $K^{\#}$ has range K and the diagram

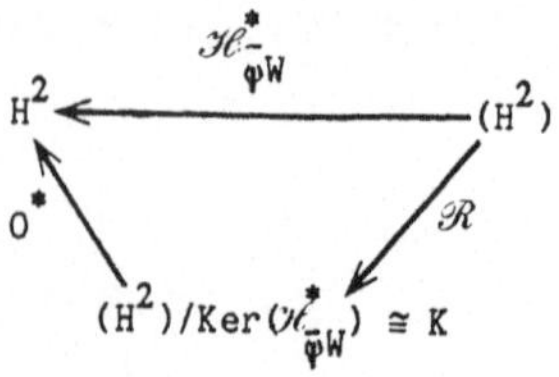

To conform to systems theory, one should regard $\mathcal{H}^{*}_{\bar{\varphi}W} = P_{+}(\bar{W}\varphi f)$ as a causal operator and $\mathcal{H}_{\bar{\varphi}W}$ as an anti-causal operator. The canonical state space realizations corresponding to these two Hankel operators will be exactly controllable and observable (at least when Range $(\mathcal{H}_{\bar{\varphi}W})$ is closed, which can be ensured by a Corona condition on the pair (φ, W)). Finally, $P_K \mathcal{H}^{*}_{\bar{\varphi}W}\mathcal{H}_{\bar{\varphi}W}P_K = T^{*}T$, and therefore the eigenvalue problem for $T^{*}T$ is the same as the eigenvalue problem for $\mathcal{H}^{*}_{\bar{\varphi}W}\mathcal{H}_{\bar{\varphi}W}$ considered as an operator from K to K. One should therefore work with the realizations of $\mathcal{H}_{\bar{\varphi}W}$ and $\mathcal{H}^{*}_{\bar{\varphi}W}$ instead of T and T^{*} as done in this paper. It is also clear that a large part of the state space constructions of Glover [1984] admit a generalization to this setting.

The details of these ideas will be presented elsewhere.

References

Adamjan, VM, Arov, DZ and Krein, MG
[1968] Infinite Hankel Matrices and Generalized Problems of Caratheodory, Fejer and I. Schur, Functional Anal. and Applns. 2, pp. 269-281.

Desoer, CA and Vidyasagar
[1975] Feedback Systems: Input-Output Properties, Academic Press, New York.

Fagnani, F
[1986] Problemi di Minimizzazione in H^{∞} Per Sistemi Dinamici in Dimensione Infinita, Thesis, Laurea, University of Pisa.

Flamm, DS

[1985] H^∞-Optimal Sensitivity for Delay Systems, Ph.D. Thesis Proposal, Dept. of Elec. Eng. and Computer Science, M.I.T., Cambridge, MA.

[1986] Control of Delay Systems for Minimax Sensitivity, Ph.D. Thesis LIDS-TH-1560, M.I.T., Cambridge, MA 02139

Flamm, DS and Mitter, SK

[1986] H^∞-Sensitivity Minimization for Delay Systems: Part I. Technical Report LIDS-P-1605, M.I.T., Cambridge, MA, Abridged version to appear in Systems Control Letters, 1987.

Foias, A, Tannenbaum, A and Zames, G

[1986] Weighted Sensitivity Minimization for Delay Systems, IEEE Trans. on Automatic Control, AC-31 (81), pp. 763-766.

Foures, Y and Segal, IE

[1955] Causality and Analyticity, Trans. Am. Math. Soc., 78, pp. 385-405.

Fuhrmann, P

[1981] Linear Systems and Operators in Hilbert Space, McGraw Hill, New York.

Garnett, JB

[1981] Bounded Analytic Functions, Academic Press, New York.

Glover, K

[1984] All Optimal Hankel-norm Approximations of Linear Multivariable Systems and their L^∞-Bounds, International Journal of Control, 39, pp. 1115-1193.

Hoffman, K

[1962] Banach Space of Analytic Functions, Prentice Hall, Englewood Cliffs, N.J.

Sarason, D

[1967] Generalized Interpolation in H^∞, Trans. Am. Math. Soc., 127, pp. 179-203.

[1985] Operator-theoretic aspects of the Nevanlina-Pick Interpolation Problems, in Operator and Function Theory, ed. S.C. Power, pp. 279-314.

CONTROL SYSTEM DESIGN TO MINIMIZE MAXIMUM ERRORS *

J.B. Pearson and M.A. Dahleh

Department of Electrical and Computer Engineering
Rice University
Houston, TX 77251-1892 U.S.A.

Abstract

The standard approach to classical control system design is to shape the loop gain function in order to meet specifications [1]. Generally the idea is to keep errors small and this involves high loop gain in the low frequency ranges where command inputs and plant disturbances are expected to lie and low loop gain where high frequency sensor noise is a problem. It is clear that the intuitive approach described above is motivated by trying to keep integral-square errors small. The H^∞-optimization theory introduced into control system design recently by Zames [2] furnishes a systematic way to minimize the error-signal energy where the system inputs are bounded energy signals. This is very closely related to the techniques used in classical design, but is constrained by the requirement of bounded energy signals.

In many control systems, a desirable objective is to limit the magnitudes of certain signals in the system. Using H^∞-theory, an indirect and not too effective way to do this is to include the integral square values of these signals in the performance index, and thus limit the maximum energy that these signals can have. Although classical design is a firmly entrenched tool that has been successfully used for many years, it is clearly not applicable to problems in which maximum error magnitudes must be limited or when inputs are not finite energy signals. Although these problems have been recognized as important for years, no substantial progress has been made toward their solution until recently when Vidyasagar [3] posed the problem of minimizing the maximum error magnitude when the system inputs are bounded in magnitude. This represented the first serious attempt to carefully formulate the problem and obtain solutions. This paper will report on research in progress on the discrete-time version of this problem.

1. Introduction

In recent years, the H^∞-optimization theory introduced by Zames [2] has become a popular way to formulate control system design problems. It allows both performance and robustness to be included in the design objectives and is a natural generalization of the classical trial and error methods of loop-gain shaping that have been used effectively on single-loop systems for years.

* This research was supported by NSF Grant ECS 85-05645.

NATO ASI Series, Vol. F34
Modelling, Robustness and Sensitivity Reduction in Control Systems. Edited by R. F. Curtain

These loop-gain shaping ideas, i.e., of keeping the loop gain high at low frequencies to insure small errors to command inputs and low frequency disturbances and the loop-gain low at high frequencies to insure small errors to high frequency sensor noise, can be explained as methods to keep integral square errors small when the inputs are bounded energy signals. For the more realistic assumption of inputs that are bounded in amplitude, the above methods are not applicable, and the problem is more appropriately posed in the time domain, rather than the frequency domain. In [3] Vidyasagar posed the problem :

> Suppose one is given a (possibly unstable) plant P, which is being subjected to a disturbance d at its output. Suppose, in addition that the disturbance d can be thought of as the output of a system W, which is in turn driven by an input v that is bounded in time by 1, but is otherwise arbitrary. The objective is to design a controller C that stabilizes the plant P and at the same time optimally rejects the disturbance; in other words, C stabilizes P, and results in the smallest possible maximum output amplitude in response to the disturbance.

A more general version of this problem is addressed in this paper, which represents a report of ongoing research conducted at Rice University by the authors.

In this paper we will be concerned only with the discrete-time case. In Section 2, we will present some background material concerning duality in optimization problems and the characterization of certain bounded linear functionals. In Section 3, we will formulate the problem, in Section 4 we will discuss the solution of the SISO case, and in Section 5 we will discuss the current status of the MIMO case and, in particular, the mixed sensitivity problem.

2. Mathematical Preliminaries

Let X be a normed linear space. The space of all bounded linear functionals on X is denoted by X^*. Let $x \in X$, $r \in X^*$, then the notation $<x,r>$ means the value of the linear functional r at x. The induced norm on X^* is defined as :

$$\|r\| = \underset{x \in BX}{supremum} \; | <x,r> | \; .$$

DEFINITION: Let S be a subspace of X. The annihilator subspace of S, denoted by $S^\perp$ is defined as follows :

$$S^\perp = \left\{ r \in X^* \mid <x,r> = 0 \quad \text{for all} \quad x \in S \right\}.$$

DEFINITION: A vector $r \in X^*$ is said to be aligned with a vector $x \in X$ if $<x,r> = \|x\| \; \|r\|$.

Next we will state a very important theorem that will play the major part of our analysis. The proof of this theorem can be found in many functional analysis and optimization textbooks, e.g. [4].

Theorem 1 :

1. Let x be an element in a real normed linear space X and let μ denote its distance from the subspace S. Then,

$$\mu = \underset{k \in S}{infimum} \; \|x-k\| = \underset{r \in BS^\perp}{maximum} <x,r>$$

where the maximum is achieved for some $\bar{r}$ in $BS^{\perp}$, with $\|\bar{r}\| = 1$.

2. If the infimum on the left is achieved for some $\bar{k}$ in S, then $\bar{r}$ is aligned with $x - \bar{k}$.

The above theorem states the equivalence of two optimization problems, one in X called the *primal problem*, and the other in X^* called the *dual problem*. $\bar{r}$ that solves the dual problem is called an *extremal functional*, and will always exist. In many situations, the problem can be set in the dual space of some particular space. Such problems always have solutions, as will be shown below. The duality theorem for these problems is very similar to the above theorem and is stated below for completeness.

Theorem 2 :

1. Let S be a subspace of a real normed linear space X. Let $x^* \in X^*$ be a distance μ from $S^{\perp}$. Then,

$$\mu = \min_{r^* \in S^{\perp}} \|x^* - r^*\| = \sup_{x \in BS} \langle x, x^* \rangle$$

where the minimum on the left is achieved for some $r_0^* \in S^{\perp}$.

2. If the supremum on the right is achieved for some $x_0 \in BS$, then $x^* - r_0^*$ is aligned with x_0.

A special case of the above theorem is the case when S is finite dimensional. In this situation, the supremum on the right will always exist, and hence, both problems have solutions [4].

Let l^1 denote the space of all sequences $h = \{h(k)\}$ such that

$$\|h\|_1 = \sum_{k=0}^{\infty} |h(k)| < \infty$$

and l^{∞} denote the space of all sequences $f = \{f(k)\}$ such that

$$\|f\|_{\infty} = sup\, |f(k)| < \infty .$$

Given a sequence h in l^1, we can define the Z-transform of h as

$$\hat{H}(z) = \sum_{k=0}^{\infty} h(k)\, z^k .$$

It is a well known result that if $\hat{H}(z)$ is the Z-transform of a sequence h; the pulse response of a linear system, then $\hat{H}(z)$ is BIBO-stable if, and only if, h is a l^1 sequence.

Let A be the space of all matrices whose elements are BIBO-stable functions. For the rest of this paper, $\hat{H}$ will denote an element in A, $H = (h_{ij})$ is its pulse response matrix, i.e., $H \in l^1_{mn}$. Let l_n^{∞} denote the space of all vectors of bounded sequences with a norm defined by:

$$\|f\|_{\infty} = \max_j \|f_j\|_{\infty} \qquad f = (f_1, \cdots, f_n)^t .$$

We can view A as the space of bounded linear shift-invariant operators on l_n^{∞}. Let $\hat{H}(z) \in A$, $f \in l_n^{\infty}$, then:

$$\hat{H} : l_n^{\infty} \rightarrow l_m^{\infty}$$
$$\hat{H} f = H * f$$

where $*$ denotes the convolution operator. The induced norm on A is given by:

$$\|\hat{H}\|_A = \max_i \sum_{j=1}^{n} \|h_{ij}\|_1 .$$

Hence, A and l^1_{mn} are identified with each other with the above norm, i.e.,

$$\|\hat{H}\|_A = \|H\|_1 .$$

The following theorem is a straightforward generalization of the result in [4].

Theorem 3 :

1. Every bounded linear functional on l^1_{mn} is representable uniquely in the form :

$$f(H) = \sum_{i=1}^{n} \sum_{j=1}^{m} \sum_{k=0}^{\infty} y_{ij}(k)\, h_{ij}(k)$$

where $Y = (y_{ij}) \in l^\infty_{mn}$. Furthermore, every element in l^∞_{mn} defines a member of $(l^1_{mn})^*$ in this way, and we have :

$$\|f\| = \sum_{i=1}^{m} \max_j \|y_{ij}\|_\infty \triangleq \|Y\|_\infty \qquad j = 1, \cdots n .$$

2. $|f(H)| = |<H,Y>| \leq \|Y\|_\infty \|H\|_1$.

3. Equality in the above holds when H is aligned with Y. Then, H, Y satisfy the following conditions :

   a. $h_{ij}(k) = 0$ whenever $y_{ij}(k) < \max_j \|y_{ij}\|_\infty$

   b. $h_{ij}(k)\, y_{ij}(k) \geq 0$

   c. $\sum_{j=1}^{n} \|h_{ij}\|_1 = \text{constant for all } i$

   d. If $y_{ij}(k)$ is identically equal to zero for a fixed i, then $h_{ij}(k)$ can be any sequence such that :

$$\sum_{j=1}^{n} \|h_{ij}\|_1 \leq \text{constant}.$$

It should be noted that the space l^1_{mn} is not reflexive, i.e., $(l^\infty_{mn})^* \neq l^1_{mn}$. However, there exists a subspace of l^∞_{mn} whose dual is exactly l^1_{mn}. Define c^0, a subspace of l^∞ as follows :

$$c^0 = \left\{ x \in l^\infty \mid \lim_{k\to\infty} x(k) = 0 \right\}.$$

As before, c^0_{mn} will denote the space of all matrices whose elements are in c^0. It can be shown [4] that $(c^0_{mn})^* = l^1_{mn}$, where linear functionals are defined as in the above theorem.

3. Problem Definition

As previously stated, our objective is to stabilize a given system and minimize the effect of a class of persistent disturbances. As usual, we start with an admissible system [5],

$$y = -\hat{P}_{11}\, u + \hat{P}_{12}\, e$$
$$z = \hat{P}_{21}\, u + \hat{P}_{22}\, e$$

where y is the measured output, z the regulated output, u the control input and e the exogenous input. Using feedback (and feedforward)

$$u = \hat{C}\, y$$

and incorporating the YJBK parametrization of all stabilizing compensators [6,7], we obtain the transfer function between e and z in the standard form:

$$\hat{\Phi}_1 = \hat{H}_1 - \hat{U}_1 \hat{Q}_1 \hat{V}_1 .$$

In the above model, the exogenous signals are assumed to be of the form:

$$e = W_2 * d$$

where d is a bounded function with $\|d\|_\infty \leq 1$. Our objective is to minimize the maximum amplitude of the regulated output (possibly weighted) due to the described exogenous inputs. More precisely, we want

$$\inf_{\hat{Q}} \sup_{d \in Bl^\infty} \| \hat{W}_1 \hat{W}_2 \hat{\Phi}_1 d \|_\infty = \inf_{\hat{Q}} \| \hat{\Phi} \|_A$$

where $\hat{\Phi} = \hat{W}_1 \hat{W}_2 \hat{\Phi}_1 = \hat{H} - \hat{U} \hat{Q} \hat{V}$. In the following section, we will discuss the solution of this problem beginning with the SISO case.

4. Solution of SISO Case [8]

In this case, the objective function can be written as

$$\hat{\Phi} = \hat{H} - \hat{U} \hat{Q}$$

where $\hat{H}, \hat{U} \in A$, and we want to find $\hat{Q} \in A$ so that $\| \hat{\Phi} \|_A$ is minimized. Define $\hat{K} = \hat{U} \hat{Q}$ and assume that $\hat{U}$ has n distinct zeros inside the open unit disk a_i, $i = 1, 2, ..., n$. The problem is to find a stable rational function $\hat{K}$ such that $\hat{K}(a_i) = 0$. For simplicity of exposition, assume the a_i are real.

Let $\hat{K}(z) = \sum_{i=0}^{\infty} k_i z^i$; then $\hat{K}(a_j) = 0$ if, and only if,

$$\sum_{i=1}^{\infty} k_i (a_j)^i = 0$$

if, and only if,

$$< k, \underline{a}_j > = 0$$

where $\underline{a}_j = 1, a_j^1, a_j^2, a_j^3 \cdots$.

Let S be the subset of l^1 defined as

$$S = \{ k \in l^1 \mid < k, \underline{a}_j > = 0 \quad j = 1, ..., n \} .$$

Then the optimization problem can be stated as follows:

$$\inf_{\hat{K} \in S} \| \hat{H} - \hat{K} \|_A . \qquad \text{(OPT)}$$

The solution of (OPT) follows directly from Theorem 1. Define $S^\perp$ as the annihilator subspace of S. Clearly, $S^\perp$ is the subspace of all possible linear combinations of the sequences $\underline{a}_j$. Therefore, any $r \in S^\perp$ can be written as

$$r = \sum_{j=1}^{n} \alpha_j \underline{a}_j ,$$

and therefore,

$$<h,r> \; = \; \sum_{j=1}^{n} \alpha_j <h, \underline{a}_j> \; = \; \sum_{j=1}^{n} \alpha_j \hat{H}(a_j) \; .$$

Thus, we have

Theorem 4 : [8]

$$\mu_0 \; = \; \inf_{\hat{K} \in S} \| \hat{H} - \hat{K} \|_A \; = \; \max_{\alpha_j} \sum_{j=1}^{n} \alpha_j \hat{H}(a_j)$$

such that

$$\left\| \sum_{j=1}^{n} \alpha_j \underline{a}_j \right\|_\infty \; \le \; 1 \; .$$

It is clear from the above constraints, which can be rewritten as

$$| \sum_{j=1}^{n} \alpha_j a_j^i | \; \le \; 1 \qquad i = 0, 1, 2, \cdots$$

that since $|a_j| < 1$, there will be some value of $i = N$ such that if the constraints are satisfied for all $i \le N$, they will be satisfied for all i. Therefore, Theorem 4 describes a simple linear programming problem with a finite number of constraints that can be solved for μ_0.

Also since the ith constraint is exactly the ith coefficient of the linear functional r, the $\bar{r}$ that solves the dual optimization problem has the properties

1. $\|\bar{r}\| = 1$, and
2. $|\bar{r}_i| < 1$ for all $i > N$.

These properties are important in constructing the solution of (OPT) which is discussed next.

Let $\bar{r}$ denote the extremal functional of the dual problem. Suppose $\hat{K}_0$ is a minimizer of (OPT), i.e.,

$$\inf_{\hat{K} \in S} \| \hat{H} - \hat{K} \|_A \; = \; \| \hat{H} - \hat{K}_0 \|_A \; = \; \mu_0 \; .$$

Then the sequence

$$\phi \; = \; h \; - \; k_0 \in l^1$$

is aligned with $\bar{r} \in l^\infty$, i.e.,

$$<\phi, \bar{r}> \; = \; \|\phi\|_1 \; \|\bar{r}\|_\infty \; = \; \mu_0$$

which is true if, and only if,

$$\sum_{i=0}^{\infty} \phi_i \bar{r}_i \; = \; \|\phi\|_1 \; \|\bar{r}\|_\infty$$

if, and only if,

$$\phi_i \; = \; 0 \quad \text{when} \quad |\bar{r}_i| \; \ne \; \|\bar{r}\|_\infty$$

and

$$\phi_i \bar{r}_i \; \ge \; 0 \; .$$

There are many choices of ϕ that will satisfy the alignment conditions, but not all are admissible, i.e., $h - \phi \in S$. Hence, ϕ_i must satisfy

$$\sum_{i=0}^{\infty} \phi_i a_j^i \; = \; \hat{H}(a_j) \qquad j \; = \; 1, ..., n \; .$$

The following theorem summarizes the above results.

Theorem 5 : [8]

(OPT) has a solution $k_0 \in S$ if, and only if, the following conditions are satisfied for $\phi = h - k_0$:

1. $\phi_i = 0$ whenever $|\bar{r}_i| \neq 1$
2. $\phi_i \bar{r}_i \geq 0$
3. $\sum_{i=0}^{\infty} |\phi_i| = \mu_0$
4. $\sum_{i=0}^{\infty} \phi_i a_j^i = \hat{H}(a_j) \qquad j = 1, ..., n.$

This theorem provides a set of linear equations, that are necessary and sufficient for the existence of a minimizer for (OPT). If a sequence ϕ satisfies these conditions, the solution k_o is given by $h - \phi$. Since there are only finitely many non-zero ϕ_i's, $\hat{K}_o$ will be rational whenever $\hat{H}$ is rational.

It is a simple excercise to prove that solutions to (OPT) will always exist. This is done by utilizing the sub-space c^o and Theorem 2 of Section 2. Hence, Theorems 4 and 5 will always provide equations which can be solved for μ_0 and ϕ_i.

5. The MIMO Case

In the MIMO case, the objective function is

$$\hat{\Phi} = \hat{H} - \hat{U}\hat{Q}\hat{V}$$

where $\hat{H}, \hat{U}$ and $\hat{V}$ are stable rational matrices of appropriate dimensions. In [9] the problem has been solved where $\hat{U}$ and $\hat{V}$ are assumed to have row rank and column rank, respectively. Here there is no loss of generality to assume that $\hat{U}$ and $\hat{V}$ are inner and co-inner respectively, and thus, $\hat{U}$ and $\hat{V}$ have a finite number of zeros in the open unit disk. We then replace $\hat{U}\hat{Q}\hat{V}$ by $\hat{K}$ and determine conditions under which a stable $\hat{K}$ always produces a stable $\hat{Q}$. These conditions are then interpreted as bounded linear functionals annihilating the space of allowable $\hat{K}$'s. In this case, the problem reduces to the solution of a linear programming problem with infinitely many constraints as in the SISO case. It can be established here too that only finitely many constraints need be considered and so the problem is, in principle, as simple as the SISO problem.

Unfortunately, the row-rank, column-rank case is not very interesting and even the simplest practical problems fall outside its purview. In particular, the mixed sensitivity problem [10] which can be stated as follows. Given the sensitivity function $\hat{S}$ and complementary sensitivity function $1-\hat{S}$ of a system. Determine a controller $\hat{C}$ that minimizes the weighted norm

$$\|\hat{\Phi}\| = \left\| \begin{bmatrix} \hat{W}_1 \hat{S} & \hat{W}_2(1-\hat{S}) \end{bmatrix} \right\| .$$

For SISO systems

$$\hat{\Phi} = \begin{bmatrix} \hat{w}_1 \hat{h}_1 & \hat{w}_2 \hat{h}_2 \end{bmatrix} - \hat{q}_1 \begin{bmatrix} \hat{w}_1 \hat{p}_1 & \hat{w}_2 \hat{p}_2 \end{bmatrix}$$

where $\hat{h}_1, \hat{h}_2, \hat{p}_1, \hat{p}_2$ are polynomials. Choose $\hat{w}_1 = \hat{n}_1 / \hat{d}_1$ and $\hat{w}_2 = \hat{n}_2 / \hat{d}_2$ as arbitrary stable functions such that $\hat{d}_1$ and $\hat{d}_2$ are coprime. Also, we will assume that $(\hat{d}_i, \hat{p}_i)$ are coprime.

Define $\underline{\hat{v}}_1 = \hat{p}_1 \hat{n}_1 \hat{d}_2$, $\underline{\hat{v}}_2 = \hat{p}_2 \hat{n}_2 \hat{d}_1$, $\hat{q} = \dfrac{\hat{q}_1}{\hat{d}_1 \hat{d}_2}$. Then

$$\hat{\Phi} = \begin{bmatrix} \hat{w}_1 \hat{h}_1 & \hat{w}_2 \hat{h}_2 \end{bmatrix} - \hat{q} \begin{bmatrix} \hat{\underline{v}}_1 & \hat{\underline{v}}_2 \end{bmatrix}.$$

Now let $\hat{k}_1 = \hat{q}\,\hat{\underline{v}}_1$, $\hat{k}_2 = \hat{q}\,\hat{\underline{v}}_2$ and assume that $\hat{\underline{v}}_1$ has N_1 distinct zeros a_{1l} and $\hat{\underline{v}}_2$ has N_2 distinct zeros a_{2l} inside the open unit disk. Then we can state the following result.

Lemma

With $\hat{k}_1 = \hat{q}\,\hat{\underline{v}}_1$ and $\hat{k}_2 = \hat{q}\,\hat{\underline{v}}_2$, $\hat{q}$ is stable if, and only if, $\hat{k}_1$ and $\hat{k}_2$ are stable and

1. $\hat{k}_1(a_{1l}) = 0 \qquad l = 1, ..., N_1$
2. $\hat{k}_2(a_{2l}) = 0 \qquad l = 1, ..., N_2$
3. $\hat{k}_1\,\hat{v}_2 - \hat{k}_2\,\hat{v}_1 = 0$ where $\hat{\underline{v}}_i = \hat{r}\,\hat{v}_i$ and $\hat{v}_1, \hat{v}_2$ are coprime.

Conditions 1 and 2 in the Lemma are those that we have seen before and describe $N_1 + N_2$ linear functionals annihilating the subspace of admissible $\hat{k}_i$'s. Condition 3, however, is new and results in our dual space being infinite dimensional.

Recall that $\hat{v}_1$ and $\hat{v}_2$ are polynomials $\hat{v}_i = v_{io} + v_{i1}\,z + \cdots + v_{iN}\,z^N \quad i = 1, 2$. Define the l^∞-sequences

$$\begin{aligned} s_{10} &= (v_{10},\, 0, 0, \cdots) \\ s_{11} &= (v_{11}, v_{10,}\, 0, \cdots) \\ s_{1N} &= (v_{1N}, ..., v_{10,}\, 0, \cdots) \\ s_{1j} &= (0, 0, ..., v_{1N}, ..., v_{10}, 0, \cdots) \qquad j \geq N. \end{aligned}$$

Then define

$$D_j = \begin{bmatrix} s_{2j} & -s_{1j} \end{bmatrix} \qquad j = 0, 1, 2, \cdots$$

Condition 3 in the Lemma can be written as

$$< K, D_j > \; = 0 \qquad j = 0, 1, 2, \cdots$$

where K represents the matrix sequence

$$K = \begin{bmatrix} k_1 & k_2 \end{bmatrix}.$$

From the above, with S the subspace of admissible K's, the annihilator subspace $S^\perp$ is the linear span of the D_j's and the sequences $[\,\underline{a}_{1l}\;\; 0\,]$ and $[\,0\;\; \underline{a}_{2l}\,]$. Thus, an element $G \in S^\perp$ can be written as

$$G = \sum_{j=0}^{\infty} \alpha_j D_j + \sum_{l=1}^{N_1} \beta_l F_{1l} + \sum_{l=1}^{N_2} \gamma_l F_{2l}$$

where $F_{1l} = [\,\underline{a}_{1l}\;\; 0\,]$, $F_{2l} = [\,0\;\; \underline{a}_{2l}\,]$.

In general, the infinite summation $\sum_{j=0}^{\infty} \alpha_j D_j$ will result in a dual problem with infinitely many variables in both the objective function and the constraints. However, in this particular case

$$\hat{t} = \hat{w}_1\,\hat{h}_1\,\hat{v}_2 - \hat{w}_2\,\hat{h}_2\,\hat{v}_1$$

is a polynomial of degree M, i.e.,

$$\hat{t} = t_0 + t_1\,z + \cdots + t_M\,z^M$$

and this will result in a finite number of variables in the objective function.

We now calculate $< H, G >$ for $G \in S^\perp$ and $H = [h_1\, h_2]$. From above

$$<H,G> \; = \; \sum_{j=0}^{\infty} \alpha_j <H,D_j> + \sum_{l=1}^{N_1} \beta_l \hat{h}_1(a_{1l}) + \sum_{l=1}^{N_2} \gamma_l \hat{h}_2(a_{2l}) \; .$$

Since

$$\begin{aligned} <H,D_j> &= j^{\text{th}} \text{ coefficient of } (\hat{h}_1 \hat{v}_2 - \hat{h}_2 \hat{v}_1) \\ &= t_j \qquad j = 0, \ldots, M \;, \end{aligned}$$

it follows that

$$<H,G> \; = \; \sum_{j=0}^{M} \alpha_j t_j + \sum_{l=1}^{N_1} \beta_l \hat{h}_1(a_{1l}) + \sum_{l=1}^{N_2} \gamma_l \hat{h}_2(a_{2l})$$

and the dual problem involves maximization of this function subject to

$$\|G\|_\infty \leq 1 \; .$$

We summarize the results as follows.

Theorem 6 :

The solution of the mixed sensitivity problem is obtained by solving the problem

$$\mu_0 \; = \; \max_{\alpha_j, \beta_j, \gamma_j} \left[\sum_{j=0}^{M} \alpha_j t_j + \sum_{j=1}^{N_1} \beta_j \hat{h}_1(a_{1j}) + \sum_{j=1}^{N_2} \gamma_j \hat{h}_2(a_{2j}) \right]$$

subject to

$$\left| v_{20} \alpha_k + \cdots + v_{2N} \alpha_{k+N} + \sum_{j=1}^{N_1} \beta_j (a_{1j})^k \right| \leq 1$$

$$\left| - v_{10} \alpha_k - \cdots - v_{1N} \alpha_{k+N} + \sum_{j=1}^{N_2} \gamma_j (a_{2j})^k \right| \leq 1$$

$$k = 0, 1, 2, \cdots .$$

Our results to date on this problem have been to establish conditions under which truncated versions have solutions. In general, this truncated problem will have solutions for all $k \geq k_{\max}$ where $k_{\max}$ is easily determined. We can then obtain solutions ϕ of norm μ_k, such that for $k_1 > k_2$, $\mu_{k1} \leq \mu_{k2}$. The problem can be solved iteratively. Some computational experience is being currently obtained and will be reported elsewhere.

References

[1] Doyle, J.C. and Stein, G., "Multivariable feedback design: Concepts for a classical / modern synthesis," *IEEE Trans. Auto. Control*, **AC-26**, pp.4-16 (Feb. 1981).

[2] Zames, G., "Feedback and optimal sensitivity: Model reference transformations, multiplicative seminorms, and approximate inverses," *IEEE Trans. on Automatic Control*, **AC-26**, pp.301-320 (Apr. 1981).

[3] Vidyasagar, M., "Optimal rejection of persistent, bounded disturbances," *IEEE Trans. on Automatic Control*, **AC-31**, pp.527-534 (1986).

[4] Luenberger, D.G., *Optimization by vector space methods*, New York: John Wiley and Sons (1969).

[5] Cheng, L. and Pearson, J.B., ''Synthesis of linear multivariable regulators,'' *IEEE Trans. on Automatic Control,* **AC-26**, pp.194-202 (Feb. 1981).

[6] Youla, D.C., Bongiorno, J.J., and Jabr, H.A., ''Modern Wiener-Hopf design of optimal controllers, Part II: The multivariable case,'' *IEEE Trans. on Automatic Control,* **AC-21**, pp.319-338 (Jun. 1976).

[7] Kucera, V., Chapter 5, section 3, Theorems 6 and 9 in *Discrete Linear Control*, New York: John Wiley and Sons (1979).

[8] Dahleh, M.A. and Pearson, J.B., '' l^1-optimal feedback controllers for discrete-time systems,'' (Technical Report 8513, Department of Electrical and Computer Engineering, Rice University) Proceedings ACC, Seattle, WA, pp. 1964-1968 (Jun. 1986).

[9] Dahleh, M.A. and Pearson, J.B., '' l^1-optimal feedback controllers for MIMO discrete-time systems,'' Technical Report 8602, Department of Electrical and Computer Engineering, Rice University (Feb. 1986).

[10] Kwakernaak, H., ''Minimax frequency domain performance and robustness optimization in linear feedback systems,'' *IEEE Trans. A-C,* **AC-30**(10), pp.994-1004 (Oct. 1985).

Implications of a Characterization Result on Strong and Reliable Decentralized Control

A. Bulent Ozguler
Department of Mathematics
Bilkent University
P.O.B.8 Maltepe
06572 Ankara
Turkey

Muzaffer Hiraoglu
Dept. of Electrical Eng.
Bogazici University
P.O.B.2 Bebek
80815 Istanbul
Turkey

0.ABSTRACT

A number of special purpose decentralized control problems are defined and examined for a two-by-two plant. Using a characterization of the set of all diagonal stabilizing compensators it is shown that reliable diagonal stabilization can equivalently be viewed either as a strong diagonal stabilization problem or simultaneous stabilization problems for suitably defined plants. Various solvability conditions are determined for strong and reliable stabilization and robust reliable stabilization of a scalar plant is shown to be equivalent to diagonal reliable stabilization problem of special two-by-two plant.

1.INTRODUCTION

In this paper, we define and examine a number of constrained decentralized control problems for linear systems. The discussion is limited to two-by-two systems for the purpose of stating definitive results and for the sake of simplicity; although, some of the results can be generalized either to diagonal control of a general square system or to decentralized control of a two-channel system. (See GUCLU and OZGULER[1986], HIRAOGLU[1986], and OZGULER[1986].)

Let $Z = [z_{ij}]$ be the two-by-two transfer matrix of a strictly causal system called <u>plant</u> and let C be the transfer matrix

NATO ASI Series, Vol. F34
Modelling, Robustness and Sensitivity Reduction in Control Systems. Edited by R. F. Curtain

of another two-by-two (causal) system called compensator. Following DESOER et. al.[1980], we say that the pair (Z,C) is internally stable iff the poles of the four-by-four transfer matrix

$$\begin{bmatrix} (I+ZC)^{-1}Z & (I+ZC)^{-1}ZC \\ (I+CZ)^{-1}CZ & (I+CZ)^{-1}C \end{bmatrix}$$

are all in a prespecified region (usually the left half plane or the unit disk) of the complex plane. Such a C, which always exists, is called a stabilizing compensator for Z.
We can now formally define the problems that we consider in the subsequent sections.

(1.1) DEFINITIONS. (a) Diagonal stabilization problem (DSP) is said to be solvable for Z iff there exists a diagonal stabilizing compensator for Z, i.e., there exists a transfer matrix C such that $C = \text{diag}\{c_1, c_2\}$ and (Z,C) is internally stable. (b) Diagonal strong stabilization problem (DSSP) is said to be solvable for Z iff there exists a compensator C that solves DSP and such that all poles of C are in the stability region of the complex plane. (b) Diagonal reliable stabilization problem (DRSP) is said to be solvable for Z iff there exists a compensator $C = \text{diag}\{c_1, c_2\}$ that solves DSP and such that (z_{11}, c_1) and (z_{22}, c_2) are internally stable pairs.

It is well known that DSP is solvable for a given system iff it is free of decentralized (in our case, diagonal) fixed modes(WANG and DAVISON [1973]). There are various procedures for determining a solution to DSP like those of WANG and DAVISON [1973], CORFMAT and MORSE [1976], VIDYASAGAR and VISWANADHAM [1982a], or GUCLU and OZGULER [1986]. Among these the procedure of the last, although restricted to diagonal feedback, yields an explicit expression for the compensator through the solution of a nonlinear but easily solvable equation over the ring of polynomials or over the ring of stable transfer functions. Our approach here to DSSP and DRSP is actually based on this explicit synthesis procedure.

The motivation for considering DSSP comes from the fundamental work of YOULA et. al.[1974], where the problem has been solved in the centralized case for a general multivariable plant. It is known that a strong stabilization scheme exhibits superior sensitivity properties compared to a scheme via an unstable compensator in some cases. This advantage of strong stabilization is also becoming realized in the context of stabilization using H -techniques.(See FREUDENBERG and LOOZE [1986]). A further, perhaps more important, property of strong stabilization is that it is an integral part in various other seemingly unrelated control problems such as simultaneous stabilization (VIDYASAGAR and VISWANADHAM [1982b]) and robust reliable stabilization (GHOSH [1984]). This paper is yet another step in the direction of indicating the role of strong stabilization in other control problems.

A diagonal reliable stabilization scheme on the other hand has the following properties: (i) In the case of an interconnection breakdown in the plant corresponding to the case $z_{12}=0$ or $z_{21}=0$, the diagonal subplants are internally stable, (ii) the overall plant also remains internally stable, provided the broken plant is free of decentralized fixed modes.(iii) In case of a feedback interconnection failure corresponding to the case $c_1 = 0$ or $c_2 = 0$, the instability of a reliably stabilized plant is at most as bad as the instability of the original open loop plant; in other words, no new unstable poles are introduced to the plant by such a feedback failure but the original unstable poles of the plant.It is easy to construct examples to the effect that an arbitrary solution to DSP does not have properties (i) - (iii). In fact, as we illustrate in Sections 3 and 4, reliable diagonal stabilization schemes exist only under rather severe constraints on the plants. One further motivation for considering DRSP is its close relation to a recently considered robust reliable stabilization problem of GHOSH [1984].

The paper is organized as follows.

In Section 2, we set up the necessary background for a "characteristic function representation" of a plant Z over the ring of stable transfer functions and restate the result of GUCLU and OZGULER [1986] for DSP in the language of stable transfer functions. We also state a characterization result for the solutions of DSP [Theorem(2.6)]. In this section, we also obtain a solvability condition for DSSP in the form of a nonlinear equation over the ring of stable transfer functions[Corollary(2.12)]. This equation in general seems to be difficult to solve. However, it is possible to prove in Theorem (2.14) that DSSP is solvable for minimum phase plants. In Section 3, we examine DRSP and show that it can be viewed as a DSSP or as a simultaneous DSP for suitably defined plants in Theorem (3.4). This also yields for DRSP solvability conditions in terms of equations over the ring of stable transfer functions.We also show in Section 3 that a plant which is stable at its diagonal can be reliably stabilized iff the multiple of the off-diagonal subplants has parity interlacing property [Theorem(3.13)]. Finally, in Section 4, robust reliable stabilization of a scalar plant is examined in the light of the results of Section 3 and a characterization result is proved in Theorem (4.3).

2.DIAGONAL STABILIZATION

Let R(z) denote the set of real rational functions of z . The set of transfer functions form a subring of R(z) and consist of elements of R(z) with no pole at infinity. Units in the ring of transfer functions are <u>biproper</u> rational functions which have no pole or zero at infinity. Thus, a biproper rational function is such that it is a nonzero transfer function with its inverse also being a transfer function. In order to incorporate a general concept of stability to transfer functions we fix a conjugate symmetric region of the complex plane intersecting the real axis and call a transfer function stable iff all its poles are in this stability region. We denote the set of stable transfer functions by R_s following MORSE[1976]. The set R_s is also a subring of R(z) and is actually a Euclidean Domain. Consequently, any finite number of stable transfer functions (and also matrices with entries in R_s) can be assigned a greatest common divisor (GCD) and a least common multiple (LCM) which are unique upto multiplications by units. (For matrices, under mild conditions, uniqueness upto multiplications by unimodular matrices can also be assured.) A further nice consequence of R_s being a Euclidean Domain is that various calculations over this ring can be reduced to solving a finite number of equations over the base ring R. Finally, Smith canonical forms of matrices with entries in R_s can be defined and computed, MCDUFFEE[1945].

Any rational function a in R(z) has a representation of the form a=p/q with p, q in R_s, q nonzero, and such that GCD{p,q} is a unit (this latter fact is also expressed by saying either "{p,q} is coprime" or "GCD{p,q}=1" meaning that a GCD is one). If a is a transfer function, then q in this representation is biproper. Conversely, if a rational function a has a coprime fractional representation p/q with q biproper, then a is a transfer function, i.e., a is proper. Also note that, if a is in R_s , then in any coprime fractional representation of a , q is a unit. An easy way to see this fact is to note that "{p,q} is coprime" is equivalent to "px+qy=1 for some x, y in R_s". Now, a=p/q implies that q(ax+y)=1 , where ax+y is in R_s. Consequently, q is a unit.

A <u>minimum phase</u> transfer function is one with all its finite zeros in the stability region. The following fact is worth noting: If p, q are in R_s with p minimum phase and q biproper, then {p,q} is coprime. In fact, any divisor of p and q is such that it is stable, biproper, and minimum phase; consequently it is a unit.

Let each entry of the plant $Z=[z_{ij}]$ of (1.1) be represented in coprime fractions $z_{ij}=n_{ij}/m_{ij}$. Also let $\det Z=n_0/m_0$ be a coprime fractional represantation of the determinant of Z and define

$$m:=\mathrm{LCM}\{m_{11},m_2,m_{21},m_{22},m_0\}$$

and note that m is biproper as each m_{ij} and m_0 is. Then,

$$n_1:=mz_{11},\ n_2:=mz_{12},\ n_3:=mz_{21},\ n_4:=mz_{22}$$

are all in R_s and further,

$$d:=m\,\det Z=m(z_{11}z_{22}-z_{12}z_{21})$$

is also in R_s. Further, these equalities also yield

$$dm=n_1n_4-n_2n_3\ . \tag{2.1}$$

We have thus obtained the following representation of Z :

$$Z=(1/m)\begin{bmatrix} n_1 & n_2 \\ n_3 & n_4 \end{bmatrix}, \tag{2.2}$$

which has the properties (i) <u>equality</u> (2.1) <u>holds for some</u> d <u>in</u> R_s and (ii) GCD{m,n_1,n_2,n_3,n_4,d} <u>is a unit</u>, this latter fact being a direct consequence of the above definitions. Conversely, if Z is represented as in (2.2) for some m, n_1, n_2, n_3, and n_4 in R_s which satisfy (i) and (ii), then it can easily be shown that they also satisfy $m = LCM\{m_{11}, m_{12}, m_{21}, m_{22}, m_0\}$. Noting the analogy between (2.2) and a similar representation for Z , where this time m is the characteristic polynomial of Z and n_1, n_2, n_3, n_4, d are polynomials, we call m in (2.2) <u>a characteristic function of</u> Z and the representation (2.2) <u>a characteristic function representation of</u> Z .These names can further be justified by noting that if c is the characteristic polynomial of Z then, m=bc ,where b is a stable and minimum phase rational function with precisely "deg c" poles at infinity. (This fact again easily follows by the definition of m .) The characteristic function representation (2.2) of Z will be convenient in examining the diagonal control problems defined in the previous section.

Our first result concerns the diagonal stabilization problem (DSP) of Definition (1.2). To this end, suppose that

(2.3) $GCD\{m, n_1, n_4, d\} = 1$.

It follows that there exist unimodular matrices U and V (matrices such that det U and det V are units in R_s) over R_s such that

$$(2.4)\quad \begin{bmatrix} u_1 & u_2 \\ u_3 & u_4 \end{bmatrix} \begin{bmatrix} m & n_4 \\ n_1 & d \end{bmatrix} \begin{bmatrix} v_1 & v_3 \\ v_2 & v_4 \end{bmatrix} = \begin{bmatrix} 1 & 0 \\ 0 & D \end{bmatrix} \qquad =$$

where diag{1,D} is the Smith cannonical form of the second matrix. It can further be assured that det U = det V = 1 by multiplying D by a unit if necessary. Then, we have that

(2.5) $D = md - n_1n_4 = - n_2n_3$.

Although the elements u_i, v_j satisfying (2.3) and $u_1u_4 - u_2u_3 = v_1v_4 - v_2v_3 = 1$ are by no means unique, let us fix one set of such u_i, v_j $(i,j=1,2,3,4)$.

(2.6) THEOREM. DSP has a solution iff (2.3) holds. If (2.3) holds, then any solution $C = \mathrm{diag}\{c_1, c_2\}$ is of the form

(2.7a) $c_1 = (k_1u_2+k_2u_4)/(k_1u_1+k_2u_3)$,

(2.7b) $c_2 = (l_1v_2+l_2v_4)/(l_1v_1+l_2v_3)$,

where k_1,k_2,l_1,l_2 are elements in R_s satisfying

(2.8) $k_1l_1 - n_2n_3k_2l_2 = 1$,

and u_i, v_j are chosen to satisfy (2.4). Conversely, for any k_1, k_2,l_1, and l_2 satisfying (2.8), $C = \mathrm{diag}\{c_1,c_2\}$, where c_1,c_2 are defined by (2.7), is a solution to the problem.

PROOF. The existence part of the claim is a direct consequence of the result of VIDYASAGAR and VISWANADHAM [1982] specialized to N=2. Also see the main result of GUCLU and OZGULER [1985] for an alternative simple procedure for the synthesis part. The characterization part follows by OZGULER [1986] and HIRAOGLU [1986]. #

(2.9) REMARK. The result of Theorem (2.6) is to be compared to the corresponding result for internally stabilizing compensators of a scalar plant $Z=n/m$. Let u_1,u_2 be in R_s satisfying $mu_1+nu_2=1$. The set of all c internally stabilizing Z is then given by $c=(u_2-km)/(u_1+kn)$, where k

is an element in R_s. Thus, the characterizing element k is free in R_s in the case of a scalar plant. The compensators (2.7) however are characterized by $\{k_1,k_2,l_1,l_2\}$ in R_s^4 satisfying the constraint (2.8). An important special case of diagonal stabilization is the case $z_{12}z_{21}=0$ in (1.1). In this case, the problem reduces to stabilization of two scalar plants z_{11} and z_{22}. The characterization (2.7) on the other hand yields for this special case

$$c_1 = (u_2+k_3u_4)/(u_1+k_3u_3),$$

$$c_2 = (v_2+l_3v_4)/(v_1+l_3v_3),$$

where k_3, l_3 are free in R_s. This is because $n_2n_3=0$ in (2.8) and hence k_1, l_1 are units in R_s.

(2.10) REMARK. One way of viewing the characterization result for scalar plants is that the set of compensators are described in terms of the set of compensators $\{k\}$ for the subsidiary plant $Z_0=0$. Note that any stable transfer function can be considered as a stabilizing compensator for $Z_0=0$. Extending this interpretation to DSP for a two by two plant, we see that the set of compensators $\mathrm{diag}\{c_1,c_2\}$ is characterized in terms of the set of all diagonal compensators $\{k_2/k_1,l_2/l_1\}$ of the subsidiary two by two, stable, off-diagonal plant

$$Z_0 = \begin{bmatrix} 0 & n_2 \\ n_3 & 0 \end{bmatrix}.$$

In fact, for Z_0, we have $m=1$, $n_1=0$, $n_4=0$, $d=-n_2n_3$ and hence the set of diagonally stabilizing compensators of Z_0 are determined by all solutions of (2.8), by the first part of Theorem (2.6) and by (2.4).

These considerations in Remarks (2.9) and (2.10) indicate that the characterization of Theorem (2.6) is a natural generalization of the scalar case characterization of DESOER et.al.[1980].

Let $\mathrm{diag}\{x_2/x_1, y_2/y_1\}$ be a solution to DSP. Then, $c_1=x_2/x_1$ and $c_2=y_2/y_1$ are given by (2.7) for some $\{k_1,l_1,k_2,l_2\}$ satisfying (2.8). It is straightforward to check that

(2.11) $mx_1y_1+n_1x_2y_1+n_4x_1y_2+dx_2y_2 = u$

for some unit u in R_s. Conversely, given a quadruple $\{x_1,y_1,x_2,y_2\}$ satisfying (2.11) for some unit u, let $u_1:=x_1/u$, $u_2:=x_2/u$, $v_1:=y_1/u$, $v_2:=y_2/u$, and also let $u_3:=-(n_1v_1+dv_2)$, $u_4:=mv_1+n_4v_2$, $v_3:=-(u_1n_4+u_2d)$, and $v_4:=u_1m+u_2n_1$. It can be checked that (2.4) holds with $D=-n_2n_3$. Consequently, choosing $k_1=l_1=1$ and $k_2=l_2=0$ in Theorem (2.6), we have that $\mathrm{diag}\{x_2/x_1,y_2/y_1\}$ is a solution to DSP. This yields the following corollary to Theorem (2.6).

(2.12) COROLLARY. <u>There exists a solution to DSSP iff there exist elements</u> x,y <u>and a unit</u> u <u>in</u> R_s <u>such that</u>

(2.13) $m+n_1x+n_4y+dxy = u.$

<u>A compensator</u> $\mathrm{diag}\{x,y\}$ <u>with</u> x, y <u>in</u> R_s <u>solves DSSP iff</u> (2.3) <u>holds for some unit</u> u <u>in</u> R_s.

PROOF. The existence part is actually implied by the second statement of the corollary. It is separately stated only for emphasis. To see the claim of the second statement, note that a compensator $\mathrm{diag}\{x_2/x_1, y_2/y_1\}$ solves DSSP iff x_1 and y_1 are units and (2.11) holds for some unit u. Setting $x:=x_2/x_1$ and $y:=y_2/y_1$ in (2.11) proves the claim.#

(2.13) REMARK. Given a scalar plant $Z=n/m$, it admits a strong stabilizing compensator iff $m+nx=u$ for some unit u and element x in R_s, by the result of YOULA et.al.[1973] restated in VIDYASAGAR and VISWANADHAM [1982] in the language of stable rational fractions. Equation (2.13) on the other hand can be rewritten as $(m+n_1x)+(n_4+dx)y = u$ or as $(m+n_4y)+(n_1+dy)x = u$. Consequently, fixing one of the compensators x or y, the problem becomes that of determining a strong stabilizing compensator for the scalar plant $h(x) := (n_4+dx)/(m+n_1x)$ or $g(y) := (n_1+dy)/(m+n_4y)$. This allows us to restate the existence condition for DSSP as follows: DSSP is solvable iff there exists x in R_s such that either $h(x)$ or $g(x)$ is a coprime fraction and has the <u>parity interlacing property</u>. (See YOULA et.al.[1973] for a definition.)

By our Remark (2.13), determining a solvability condition for DSSP in terms of $\{m,n_1,n_4,d\}$ thus turns out to be nontrivial. There are, however, a few cases where this is possible. Among these cases, the trivial ones are obtained when (i) $d = 0$, (ii) $n_2n_3 = 0$, (iii) $\{m,n_1\}$ is coprime and z_{11} has parity interlacing property, and (iv) $\{m,n_4\}$ is coprime and z_{22} has parity interlacing property.

(i) [d=0] In this case, det Z = 0 and we have a singular plant. Let $n := GCD\{n_1, n_4\}$. Then, by equality (2.13), $m+n(n_{10}x+n_{40}y) = u$, where $n_{10} := n_1/n$ and $n_{40} := n_4/n$ with $\{n_{10}, n_{40}\}$ coprime. Clearly, in this case the problem is solvable iff $\{m,n\}$ is coprime and n/m has the parity interlacing property. In fact, the "only if" part being clear, suppose that $m+nt=u$ for some t and unit u in R_s. By coprimeness of $\{n_{10}, n_{40}\}$ there exist x, y in R_s such that $t=n_{10}x+n_{40}y$ and we have by (2.12) that diag$\{x,y\}$ is a solution to the problem. It can be shown that in any characteristic function representation of Z, $GCD\{n_1, n_4, d\}$ is the smallest invariant numerator in the Smith McMillan form of Z over R_s. The condition that $\{m,n\}$ is coprime and n/m has parity interlacing property is precisely the solvability condition for a two by two plant to have a (central) strong stabilizing compensator (see VIDYASAGAR and VISWANADHAM [1982]). Therefore, if $d=0$, DSSP is solvable for Z iff Z is (central) strong stabilizable.

(ii) $[n_2n_3=0]$ This corresponds to a lower or upper triangular plant. The equation (2.13) seperates as $(m_{11}+n_{11}x)(m_{22}+n_{22}x) = u$ provided the plant is free of decentralized fixed modes. Clearly, in this case, the plant Z is diagonally strong stabilizable iff both of the scalar plants z_{11} and z_{22} are strong stabilizable.

(iii) & (iv) We examine only one of these cases as the other one is analogous. Note that, if (iii) holds then diag$\{x,0\}$, where x is such that $m+n_1x=u$ for some unit u, is a solution to the problem. Therefore, in cases (iii) and (iv) the plant is strong stabilizable using only one of the channels.

A less trivial result is the following.

(2.14) THEOREM. _Let_ Z _be nonsingular and let_ d _be minimum phase. Then,_ DSSP _is solvable for_ Z.

PROOF. Let $GCD\{n_4,d\}=:e$ and note that e is minimum phase as d is. Let $d=d_0e$, $n_4=n_{40}e$ for coprime elements d_0 and n_{40} in R_S. AS d_0 is minimum phase, there exists x such that $n_{40}+d_0x=v$ for some unit v. Since ve is minimum phase and $m+n_1x$ is biproper, the pair $\{m+n_1x,ve\}$ is coprime and there exists y such that $m+n_1x+vey=u$ for some unit u. Consequently, $m+n_1x+n_4y+dxy=u$ and by Corollary (2.12), $diag\{x,y\}$ is a solution to DSSP.

(2.15) REMARK. Since $\det Z = d/m$, d being minimum phase is equivalent to $\det Z$ being minimum phase.

3. RELIABLE DIAGONAL STABILIZATION

In this section, we show that DRSP can equivalently be viewed either as a DSSP or as a simultaneous DSP. We express solvability conditions in terms of equations over R_S and relate these to problem data in some significant special cases.

Let Z be in a characteristic function representation (2.2) and suppose Z is diagonally stabilizable. Let (2.4) hold with $D=-$

n_2n_3 for a fixed set of elements u_i, v_i $(i=1,2,3,4)$ in R_s. Also let

$$(3.1) \qquad m_{11}a_1+n_{11}b_1=1 \ , \ m_{22}a_2+n_{22}b_2=1$$

for a fixed set of elements a_i, b_i $(i=1,2)$ in R_s. Note that such elements exist as $\{m_{11},n_{11}\}$ and $\{m_{22},n_{22}\}$ are coprime pairs. In order to state our first result, we now define the following subsidiary transfer matrices:

$$(3.2) \qquad T := (1/m_0) \begin{bmatrix} (dm_{22}-n_1n_{22})b_1 & n_2 \\ n_3 & (dm_{11}-n_4n_{11})b_2 \end{bmatrix}$$

where $m_0:=ma_1a_2+n_1b_1a_2+n_4a_1b_2+db_1b_2$,

$$(3.3a) \qquad T_d:=\mathrm{diag}\{(dm_{11}-n_4n_{11})v_2/(m_{11}u_1+n_{11}u_2),(dm_{22}-n_1n_{22})u_2/(m_{22}v_1+n_{22}v_2)\},$$

$$(3.3b) \qquad T_0 := \begin{bmatrix} 0 & n_2 \\ n_3 & 0 \end{bmatrix}.$$

(3.4) THEOREM. *The following statements are equivalent*:

(i) *DRSP is solvable for* Z.

(ii) *DSSP is solvable for* T.

(iii) *Simultaneous DSP is solvable for* $[T_0,T_d]$.

(iv) *There exist elements* x, y *and a unit* u_0 *in* R_s *such that*

$$(3.5) \qquad u_0=m_0+(dm_{22}-n_1n_{22})b_1x+(dm_{11}-n_4n_{11})b_2y+(dm_{22}-n_1n_{22})m_{11}xy.$$

(v) <u>There exist elements</u> x_1, y_1, x_2, y_2 <u>and units</u> u, v <u>in</u> R_s <u>such that</u>

(3.6a) $x_1y_1 - n_2n_3x_2y_2 = 1,$

(3.6b) $(m_{11}u_1 + n_{11}u_2)x_1 + (dm_{11} - n_4n_{11})v_2x_2 = u,$

(3.6c) $(m_{22}v_1 + n_{22}v_2)y_1 + (dm_{22} - n_1n_{22})u_2y_2 = v.$

PROOF. Let us first make a few observations on the matrices T and T_d. Let $d_1 := GCD\{m, n_1\}$ and $d_2 := GCD\{m, n_4\}$. Note that $m = m_{11}d_1 = m_{22}d_2$, $n_1 = n_{11}d_1$, $n_4 = n_{22}d_2$ and hence with

$$N_0 := \begin{bmatrix} -n_2n_3b_1/d_2 & n_2 \\ n_3 & -n_2n_3b_2/d_1 \end{bmatrix}$$

we have $T = N_0/m_0$, where we used $md = n_1n_4 - n_2n_3$. Using (3.1) it is not difficult to verify that

(3.7) $\det N_0 = m_0m_{11}(dm_{22} - n_1n_{22}),$

i.e., m_0 divides the determinant of the numerator matrix of T. Further, one can show by making use of various coprimeness conditions that

$m_0, (dm_{22} - n_1n_{22})b_1, (dm_{11} - n_4n_{11})b_2, n_2, n_3, (dm_{22} - n_1n_{22})m_{11}$

is coprime. Consequently, T in (3.2) is in characteristic function representation with m_0 its characteristic function. On the other hand, using (2.4), it further follows that

$\{m_{11}u_1 + n_{11}u_2, (dm_{11} - n_4n_{11})v_2\},$

$\{m_{22}v_1 + n_{22}v_2, (dm_{22} - n_1n_{22})u_2\},$

are coprime pairs.Consequently, the diagonal elements of T_d in (3.3a) are in coprime fractional representations.

[(ii) iff (iii)] By Corollary (2.12), $diag\{x, y\}$ strong stabilizes T iff $m + n_1x + n_4y + dxy$ is a unit. By (3.2) and (3.7), the relevant quantities are $m=m_0, n_1=(dm_{22}-n_1n_{22})b_1, n_4=(dm_{11}-n_4n_{11})b_2$, and $d=(dm_{22}-n_1n_{22})m_{11}$.

[(iii) iff (iv)] Let $\mathrm{diag}\{x_2/x_1, y_2/y_1\}$ simultaneously stabilize $[T_0, T_d]$. Then, by (2.11) applied to T_0, we have

(3.8) $\quad x_1y_1-n_2n_3x_2y_2=u_0,$

for some unit u_0. Further, as the diagonal entries of T_d are stabilized by x_2/x_1 and y_2/y_1, we also have (3.6b) and (3.6c) for some units u and v. Normalizing (3.8) by replacing x_1/u_0 with x_1 and x_2/u_0 with x_2, we obtain (3.6a). Under this normalization (3.6b,c) still hold with u/u_0 replacing u. Conversely, if (3.6) hold for some units u and v, then it is obvious that $\mathrm{diag}\{x_2/x_1, y_2/y_1\}$ stabilizes both of T_0 and T_d.

[(i) iff (iv)] By Definition (1.2), $C=\mathrm{diag}\{c_1, c_2\}$ solves DRSP for Z iff (Z,C), (z_{11}, c_1), and (z_{22}, c_2) are all internally stable pairs. Let x_2/x_1 and y_2/y_1 be coprime fractional representations of c_1 and c_2, respectively. Then, C is a reliable stabilizer of Z iff

(3.9a) $\quad m_{11}x_1+n_{11}x_2=u,$

(3.9b) $\quad m_{22}y_1+n_{22}y_2=v,$

(3.9c) $\quad mx_1y_1+n_1x_2y_1+n_4x_1y_2+dx_2y_2=u_0,$

for some units u, v, u_0, where the last equality is by (2.11). Letting $k_1:=x_1/u$, $k_2:=x_2/u$, $l_1:=y_1/v$, $l_2:=y_2/v$ and comparing (3.9a,b) and (3.1), we have

$$k_1=a_1-n_{11}y,\ k_2=b_1+m_{11}y$$

$$l_1=a_2-n_{22}x,\ l_2=b_2+m_{22}x.$$

Substituting into (3.9c), we obtain (3.5) with u_0/uv

replacing u. Conversely, if (3.5) holds, then the compensator $\mathrm{diag}\{(b_1+m_{11}y)/(a_1-n_{11}y),\ (b_2+m_{22}x)/(a_2-n_{22}x)\}$ is easily checked to reliably stabilize Z.

[(i) iff (v)] Let $C=\mathrm{diag}\{c_1,\ c_2\}$ be a solution to DRSP for Z. Since Z solves DSP in particular, by Theorem (2.6), there exist k_1, k_2, l_1, and l_2 satisfying (2.8) such that c_1 and c_2 are given by (2.7). By internal stability of (z_{11},c_1) and (z_{22},c_2), we further have

$$m_{11}(k_1u_1+k_2u_3)+n_{11}(k_1u_2+k_2u_4)=u,$$

$$m_{22}(l_1v_1+l_2v_3)+n_{22}(l_1v_2+l_2v_4)=v,$$

for some units u and v. These equalities yield

(3.10a) $(m_{11}u_1+n_{11}u_2)k_1+(dm_{11}-n_4n_{11})v_2k_2=u,$

(3.10b) $(m_{22}v_1+n_{22}v_2)l_1+(dm_{22}-n_1n_{22})u_2l_2=v,$

on substituting $u_3=-(n_1v_1+dv_2)$, $u_4=mv_1+n_4v_2$, $v_3=-(u_1n_4+u_2d)$, $v_4=mu_1+u_2n_1$ which follow by (2.4). Clearly, (2.8) and (3.10) yield equations (3.6) with $x_i=k_i$, $y_i=l_i$ $(i=1,2)$. Conversely, if (3.6) hold for some x_1, x_2, y_1, y_2, let

$$c_1:=(x_1u_2+x_2u_4)/(x_1u_1+x_2u_3),$$

$$c_2:=(y_1v_2+y_2v_4)/(y_1v_1+y_2v_3).$$

Then, by Theorem (2.6), $C:=\mathrm{diag}\{c_1,c_2\}$ internally stabilizes Z and it can further be verified employing (2.4) that (z_{11},c_1) and (z_{22},c_2) are also internally stable pairs. Consequently, C is a solution to DRSP. #

(3.11) REMARK. It is possible to strengthen the result of Theorem (3.4), obtaining a characterization of the set of all solutions to DRSP in terms of the set of all solutions to either DSSP for T or simultaneous DSP for $[T_0,T_d]$. A closer inspection of the proof of Theorem (3.4) reveals that $C=\mathrm{diag}\{c_1,c_2\}$ is a solution to DRSP iff $c_1=(b_1+m_{11}y)/(a_1-n_{11}y)$, $c_2=(b_2+m_{22}x)/(a_2-n_{22}x)$, where $\mathrm{diag}\{x,y\}$ is a solution to DSSP for T, equivalently, iff $c_1=(x_1u_2+x_2u_4)/(x_1u_1+x_2u_3)$, $c_2=(y_1v_2+y_2v_4)/(y_1v_1+y_2v_3)$, where $\mathrm{diag}\{x_2/x_1,y_2/y_1\}$ is a solution to simultaneous DSP for $[T_0,T_d]$.

The merit of Theorem (3.4), which is basically a "problem transformation result" is that it makes the known results on DSSP or simultaneous DSP readily available in solving DRSP. In the rest of this section and in the next we will make use of the existing results on strong stabilization in order to obtain solutins in terms of plant parameters z_{ij} to some special DRSP's.

Let Z be a nonsingular plant with <u>stable diagonal entries</u>, i.e., with z_{11} and z_{22} in R_s. A diagonal stabilizing compensator for Z exists provided Z is free of decentralized fixed modes. Such a compensator might, however, destroy the stability of diagonal subplants while achieving overall closed loop stability. The question, we then ask, is this: <u>When can a plant stable at its diagonal be reliably stabilized via a diagonal compensator</u>. We show below that the answer to this question is affirmative only under serious constraints on z_{ij}'s.

Let us represent Z with each of its entries in coprime fractions as

$$(3.12)\quad Z = \begin{bmatrix} n_{11} & n_{12}/m_{12} \\ n_{21}/m_{21} & n_{22} \end{bmatrix},$$

where $m_{11}=m_{22}=1$ by the hyphothesis that Z is stable at its diagonal. Let $d_1=GCD\{n_{12},m_{21}\}$, $d_2=GCD\{n_{21},m_{12}\}$ so that $n_{12}=d_1l_{12}$, $m_{21}=d_1k_{21}$, $n_{21}=d_2l_{21}$, $m_{12}=d_2k_{12}$ for some l_{12}, k_{12}, l_{21}, and k_{12} satisfying $GCD\{l_{12},k_{21}\}=GCD\{l_{21},k_{12}\}=1$. Defining $l:=l_{12}l_{21}$ and $k:=k_{12}k_{21}$, we have

$$z_{12}z_{21}=n_{12}n_{21}/m_{12}m_{21}=l/k,$$

where $GCD\{l,k\}=1$. Further note that k is a denominator for detZ since

$$\det Z=(n_{11}n_{22}k-l)/k,$$

where $GCD\{n_{11}n_{22}k-l,k\}=GCD\{l,k\}=1$. By definition, the characteristic function of Z is

$$m:=LCM\{m_{12},m_{21},k\},$$

so that $m=m_{12}s_{12}=m_{21}s_{21}=ks$ for some s_{12}, s_{21}, and s satisfying $GCD\{s_{12},s_{21},s\}=1$. Further,

$$n_1=n_{11}m,\ n_4=n_{22}m,\ d=s(n_{11}n_{22}k-l).$$

By coprimeness of $\{l,k\}$, $GCD\{k,n_{11}k,n_{22}k,n_{11}n_{22}k-l\}=1$. Therefore,

$$s=GCD\{m,n_1,n_4,d\}.$$

We now show that "d_1d_2 is an associate of s". In fact, by $ks=m_{12}s_{12}=m_{21}s_{21}$, we have $d_2k_{12}s_{12}=d_1k_{21}s_{21}=sk_{12}k_{21}$. This implies that $d_2s_{12}=sk_{21}$ and $d_1s_{21}=sk_{12}$. By coprimeness of $\{d_2,k_{21}\}$ and $\{d_1,k_{12}\}$, it now follows that $s=d_1e_2=d_2e_1$, $s_{12}=k_{21}e_1$, and $s_{21}=k_{12}e_2$ for some e_1 and e_2. Note that $GCD\{s_{12},s_{21},s\}=1$ implies that $GCD\{e_1,e_2\}=1$. We thus have

$d_2e_1=d_1e_2$,

where $GCD\{d_1,d_2\}=GCD\{e_1,e_2\}=1$. Consequently, d_1 and e_1, and hence, s and d_1d_2 are associates as claimed above.

(3.13) THEOREM. DRSP is solvable for Z of (3.12) iff (i) Z is free of decentralized fixed modes and (ii) the transfer function $z_{12}z_{21}$ has parity interlacing property.

PROOF. Let $a_1=a_2=1$, $b_1=b_2=1$ in (3.1). Then, by Theorem (3.4), DRSP is solvable for Z of (3.12) iff there exist x, y, and a unit u_0 such that

(3.14) $m+(d-n_1n_{22})xy=u_0$.

Substituting $m=sk$, $d=s(n_{11}n_{22}k-l)$, $n_1=n_{11}sk$, we have

(3.15) $s(k-lxy)=u_0$.

It follows that, s and k-lxy are both units. Since the unstable decentralized fixed modes of Z are precisely the unstable zeros of s, condition (i) follows. On the other hand, the fact that k-lxy is a unit implies that {k,l} is coprime and that l/k has the parity interlacing property. Since $z_{12}z_{21}=l/k$, the second condition also follows. Conversely, by (i), s is a unit and hence $l/k=n_{12}n_{21}/m_{12}m_{21}$ are both coprime fractional representations of $z_{12}z_{21}$. Now, (ii) implies that, there exist x and a unit u such that $k-lx=u$. Setting $u_0:=us$, $y:=1$ we satisfy (3.15) and hence (3.14). Therefore, by Theorem(3.4), DRSP is solvable for Z of (3.12). In fact, the set of all solutions to DRSP in this particular case is given,by Remark(3.11), as $\{C=diag\{c_1,c_2\}$: $c_1=y/(1-n_{11}y)$, $c_2=x/(1-n_{22}x)$; where x, y are solutions to $k-lxy=u$ for some unit u$\}$. #

(3.16) REMARK. If Z of (3.12) is singular, then $n_{11}n_{22}k=1$, implying by GCD{k,1}=1, that k is a unit. This in turn implies that m_{12} and m_{21} are divisors of s since $ks=m_{12}s_{12}=m_{21}s_{21}$. Hence, $m=LCM\{m_{12},m_{21},k\}$ is a divisor of s. Consequently, all unstable zeros of the characteristic function are unstable fixed modes of Z. Therefore, in the singular case, diagonal stabilizability implies that DRSP is solvable.

4.ROBUST RELIABLE STABILIZATION OF A SCALAR PLANT

Consider the following problem of GHOSH [1984]: Given a scalar plant t = n/m in coprime fractional represantation, determine a pair of compensators $[c_1,c_1]$ such that (i) (t,c_1), (t,c_2), (t,c_1+c_2) are internally stable pairs and (ii) (t,k_1), (t,k_2), (t,k_1+k_2) are also internally stable for all k_1 and k_2 in some open neighborhood of c_1 and c_2, respectively, with respect to Rat n topology on rational functions(BROCKETT [1976]).

It has been shown in GHOSH [1984] that such a robust, reliable stabilization scheme is possible iff t has parity interlacing property. Moreover, the question of whether a given compensator c_1 admits a compensator c_2 such that $[c_1,c_2]$ is a solution to robust reliable stabilization problem is also answered in GHOSH [1984].

Here we formulate the problem as a DRSP of a two-by-two plant with identical entries t and employ the results of Section 3 to obtain a characterization of all robust reliable stabilizers of t. In achieving this, we use the following result of GHOSH [1984]which basically converts the problem to an algebraic one.

(4.1) LEMMA. *A pair of compensators* $c_1=x_2/x_1$, $c_2=y_2/y_1$ *is a robust reliable stabilizer of* $t=n/m$ *iff*

(i) $mx_1+nx_2=u$,

(ii) $my_1+ny_2=v$,

(iii) $mx_1y_1+n(x_1y_2+x_2y_1)=u_0$,

for some units u, v, u_0.

The following lemma is an immediate consequence of Definition (1.1) and Lemma (4.1).

(4.2) LEMMA. *The robust reliable stabilization problem is solvable for* $t=n/m$ *iff DRSP is solvable for*

$$Z = (1/m)\begin{bmatrix} n & n \\ n & n \end{bmatrix}.$$

Moreover, a pair $[c_1,c_2]$ *is a robust reliable stabilizer of* t *iff* $\mathrm{diag}\{c_1,c_2\}$ *is a solution to DRSP for* Z.

We can now state the main result of this section.

(4.3) THEOREM. *The robust reliable stabilization problem is solvable iff* $t=n/m$ *has parity interlacing property. If the latter holds, let a unit* u *and an element* b *in* R_s *be chosen to satisfy*

(4.4) $mu+nb=1$.

Then, the set of all solutions is given by either one of the sets S_1 and S_2:

$S_1=\{\ [(b+my)/(u-ny),(b+mx)/(u-nx)] :\ x$ and y are solutions to $u(1+nb)-n^2bx-n^2by-n^2mxy=v$ for some unit $v\ \}$

$S_2=\{\ [(b+mx)/(u-nx),y/(1+n^2xy-uny)] :\ x$ and y are solutions to $m+mn^2xy-n^2by=v$ for some unit $v\ \}$.

PROOF. The first statement is by GHOSH [1984]. Note that if $t=n/m$ with $\{n,m\}$ coprime has parity interlacing property, then there exist u and b satisfying (4.4) by YOULA et. al. [1974]. By Lemma (4.2), we only need to describe the set of all solutions to DRSP for Z to obtain a characterization of the set of all solutions to the problem. This can be done in two alternative ways by the results of Theorem (3.4 iv, v) and Remark (3.11). One only needs to note that choices $a_1=a_2=u$ and $b_1=b_2=b$ satisfy (3.1) and that (2.4) in this special case becomes

$$\begin{bmatrix} u & b \\ -n & m \end{bmatrix} \begin{bmatrix} m & n \\ n & 0 \end{bmatrix} \begin{bmatrix} 1 & -un \\ 0 & 1 \end{bmatrix} = \begin{bmatrix} 1 & 0 \\ 0 & -n^2 \end{bmatrix} .$$

#

(4.5) REMARK. Some elements in S_1 and S_2 are easily determined as follows. Let $y=0$ and consider the equation $u(1+nb)-n^2bx=v$. If z is any real unstable zero of nb, then $u(z)[1+n(z)b(z)]=u(z)$. Since u is a unit, the signs of $u(z)$ for all real z are the same. Further, $\{u(1+nb),nnb\}$ is coprime. Hence, by VIDYASAGAR and VISWANADHAM [1982b], there exist a unit v and an element x in R_s satisfying $u(1+nb)-n^2bx=v$. This yields a pair $[b/u,(b+mx)/(u-nx)]$ in S_1.

Similarly, let x=0 and consider $m-n^2by=0$. By (4.4), the signs of m(z) are the same for all real unstable zeros of nb. Since $\{m,n^2b\}$ is coprime by (4.4), it follows that a unit v and an element y satisfying $m-n^2by=v$ exist. This yields a pair $[b/u,y/(1-uny)]$ in S_2.

REFERENCES

Brockett RW (1976) Some geometric questions in the theory of linear systems. IEEE Trans. on Aut. Control AC-21:449-455

Corfmat JP, Morse AS (1976) Decentralized control of linear multivariable systems. Automatica 12:479-496

Desoer CA, Liu RW, Murray J, Saeks R (1980) Feedback system design: The fractional represantation approach to analysis and synthesis. IEEE Trans. on Aut. Control AC-25:399-412

Freudenberg JS, Looze DP (1986) An analysis of H - optimization design methods. IEEE Trans. on Aut. Control AC-31:1-10

Ghosh BK (1984) A robust reliable stabilization scheme for single input single output systems using transcendental methods. Systems and Control Letters 5:111-115

Guclu AN, Ozguler AB (1986) Diagonal Stabilization of linear multivariable systems. Int. J. Control 43:965-980

Hiraoglu M. (1986) Decentralized stabilization with controller constraints. M.Sc. thesis submitted to Bosphorous University, Istanbul, Turkey

McDuffee CC (1956) The Theory of Matrices. Chelsea Newyork

Morse AS "System invariants under feedback and cascade control" in Lecture Notes in Economics and Mathematical Systems. Springer New York

Ozguler AB (1986) A characterization of the set of all diagonal compensators for a two-by-two plant, preprint. Bilkent University, P.O.B.8, Maltepe, 06572, Ankara, Turkey

Vidyasagar M, Viswanadham N(1982a)Algebraic characterization of decentralized fixed modes and pole assignment. University of Waterloo Report 82:06

Vidyasagar M, Viswanadham N (1982b) Algebraic design techniques for reliable stabilization. IEEE Trans. on Aut. Control AC-27:1085-1095

Wang SH, Davison EJ (1973) On the stabilization of multivariable decentralized control systems. IEEE Trans. on Aut. Control

Youla DC, Bongiorno JJ, Lu CN (1974) Sinlgle loop feedback stabilization of linear multivariable systems. Automatica 12:159-173

Sensitivity Minimization as a Nevanlinna–Pick Interpolation Problem

Joseph A. Ball and D. William Luse
Department of Mathematics and Department of Electrical Engineering
Virginia Polytechnic Institute and State University
Blacksburg, VA 24061
USA

Using known results on matrix Nevanlinna–Pick interpolation combined with a spectral factorization algorithm of Youla, we present an algorithm for obtaining all stable closed loop sensitivities having H^∞–norm within a prescribed (suboptimal) tolerance for a given open loop plant P(s). Other desirable characteristics for the associated stabilizing compensator (such as suitable roll-off at ∞) can then be sought within this class.

1. INTRODUCTION

We consider the closed loop system Σ given by Figure 1, where the plant P and the compensator C are given by their transfer functions P(s) and C(s) which are rational

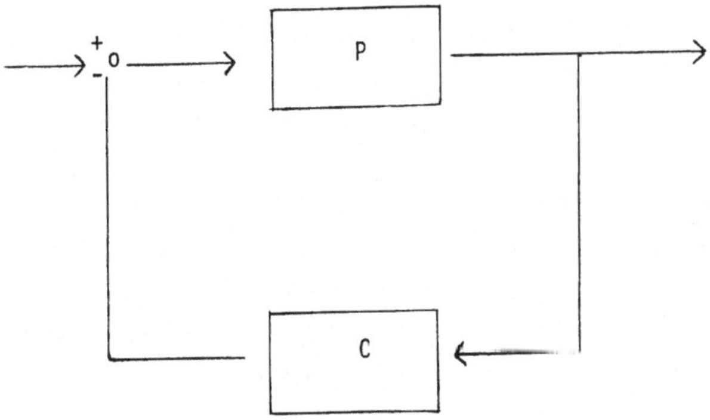

Figure 1

NATO ASI Series, Vol. F34
Modelling, Robustness and Sensitivity Reduction
in Control Systems. Edited by R. F. Curtain

matrix functions (of sizes $m \times n$ and $n \times m$) in the transform variable s. The closed-loop system is said to be (internally) stable if the four matrix functions $(I+PC)^{-1}$, $C(I+PC)^{-1}$, $(I+PC)^{-1}P$ and $(I+CP)^{-1}$ are all in the class $\mathcal{S}$ of stabie functions (i.e. proper rational matrix functions with no poles in the closed right half plane). The sensitivity of the system is taken to be $S=(I+PC)^{-1}$ while the weighted sensitivity is $X = W_1 SW_2$, where the weight functions W_1 and W_2 are assumed to be minimum phase (i.e. analytic and invertible in the closed right half plane). The sensitivity minimization problem, first introduced by Zames [1] (see also [2,3,4,5]), is to find a compensator C which minimizes the H^{∞}-norm of the weighted sentivity while maintaining stability of the closed loop system. As has often been pointed out [3,6,7,8], the mathematical problem as formulated above does not capture all engineering design constraints in practice. To get more flexibility which can then be used at later stages to meet other design criteria, we propose choosing a tolerance level σ slightly larger than $\sigma_0 = \min.\{\|X\|: \text{the system } \Sigma \text{ is stable}\}$, and then parametrizing the larger class of sensitivities $\{X: \|X\| \leq \sigma\}$ (i.e. all "suboptimal" solutions). One can then solve for C in terms of X, and pick out those C's which are strictly proper (e.g.) while maintaining simultaneously the bound σ on $\|X\|$.

When one uses the Youla-Jabr-Bongiorno (see [9,10]) parametrization of all the stabilizing compensators, one can view the associated sensitivities as the class of all stable matrix functions satisfying some interpolation conditions (see [11] for the scalar case). The sensitivity minimization problem then becomes a matrix Nevanlinna-Pick-Caratheodory-Fejer interpolation problem, for which the mathematical theory is well developed (see [12,13,14,15]). From the general theory (see eg. [14]) it is known that there is a linear fractional parametrization of all suboptimal solutions. Explicit state space formulas for the linear fractional map were obtained in [16] (see [17] for the scalar case), but there the input interpolation data are assumed to be in the very detailed form of zero chains (see [18]). Here, we reduce the problem to a matrix Wiener-Hopf factorization (as was done in [19,20] in the context of the Nehari problem), and then use a factorization algorithm of Youla [21] to solve directly in terms of matrix polynomial coefficients. This work can be seen as complementary to that of [22,23] where algorithms are given for finding the optimal (or even "superoptimal") solution of a Nevanlinna-Pick problem; there also one has a choice as to how the input data can be expressed.

Other authors use inner-outer factorization to convert the Nevanlinna-Pick interpolation problem to a matrix Nehari problem [24]. If one assumes that the initial plant is given in terms of a state space realization, then each step in the above procedure (coprime factorization and Bezout identities for the YJB parametrization, inner-outer factorization, solution of Nehari problem) can be done with state space algorithms (see

[19,20,25,26,27] and [4,5] for surveys). There also has been work on a more general version of the problem more complicated than that discussed here [4,5,28].

2. REDUCTION TO NEVANLINNA-PICK INTERPOLATION

We consider the closed loop system Σ as in Figure 1. Let $P = N_pD_p^{-1} = \tilde{D}_p^{-1}\tilde{N}_p$ be right and left coprime factorizations of P; here N_p, D_p, $\tilde{D}_p$, $\tilde{N}_p$ are stable rational matrix functions of sizes m×n, n×n, m×m and m×n respectively. It is known (see [25,5]) that there exist stable rational matrix functions U_p, V_p, $\tilde{U}_p$ and $\tilde{V}_p$ of the appropriate sizes so that the Bezout identities

$$\begin{bmatrix} \tilde{D}_p & -\tilde{N}_p \\ U_p & V_p \end{bmatrix} \begin{bmatrix} \tilde{V}_p & N_p \\ -\tilde{U}_p & D_p \end{bmatrix} = \begin{bmatrix} I & 0 \\ 0 & I \end{bmatrix}$$

is satisfied. Then the stabilizing compensators C for Σ are known [4,5,9,10] to be given by

$$C = (-D_pQ + \tilde{U}_p)(N_pQ + \tilde{V}_p)^{-1}$$

where Q is an arbitrary function in the class $\mathcal{S}_{nxm}$ of stable nxm matrix functions. One can then compute (see [4,5])

$$S = \tilde{V}_p\tilde{D}_p + N_pQ\tilde{D}_p \tag{2.1}$$

and the weighted sensitivity X has the form

$$X = W_1\tilde{V}_p\tilde{D}_pW_2 + W_1N_pQ\tilde{D}_pW_2 \tag{2.2}$$

Since Q is an arbitrary function of class $\mathcal{S}_{nxm}$, formulas (1.1) and (1.2) exhibit the set of all possible (weighted) sensitivities associated with stabilizing compensators as the set of all solutions of a matricial interpolation problem. For the square plant case, by writing out in detail the RHP zeros and associated left zero chains for W_1N_p and the RHP zeros and right zero chains for $\tilde{D}_pW_2$, one can convert the form (2.2) for X to an equivalent set of matrix interpolation conditions of Nevanlinna-Pick-Caratheodory-Fejer type as studied in [14,15]. In terms of all these data one can also compute explicitly the linear fractional parametrizer of all interpolants X(s) satisfying $\|X\|_\infty \leq \sigma$ for some tolerance σ (see [16,17]). However, the zeros and zero chains may be difficult to compute; we discuss instead an alternative factorization route which works more directly with coefficient data of D_p, N_p and $\tilde{V}_p$.

3. THE SUBOPTIMAL SENSITIVITY MINIMIZATION PROBLEM

We now consider the problem of characterizing those sensitivities S for stable closed loop systems which satisfy an additional constraint of the form

(3.1) $\|S\|_\infty \le \sigma$

for some prescribed tolerance σ. Our analysis handles weighted sensitivities equally well; for notational simplicity we discuss only the unweighted version. From (2.2) the stable sensitivities are characterized as those functions S of the form

$$S = \tilde{V}_p \tilde{D}_p + N_p Q \tilde{D}_p$$

where $Q \in \mathcal{S}_{n\times m}$ is arbitrary. In practice it is easy to get one stabilizing compensator C_1 (eg. by the root-locus technique, or through LQ theory) and thus one stable sensitivity $\mathcal{S}_1 = (I+PC_1)^{-1}$. The form of the general stable sensitivity can then be taken to be

(3.2) $S = S_1 + N_p Q \tilde{D}_p,\ Q \in \mathcal{S}_{n\times m}$.

To impose the additional constraint (3.1) we follow the approach from [14]. Let H^2_k denote the Hardy space of $\mathbb{C}^k$-valued functions over the right half plane (RHP) (see [13]); we often consider H^2_k as a subspace of L^2_k(square integrable $\mathbb{C}^k$-valued functions on the jw-axis). As is customary in the subject [4,5], we assume that $\tilde{D}_p$ and N_p have no zeros on the jw-axis, i.e. that the plant has no jw-axis poles or zeros (including ∞); this assumption can be removed at a later stage. Then results from [14] can be summarized as follows.

THEOREM 3.1. There exists a matrix function S of the form (3.2) with $\|S\|_\infty \le \sigma$ if and only if $\|\Gamma\| \le \sigma$, where Γ: $\tilde{D}_p^{-1} H^2_n \to L^2_m \ominus N_p\, H^2_n$ is the operator defined by

$$\Gamma(f) = P(S_1 f)$$

for $f \in \tilde{D}_p^{-1} H^2_n$, where P is the orthogonal projection of L^2_m onto the orthogonal complement of $N_p H^2_n$ in L^2_m. Moreover, for $\sigma > \|\Gamma\|$ there exists a rational matrix function $\theta(s) = \begin{bmatrix} \theta_{11}(s) & \theta_{12}(s) \\ \theta_{21}(s) & \theta_{22}(s) \end{bmatrix}$ such that the stable sensitivities S(s) with $\|S\|_\infty \le \sigma$ are precisely those matrix functions of the form

(3.3) $S(s) = [\theta_{11}(s)G(s) + \theta_{12}(s)][\theta_{21}(s)G(s)+\theta_{22}(s)]^{-1}$, for some $G \in \mathcal{S}_{p\times m}$ (where $p=\min\{n,m\}$) with $\|G\|_\infty \le 1$. The function $\theta(s)$ can be taken to be any matrix function such that

(i) $\theta H^2_{p+m} = LH^2_{p+m}$

where

$$L(s) = \begin{bmatrix} S_1(s)\tilde{D}_p^{-1}(s) & N_p(s) \\ \tilde{D}_p^{-1}(s) & 0 \end{bmatrix}$$

and

(ii) $\theta(-\bar{s})^* J_\sigma \theta(s) = J$

where

$$J_\sigma = \begin{bmatrix} I_m & 0 \\ 0 & -\sigma^2 I_m \end{bmatrix}, \quad J = \begin{bmatrix} I_p & 0 \\ 0 & -I_m \end{bmatrix}$$

and $\theta_{11}, \theta_{12}, \theta_{21}, \theta_{22}$ have sizes m×p, m×m, m×p and m×m respectively.

The next lemma indicates how to reduce the computation of θ to a signed Wiener-Hopf factorization problem.

LEMMA 3.2. The matrix function $\theta(s)$ satisfies conditions (i) and (ii) of Theorem 3.1 if and only if

(3.4) $\theta = LX^{-1}$

where

(i) $L(-\bar{s})^* J_\sigma L(s) = X(-\bar{s})^* JX(s)$

and

(ii) both X and X^{-1} are in $\mathcal{S}_{(p+m)x(p+m)}$.

PROOF. Simply check that (i) and (ii) in Lemma 3.2 are equivalent to (i) and (ii) in Theorem 3.1 when X and θ are related as in (3.4). The same observation was a key point in [19,20].

To solve the factorization problem for X in Lemma 3.2, we adapt an algorithm of Youla [21] to the indefinite case. Then we can compute $\theta = LX^{-1}$.

As an application of the general linear fractional formula (3.3) for all stable sensitivities with norm $\le \sigma$, we can solve for a <u>strictly proper</u> stabilizing compensator C_0 while maintaining that the norm $\|S_0\|_\infty$ of the corresponding sensitivity $S_0 = (I+PC_0)^{-1}$ be less than σ. To do this, assuming that the plant P has no poles or zeros at ∞, note that strict properness of C_0 is equivalent to $S_0(\infty) = I$. One can back-solve (3.3) to determine the value of the corresponding $G_0(\infty)$ compatible with $S_0(\infty) = I$. When $\|G_0(\infty)\| \le 1$, choose any $G_o \in \mathcal{S}_{m\times n}$ with the desired value $G_0(\infty)$ at ∞ and with $\|G_0\|_\infty \le 1$. Thus the linear-fractional formula (3.3) gives S_0, from which one can back-solve $S_0=(I+PC_0)^{-1}$ for C_0. We illustrate with a couple of examples.

EXAMPLES 1. ("tall plant" case). We let $P(s) = \begin{bmatrix} \frac{s-2}{s-1} \\ \frac{s+3}{s-1} \end{bmatrix}$. Then P has right coprime factorization $P=ND^{-1}$ with

$$N(s) = \begin{bmatrix} \frac{s-2}{s+1} \\ \frac{s+3}{s+1} \end{bmatrix}, \quad D(s) = \left[\frac{s-1}{s+2}\right]$$

and left coprime factorization $P=\tilde{D}^{-1}\tilde{N}$ with

$$\tilde{D}(s) = \begin{bmatrix} \frac{s-1}{s+1} & 0 \\ 4 & 1 \end{bmatrix}, \quad \tilde{N}(s) = \begin{bmatrix} \frac{s-2}{s+1} \\ 5 \end{bmatrix}$$

One stabilizing compensator is $C_1 = [0 \quad 1]$ with corresponding sensitivity

$$\begin{aligned} S_1 &= [I + PC_1]^{-1} \\ &= \left\{ I + \begin{bmatrix} 0 & \frac{s-2}{s-1} \\ 0 & \frac{s+3}{s-1} \end{bmatrix} \right\}^{-1} \\ &= \begin{bmatrix} 1 & \frac{s-2}{s-1} \\ 0 & \frac{2s+2}{s-1} \end{bmatrix}^{-1} \\ &= \begin{bmatrix} \frac{2s+2}{s-1} & -\frac{s-2}{s-1} \\ 0 & 1 \end{bmatrix} \cdot \frac{s-1}{2s+2} \\ &= \begin{bmatrix} 1 & -\frac{s-2}{2s+2} \\ 0 & \frac{s-1}{2s+2} \end{bmatrix}. \end{aligned}$$

One then can compute

$$\begin{aligned} S_1\tilde{D}^{-1} &= \begin{bmatrix} 1 & -\frac{s-2}{2(s+1)} \\ 0 & \frac{s-1}{2(s+1)} \end{bmatrix} \begin{bmatrix} 1 & 0 \\ -4 & \frac{s-1}{s+1} \end{bmatrix} \frac{s+1}{s-1} \\ &= \begin{bmatrix} 3 & -\frac{s-2}{2(s+1)} \\ -2 & \frac{s-1}{2(s+1)} \end{bmatrix} \end{aligned}$$

and thus

$$L = \begin{bmatrix} S_1\tilde{D}^{-1} & N \\ \tilde{D}^{-1} & 0 \end{bmatrix}$$

$$= \begin{bmatrix} 3 & -\frac{s-2}{2(s+1)} & \frac{s-2}{s+1} \\ -2 & \frac{s-1}{2(s+1)} & \frac{s+3}{s+1} \\ \frac{s+1}{s-1} & 0 & 0 \\ \frac{-4(s+1)}{s-1} & 1 & 0 \end{bmatrix}$$

$$= \frac{1}{s-1} \begin{bmatrix} 3(s-1) & -\frac{(s-1)(s-2)}{2} & (s-1)(s-2) \\ -2(s-1) & \frac{(s-1)^2}{2} & (s-1)(s+3) \\ s+1 & 0 & 0 \\ -4(s+1) & (s+1)(s-1) & 0 \end{bmatrix}.$$

$$\cdot \operatorname{diag.} \left(1, \frac{1}{s+1}, \frac{1}{s+1}\right).$$

To apply the Youla algorithm [21], it is convenient to work with matrix polynomials; we therefore consider the middle factor L_0 in the above representation for L

$$L_0 = \begin{bmatrix} -3+3s & -1+1.5s-.5s^2 & 2-3s+s^2 \\ 2-2s & .5-s+.5s^2 & -3+2s+s^2 \\ 1+s & 0 & 0 \\ -4-4s & -1+0s+s^2 & 0 \end{bmatrix}$$

and use the Youla algorithm to find a matrix polynomial Y having no zeros in RHP such that

$$L_0^* J_\sigma L_0 = Y^* J Y.$$

We then set $\theta = L_0 Y^{-1}$; one can easily check that this agrees with $\theta = LX^{-1}$ as required by Lemma 3.2; a priori one may have to adjust Y with a unimodular polynomial factor but this turns out not to be necessary in this example. The result for $\sigma^2 = 2$ is

$$\theta(s) = \frac{1}{d(s)} [N_{ij}(s)]_{\substack{1 \le i \le 4 \\ 1 \le j \le 3}}$$

where

$N_{11} = (-1.34)(s-1)(s-3)$

$N_{12} = (1.17)(s-1)(s+4)$

$N_{13} = (.655)(s-1)(s+2.55)$

$N_{21} = -(.447)(s-1)(s+4)$

$N_{22} = -(.975)(s-1)(s+4)$

$N_{23} = -(.436)(s-1)(s+2.55)$
$N_{31} = -(.67)(s-.606)(s-4.62)$
$N_{32} = (1.07)(s-.763)(s+3.607)$
$N_{33} = (.218)(s-1)(s+2.55)$
$N_{41} = (.224)(s+1.85)(s-7)$
$N_{42} = -(.195)(s+2.01)(s+6.76)$
$N_{43} = -(.873)(s+1)(s+2.55)$
and

$$d(s) = (s+1)(s+2.55).$$

To find a strictly proper stabilizing compensator C_0 with $\|S_0\|_\infty \le \sigma = \sqrt{2}$, we seek a parametrizer $G_0 \in S_{1x2}$ so that the corresponding S_0 given by (3.3) satisfies $S_0(\infty) =$ I. Thus we must solve

$$[\Theta_{11}(\infty)-\Theta_{21}(\infty)]G_0(\infty) = -\Theta_{12}(\infty)+\Theta_{22}(\infty)$$

where

$$\Theta_{11}(\infty)-\Theta_{21}(\infty) = \begin{bmatrix} -.671 \\ -.671 \end{bmatrix}$$

and

$$-\Theta_{22}(\infty)+\Theta_{22}(\infty) = \begin{bmatrix} -.0976 & -.436 \\ -.0976 & -.436 \end{bmatrix}$$

Despite the appearance of being overdetermined we can solve for $G_0(\infty) = \begin{bmatrix} G_1 \\ G_2 \end{bmatrix}$ with $G_1 = .0976/.671$ and $G_2 = .436/.671$. Note also that $\|G_0(\infty)\| < 1$. For convenience we choose $G_0(s)$ to be the constant $\begin{bmatrix} G_1 \\ G_2 \end{bmatrix}$. Then from (3.3) we get the corresponding sensitivity $S_0(s)$ to be

$S_0 = [S_{0,ij}]_{1 \le i,j \le 2}$ where

$$S_{0,11} = \frac{(s+7.06)(s+.601)}{(s+5.86)(s+1)}$$

$$S_{0,12} = \frac{2.40(s-2)}{(s+5.86)(s+1)}$$

$$S_{0,21} = \frac{.807(s+3)}{(s+5.86)(s+1)}$$

$$S_{0,22} = \frac{(s+4.48)(s-.853)}{(s+5.86)(s+1)}$$

We compute next $PC_0 = S_0^{-1} - I$ to get

$$PC_0 = \frac{1}{(s-1)(s+5.44)} \begin{bmatrix} -.807(s-2) & 3.23(s-2) \\ -.807(s+3) & 3.23(s+3) \end{bmatrix}$$

Let $C_0(s) = [c_1(s) \quad c_2(s)]$; we must solve for c_1 and c_2:

$$PC_0 = \begin{bmatrix} \frac{s-2}{s-1} \\ \frac{s+3}{s-1} \end{bmatrix} [c_1 \quad c_2].$$

This yields

$$c_1 = \frac{-.807}{s+5.44}$$

$$c_2 = \frac{3.27}{s+5.44}$$

and $C_0 = [c_1, c_2]$ is a strictly proper stabilizing compensator as desired.

EXAMPLE 2 ("fat plant" case).

We take $P(s) = \begin{bmatrix} \frac{s-2}{s-1} & \frac{s+3}{s-1} \end{bmatrix}$.

Then P has right and left coprime factorizations $P = ND^{-1} = \tilde{D}^{-1}\tilde{N}$ where

$$N(s) = \begin{bmatrix} \frac{s-2}{s+1} & 5 \end{bmatrix}$$

$$D(s) = \begin{bmatrix} \frac{s-1}{s+1} & \frac{4}{s+1} \\ 0 & \frac{1}{s+1} \end{bmatrix}$$

$$\tilde{D}(s) = \frac{s-1}{s+1}$$

$$\tilde{N}(s) = \begin{bmatrix} \frac{s-2}{s+1} & \frac{s+3}{s+1} \end{bmatrix}.$$

One stabilizing compensator is $C_1 = \begin{bmatrix} 0 \\ 1 \end{bmatrix}$ with associated sensitivity

$$S_1 = \frac{1}{1+PC_1} = \frac{s-1}{2(s+1)}.$$

We compute

$$L = \begin{bmatrix} S_1\tilde{D}^{-1} & N \\ \tilde{D}^{-1} & 0 \end{bmatrix}$$

$$= \begin{bmatrix} \frac{s-1}{2(s+1)}\,\frac{s+1}{s-1} & \frac{s-2}{s+1} & \frac{5}{s+1} \\ \frac{s+1}{s-1} & 0 & 0 \end{bmatrix}$$

$$= \frac{1}{s-1} \begin{bmatrix} \frac{s-1}{2} & (s-2)(s-1) & 5(s-1) \\ s+1 & 0 & 0 \end{bmatrix} \cdot$$

$$\operatorname{diag}(1, \frac{1}{s+1}, \frac{1}{s+1}).$$

We take L_0 to the the middle matrix polynomial factor

$$L_0(s) = \begin{bmatrix} \frac{s-1}{2} & (s-2)(s-1) & 5(s-1) \\ s+1 & 0 & 0 \end{bmatrix}$$

and use the Youla algorithm [20] to find a matrix polynomial Y(s) with no zeros in RHP such that

$$L_0^* J_\sigma L_0 = Y^* J Y$$

(with $\sigma^2 = 1.3$) and then set $\Theta = L_0 Y^{-1}$; again one can check that this agrees with $\Theta = LX^{-1}$ required by Lemma 3.2. The result is

$$\Theta(s) = \frac{1}{s+1} \begin{bmatrix} 1.11(s-1) & .488(s-1) \\ .423(s+1) & .976(s+1) \end{bmatrix}$$

We then seek $g(\infty)$ such that

$$[\Theta_{11}(\infty) - \Theta_{21}(\infty)]g(\infty) = -\Theta_{12}(\infty) + \Theta_{22}(\infty).$$

From the above form of Θ we see that

$$\Theta_{11}(\infty) - \Theta_{21}(\infty) = .685$$

$$-\Theta_{12}(\infty) + \Theta_{22}(\infty) = .488$$

and thus we can take $g(s) = .488/.685$. The corresponding sensitivity $S_0(s)$ from (3.3) (with G=g) works out to be

$$S_0(s) = (-1.28+1.28s)(1.28s+1.28s)^{-1}$$

$$= \frac{s-1}{s+1}.$$

Thus

$$1+PC_0 = S_0^{-1} = \frac{s+1}{s-1} = 1 + \frac{2}{s-1}$$

so $PC_0 = \frac{2}{s-1}$ and

$$\begin{bmatrix} \frac{s-2}{s-1} & \frac{s+3}{s-1} \end{bmatrix} \begin{bmatrix} c_1 \\ c_2 \end{bmatrix} = \frac{2}{s-1}$$

$$c_1(s-2) + c_2(s+3) = 2.$$

This last equation has many solutions; in particular it is possible to solve with improper C_0's. One choice of strictly proper $C_0 = [c_1, c_2]^t$ is

$$c_1 = \frac{2}{5} \frac{1}{s+2}$$

$$c_2 = \frac{8}{5} \quad \frac{1}{s+2}.$$

REFERENCES

1. G. Zames, Feedback and optimal sensitivity: model reference transformations, multiplicative seminorms and approximate inverses, IEEE Trans. Auto. Control AC-26 (1981), 301-320.

2. G. Zames and B.A. Francis, Feedback, minimax sensitivity and optimal robustness, IEEE Trans. Auto. Control AC-28 (1983), 585-601.

3. B.A. Francis, J.W. Helton and G. Zames, H^∞-optimal feedback controllers for linear multivariable systems. IEEE Trans. Auto. Control AC-29 (1984), 888-900.

4. B.A. Francis and J.C. Doyle, Linear control theory with an H^∞-optimality criterion, SIAM J. Control and Opt., to appear.

5. M. Vidyasagar, Control System Synthesis: A Factorization Approach, MIT Press (Cambridge, MA) 1985.

6. J.W. Helton, Worst case analysis analysis in the frequency domain: an H^∞-approach to control, IEEE Trans. Auto. Control, AC-30(1985), 1154-1170.

7. Y.K. Foo and I. Postlethwaite, An H^∞-minimax approach to the design of robust control systems, Systems and Control Letters 5 (1984), 81-88.

8. H. Kwakernaak, A polynomial approach to minimax frequency domain optimization of multivariable feedback systems, Int. J. Control (1986), 117-159.

9. D.C. Youla, H. Jabr and J. Bongiorno, Modern Wiener-Hopf design of optimal controllers, I and II, IEEE Trans. Auto. Control AC-21 (1977), 3-13 and 319-338.

10. C. Desoer, R.W. Liu, J. Murray and R. Saeks, Feedback system design: the fractional representation approach to analysis and synthesis, IEEE Trans. Automatic Control AC-25 (1980), 399-412.

11. D.C. Youla, J. Bongiorno and Y. Lu, Single loop feedback stabilization of linear multivariable dynamic plants, Automatica 10 (1974), 150 173.

12. D. Sarason, Generalized interpolation in H^∞. Trans. AMS 127 (1967), 179-203.

13. M. Rosenblum and J. Rovnyak, Hardy Classes and Operator Theory, Oxford Univ. Press (1985), Oxford.

14. J.A. Ball and J.W. Helton, A Beurling-Lax theorem for the Lie group U(m,n) which contains must classical interpolation theory, J. Operator Theory 9 (1983), 107-142.

15. J.A. Ball and J.W. Helton, Beurling-Lax representations using classical Lie groups III: groups preserving two bilinear forms, Amer. J. Math. 108 (1986), 95-174.

16. J. A. Ball and A.C.M. Ran, Local inverse spectral problems for rational matrix functions, Integral Equations and Operator Theory, to appear.

17. L.B. Golinsky, On a generalization of the matrix Nevanlinna-Pick problem, Proc. of the Armenian Academy of Science 18 (1983), 108-205 [in Russian].

18. H. Bart, I. Gohberg and M.A. Kaashoek, Minimal Factorization of Matrix and Operator Functions, OT 1 Birkhauser (Basel), 1979.

19. J.A. Ball and A.C.M. Ran, Hankel norm approximation of a rational matrix function in terms of its realization, in Modelling, Identification and Robust Control (ed. C.I. Byrnes and A. Lindquist) North-Holland (Amsterdam) (1986), pp. 285-296.

20. J.A. Ball and A.C.M. Ran, Optimal Hankel norm model reductions and Wiener-Hopf factorization I: The canonical case, SIAM J. Control and Opt., to appear.

21. D.C. Youla, On the factorization of rational matrices, IRE Trans. on Information Theory, July (1961), 172-189.

22. A.C. Allison and N.J. Young, Numerical algorithms for the Nevanlinna-Pick problem, Numerische Mathematik 42 (1983), 125-145.

23. N.J. Young, The Nevanlinna-Pick problem for matrix valued functions, J. Operator Theory 15 (1986), 239-265.

24. V.M. Adamjan, D.Z. Arov and M.G. Krein, Infinite Hankel block matrices and related extension problems, Transl. Amer. Math. Soc. (2), 111 (1978), 133-156.

25. A.N. Nett, C.A. Jacobson and M.J. Balas, Connection between state space and doubly coprime fractional representation, IEEE Trans. Ant. Control AC-29 (1984), 831-832.

26. J.A. Ball and A.C.M. Ran, Global inverse spectral problems for rational matrix functions, Linear Alg. and Appl., to appear.

27. K. Glover, All optimal Hankel-norm approximations of linear multivariable systems and their L^∞-error bounds, Int. J. Control 39 (1984), 1115-1193.

28. J.A. Ball and N. Cohen, Sensitivity minimization in H^∞-norm: parametrizatoin of all suboptimal solutions, Int. J. Control, to appear.

THE SPECTRAL FACTORIZATION PROBLEM FOR SISO DISTRIBUTED SYSTEMS

F.M. Callier and J. Winkin

Department of Mathematics
Facultés Universitaires N.D. de la Paix
8, Rempart de la Vierge
B-5000 Namur BELGIUM

ABSTRACT. We study spectral factorization for singlevariable linear distributed systems, viz. those modelled by transfer functions (TF's) in the algebra $\hat{B}(\sigma_0)$, [11], [26], where a) stability is guaranteed with a margin $|\sigma_0|$, and b) delay is allowed in the direct input-output transmission. This enables us to generalize the graph metric, [16], to such systems. The results are particularized to TF's of exponential order, viz. those in the subalgebra $\hat{F}(\sigma_0)$, [14]. The theory is illustrated by an application to robust feedback control.

1. NOTATIONS AND ABBREVIATIONS

1.1. The following notations are used throughout : $\mathbb{R}$, ($\mathbb{R}_-$, $\mathbb{R}_+$ resp.), denotes the set of real, (nonpositive real, nonnegative real resp.), numbers; $\mathbb{C}$ denotes the field of complex numbers. For any σ, σ' $\in \mathbb{R}$, $\mathbb{C}_{\sigma+} := \{s \in \mathbb{C} : \text{Re } s \geq \sigma\}$, $\mathring{\mathbb{C}}_{\sigma+} := \{s \in \mathbb{C} : \text{Re } s > 0\}$, $\mathbb{C}_{\sigma-} := \mathbb{C}\backslash\mathring{\mathbb{C}}_{\sigma+}$ and $\mathring{\mathbb{C}}_{\sigma-} := \mathbb{C}\backslash\mathbb{C}_{\sigma+}$; the vertical strip $\{s \in \mathbb{C} : \text{Re } s \in [\sigma,\sigma']\}$, ($\{s \in \mathbb{C} : \text{Re } s \in (\sigma,\sigma')\}$ resp.), is denoted by $S(\sigma,\sigma')$, ($\mathring{S}(\sigma,\sigma')$ resp.); $S_\sigma := S(\sigma,-\sigma)$ and $\mathring{S}_\sigma := \mathring{S}(\sigma,-\sigma)$. For any $s \in \mathbb{C}$, $\bar{s}$ denotes the complex conjugate of s. LTD, (LTD^-, LTD^+ resp.), denotes the set of $\mathbb{C}$-valued Laplace-transformable distributions with support on $\mathbb{R}$, ($\mathbb{R}_-$, $\mathbb{R}_+$ resp.); if $f,g \in \text{LTD}$, $(f*g)(\cdot) := \int_{-\infty}^{+\infty} f(\cdot-u).g(u).du$ is the convolution product of f and g. The (two-sided) Laplace transform as well as corresponding sets of Laplace transforms are denoted by a hat; i.e., for any $f \in \text{LTD}$, $\hat{f}(s) := \int_{-\infty}^{+\infty} f(t).\exp(-st).dt$, (sometimes we also use $L[f] := \hat{f}$), and e.g. $\widehat{\text{LTD}} := \{\hat{f} : f \in \text{LTD}\}$. $\dotplus$ denotes the direct sum of linear subspaces, (sometimes it will be used to denote a disjoint sum of sets).

1.2. SISO, MIMO, s.t., a.e., w.l.g., u.th.c., TF, exp. are abbreviations for resp. "single-input single-output", "multiple-input multiple-output", "such that", "almost everywhere", "without loss of generality", "under these conditions", "transfer function", "exponential(ly)".

NATO ASI Series, Vol. F34
Modelling, Robustness and Sensitivity Reduction in Control Systems. Edited by R. F. Curtain

2. INTRODUCTION

The main objective of this paper is to solve the spectral factorization problem for SISO linear distributed systems, namely those modelled by TF's in the algebra $\hat{B}(\sigma_0)$, [11], [26], where (i) stability is guaranteed with a margin $|\sigma_0|$, and (ii) delay is allowed in the direct input-output transmission, (Section 3). Such a problem plays a fundamental role in Wiener-Hopf type integral equation theory, e.g. [5], and in LQ-optimal control theory, e.g. [33]. Here it will be used to extend to distributed systems an important tool for the study of feedback stability robustness, viz. the graph metric, [16]-[17], (Section 4). The results will be particularized to TF's of exponential order, i.e. those in the subalgebra $\hat{\mathbf{F}}(\sigma_0)$, [14], (Section 5). The theory will be illustrated by an application to robust feedback stabilization of a heat flow in a finite rod, (Section 6).

To set the stage, we shortly revisit an example from mathematical physics, [5, pp. 263-264]. Let the (two-sided) autocorrelation function

$$f(t) = \delta(t) - (4\pi\, \mathrm{ch}(t/2))^{-1} = f(-t) \quad , \quad t \in \mathbb{R} , \tag{2.1}$$

where $\delta(\cdot)$ denotes the Dirac delta distribution. The corresponding power spectral density is

$$\hat{f}(j\omega) = 1 - (2 \,.\, \cos(\pi j\omega))^{-1} = \hat{f}(-j\omega) \; , \quad \omega \in \mathbb{R} . \tag{2.2}$$

It follows that, with $\alpha := 3^{-1}$ and $\beta := 2^{-1}$, $\hat{f}(s) = \sin(2^{-1}\pi(\alpha+s))\,.$ $\sin(2^{-1}\pi(\alpha-s))\,.\,(\sin(2^{-1}\pi(\beta+s))\,.\,\sin(2^{-1}\pi(\beta-s)))^{-1}$. Hence,

$$\hat{f}(s) = \frac{\alpha^2-s^2}{\beta^2-s^2} \cdot \prod_{n=1}^{\infty} \frac{(1-(\frac{\alpha+s}{2n})^2)\,.\,(1-(\frac{\alpha-s}{2n})^2)}{(1-(\frac{\beta+s}{2n})^2)\,.\,(1-(\frac{\beta-s}{2n})^2)} = \hat{f}(-s) \; , \tag{2.3}$$

s.t. $\hat{f}$ has its zeros at $2n\pm\alpha$ and its poles at $2n\pm\beta$, $n=0,\pm1,\ldots$, [12, p. 356].

By symmetric extraction of elementary factors

$$\hat{f}_n(s) := \frac{(2n+\alpha+s)\,.\,(2(n+1)-\alpha+s)}{(2n+\beta+s)\,.\,(2(n+1)-\beta+s)} \tag{2.4a}$$

from $\hat{f}$, (2.3) , we get an outer spectral factor $\hat{r}$ of $\hat{f}$, viz.

$$\hat{r}(s) := \prod_{n=0}^{\infty} \hat{f}_n(s) \, . \tag{2.4b}$$

$\hat{r}$ is the TF of a stable and minimum phase system whose output power spectral density is $\hat{f}(s) = \hat{r}(-s)\,.\,\hat{r}(s)$, [2]. The corresponding exp.

stable impulse response is, by residue calculus, e.g. [12],

$$r(t) = \delta(t) + \sum_{n=0}^{\infty} a_n \cdot \exp(-(2^{-1}+n)t) , \quad t \geq 0 , \tag{2.5}$$

where the coefficients a_n depend on Euler's gamma fucntion, [12]. Observe that $r(\cdot)$ involves a direct input-output transmission without delay. However it is not possible to restrict oneself to finitely many delays under feedback : e.g., with $\hat{f}(s) = 1 - \exp(-s)$, $(1 + \hat{f}(s))^{-1} = 2^{-1} \Sigma_{n=0}^{\infty} 2^{-n} \exp(-ns)$. In addition such TF's occur frequently e.g. in transmission line theory, e.g. [36, pp. 99,104].

The following definitions and related properties, [11], [26], are then in order. $L_p(\mathbb{R}_+)$, or for short L_p , denotes for $p \in [1,\infty)$, (for $p=\infty$ resp.), the Banach space of p-th-power absolutely integrable, (essentially bounded resp.), functions with support on $\mathbb{R}_+$. For $\sigma \in \mathbb{R}$ and $p \in [1,\infty]$, $L_{p,\sigma} := \{f(\cdot) : \exp(-\sigma\cdot) f(\cdot) \in L_p\}$. $f \in LTD^+$ is said to belong to $\mathbf{A}(\sigma)$ iff, for $t < 0$, $f(t) = 0$, and, for $t \geq 0$, $f(t) = f_a(t) + f_{sa}(t)$, where the regular functional part $f_a(\cdot) \in L_{1,\sigma}$ and the singular atomic part $f_{sa}(\cdot) := \Sigma_{i=0}^{\infty} f_i \; \delta(\cdot - t_i)$ s.t. $t_0 = 0$, $t_i > 0$ for $i=1,2,\ldots$, and $f_i \in \mathbb{C}$ for $i=0,1,\ldots$, with $\Sigma_{i=0}^{\infty} |f_i| \exp(-\sigma t_i) < \infty$.

$$\mathbf{A}(\sigma) \text{ is a commutative convolution Banach algebra,} \tag{2.6}$$

e.g. [13], [8], with unit element $\delta(\cdot)$ and with norm defined by

$$\| f \|_{\mathbf{A}(\sigma)} := \int_0^{\infty} \exp(-\sigma t) \cdot |f_a(t)| \cdot dt + \sum_{i=0}^{\infty} \exp(-\sigma t_i) \cdot |f_i| . \tag{2.7}$$

For $\sigma_0 \in \mathbb{R}$, $f \in LTD^+$ is said to belong to the convolution algebra $\mathbf{A}_-(\sigma_0)$ iff $f \in \mathbf{A}(\sigma)$ for some $\sigma < \sigma_0$: $\hat{\mathbf{A}}_-(\sigma_0)$ is a good class of σ_0-stable, (i.e. holomorphic in $\mathbb{C}_{\sigma_0}^+$), TF's, [14, Cor. 2.1]. $\hat{\mathbf{A}}_-^{\infty}(\sigma_0)$ is the multiplicative subset of $\hat{\mathbf{A}}_-(\sigma_0)$ of elements that are bounded away from zero at infinity in $\mathbb{C}_{\sigma_0}^+$. $\hat{\mathbf{B}}(\sigma_0)$ is the TF algebra of fractions of $\hat{\mathbf{A}}_-(\sigma_0)$ with respect to $\hat{\mathbf{A}}_-^{\infty}(\sigma_0)$, i.e. $\hat{\mathbf{B}}(\sigma_0) := [\hat{\mathbf{A}}_-(\sigma_0)].[\hat{\mathbf{A}}_-^{\infty}(\sigma_0)]^{-1}$, or equivalently, $\hat{f} \in \hat{\mathbf{B}}(\sigma_0)$ iff $\hat{f} = \hat{n}\, \hat{d}^{-1}$ for some $\hat{n} \in \hat{\mathbf{A}}_-(\sigma_0)$ and $\hat{d} \in \hat{\mathbf{A}}_-^{\infty}(\sigma_0)$. W.l.g. the denominator $\hat{d}$ may be chosen rational. $\mathbf{I}(\sigma_0)$ denotes the subclass of those elements of $\mathbf{A}_-(\sigma_0)$ that are invertible in $\mathbf{A}_-(\sigma_0)$.

$$f \in \mathbf{I}(\sigma_0) , \quad \text{i.e.} \quad f*g = \delta = g*f \quad \text{for some} \quad g \in \mathbf{A}_-(\sigma_0) , \tag{2.8a}$$

iff

$$\inf\{|\hat{f}(s)| : s \in \mathbb{C}_{\sigma_0}^+\} > 0 . \tag{2.8b}$$

Hence $\hat{\mathbf{I}}(\sigma_0) := \{\hat{f} \in \hat{\mathbf{A}}_-(\sigma_0) : \hat{f}$ is invertible in $\hat{\mathbf{A}}_-(\sigma_0)\} \subset \hat{\mathbf{A}}_-^{\infty}(\sigma_0)$.

See [11], [26]-[27] for more details about this framework, summarized by $(G,H,I,J) := (\hat{B}(\sigma_0),\hat{A}_-(\sigma_0),\hat{A}_-^\infty(\sigma_0),\hat{I}(\sigma_0))$ in accordance with the theory in [29]. We are now ready to start ...

3. THE SPECTRAL FACTORIZATION PROBLEM

The main objective of this section is to establish a necessary and sufficient condition for the existence of a solution to this problem.
From now on σ_0 is nonpositive real, unless otherwise stated. Moreover we adopt the following convention : to any $f = f_a + f_{sa} := f_a(\cdot) + \Sigma_{i=-\infty}^{\infty} f_i\, \delta(\cdot-t_i) \in LTD$, (where f_a is a $\mathbb{C}$-valued function; $f_i \in \mathbb{C}$ for $i=0,\pm1,\ldots$; $t_0 = 0$, $t_i > 0$ for $i=1,2,\ldots$ and $t_i < 0$ for $i=-1,-2,\ldots$)[1] , we associate $f^+ := f_a^+ + f_{sa}^+ \in LTD^+$ and $f^- := f_a^- + f_{sa}^- \in LTD^-$ s.t. $f_a^+(0) = f_a^-(0) := 2^{-1}.f_a(0)$, $f_a^+(t) := f_a(t)$ for $t>0$, ($f_a^-(t) := f_a(t)$ for $t<0$ resp.), and $f_{sa}^+(t) := 2^{-1}.f_0.\delta(t) + \Sigma_{i=1}^{\infty} f_i.\delta(t-t_i)$ for $t\geq0$, ($f_{sa}^-(t) := 2^{-1}.f_0.\delta(t) + \Sigma_{i=-\infty}^{-1} f_i.\delta(t-t_i)$ for $t\leq0$). Obviously $f = f^- + f^+$ on $\mathbb{R}$.

<u>Definitions 3.1 :</u> Let $\sigma_0 \leq 0$ and let $f = f_a + f_{sa} := f_a(\cdot) + \Sigma_{i=-\infty}^{+\infty} f_i.\delta(\cdot-t_i) \in LTD$ be s.t. the following assumptions hold :

(A.1) $\hat{f}^+$ is σ_0-stable, i.e. $\hat{f}^+ \in \hat{A}_-(\sigma_0)$,

(A.2) $\hat{f}$ is parahermitian (p.h.) on S_{σ_0} ,
i.e. $\hat{f}(s) = \hat{f}_*(s) := \bar{\hat{f}}(-\bar{s})$ for all $s \in \mathbb{C}$ s.t. $|\mathrm{Re}\, s| \leq -\sigma_0$,

(A.3) $\hat{f}$ is nonnegative (n.n.) on the imaginary axis,
i.e. $\hat{f}(j\omega) \geq 0$ for all $\omega \in \mathbb{R}$.

U.th.c.

$\hat{r} \in \hat{A}_-(\sigma_0)$ is said to be <u>spectral factor</u> of $\hat{f}$
iff

$$\hat{f}(s) = \hat{r}_*(s) . \hat{r}(s) \quad \text{on } S_{\sigma_0} , \tag{3.1a}$$

and

$$\inf\{|\hat{r}(s)| : s \in \mathring{\mathbb{C}}_{(-\sigma_0)^+}\} > 0 . \tag{3.1b}$$

Such $\hat{r}$ is said to be <u>σ_0-outer</u>

---

[1] These specifications will be assumed to hold in the sequel.

iff

$$\hat{r}^{-1} \in \hat{\mathbf{A}}_-(\sigma_0) \;.\square \tag{3.2}$$

Comments 3.1 : α) The time-domain meaning of frequency-domain assumptions (A.1)-(A.2) is that

$$\int_{-\infty}^{+\infty} | f_a(t)| \,.\, \exp(-\sigma \,| t|) \,.\, dt \;<\; \infty \;, \tag{3.3a}$$

and

$$\sum_{i=-\infty}^{+\infty} | f_i| \,.\, \exp(-\sigma \,| t_i|) \;<\; \infty \tag{3.3b}$$

for some $\sigma < \sigma_0$, and that

$$f_a(t)=\overline{f}_a(-t) \text{ a.e. on } \mathbb{R} \;,\; t_{-i}=-t_i<0 \text{ and } f_{-i}=\overline{f}_i \quad i=0,1,2,\ldots \;. \tag{3.3c}$$

Identity (3.1a) then reads :

$$f(t) = \overline{f}(-t) = \int_0^{+\infty} \overline{r}(u) \,.\, r(t+u) \,.\, du \quad \text{a.e. on } \mathbb{R} \;, \tag{3.4}$$

since $\hat{r}_*\hat{r} = L[\overline{r}(-\cdot)*r(\cdot)]$ and $r(\cdot)$ has its support on $\mathbb{R}_+$, (see (3.8)). In a stochastic context, $f(\cdot)$ is the autocorrelation of the random process $(r(t))_{t\geq 0}$, e.g. [1, p. 190].

β) The expression "(outer) spectral factor" is borrowed from stochastic process theory, where such factor $\hat{r}$ is the TF of a proper, stable (and minimum phase) system whose output power spectral density is $\hat{f}$, e.g. [2, pp. 90-93]. The term "outer" is used by analogy with the discrete time case, e.g. [3], [4, Th. 2.8 and Th. 11.6]. Condition (3.1b) stems from the concept of "proper factorization" in [5, pp. 182-183], while (3.2) is needed to establish the existence of normalized coprime fractions of a given TF, (see Theorem 4.1).

γ) If $\sigma_0 = 0$, (3.1b) reads $\inf\{| \hat{r}(s)| \,:\, \mathrm{Re}\, s > 0\} > 0$, whence our definition of spectral factor is as in the lumped case, e.g. [6].

Observe that, in view of (2.8), (3.2) holds iff $\inf\{| \hat{r}(s)| : s \in \mathbb{C}_{\sigma_0}^{+}\} > 0$, or equivalently $\inf\{| \hat{r}_*(s)| \;:\; s \in \mathbb{C}_{(-\sigma_0)}^{-}\} > 0$. Hence a necessary condition for $\hat{f}$ to admit a σ_0-outer spectral factor is that $\hat{f}$ is bounded away from zero at infinity and on S_{σ_0} , (3.6). It turns out to be sufficient too.

Theorem 3.1 : [Existence of a σ_0-outer spectral factor]. Let $\sigma_0 \leq 0$ and let $f = f_a + f_{sa} := f_a(\cdot) + \sum_{i=-\infty}^{+\infty} f_i.\delta(\cdot-t_i) \in \mathrm{LTD}$ s.t. (A.1)-(A.3) hold.

U.th.c.

there exists an $\hat{r} \in \hat{\mathbf{A}}_-(\sigma_0)$ s.t.

$\hat{f}$ is a σ_0-outer spectral factor of $\hat{f}$ (3.5)

iff

$$\inf\{|\hat{f}(s)| \; : s \in S_{\sigma_0}\} > 0 \;.\square \tag{3.6}$$

We now face the proof of sufficiency of (3.6).
We shall need results in [7, Ch. VII, § 1.], developed in the context of Banach algebra theory, e.g. [13], [8]. However $\mathbf{A}_-(\sigma_0)$ is not a Banach algebra, but so is $\mathbf{A}(\sigma)$. This motivates the following definitions, generalizing those in [9, pp. 125-127].

<u>Definitions 3.2 :</u> Let $\sigma \leq 0$. $LA(\sigma) := L(\sigma) \dot{+} A(\sigma)$ denotes the class of all $f = f_a + f_{sa} := f_a(\cdot) + \Sigma_{i=-\infty}^{+\infty} f_i \,.\, \delta(\cdot - t_i)$ in LTD s.t. $f^+(\cdot)$, $f^-(-\cdot) \in \mathbf{A}(\sigma)$, or equivalently s.t. $f_a \in L(\sigma)$, i.e. (3.3a) holds, and $f_{sa} \in A(\sigma)$, i.e. (3.3b) holds. $LA^+(\sigma) := L^+(\sigma) \dot{+} A^+(\sigma)$, ($LA^-(\sigma) := L^-(\sigma) \dot{+} A^-(\sigma)$ resp.), denotes the subclass of those elements of $LA(\sigma)$ that have their support on $\mathbb{R}_+$ ($\mathbb{R}_-$ resp.).$\square$

Observe that

$$LA^+(\sigma) = \mathbf{A}(\sigma) \;, \tag{3.7a}$$

$$f \in LA^-(\sigma) \;\text{ iff }\; f(-\cdot) \in LA^+(\sigma) \;, \tag{3.7b}$$

and

$$LA(\sigma) = LA^-(\sigma) + LA^+(\sigma) \;. \tag{3.7c}$$

Let $f := f_a + \Sigma_{k=-\infty}^{+\infty} f_k \,.\, \delta(\cdot - t_k)$ and $g := g_a + \Sigma_{\ell=-\infty}^{+\infty} g_\ell \,.\, \delta(\cdot - \tau_\ell)$ belong to $LA(\sigma)$. We equip $LA(\sigma)$ with the convolution product

$$(f*g)(t) := \int_{-\infty}^{+\infty} f(t-u) \,.\, g(u) \,.\, du \;, \tag{3.8a}$$

and with the norm

$$\|f\|_{1,\sigma} := \|f^-(-\cdot)\|_{\mathbf{A}(\sigma)} + \|f^+(\cdot)\|_{\mathbf{A}(\sigma)} \;. \tag{3.9a}$$

It follows that

$$(f*g)(t) = (f*g)_a(t) + \sum_{k,\ell=-\infty}^{+\infty} f_k \,.\, g_\ell \,.\, \delta(t - t_k - \tau_\ell) \;, \tag{3.8b}$$

where

$$(f*g)_a(t) := (f_a*g_a)(t) + \sum_{k=-\infty}^{+\infty} f_k \,.\, g_a(t-t_k) + \sum_{\ell=-\infty}^{+\infty} g_\ell \, f_a(t-\tau_\ell) \tag{3.8c}$$

is the functional part of $(f*g)(t)$; and

$$\|f\|_{1,\sigma} = \int_{-\infty}^{+\infty} |f_a(t)| \exp(-\sigma\,|t|)\, dt + \sum_{i=-\infty}^{+\infty} |f_i| \exp(-\sigma\,|t_i|) \;. \tag{3.9b}$$

Moreover

$$\| f*g\|_{1,\sigma} \leq \| f\|_{1,\sigma} \cdot \| g\|_{1,\sigma} \cdot \qquad (3.10)$$

(3.7)-(3.10) induce important properties of distributions in $LA(\sigma)$, that are collected in the following

Fact 3.1 : [Properties of distributions in $LA(\sigma)$]. Let $\sigma \leq 0$.
U.th.c.

a) $LA(\sigma)$ and $A(\sigma)$ are commutative convolution Banach algebras with unit element $\delta(\cdot)$ and with norm $\| \cdot\|_{1,\sigma}$.

b) If $f = f_a + f_{sa} \in LA(\sigma)$, then

(i) $|\hat{f}_a(\sigma_1+j\omega_1)| \to 0$ as $|\omega_1| \to \infty$, uniformly in $\sigma_1 \in [\sigma,-\sigma]$, (3.11a)

(ii) $\hat{f}_{sa}$ is analytic almost periodic in S_σ , (3.11b)

(iii) $\hat{f}$ is uniformly continuous in S_σ , (3.12a)

(iv) $\hat{f}$ is holomorphic in $\overset{o}{S}_\sigma$. (3.12b)

c) $f \in LA(\sigma)$ is invertible in $LA(\sigma)$ (3.13a)

<=>

$\inf\{|\hat{f}(s)| : s \in S_\sigma\} > 0$, (3.13b)

=>

$\hat{f}$ is bounded away from zero at infinity in S_σ , (3.14a)

<=>

f_{sa} is invertible in $A(\sigma) \subset LA(\sigma)$, (3.14b)

<=>

$\hat{f}_{sa}(s) \neq 0$ for all $s \in S_\sigma$.□ (3.14c)

Comments 3.2 : α) It follows by (3.11b) that $\hat{f}_{sa}$ is the uniform limit of a sequence of exp. polynomials, viz. $(\Sigma_{i=-n}^{n} f_i \exp(-st_i))_{n=0}^{\infty}$, e.g. [10].
β) (3.14a) holds iff there exist $\eta > 0$ and $\rho > 0$ s.t., for all $s \in S_\sigma$ with $|\mathrm{Im}\, s| \geq \rho$, $|\hat{f}(s)| \geq \eta$, [11, p. 652].
A proof of Fact 3.1 is available in Appendix A.
Theorem 3.1 will follow by the two lemmas below.

Lemma 3.1 : [Transformation of the assumptions]. Let $\sigma_0 \leq 0$ and let $f = f_a + f_{sa} := f_a(\cdot) + \Sigma_{i=-\infty}^{+\infty} f_i \cdot \delta(\cdot-t_i) \in LTD$.
U.th.c.

a) (A.1)-(A.3) hold (3.15)

iff

there exists a $\sigma < \sigma_0$ s.t.

(B.1) $\hat{f} \in \hat{LA}(\sigma)$,

(B.2) $\hat{f}$ is p.h. on S_σ ,

s.t. $\hat{f}^+ = (\hat{f}^-)_* \in \hat{\mathbf{A}}(\sigma)$, (3.16)

(B.3) $\hat{f}$ is n.n. on the imaginary axis.

b) $\hat{f}^+$ and $(\hat{f}^-)_* \in \hat{\mathbf{A}}_-(\sigma_0)$ s.t. (3.6) holds (3.17a)

iff

there exists a $\sigma<\sigma_0$ s.t. $\hat{f} \in \hat{LA}(\sigma)$ is invertible in $\hat{LA}(\sigma)$.□ (3.17b)

<u>Proof :</u> a) If : Obvious. Only if : It follows from (A.1)-(A.2) that $\hat{f}^+ \in \hat{\mathbf{A}}(\sigma_1)$ for some $\sigma_1 < \sigma_0$, and $\hat{f}^+ = (\hat{f}^-)_*$ on $\mathbb{C}_{\sigma_0^+}$. Since $\hat{f}^+$ is holomorphic in $\mathbb{C}_{\sigma_2}$ for all σ_2 s.t. $\sigma_1 < \sigma_2 < \sigma_0$, [11, p. 652], we obtain by analytic continuation, e.g. [12, § 6.1], that $\hat{f}^+ = (\hat{f}^-)_*$ on $\mathbb{C}_{\sigma_2^+}$, for any such σ_2 . Hence (B.1)-(B.2) hold for some $\sigma < \sigma_0$, since $\hat{\mathbf{A}}(\sigma_1) \subset \hat{\mathbf{A}}(\sigma_2)$ and $(\hat{f}^-)_* = \mathbf{L}[\bar{f}^-(-\cdot)]$.

b) If : Obvious in view of (3.7) and (3.13). Only if : It follows by (3.7) that $\hat{f} \in \hat{LA}(\sigma)$ for some $\sigma < \sigma_0$. Thus (3.12a) holds s.t., by (3.6), we obtain that w.l.g.[2] $\hat{f}$ is invertible in $\hat{LA}(\sigma)$, in view of (3.13).□

<u>Lemma 3.2 :</u> [Fundamental lemma]. Let $\sigma \leq 0$ and let $f = f_a + f_{sa} := f_a(\cdot) + \Sigma_{i=-\infty}^{+\infty} f_i \cdot \delta(\cdot-t_i) \in LTD$ s.t. (B.1)-(B.3) hold.

U.th.c.

there exists an $\hat{r} \in \hat{\mathbf{A}}(\sigma) = \hat{LA}^+(\sigma)$ s.t.

a) $\hat{f}(s) = \hat{r}_*(s) \cdot \hat{r}(s)$ on S_σ , (3.18a)

b) $\hat{r}$ is invertible in $\hat{\mathbf{A}}(\sigma)$, (3.18b)

iff

$\hat{f}$ is invertible in $\hat{LA}(\sigma)$.□ (3.19)

<u>Proof of Theorem 3.1 :</u> If : By Lemma 3.1, (B.1)-(B.3) and (3.19) hold for some $\sigma < \sigma_0$, whence, by Lemma 3.2, there exists an $\hat{r} \in \hat{\mathbf{A}}(\sigma) \subset \hat{\mathbf{A}}_-(\sigma_0)$ s.t. (3.5) holds.□

<u>Proof of Lemma 3.2 :</u> Only if : By (3.7), $\hat{r}$ is invertible in $\hat{LA}^+(\sigma) \subset \hat{LA}(\sigma)$ iff $\hat{r}_*$ is invertible in $\hat{LA}^-(\sigma) \subset \hat{LA}(\sigma)$. If : Observe that, in view of (3.13)-(3.14), $\hat{f}_{sa} \in \hat{A}(\sigma)$ is invertible in $\hat{A}(\sigma) \subset \hat{A}(0)$. In

[2] By increasing σ slightly to $\sigma' \in (\sigma,\sigma_0)$, if necessary.

addition, $\hat{f}$ belongs to $\hat{LA}(\sigma) \subset \hat{LA}(0)$ and is p.h.n.n. on the imaginary axis, s.t. $\inf\{\hat{f}(j\omega) : \omega \in \mathbb{R}\} > 0$. Furthermore, $\hat{f}$ is the product of two uniformly continuous, (3.12a), real and positive functions, viz. $j\omega \to \hat{f}_{sa}(j\omega)$ and $j\omega \to 1 + \hat{f}_a(j\omega)\,\hat{f}_{sa}^{-1}(j\omega)$, that are bounded away from zero at infinity, (3.11a). It follows that the indices of $\hat{f}$, [7, pp. 158-159][3], viz. $\nu(\hat{f}) := \lim_{\ell\to\infty} (2\,\ell)^{-1}.[\arg \hat{f}_{sa}(j\omega)]_{\omega=-\ell}^{\ell}$ and $n(\hat{f}) := (2\,\pi)^{-1}$. $[\arg (1 + \hat{f}_a(j\omega)\,\hat{f}_{sa}^{-1}(j\omega))]_{\omega=-\infty}^{+\infty}$, are equal to zero. Applying [7, Th. 1.1, p. 160], we then obtain that there exists an $\hat{r} \in \hat{\mathbf{A}}(0)$ s.t. $\hat{f}(j\omega) = \hat{r}_*(j\omega)\,.\,\hat{r}(j\omega)$ for all $\omega \in \mathbb{R}$, and $\hat{r}$ is invertible in $\hat{\mathbf{A}}(0)$. Now, by (3.12), (3.7a-b), and Appendix A(4)-(5) with $(\hat{f}^+,\sigma)$ replaced by $(\hat{r},0)$, $\hat{r}_*(s)^{-1}\,.\,\hat{f}(s)$ is continuous in $S(\sigma,0)$ and holomorphic in $\overset{\circ}{S}(\sigma,0)$. Hence, by e.g. [12, p. 312], $\hat{r}$ can be extended to a function continuous in $\mathbb{C}_{\sigma+}$ and holomorphic in $\overset{\circ}{\mathbb{C}}_{\sigma+}$, s.t. $\hat{r}(s) = \hat{r}_*(s)^{-1}\,.\,\hat{f}(s)$ on $S(\sigma,0)$ and $\hat{r} \in \hat{\mathbf{A}}(\sigma)$. By similar arguments, $\hat{r}^{-1} \in \hat{\mathbf{A}}(\sigma)$, since $\hat{r}^{-1}(j\omega) = \hat{r}_*(j\omega)\,.\,\hat{f}(j\omega)^{-1}$ for all $\omega \in \mathbb{R}$. Finally, by (B.2), $\hat{r}(s) = \hat{r}_*(s)^{-1}\,.\,\hat{f}(s)$ on $S(0,-\sigma)$.□

Comments 3.3 : α) If $|\sigma|$ is sufficiently small, then we can find a domain, (i.e. an open connected set), $D \subset \mathbb{C}$ s.t. D contains the spectrum of f, viz. $\sigma(f) := \{\lambda \in \mathbb{C} : (\lambda\delta - f)$ is not invertible in $LA(\sigma)\}$ = the closure of $\{\hat{f}(s) : s \in S_\sigma\}$, e.g. [8, p. 234], and s.t. $\lambda \to \log \lambda$ is holomorphic in D, (see (3.12a), (3.13b)). $\hat{r}$ in Lemma 3.2 then reads

$$\hat{r}(s) = \exp(\log \hat{f}(s))^+ ,$$

with

$$\log \hat{f}(s) = (2\,\pi\, j)^{-1}\,.\int_\Gamma \log \lambda\,.\,(\lambda - \hat{f}(s))^{-1}\,.\,d\lambda ,$$

where Γ is any contour that surrounds $\sigma(f)$ in D, (generalized Cauchy formula), e.g. [8, p. 243].

β) Any σ_0-outer spectral factor $\hat{r} = \hat{r}_a + \hat{r}_{sa} \in \hat{\mathbf{A}}_-(\sigma_0)$ of a given $\hat{f} = \hat{f}_a + \hat{f}_{sa}$, is unique up to multiplication by a $c \in \mathbb{C}$ s.t. $|c| = 1$. Any such $\hat{r}$ reads $\hat{r} = \hat{r}_1\,.\,\hat{r}_2$, where $\hat{r}_1 = \hat{r}_{sa}$, ($\hat{r}_2 = 1 + \hat{r}_a\,\hat{r}_{sa}^{-1}$ resp.), is a σ_0-outer spectral factor of $\hat{f}_{sa}$, ($1 + \hat{f}_a\,\hat{f}_{sa}^{-1}$ resp.).

γ) Let $f = f_a + f_{sa} \in LTD$ be s.t. (A.1)-(A.3) hold and s.t. $\hat{f}$ is bounded away from zero at infinity in S_{σ_0}. It follows that $\hat{f}$ may only have finitely many zeros there, e.g. [14, Comment 4.1 (β)] : they are distributed symmetrically with respect to the imaginary axis, s.t. any

[3] Where the real line, the upper (lower) half-plane are replaced by the imaginary axis, $\mathbb{C}_+$, ($\mathbb{C}_-$ resp.).

$j\omega$-axis zero of $\hat{f}$ has an even multiplicity. Then there exists an $\hat{r} \in \hat{\mathcal{A}}_-(\sigma_0)$ s.t. $\hat{r}$ is a spectral factor of $\hat{f} = (\hat{f}_a \hat{f}_{sa}^{-1} + 1) \cdot \hat{f}_{sa}$, by Theorem 3.1 (applied to $\hat{f}_{sa}$) and [5, Th. 3.2] (applied to $(1+\hat{f}_a\hat{f}_{sa}^{-1})$). Any such $\hat{r}$ reads $\hat{r} = \hat{q} \cdot \hat{w}$, where $\hat{q}$ is a σ_0-stable rational biproper function s.t. $\hat{q}_* \hat{q}$ collects all the zeros of $\hat{f}$, and where $\hat{w} \in \hat{\mathcal{A}}_-(\sigma_0)$, (by [11, Cor. 2.2C]), is a σ_0-outer spectral factor of $\hat{q}_*^{-1} \hat{f} \hat{q}^{-1}$.

4. THE GRAPH METRIC FOR DISTRIBUTED SYSTEMS

We can now extend the graph metric (for possibly unstable lumped systems), [16], [17], to distributed systems. We first establish the existence of normalized coprime fractions. An upper bound for the graph metric is then derived. Finally, proper rational TF's are shown to be dense in the subclass of TF's with constant direct transmission.

Recall that we take a real $\sigma_0 \leq 0$, and that a pair $(\hat{n}_1,\hat{d}_1) \in \hat{\mathcal{A}}_-(\sigma_0) \times \hat{\mathcal{A}}_-^\infty(\sigma_0)$ is said to be a <u>σ_0-coprime fraction</u> of $\hat{p} \in \hat{\mathcal{B}}(\sigma_0)$ iff

$$\text{(i)}\quad \hat{p} = \hat{n}_1 \hat{d}_1^{-1} \quad \text{and} \quad \text{(ii)}\quad (\hat{n}_1,\hat{d}_1) \text{ is } \sigma_0\text{-coprime}, \tag{4.1a}$$

i.e. $\hat{u}\hat{n}_1 + \hat{v}\hat{d}_1 = 1$ for some $\hat{u}, \hat{v} \in \hat{\mathcal{A}}_-(\sigma_0)$, or equivalently, [11],

$$\inf\{|(\hat{n}_1(s),\hat{d}_1(s))| : s \in \mathbb{C}_{\sigma_0+}\} > 0 . \tag{4.1b}$$

<u>Definition 4.1 :</u> Let $\sigma_0 \leq 0$ and $\hat{p} \in \hat{\mathcal{B}}(\sigma_0)$. A σ_0-coprime fraction $(\hat{n},\hat{d}) \in \hat{\mathcal{A}}_-(\sigma_0) \times \hat{\mathcal{A}}_-^\infty(\sigma_0)$ of $\hat{p}$ is said to be <u>normalized</u> iff

$$\hat{n}_*(s-\sigma_0).\hat{n}(s+\sigma_0) + \hat{d}_*(s-\sigma_0).\hat{d}(s+\sigma_0) = 1 \text{ , for all } s=j\omega,\ \omega\in\mathbb{R}\ .\square \tag{4.2}$$

<u>Comments 4.1 :</u> α) In the MIMO Lumped case normalized σ_0-coprime fractions, (for short σ_0-NCF, or NCF when $\sigma_0 = 0$), can be computed by stabilizing the plant by a normalized LQ-optimal control law, (Appendix C), where an <u>exp. weight</u>, viz. $\exp(-2\sigma_0 t)$, has been introduced in the cost. The following equivalences are then not surprising : $(\hat{n},\hat{d}) \in \hat{\mathcal{A}}_-(\sigma_0) \times \hat{\mathcal{A}}_-^\infty(\sigma_0)$ is a σ_0-NCF of $\hat{p} \in \hat{\mathcal{B}}(\sigma_0)$ iff $(\underset{\sim}{\hat{n}}(\cdot),\underset{\sim}{\hat{d}}(\cdot)) := (\hat{n}(\cdot+\sigma_0),(\hat{d}(\cdot+\sigma_0)) \in \hat{\mathcal{A}}_-(0) \times \hat{\mathcal{A}}_-^\infty(0)$ is a NCF of the <u>shifted plant</u> $\underset{\sim}{\hat{p}}(\cdot) := \hat{p}(\cdot+\sigma_0) \in \hat{\mathcal{B}}(0)$, or equivalently, $(\hat{n}(\cdot),\hat{d}(\cdot)) := (\underset{\sim}{\hat{n}}(\cdot-\sigma_0),\underset{\sim}{\hat{d}}(\cdot-\sigma_0))$ is a σ_0-NCF of $\hat{p}(\cdot) = \underset{\sim}{\hat{p}}(\cdot-\sigma_0)$.

β) Definition 4.1 is thus the correct generalization of the concept of NC Factorization for lumped systems, [16, Def. 3.1], to our framework, where stability is guaranteed with a margin $|\sigma_0| \geq 0$ (instead of $|\sigma_0| = 0$).

Examples show that the characterization " $\hat{n}_* \hat{n} + \hat{d}_* \hat{d} = 1$ on S_{σ_0} " does not always insure the existence of such σ_0-coprime fractions.

γ) Let $\hat{u}$, $\hat{v} \in \hat{\mathbf{A}}_-(\sigma_0)$. The $H^\infty(\mathbb{C}_{\sigma_0}^+)$-norm of $\hat{u}$, $(\hat{u},\hat{v})$ resp., reads :

$$\|\hat{u}\| := \sup\{|\hat{u}(s)| : s \in \mathbb{C}_{\sigma_0}^+\} = \sup\{|\hat{u}(\sigma_0+j\omega)| : \omega \in \mathbb{R}\} , \qquad (4.3a)$$

$$\|(\hat{u},\hat{v})\| := \sup\{(|\hat{u}(s)|^2 + |\hat{v}(s)|^2)^{1/2} ; s \in \mathbb{C}_{\sigma_0}^+\} \qquad (4.3b)$$

$= \sup\{(|\hat{u}(\sigma_0+j\omega)|^2 + |\hat{v}(\sigma_0+j\omega)|^2)^{1/2} : \omega \in \mathbb{R})$ resp., by the maximum modulus principle, e.g. [18]. Recall, [11], that $\|\hat{u}\| \le \|u\|_{A(\sigma_0)}$. If $(\hat{n},\hat{d}) \in \hat{\mathbf{A}}_-(\sigma_0)\times\hat{\mathbf{A}}_-^\infty(\sigma_0)$ is a σ_0-NCF of $\hat{p} \in \hat{\mathbf{B}}(\sigma_0)$, then the mapping $\hat{\mathbf{A}}_-(\sigma_0) \to [\hat{\mathbf{A}}_-(\sigma_0)]^2 : \hat{u} \to (\hat{n}\hat{u},\hat{d}\hat{u})$ is an isometry, by (4.2)-(4.3), i.e.

$$\|(\hat{n}\hat{u},\hat{d}\hat{u})\| = \|\hat{u}\| \text{ for any } \hat{u} \in \hat{\mathbf{A}}_-(\sigma_0) . \qquad (4.4)$$

δ) In a stochastic context, (4.2) means that the graph of the plant $\hat{p}$ (with respect to L_2), viz. $g(\hat{p}) := \{(u,y) \in L_2 x L_2 : \hat{y} = \hat{p}.\hat{u}\} = \{(d*z,n*z) : z \in L_2\}$, [17, Section 7.2], driven by white noise has a white power spectral density, e.g. [2, pp. 90-93].

In [24], the existence of NCF's is established only for proper TF's in $\hat{B}(0)$, namely $\hat{f} \in \hat{\mathbf{B}}_p(0) := \{\hat{f} \in \hat{\mathbf{B}}(0) : \hat{f}(s) \to k \in \mathbb{C}$ as $|s| \to \infty$ in $\mathbb{C}_0+\}$. We prove here the more general

Theorem 4.1 : [Existence of σ_0-NCF]. Let $\sigma_0 \le 0$.

U.th.c.

Any $\hat{p} \in \hat{\mathbf{B}}(\sigma_0)$ admits a σ_0-NCF $(\hat{n},\hat{d}) \in \hat{\mathbf{A}}_-(\sigma_0)\times\hat{\mathbf{A}}_-^\infty(\sigma_0)$.□ (4.5)

Proof : In view of Comment 4.1α), w.l.g. $\sigma_0 := 0$. Any $\hat{p} \in \hat{\mathbf{B}}(\sigma_0)$ admits a σ_0-coprime fraction $(\hat{n}_1,\hat{d}_1) \in \hat{\mathbf{A}}_-(\sigma_0)\times\hat{\mathbf{A}}_-^\infty(\sigma_0)$, [11] . Define $\hat{f} := (\hat{n}_1)_* \hat{n}_1 + (\hat{d}_1)_* \hat{d}_1$. It follows from (4.1) that (A.1)-(A.3) and (3.6) hold. Hence, by Theorem 3.1, $\hat{f}$ has a σ_0-outer spectral factor $\hat{r} \in \hat{\mathbf{A}}_-(\sigma_0)$. Therefore $(\hat{n},\hat{d}) := (\hat{n}_1\hat{r}^{-1},\hat{d}_1\hat{r}^{-1})$ is a σ_0-NCF of $\hat{p}$.□

We are now ready to introduce

Definition 4.2 : Let $\sigma_0 \le 0$, $\hat{p}_1$, $\hat{p}_2 \in \hat{\mathbf{B}}(\sigma_0)$ and $(\hat{n}_1,\hat{d}_1)$, $(\hat{n}_2,\hat{d}_2)$ resp. , $\in \hat{\mathbf{A}}_-(\sigma_0)\times\hat{\mathbf{A}}_-^\infty(\sigma_0)$ be σ_0-NCF's of $\hat{p}_1$, $\hat{p}_2$ resp..

U.th.c.

Define

$$d_{\sigma_0}(\hat{p}_1,\hat{p}_2) := \max\{\delta(\hat{p}_1,\hat{p}_2),\delta(\hat{p}_2,\hat{p}_1)\} , \qquad (4.6a)$$

with

$$\delta(\hat{p}_1,\hat{p}_2) \;:=\; \inf\{\|(\hat{n}_1-\hat{n}_2\hat{u},\hat{d}_1-\hat{d}_2\hat{u})\| : \hat{u}\in\hat{\mathbf{A}}_-(\sigma_0)\;,\;\|\hat{u}\|\le 1\}\;. \quad (4.6b)$$

The metric $d_{\sigma_0} : [\hat{\mathbf{B}}(\sigma_0)]^2 \to [0,1] : (\hat{p}_1,\hat{p}_2) \mapsto d_{\sigma_0}(\hat{p}_1,\hat{p}_2)$ is called the <u>σ_0-graph metric</u> on $\hat{\mathbf{B}}(\sigma_0)$.□

The topology induced by d_{σ_0} on $\hat{\mathbf{B}}(\sigma_0)$ is the σ_0-graph topology. More precisely, if $(\hat{p}_i)_{i=1}^{\infty}$ is a sequence in $\hat{\mathbf{B}}(\sigma_0)$, then

$$d_{\sigma_0}(\hat{p}_i,\hat{p}) \to 0 \text{ as } i\to\infty \quad (4.7)$$

iff

there exist σ_0-coprime fractions $(\hat{n},\hat{d})$ of $\hat{p}$ and $(\hat{n}_i,\hat{d}_i)$ of $\hat{p}_i$ $i=1,2,\ldots$ s.t. $\|(\hat{n}_i-\hat{n},\hat{d}_i-\hat{d})\| \to 0$ as $i\to\infty$.

Indeed, necessity holds, by the proof of [16, Th. 3.1]. Sufficiency follows by

<u>Fact 4.1 :</u> [Upper bound for the σ_0-graph metric]. Let $\sigma_0 \le 0$. Let $(\hat{n},\hat{d})$, $(\hat{n}_1,\hat{d}_1) \in \hat{\mathbf{A}}_-(\sigma_0)\mathrm{x}\hat{\mathbf{A}}_-^{\infty}(\sigma_0)$ be σ_0-coprime fractions of $\hat{p}$, resp. $\hat{p}_1 \in \hat{\mathbf{B}}(\sigma_0)$, and let

$$\hat{f}(\cdot) \;:=\; \hat{n}_*(\cdot-\sigma_0)\,.\,\hat{n}(\cdot+\sigma_0) + \hat{d}_*(\cdot-\sigma_0)\,.\,\hat{d}(\cdot+\sigma_0)\;. \quad (4.8)$$

U.th.c.

If

$$\delta \;:=\; \|(\hat{n}-\hat{n}_1,\hat{d}-\hat{d}_1)\| < \eta^{-1} \;:=\; (\sup\{\hat{f}(j\omega)^{-\frac{1}{2}} : \omega\in\mathbb{R}\})^{-1}\;, \quad (4.9)$$

then

$$d_{\sigma_0}(\hat{p},\hat{p}_1) \le 2\,\eta\,\delta\,(1-\eta\,\delta)^{-1}\;.\square \quad (4.10)$$

<u>Proof :</u> Let $\hat{\underline{r}} \in \hat{\mathbf{A}}_-(0)$ be an (0-)outer spectral factor of $\hat{f}$, s.t. $\hat{r}(\cdot) := \hat{\underline{r}}(\cdot-\sigma_0) \in \hat{\mathbf{I}}(\sigma_0)$, whence

$$\eta = \|\hat{r}^{-1}\| < \infty\;. \quad (4.11)$$

Observe that $(\hat{n}',\hat{d}') := (\hat{n}\hat{r}^{-1},\hat{d}\hat{r}^{-1})$, resp. $(\hat{n}_1',\hat{d}_1') := (\hat{n}_1\hat{r}^{-1},\hat{d}_1\hat{r}^{-1})$, is a σ_0-NCF, resp. σ_0-coprime fraction, of $\hat{p}$, resp. $\hat{p}_1$. In addition, $\|(\hat{n}'-\hat{n}_1',\hat{d}'-\hat{d}_1')\| \le \eta\,.\,\delta < 1$, by (4.9) and (4.11). (4.10) follows by the proof of [16, Lemma 3.4].□

<u>Comments 4.2 :</u> α) The upper bound in (4.10) is obtained in terms of any σ_0-coprime fractions of the there involved TF's. In [16, Lemma 3.4], at least one is normalized.

β) The σ_0-graph topology is the weakest one on $\hat{\mathbf{B}}(\sigma_0)$ in which feedback stability is robust against plant and/or compensator perturbations, [19,

Th. 4.1]. In addition, the topology induced on $\hat{\mathbf{A}}_-(\sigma_0)$ by d_{σ_0} is equivalent to that induced by $\|\cdot\|$, (use the proof of [16, Lemma 2.2]).

Now recall that $\mathbf{R}(\sigma_0)$ denotes the algebra of σ_0-exp. stable, (i.e. holomorphic in $\mathbb{C}_{\sigma_0}^+$), proper rational TF's, while $\mathbf{R}^\infty(\sigma_0)$ denotes its subset of biproper, (i.e. nonzero at infinity), elements. $\mathbb{C}_p(s)$ is the algebra of proper rational TF's. It is known, [26], that $\mathbb{C}_p(s) = [\mathbf{R}(\sigma_0)] . [\mathbf{R}^\infty(\sigma_0)]^{-1}$, i.e. $\hat{f} \in \mathbb{C}_p(s)$ iff $\hat{f} = \hat{n}\,\hat{d}^{-1}$ for some $\hat{n} \in \mathbf{R}(\sigma_0)$ and $\hat{d} \in \mathbf{R}^\infty(\sigma_0)$. Recall also, [26], that

$$\hat{\mathbf{B}}(\sigma_0) = [\hat{\mathbf{A}}_-(\sigma_0)] . [\mathbf{R}^\infty(\sigma_0)]^{-1} . \tag{4.12}$$

Let $\hat{\mathbf{B}}_p(\sigma_0) \subsetneq \hat{\mathbf{B}}(\sigma_0)$ be the subclass of those TF's with direct transmission without delay : $\hat{f} \in \hat{\mathbf{B}}_p(\sigma_0)$ iff $\hat{f} \in \hat{\mathbf{B}}(\sigma_0)$ s.t. $\hat{f}(s) \to k \in \mathbb{C}$ as $|s| \to \infty$ in $\mathbb{C}_{\sigma_0}^+$, (whence the subscript p for proper). Otherwise stated, by (4.12),

$$\hat{\mathbf{B}}_p(\sigma_0) = [U\{\hat{L}_{1,\sigma} : \sigma < \sigma_0\} \dot{+} \mathbb{C}] . [\mathbf{R}^\infty(\sigma_0)]^{-1} . \tag{4.13}$$

Furthermore, by (4.13) and [11, Th. 3.3], any $\hat{f} \in \hat{\mathbf{B}}_p(\sigma_0)$ reads $\hat{f} = \hat{r} + \hat{q}$, where $\hat{r}$ is completely σ_0-exp. unstable strictly proper rational and $\hat{q} \in U\{\hat{L}_{1,\sigma} : \sigma < \sigma_0\} \dot{+} \mathbb{C}$. We can now state

<u>Theorem 4.2 :</u> [Density of $\mathbb{C}_p(s)$ in $\hat{\mathbf{B}}_p(\sigma_0)$]. Let $\sigma_0 \leq 0$.
U.th.c.

$$\mathbb{C}_p(s) \text{ is dense in } (\hat{\mathbf{B}}_p(\sigma_0), d_{\sigma_0}) , \tag{4.14}$$

or equivalently, for any $\hat{p} \in \hat{\mathbf{B}}_p(\sigma_0)$, there exists a sequence $(\hat{p}_i)_{i=1}^\infty$ in $\mathbb{C}_p(s)$ s.t. $d_{\sigma_0}(\hat{p}_i, \hat{p}) \to 0$ as $i \to \infty$.□

<u>Proof :</u> Let $(\hat{n},\hat{d}) \in [U\{\hat{L}_{1,\sigma} : \sigma < \sigma_0\} \dot{+} \mathbb{C}] x \mathbf{R}^\infty(\sigma_0)$ be a σ_0-coprime fraction of $\hat{p}$, (4.13) . By [25, Th. 6.1.1], $\mathbf{R}(\sigma_0)$ is dense in $(U\{\hat{L}_{1,\sigma} : \sigma < \sigma_0\} \dot{+} \mathbb{C} , \|\cdot\|)$, whence there exists a sequence $(\hat{n}_i)_{i=1}^\infty$ in $\mathbf{R}(\sigma_0)$ s.t. $\|\hat{n}_i - \hat{n}\| \to 0$ as $i \to \infty$. By the genericity of coprimeness, [17, Section 7.6], any $(\hat{n}_i, \hat{d})$ is w.l.g. σ_0-coprime. The conclusion follows by (4.7) , with $\hat{p}_i := \hat{n}_i\, \hat{d}^{-1}$.□

<u>Comments 4.3 :</u> α) Any $\hat{p}_{sa}(s) := \Sigma_{i=1}^{+\infty} p_i . \exp(-st_i)$ in $\hat{\mathbf{A}}_-(\sigma_0)$ can be approximated arbitrarily closely by an exponential polynomial $\hat{p}_N(s) := \Sigma_{i=1}^N p_i . \exp(-st_i)$. Indeed, $\Sigma_{i=1}^{+\infty} |p_i| . \exp(-\sigma_0 t_i) < \infty$, s.t., for all $\varepsilon > 0$, there exists N s.t. $\|\hat{p}_{sa} - \hat{p}_n\| \leq \|p_{sa} - p_n\|_{\mathbf{A}(\sigma_0)} < \varepsilon$ for any $n \geq N$. But then the delays t_i are to be known exactly. This is the case in

most applications, where typically the t_i's are equally spaced, (e.g. transmission lines).

β) (4.14) makes it possible to design finite-dimensional controllers for robust stabilization of infinite-dimensional systems, (see [20] and references therein).

5. SPECTRAL FACTORS OF EXPONENTIAL ORDER

The results in Sections 3 and 4 are now particularized to TF's of exp. order, [14]. Such TF's fit semigroup systems, e.g. [21], [22], better than those in $\hat{B}(\sigma_0)$, [23]-[24], [14]. Let $\sigma_0 \in \mathbb{R}$. A function $f : \mathbb{R} \to \mathbb{C}$ with support on $\mathbb{R}_+$ is said to be <u>σ_0-exp. stable</u> iff there exists a $\sigma < \sigma_0$ s.t. $f \in L_{\infty,\sigma}$, or equivalently,

$$|f(t)| \leq M \cdot \exp(\sigma t) \text{ a.e. on } \mathbb{R}_+ \text{, for some } \sigma<\sigma_0 \text{ and } M>0 . \quad (5.1)$$

$L_\varepsilon(\sigma_0)$ denotes the set of σ_0-exp. stable functions. The latter is a proper ideal of $\mathcal{A}_-(\sigma_0)$, [14, Th. 2.1], i.e.

$$L_\varepsilon(\sigma_0) \text{ is a linear subspace of } \mathcal{A}_-(\sigma_0) , \quad (5.2a)$$

and

$$\text{for all } u \in L_\varepsilon(\sigma_0) \text{ and } f \in \mathcal{A}_-(\sigma_0) , \; y := f * u \in L_\varepsilon(\sigma_0) . \quad (5.2b)$$

$f = f_a + f_{sa} \in \mathcal{A}_-(\sigma_0)$ is said to be σ_0-exp. stable iff $f_a \in L_\varepsilon(\sigma_0)$. The subset of σ_0-exp. stable distributions of $\mathcal{A}_-(\sigma_0)$, denoted by $E(\sigma_0)$, is a (commutative) subalgebra of $\mathcal{A}_-(\sigma_0)$, [14, Th. 3.1]. $\hat{I}_\varepsilon(\sigma_0)$ denotes the subclass of those TF's in $\hat{E}(\sigma_0)$ that are invertible in $\hat{E}(\sigma_0)$. By [14, Th. 3.2], $\hat{I}_\varepsilon(\sigma_0) = \hat{E}(\sigma_0) \cap \hat{I}(\sigma_0)$, i.e. $\hat{f} \in \hat{E}(\sigma_0)$ is invertible in $\hat{E}(\sigma_0)$ iff (2.8b) holds. Finally, $\hat{F}(\sigma_0)$ denotes the commutative algebra of fractions of $\hat{E}(\sigma_0)$ with respect to $\hat{E}^\infty(\sigma_0) := \hat{E}(\sigma_0) \cap \hat{\mathcal{A}}_-^\infty(\sigma_0)$, i.e. $\hat{F}(\sigma_0) := [\hat{E}(\sigma_0)].[\hat{E}^\infty(\sigma_0)]^{-1} = [\hat{E}(\sigma_0)].[R^\infty(\sigma_0)]^{-1} \subsetneq \hat{B}(\sigma_0)$. For further properties of $\hat{E}(\sigma_0)$ and $\hat{F}(\sigma_0)$, see [14].

<u>Definition 3.1ε</u> : Let $\sigma_0 \leq 0$ and let $f = f_a + f_{sa} := f_a(\cdot) + \Sigma_{i=-\infty}^{+\infty} f_i \cdot \delta(\cdot - t_i) \in$ LTD satisfy (A.2), (A.3) and

(A.1ε) $\hat{f}^+ \in \hat{E}(\sigma_0) \subsetneq \hat{\mathcal{A}}_-(\sigma_0)$.

U.th.c.

$\hat{r} \in \hat{E}(\sigma_0)$ is said to be an <u>$\varepsilon\sigma_0$-outer spectral factor</u> of $\hat{f}$ iff $\hat{r}$ is a spectral factor of $\hat{f}$, (3.1), s.t. $\hat{r}^{-1} \in \hat{E}(\sigma_0)$.□

Comments 5.1 : α) (A.1ε) is necessary for $\hat{f}$ to have an $\varepsilon\sigma_0$-outer spectral factor $\hat{r} \in \hat{\mathbf{E}}(\sigma_0)$. Indeed assume that $|r_a(t)| \leq K \cdot \exp(\sigma t)$ a.e. on $\mathbb{R}_+$, for some $K > 0$ and $\sigma < \sigma_0$. Observe that $f(\cdot) = \bar{r}(-\cdot)*r(\cdot)$, whence $f_a(\cdot) = \bar{r}_a(-\cdot)*r_a(\cdot) + \bar{r}_{sa}(-\cdot)*r_a(\cdot) + \bar{r}_a(-\cdot)*r_{sa}(\cdot)$, where w.l.g.[4] $\bar{r}_a(-\cdot)$, $\bar{r}_{sa}(-\cdot)$ and $r_{sa}(\cdot) \in LA(\sigma)$. It follows by Lemma 5.1 and (5.6) below, that f_a satisfies (5.4) whence $f_a^+ \in L_\varepsilon(\sigma_0)$.

β) If $\hat{r} \in \hat{\mathbf{A}}_-(\sigma_0)$ is a spectral factor of a p.h.n.n. $\hat{f}$ and if $f_a(0)$ is finite, then $r_a \in L_2$. Indeed, by (3.1a), $f(\cdot) = \bar{r}(-\cdot)*r(\cdot)$, whence, by (3.8), $f_a(0) = \int_0^\infty |r_a(u)|^2 du + 2 \Sigma_{i=0}^\infty \operatorname{Re}(r_i.\bar{r}_a(t_i))$. It follows that $\int_0^\infty |r_a(u)|^2 du \leq |f_a(0)| + 2 \Sigma_{i=0}^\infty |r_i| . |r_a(t_i)| < \infty$. If $\hat{r} \in \hat{\mathbf{E}}(\sigma_0)$, with $\sigma_0 \leq 0$, then $\hat{r}_a \in \hat{L}_\varepsilon(\sigma_0)$ s.t., by [14, Fact 2.1(a)], $r_a \in L_2$.

We come now to the main result of this section : if a p.h.n.n. $\hat{f}$ is of exp. order, then so is any of its σ_0-outer spectral factors. This is an easy consequence of the following

Theorem 3.1ε : [Existence of $\varepsilon\sigma_0$-outer spectral factors]. Let $\sigma_0 \leq 0$ and let $f = f_a + f_{sa} := f_a(\cdot) + \Sigma_{i=-\infty}^{+\infty} f_i.\delta(\cdot - t_i) \in LTD$ s.t. (A.1ε) , (A.2)-(A.3) hold.

U.th.c.

there exists an $\hat{r} \in \hat{\mathbf{E}}(\sigma_0)$ s.t.

$$\hat{r} \text{ is an } \varepsilon\sigma_0\text{-outer spectral factor of } \hat{f} , \tag{5.3}$$

iff

$$\inf\{|\hat{f}(s)| : s \in S_{\sigma_0}\} > 0 . \square \tag{3.6}$$

The proof of the latter is based on the following lemmas.

Lemma 5.1 : [Ideal property]. Let $\sigma_0 \leq 0$.

U.th.c.

The functions $g : \mathbb{R} \to \mathbb{C}$ s.t. there exist $\sigma < \sigma_0$ and $M > 0$ s.t.

$$|g(t)| \leq M . \exp(\sigma |t|) \text{ a.e. on } \mathbb{R} , \tag{5.4}$$

form a proper ideal of the convolution algebra

$$\{f \in LTD : f \in LA(\sigma) , \text{ for some } \sigma<\sigma_0\} . \square \tag{5.5}$$

Comment 5.2 : $g(\cdot)$ is s.t. (5.4) holds for some $\sigma<\sigma_0$ and $M>0$

[4] By increasing σ slightly to $\sigma' \in (\sigma,\sigma_0)$, if necessary.

iff

$$\hat{g}^+ \text{ and } (\hat{g}^-)_* \in \hat{L}_\varepsilon(\sigma_0) ; \tag{5.6}$$

$$f \in LA(\sigma) \text{ for some } \sigma<\sigma_0 \text{ iff } \hat{f}^+ \text{ and } (\hat{f}^-)_* \in \hat{\mathbf{A}}_-(\sigma_0) . \tag{5.7}$$

The proof of Lemma 5.1 is given in Appendix B.

<u>Lemma 5.2 :</u> [Regularization]. Let $\sigma_0 \leq 0$ and let $\hat{g} = \hat{g}_a+1 \in \hat{LTD}$ and $\hat{h} = \hat{h}_a+1 \in \hat{LTD}_-$ s.t. h has its support on $\mathbb{R}_-$, with

$$\text{(i)} \quad \hat{g}^+ = (\hat{g}^-)_* \in \hat{\mathbf{E}}(\sigma_0) \tag{5.8}$$

and

$$\text{(ii)} \quad \hat{h}_* \in \hat{\mathbf{A}}_-(\sigma_0) . \tag{5.9}$$

U.th.c.

If $\hat{q} := \hat{g}.\hat{h} \in \hat{LTD}_+$ s.t. q has its support on $\mathbb{R}_+$, then $\hat{q} \in \hat{\mathbf{E}}(\sigma_0)$.□ (5.10)

<u>Proof :</u> $\hat{q} = \hat{q}_a + \hat{q}_{sa} = \hat{q}_a + 1$, whence $\hat{q} \in \hat{\mathbf{E}}(\sigma_0)$ iff $\hat{q}_a \in \hat{L}_\varepsilon(\sigma_0)$. Moreover, $q_a(t) = \overset{+}{q}_a(t) = (g*f)^+_a(t) = (g_a*h_a)^+(t) + \overset{+}{g}_a(t) + \overset{+}{h}_a(t) = (g_a*h_a)^+(t) + \overset{+}{g}_a(t)$ a.e. on $\mathbb{R}_+$, since $q \in LTD_+$ and $h \in LTD_-$, (5.9). In view of (5.8)-(5.9), it follows by (5.6)-(5.7) and Lemma 5.1 that $q_a \in L_\varepsilon(\sigma_0)$.□

<u>Proof of Theorem 3.1ε :</u> Only if : Apply Theorem 3.1.

If : Again by Theorem 3.1, $\hat{f}$ has a σ_0-outer spectral factor $\hat{r} \in \hat{\mathbf{A}}_-(\sigma_0)$, s.t. $\hat{r}^{-1} \in \hat{\mathbf{A}}_-(\sigma_0)$. If $\hat{r} \in \hat{\mathbf{E}}(\sigma_0)$, then $\hat{r}^{-1}$ will belong to $\hat{\mathbf{E}}(\sigma_0)$, since $\hat{\mathbf{I}}_\varepsilon(\sigma_0) = \hat{\mathbf{E}}(\sigma_0) \cap \hat{\mathbf{I}}(\sigma_0)$. By Comment 3.3β), $\hat{r} = \hat{r}_1 \hat{r}_2$, where $\hat{r}_2 = (\hat{r}_2)_a+1 \in \hat{\mathbf{A}}_-(\sigma_0)$, ($\hat{r}_1 = (\hat{r}_1)_{sa} \in \hat{\mathbf{A}}_-(\sigma_0)$ resp.), is a σ_0-outer spectral factor of $(\hat{f}_a \hat{f}_{sa}^{-1}+1)$, ($\hat{f}_{sa}$ resp.). Hence

$$\hat{r} = \hat{r}_1 (\hat{r}_2)_a + \hat{r}_1 , \tag{5.11a}$$

with

$$\hat{r}_1 , \hat{r}_1^{-1} \in \hat{\mathbf{E}}(\sigma_0) \quad \text{s.t.} \quad \hat{r}_1 = \hat{f}_{sa}(\hat{r}_1)_*^{-1} , \tag{5.11b}$$

and

$$\hat{r}_2 , \hat{r}_2^{-1} \in \hat{\mathbf{A}}_-(\sigma_0) \quad \text{s.t.} \quad \hat{r}_2 = (\hat{f}_a \hat{f}_{sa}^{-1}+1) (\hat{r}_2)_*^{-1} . \tag{5.11c}$$

In view of (5.11) and (5.2b), it suffices to prove that $\hat{r}_2 \in \hat{\mathbf{E}}(\sigma_0)$: indeed $\hat{r}_2 \in \hat{\mathbf{E}}(\sigma_0)$ iff $(\hat{r}_2)_a \in \hat{L}_\varepsilon(\sigma_0)$, whence, by (5.2b), $\hat{r}_1 (\hat{r}_2)_a \in \hat{L}_\varepsilon(\sigma_0)$; it follows by (5.11) that $\hat{r} \in \hat{\mathbf{E}}(\sigma_0)$. Now apply Lemma 5.2 to $\hat{g} := \hat{f}_a \hat{f}_{sa}^{-1} + 1$, $\hat{h} := (\hat{r}_2)_*^{-1}$ and $\hat{q} := \hat{r}_2$ to get the conclusion. Note that $\hat{h}_* = \hat{r}_2^{-1} \in \hat{\mathbf{A}}_-(\sigma_0)$, (5.11c), and that $\hat{g}^+ = \hat{g}_*^- \in \hat{\mathbf{A}}_-(\sigma_0)$, (A.1ε)-(A.2). Finally, $\hat{g}^+ \in \hat{\mathbf{E}}(\sigma_0)$: indeed, $\hat{f}_a^+ = (\hat{f}_a^-)_* \in \hat{L}_\varepsilon(\sigma_0)$, and

$(\hat{f}_{sa}^{-1})^+ = (\hat{f}_{sa}^{-1})_*^- \in \hat{\mathcal{A}}_-(\sigma_0)$, by (A.1ε)-(A.2) and the fact that invertibility implies invertibility of the atomic part (see Lemma 3.1b and (3.13)-(3.14)); hence $\hat{g}_a^+ = (\hat{f}_a \hat{f}_{sa}^{-1})^+ \in \hat{L}_\varepsilon(\sigma_0)$, by (5.6)-(5.7) and Lemma 5.1.□

By parallel reasonings, the remainder of Sections 3 and 4 can be developed for TF's of exp. order. ($\hat{B}(\sigma_0)$, $\hat{\mathcal{A}}_-(\sigma_0)$, $\hat{\mathcal{A}}_-^\infty(\sigma_0)$, $\hat{I}(\sigma_0)$, $U\{\hat{L}_{1,\sigma} : \sigma<\sigma_0\}$, $\hat{B}_p(\sigma_0)$), σ_0-coprime, σ_0-NCF, σ_0-graph metric and d_{σ_0} resp. must then be replaced resp. by ($\hat{\mathbf{F}}(\sigma_0)$, $\hat{\mathbf{E}}(\sigma_0)$, $\hat{\mathbf{E}}^\infty(\sigma_0)$, $\hat{I}_\varepsilon(\sigma_0)$, $\hat{L}_\varepsilon(\sigma_0)$, $\hat{\mathbf{F}}_p(\sigma_0)$), $\varepsilon\sigma_0$-coprime, $\varepsilon\sigma_0$-NCF, $\varepsilon\sigma_0$-graph metric and $d_{\varepsilon\sigma_0}$. Such sharper results will be indicated by an added subscript ε . For example, Theorem 4.1ε reads :

<u>Theorem 4.1ε :</u> [Existence of $\varepsilon\sigma_0$-NCF]. Let $\sigma_0 \leq 0$.
U.th.c.
Any $\hat{p} \in \hat{\mathbf{F}}(\sigma_0)$ admits an $\varepsilon\sigma_0$-NCF $(\hat{n},\hat{d}) \in \hat{\mathbf{E}}(\sigma_0)\times\hat{\mathbf{E}}^\infty(\sigma_0)$.□

Note that, for all $\hat{p}_1,\hat{p}_2 \in \hat{\mathbf{F}}(\sigma_0) \subsetneq \hat{B}(\sigma_0)$, $d_{\sigma_0}(\hat{p}_1,\hat{p}_2) \leq d_{\varepsilon\sigma_0}(\hat{p}_1,\hat{p}_2)$, since $\hat{\mathbf{E}}(\sigma_0) \subsetneq \hat{\mathcal{A}}_-(\sigma_0)$ and any $\varepsilon\sigma_0$-NCF is a σ_0-NCF.

6. APPLICATION : ROBUST FEEDBACK CONTROL OF THE HEAT FLOW IN A FINITE ROD

Consider the diffusion of a heat flow in a rod of unit length, insulated at the ends. If $\theta(x,t)$ denotes the temperature at the point $x \in [0,1]$ at time $t \geq 0$, then the system is described by the diffusion equation

$$\frac{\partial\theta}{\partial t}(x,t) = \frac{\partial^2\theta}{\partial x^2}(x,t)+\delta(x)u(t)\ ,\ y(t)=\langle\delta(x-1),\theta(x,t)\rangle_{L^2([0,1])}=\theta(1,t)\ , \qquad (6.1a)$$

with boundary and initial conditions :

$$\frac{\partial\theta}{\partial x}(0,t) = \frac{\partial\theta}{\partial x}(1,t) = 0\ , \quad \theta(x,0) = 0\ . \qquad (6.1b)$$

By the method of images, [38, pp. 211-212], (6.1) is an equivalent description of the diffusion of a heat flow in a rod of length two, insulated at the ends, with heat source $\delta(x)$ in the middle. By [37, pp. 37-38, 101-102], (6.1) can be described by the semigroup state-space system

$$\dot{z}(t) = A\,z(t) + B\,u(t)\ , \quad y(t) = C\,z(t)\ , \quad z(0) = 0\ , \qquad (6.2)$$

with Sobolev state-space $H^{-\frac{1}{2}-\varepsilon}(0,1)$, $\varepsilon>0$, and with operator $A = \partial^2/\partial x^2$. By [37, p. 101], the TF of (6.2) reads

$$\hat{p}(s) = C\,(s\,I - A)^{-1}\,B\,. \tag{6.3}$$

However we prefer to use classical techniques, e.g. [38], [36], to compute $\hat{p}$: in our opinion (6.3) leads to more involved computations than the latter. Taking Laplace transform in (6.1a), we obtain that

$$s\,\hat{\theta}(x,s) = \frac{d^2}{dx^2}\hat{\theta}(x,s) + \delta(x) \quad \text{and} \quad \hat{y}(s) = \hat{\theta}(1,s)\,,$$

whence

$$\hat{p}(s) = (\sqrt{s}\,.\,\mathrm{sh}(\sqrt{s}))^{-1}\,. \tag{6.4a}$$

By elementary calculus of residues, e.g. [12, chap. 4],

$$\hat{p}(s) = s^{-1} + 2\sum_{n=1}^{\infty} (-1)^n\,(s + n^2\,\pi^2)^{-1}\,, \tag{6.4b}$$

or, by e.g. [12, p. 356],

$$\hat{p}(s) = (s\,.\prod_{n=1}^{\infty}\,[1 + (n\,\pi)^{-2}\,s])^{-1}\,. \tag{6.4c}$$

In view of (6.4b), $\hat{p} \in \hat{\mathbf{F}}(0)$, [14, Th. 5.1] : $\hat{p}$ is of exp. order. An ε0-coprime fraction $(\hat{n},\hat{d}) \in \hat{L}_{\varepsilon}(0)\mathrm{x}\mathbf{R}^{\infty}(0)$ of $\hat{p}$ is, with $\alpha>0$,

$$(\hat{n}(s),\hat{d}(s)) = ((s+\alpha)^{-1}+2\sum_{n=1}^{\infty}(-1)^n s(s+\alpha)^{-1}(s+(n\pi)^2)^{-1}, s(s+\alpha)^{-1})\,. \tag{6.5}$$

If $\underset{\sim}{\hat{r}} \in \hat{\mathbf{E}}(0)$ is a spectral factor of $(\hat{p}_*\hat{p}+1)^{-1}$, then, by the proof of Theorem 4.1ε , $\hat{r} := \hat{d}\,\underset{\sim}{\hat{r}}^{-1}$ is an ε0-outer spectral factor of

$$\hat{f} := \hat{n}_*\,\hat{n} + \hat{d}_*\,\hat{d} = \hat{d}_*\,(\hat{p}_*\,\hat{p} + 1)\,\hat{d}\,. \tag{6.6}$$

Furthermore $(\underset{\sim}{\hat{n}},\underset{\sim}{\hat{d}}) := (\hat{n}\hat{r}^{-1},\hat{d}\hat{r}^{-1}) = (\hat{p}\underset{\sim}{\hat{r}},\underset{\sim}{\hat{r}})$ is an ε0-NCF of $\hat{p}$. By (6.4a), (6.4c), $(\hat{p}_*\hat{p}+1)(s) = (1 - s\,.\,\sin\sqrt{s}\,.\,\mathrm{sh}\sqrt{s})\,.\,(-s\,.\,\sin\sqrt{s}\,.\,\mathrm{sh}\sqrt{s})^{-1}$, i.e.

$$(\hat{p}_*\hat{p}+1)(s) = (s^2 - \alpha_0^2)\,.\,s^{-2}\,.\prod_{n=1}^{\infty}\,[(\alpha_n^4 - s^2)\,.\,((n\pi)^4 - s^2)^{-1}]\,. \tag{6.7}$$

In (6.7), $\pm\alpha_0$, $\pm\alpha_1$, ... are the roots of the transcendental equation $z^2\,.\,\sin z = (\mathrm{sh}\,z)^{-1}$, s.t. $\alpha_i > 0$, $i=0,1,\ldots$, and $\alpha_n^2 \rightarrow (n\pi)^2$ as $n\rightarrow\infty$. By symmetric extractions we obtain a spectral factor of $(\hat{p}_*\hat{p}+1)^{-1}$, viz.

$$\underset{\sim}{\hat{r}}(s) = s\,.\,(s+\alpha_0)^{-1}\,.\prod_{n=1}^{\infty}\,[((n\pi)^2+s)\,.\,(\alpha_n^2+s)^{-1}] \in \hat{\mathbf{E}}(0)\,. \tag{6.8a}$$

An ε0-NCF of $\hat{p}$ is thus

$$(\underset{\sim}{\hat{n}}(s),\underset{\sim}{\hat{d}}(s)) = \left(\{s+\alpha_0)\,.\prod_{n=1}^{\infty}\,[(\alpha_n^2+s)\,(n\pi)^{-2}]\}^{-1}\,,\,\underset{\sim}{\hat{r}}(s)\right)\,. \tag{6.8b}$$

We now approximate the nominal model s.t. the two first dominant modes, (6.4b), are retained. We then get a lumped system with TF

$$\hat{p}_r(s) \ := \ (\pi^2 - s) \ . \ [s \ . \ (s + \pi^2)]^{-1} \ . \qquad (6.9)$$

An $\varepsilon 0$-coprime fraction of $\hat{p}_r$ is $(\hat{n}_r, \hat{d}_r) \in \mathbf{R}(0) \times \mathbf{R}^{\infty}(0)$, with

$$(\hat{n}_r(s), \hat{d}_r(s)) \ := \ ((s+\alpha)^{-1} - 2\ s\ [(s+\pi^2).(s+\alpha)]^{-1} \ , \ \hat{d}(s)) \ . \qquad (6.10)$$

In view of (4.3), (6.5) and (6.10), $\delta := \| (\hat{n}-\hat{n}_r, \hat{d}-\hat{d}_r) \| = \| \hat{n} - \hat{n}_r \|$, s.t. $\delta \leq 2 \ . \ \Sigma_{n=2}^{\infty} \| (s+n^2r^2)^{-1} \| \ . \ \| s.(s+\alpha)^{-1} \| = 2 \ \Sigma_{n=2}^{\infty} (n\pi)^{-2} \leq 2.\pi^{-2}$. In addition, with $\hat{f}_r := (\hat{n}_r)_* \ \hat{n}_r + \hat{a}_* \ \hat{a}$, $\eta_r := \sup\{\hat{f}_r(j\omega)^{-1/2} : \omega \in \mathbb{R}\} = \sup\{[(\alpha^2+\omega^2).(1+\omega^2)^{-1}]^{1/2} : \omega \in \mathbb{R}\} = \max\{1,\alpha\}$. It follows by Fact 4.1 that, if α is not too large,

$$d_{\varepsilon 0}(\hat{p}_r, \hat{p}) \ \leq \ 4 \ . \ \max\{1,\alpha\} \ . \ (\pi^2 - 2 \ . \ \max\{1,\alpha\})^{-1} \ . \qquad (6.11)$$

Then, by the proof of [34, Cor. 5.1], any finite-dimensional stabilizing compensator $\hat{c} \in \mathbb{C}_p(s)$ of $\hat{p}_r$ s.t.

$$\| T(\hat{p}_r, \hat{c}) \| \ \leq \ \pi^2 \ . \ (4 \ . \ \max\{1,\alpha\})^{-1} - 2^{-1} \qquad (6.12a)$$

is, in view of (6.11), a stabilizing compensator of $\hat{p}$. Here,

$$T(\hat{p}_r, \hat{c}) := \begin{bmatrix} -\hat{p}_r \ \hat{c} \ (1+\hat{p}_r\hat{c})^{-1} & -\hat{p}_r \ (1+\hat{c} \ \hat{p}_r)^{-1} \\ \hat{c} \ (1+\hat{p}_r\hat{c})^{-1} & (1+\hat{c}\hat{p}_r)^{-1} \end{bmatrix} = H(\hat{p}_r, \hat{c}) - \begin{bmatrix} 1 & 0 \\ 0 & 0 \end{bmatrix}, \qquad (6.12b)$$

where $H(\hat{p}_r, \hat{c})$ is the closed-loop TF matrix resulting from the application of the feedback controller $\hat{c}$ to the plant $\hat{p}_r$. Moreover,

$$\| (\hat{p}_r, \hat{c}) \| \ := \ \sup\{\bar{\sigma}[T(\hat{p}_r, \hat{c})(j\omega) \ , \ \omega \in \mathbb{R}\} \ , \qquad (6.12c)$$

where $\bar{\sigma}[\cdot]$ denotes the largest singular value of a matrix. Choosing $\alpha := 1$ to get the best upper bound in (6.11), we obtain that $(\hat{u}_r(s), \hat{v}_r(s)) := (1, 1+2(s+\pi^2)^{-1})$ is a solution of the Diophantine equation $\hat{u}_r \ \hat{n}_r + \hat{v}_r \ \hat{d} = 1$. By [29], the controller

$$\hat{c}(s) \ := \ \hat{v}_r(s)^{-1} \ . \ \hat{u}_r(s) \ = \ (s + \pi^2) \ . \ (s + \pi^2 + 2)^{-1} \qquad (6.13)$$

robustly stabilizes $\hat{p}_r$, [17, chap. 7]. In addition, (6.12) holds : indeed, $\bar{\sigma}[T(\hat{p}_r, \hat{c})(j\omega)] = [2 + 4 \ (1+\pi^2) \ (\pi^4+\omega^2)^{-1}]^{1/2}$, whence $\| T(\hat{p}_r, \hat{c}) \| = \bar{\sigma}[T(\hat{p}_r, \hat{c})(0)]$. Note that $\hat{c}$ is exp. stable, (6.13), i.e. $\hat{p}$, ($\hat{p}_r$ resp.), is strongly stabilizable, e.g. [17, § 5.3]. The nominal closed-loop system $H(\hat{p}, \hat{c})$ stability robustness is confirmed by a Nyquist-type test, e.g. [41]-[42] : see Fig. 6.1. The closed-loop behavior of the temperature θ , the plant output y resp., with initial condition $\theta(x,0) = 1$ is depicted in Fig. 6.2, Fig. 6.3 resp.

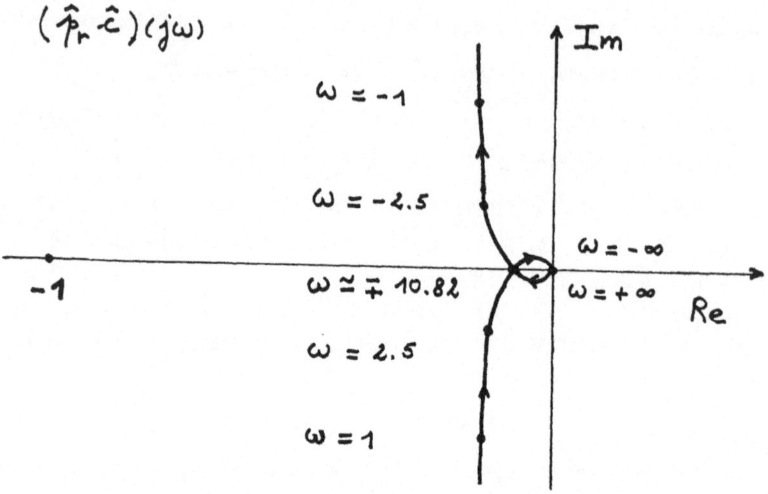

Fig. 6.1. Nyquist diagram for the approximate closed-loop system. The curve $(\hat{p}\hat{c})(j\omega)$, $\omega \in \mathbb{R}$, lies within a strip centered on $(\hat{p}_r\hat{c})(j\cdot)$, being $2\varepsilon = 4\pi^{-2}$ wide, s.t. $\| \hat{p}\hat{c}-\hat{p}_r\hat{c}\| \le \| \hat{p}-\hat{p}_r\| \le \varepsilon$. Note that $(\hat{p}_r\hat{c})\,(\pm j\infty) = 0 \angle \pm 90°$.

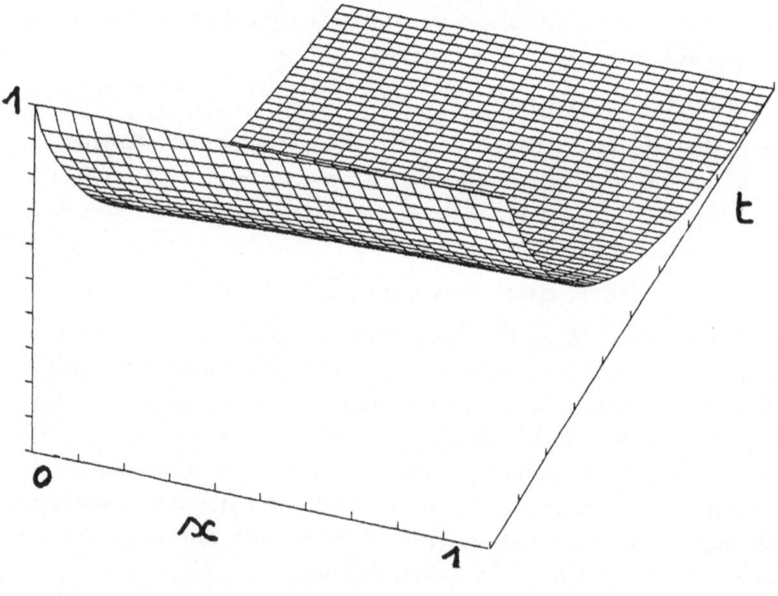

Fig. 6.2. Temperature under feedback.

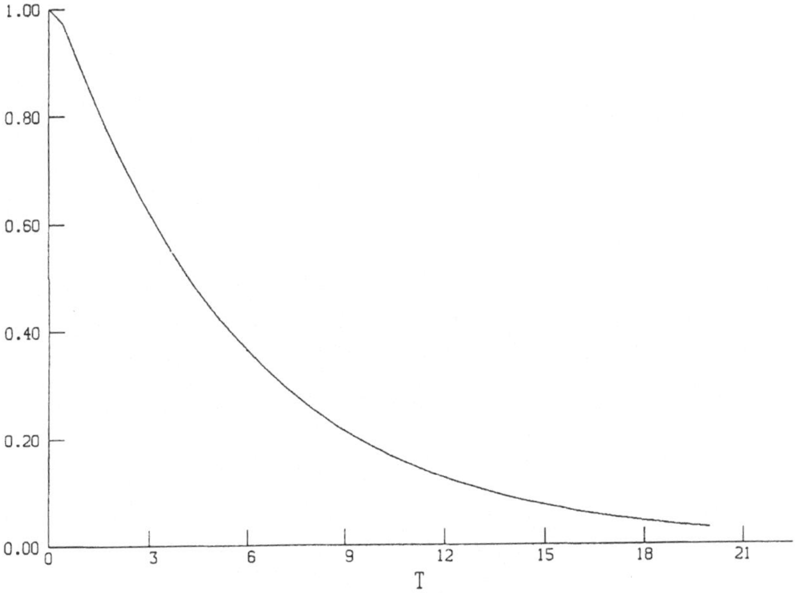

Fig. 6.3. Plant output under feedback.

7. CONCLUDING REMARKS

In this paper, we have solved the conceptual problem of spectral factorization for singlevariable linear distributed systems (of exponential order or not). As a byproduct, we have established some results dealing with the robustness of stability under feedback.

A further research step is now to study multivariable spectral factorization. The latter must then be seen as a Hilbert problem, [31].

Examples suggest that spectral factors could be approximated via symmetric extractions of elementary factors, as in the lumped case, e.g. [6]. The algorithm developed in [35] can be used to compute a spectral factor of an almost periodic parahermitian positive $\hat{f}_{sa}(s) := \Sigma_{i=-\infty}^{+\infty} f_i \cdot \exp(-st_i)$, when the delays t_i are equally spaced.

ACKNOWLEDGEMENTS

The authors gratefully acknowledge J. Bontsema from University of Groningen (The Netherlands) for simulating the example (Figs. 6.2 and 6.3).

APPENDIX A : PROOF OF FACT 3.1

a) $LA(\sigma)$ is isometrically isomorphic, [15, p. 9], to the convolution Banach algebra $L_0(\exp(-\sigma|\xi|)) + A_0(\exp(-\sigma|\xi|))$, a Banach subalgebra of $S_0(\exp(-\sigma|\xi|))$, [15, pp. 144, 153]. Moreover, in view of (3.9b),

$$A(\sigma) \text{ is a closed subalgebra of } LA(\sigma) . \qquad (1)$$

b) (3.11)-(3.12) follow by (3.7) . Indeed, the following hold for any $f^+ \in \mathbf{A}(\sigma)$, [11, p. 652] :

(i) $|\hat{f}^+(\sigma_1+j\omega_1)| \to 0$ as $|\omega_1| \to \infty$ uniformly in $\sigma_1 \in [\sigma,\sigma']$,

for any $\sigma' \geq \sigma$, (Riemann-Lebesgue). (2)

(ii) $\hat{f}^+_{sa}$ is analytic almost periodic in $S(\sigma,\sigma')$, for any $\sigma' \geq \sigma$. (3)

(iii) $\hat{f}^+$ is uniformly continuous in $\mathbb{C}_{\sigma^+}$, (4)

by (2)-(3), since $|\hat{f}^+(\sigma_1+j\omega_1)| = |\hat{f}^+_a(\sigma_1+j\omega_1)+f_0+\sum_{i=1}^{+\infty} f_i.\exp(-(\sigma_1+j\omega_1)t_i)| \to |f_0|$ as $\sigma_1 \to \infty$ uniformly in $\omega \in \mathbb{R}$.

(iv) $\hat{f}^+$ is holomorphic in $\mathring{\mathbb{C}}_{\sigma^+}$. (5)

c) <u>(3.13a) <=> (3.13b) :</u> see [24, Appendix], where $A_2(\sigma) := LA(\sigma)$.
<u>(3.13b) => (3.14a) :</u> Obvious, by Comment 3.2β).

Recall now that $\hat{f}_{sa}$ is almost periodic in S_σ , (3.11b), or equivalently, by e.g. [10, p. 73, Corollary p. 75],

for any $\varepsilon > 0$ there exists $\ell > 0$ s.t. each interval of length ℓ on the imaginary axis contains at least one ε-translation number $j\tau$, i.e. one point $j\tau$ s.t. $|\hat{f}_{sa}(\sigma_1+j(\omega+\tau))-\hat{f}_{sa}(\sigma_1+j\omega)| < \varepsilon$ for all $\sigma_1 \in [\sigma,-\sigma]$ and $\omega \in \mathbb{R}$. (6)

Moreover, by (1) and (3.13),

$$f_{sa} \text{ is invertible in } A(\sigma) \subset LA(\sigma) \qquad (3.14b)$$

iff

$\inf\{|\hat{f}_{sa}(s)| : s \in S_\sigma\} > 0 .$ (7)

(3.14a) <=> (7) : <= : Obvious, by (3.11a).

=> : By (3.11a), there exists $\rho_1 > 0$ s.t. $\inf\{|\hat{f}_{sa}(s)| : s \in S_\sigma$ with $|\mathrm{Im}\, s| \geq \rho_1\} > 0$. (7) follows by (6).

(7) <=> (3.14c) : => : Obvious.

<= : Assume by contradiction that there exists a sequence $(s_k)_{k=1}^{\infty}$ in S_σ s.t. $|\mathrm{Im}\, s_k| \to \infty$ and $|\hat{f}_{sa}(s_k)| \to 0$ as $k \to \infty$. Then $(\mathrm{Re}\, s_k)_{k=1}^{\infty}$ is bounded s.t., by (3.12a), w.l.g. $|\mathrm{Re}\, s_k - \sigma_1| \to 0$, $|\mathrm{Im}\, s_k| \to \infty$ and $|\hat{f}_{sa}(\sigma_1 + j\,\mathrm{Im}\, s_k)| \to 0$ as $k \to \infty$, for some $\sigma_1 \in [\sigma,-\sigma]$. In view of (3.14c), (6) and (3.12b), it follows by [10, Th. 3.6], that $\hat{f}_{sa}(\sigma_1+\cdot)$ vanishes on the imaginary axis. This contradicts (3.14c).

APPENDIX B : PROOF OF LEMMA 5.1

In view of (5.2a) and (5.6)-(5.7), the set $\{g \in LTD$: g is a function s.t. (5.4) holds for some $\sigma<\sigma_0$ and $M>0\}$ is a linear subspace of $\{f \in LTD : f \in LA(\sigma)$ for some $\sigma<\sigma_0\}$, (5.5). We still have to prove that : for any function $g : \mathbb{R} \to \mathbb{C}$ s.t.

$|g(t)| \leq M . \exp(\sigma|t|)$ a.e. on $\mathbb{R}$, for some $\sigma<\sigma_0$ and $M>0$, (1)

and for any distribution $f \in LTD$ s.t.

$f \in LA(\sigma)$ for some $\sigma < \sigma_0$, (2)

$h := f*g$ is a $\mathbb{C}$-valued function with support on $\mathbb{R}$ s.t.

$|h(t)| \leq L.\exp(\sigma'|t|)$ a.e. on $\mathbb{R}$, for some $\sigma'<\sigma_0$ and $L>0$. (3)

In view of (5.6), it is necessary and sufficient to prove that $\hat{h}^+$ and $\hat{h}^-_*$ $\in \hat{L}_\varepsilon(\sigma_0)$. Observe that $\hat{h} = \hat{f}^- \hat{g}^- + \hat{f}^+ \hat{g}^+ + \hat{f}^- \hat{g}^+ + \hat{f}^+ \hat{g}^-$. It follows from (5.6), (5.7) and (5.2b) that $\hat{f}^+ \hat{g}^+$ and $(\hat{f}^- \hat{g}^-)_* = \hat{f}^-_* \hat{g}^-_* \in \hat{L}_\varepsilon(\sigma_0)$. We now establish that

$(\hat{f}^- \hat{g}^+)^+$ and $(\hat{f}^- \hat{g}^+)^-_* \in \hat{L}_\varepsilon(\sigma_0)$; (4)

the reasoning for $\hat{f}^+ \hat{g}^-$ is entirely analogous. In the sequol, 1(t) denotes the unit step function : 1(t) = 0 if t<0 and 1(t) = 1 if $t\geq 0$. Note that, (3.8), a.e. on $\mathbb{R}$, $(f^-*g^+)(t) = \int_{-\infty}^{+\infty} f^-(u).g^+(t-u).du = 2^{-1} f_0\, g(t)\, 1(t) + \int_{-\infty}^{0} f_a(u).g(t-u).1(t-u).du + \Sigma_{i=-\infty}^{-1} f_i.g(t-t_i).1(t-t_i)$. Hence, for almost all $t>0$, $|(f^-*g^+)(t)| \leq M . (\int_{-\infty}^{0} |f_a(u)|.\exp(\sigma u).du + 2^{-1} |f_0| + \Sigma_{i=-\infty}^{-1} |f_i| .\exp(\sigma t_i)).\exp(\sigma t) \leq M . \|f\|_{1,\sigma} . \exp(\sigma t)$, with $M . \|f\|_{1,\sigma} < +\infty$; these inequalities follow by (1)-(2) and the fact that $\sigma<0$

and $t_i<0$, $i=-1,-2,\ldots$. For the same reasons, for almost all $t<0$, $|(f\bar{*}g^+)(t)| \leq M . (\int_{-\infty}^t |f_a(u)| .\exp(\sigma(u-t)).du + \Sigma_{i=-\infty}^{-1} |f_i| .\exp(\sigma(t_i-t))) \leq M . \|f\|_{1,\sigma} . \exp(-\sigma t)$. Thus (4) is established, in view of (5.6).

APPENDIX C : LQ-OPTIMAL CONTROL AND NORMALIZED COPRIME FRACTIONS

Let $P(s) \in \mathbb{R}_p(s)^{n_o x n_i}$ be the strictly proper TF matrix of a lumped linear time-invariant MIMO system. Let $N(s) \in \mathbb{R}[s]^{n_o x n_i}$ and $D(s) \in \mathbb{R}[s]^{n_i x n_i}$ be right coprime (r.c.) polynomial matrices s.t. $P = N D^{-1}$, with w.l.g. D column-reduced (c.r.), e.g. [28]. Moreover, with $\delta_{cj}[\cdot]$ denoting the j-th column-degree of a polynomial matrix, $\delta_{cj}[N] < \delta_{cj}[D]$, $j=1,\ldots,n_i$.

A pair $(N_r,D_r) \in \mathbf{R}(0)^{n_o x n_i} x \mathbf{R}(0)^{n_i x n_i}$, with N_r strictly proper and D_r biproper, (i.e. $\det D_r(\infty) \neq 0$, s.t. w.l.g. $D_r(\infty) = I$), is a normalized right coprime fraction (NRCF) of P iff, by definition, [16]-[17],

(F) $P = N_r D_r^{-1}$, (1a)

(RC) $U_r N_r + V_r D_r = I$, for some $[U_r,V_r] \in \mathbf{R}(0)^{n_i x (n_o+n_i)}$, (1b)

(N) $(N_r)_* N_r + (D_r)_* D_r = I$. (1c)

(1) holds iff

$$(D_r,N_r) = (DW^{-1},NW^{-1}) , \tag{2a}$$

where $W \in \mathbb{R}[s]^{n_i x n_i}$ is a stable spectral factor of $N_*N + D_*D := N^T(-s) N(s) + D^T(-s) D(s)$, i.e.

$$H := N_* N + D_* D = H_* = W_* W , \tag{2b}$$

with

$$\det W(s) \neq 0 , \text{ for all } s \in \mathbb{C}_+ . \tag{2c}$$

Indeed, this is an easy consequence of [40, Th. 1]. Note that any W s.t. (2b)-(2c) hold, is c.r. with $\delta_{cj}[W] = \delta_{cj}[D]$, $j=1,\ldots,n_i$, [6] .

If $[A,B,C,0]$ is a minimal realization of P , let $\mathbf{M}$ be a polynomial matrix s.t. $\mathbf{M}$ and D are r.c., $\delta_{cj}[\mathbf{M}] < \delta_{cj}[D]$, $j=1,\ldots,n_i$, $N := C\mathbf{M}$, and $\mathbf{M} D^{-1} = (s I - A)^{-1} B$.

(1) holds iff (D_r,N_r) takes the form

$$\begin{bmatrix} D_r(s) \\ N_r(s) \end{bmatrix} = \begin{bmatrix} F \\ C \end{bmatrix} [s I - (A + BF)]^{-1} B + \begin{bmatrix} I \\ 0 \end{bmatrix} , \tag{3a}$$

where [A,B,C,0] is any minimal realization of P , and F is the gain matrix of the stabilizing linear state feedback control law $u = F x$, which minimizes $\int_0^\infty (x^T C^T C x + u^T u).dt$ for every initial state, (normalized LQ-optimal control); i.e., by [33],

$$F = -X^{-1} Y = -X^T Y , \qquad (3b)$$

with (X,Y) the (unique) constant solution of the Diophantine equation

$$X . D(s) + Y . \mathbf{M}(s) = W(s) , \qquad (3c)$$

where W is a stable spectral factor of $H = \mathbf{M}_* C^T C \mathbf{M} + D_* D$, (2b)-(2c). Indeed, sufficiency follows by equivalence (1)-(2); necessity is a consequence of (1)-(2) and [40, Th. 2].
By duality, normalized left coprime fractions (NLCF) are related to stable (left) spectral factorization, and to minimum-variance linear estimation : the intensity of the joint input-output additive white noise is then the unity matrix.

See e.g. [6], [35], [39] and the references therein for computational algorithms of spectral factorization.

REFERENCES

[1] Maisel, L., 1971, Probability, Statistics and Random Processes (NY : Simon and Schuster).
[2] Kwakernaak, H., and Sivan, R., 1972, Linear Optimal Control Systems (NY : Wiley - Interscience).
[3] Jonckheere, E.A., and Helton, J.W., 1985, IEEE TAC-30, 1192.
[4] Duren, P.L., 1970, Theory of H^p spaces (NY : Academic Press).
[5] Kreĭn, M.G., 1962, Amer. Math. Soc. Transl., (2) Vol. 22, 163.
[6] Callier, F.M., 1985, IEEE TAC-30, 453.
[7] Gohberg, I.C., and Fel'dman, I.A., 1974, Convolution Equations and Projection Methods for their Solution, Vol. 41 (Providence, RI).
[8] Rudin, W., 1970, Functional Analysis (NY : McGraw Hill).
[9] Willems, J.C., 1971, The Analysis of Feedback Systems, Research Mon. 62 (Cambridge : MIT Press).
[10] Corduneanu, C., 1961, Almost Periodic Functions (NY : Interscience).
[11] Callier, F.M., and Desoer, C.A., 1978, IEEE CAS-25, 651, (correction : 1979, Ibid., Vol. 26, 360).

[12] Marsden, J.E., 1973, Basic Complex Analysis (NY : W.H. Freeman and Co.).

[13] Gelfand, I.M., Raĭkov, D.A., and Shilov, G.E., 1964, Commutative Normed Rings (NY : Chelsea); esp. § 13.

[14] Callier, F.M., and Winkin, J., 1986, Int. J. Control, Vol. 43, 1353.

[15] Hille, E., and Phillips, R.S., 1957, Functional Analysis and Semigroups, Vol. 31 (Providence, RI).

[16] Vidyasagar, M., 1984, IEEE TAC-29, 403.

[17] Vidyasagar, M., 1985, Control System Synthesis : A Factorization Approach (Cambridge : MIT Press).

[18] Rudin, W., 1974, Real and Complex Analysis, 2nd Ed. (NY : McGraw Hill).

[19] Vidyasagar, M., Schneider, H., and Francis, B.A., 1982, IEEE TAC-27, 880.

[20] Curtain, R.F., and Glower, K., 1986, Syst. Control Letter, Vol. 7, 41.

[21] Curtain, R.F., and Pritchard, A.J., 1978, Infinite Dimensional Linear System Theory (NY : Springer-Verlag).

[22] Pazy, A., 1983, Semigroups of Linear Operators and Applications to Partial Differential Equations (NY : Springer-Verlag).

[23] Jacobson, C.A., 1984, M.Sc. Thesis, ECSE Dept., Rensselaer Polytechnic Institute, Troy, NY.

[24] Jacobson, C.A., 1986, Ph.D. Thesis, ECSE Dept., Ren. Poly. Ins., Troy, NY.

[25] Nett, C.N., 1984, M.Sc. Thesis, ECSE Dept., Ren. Poly. Ins., Tray, NY.

[26] Callier, F.M., and Desoer, C.A., 1980, IEEE CAS-27, 320.

[27] Callier, F.M., and Desoer, C.A., 1980, Ann. Soc. Sc. Bruxelles, Vol. 94, 7.

[28] Callier, F.M., and Desoer, C.A., 1982, Multivariable Feedback Systems (NY : Springer-Verlag).

[29] Desoer, C.A., Liu, R.W., Murray, J., and Saeks, R., 1980, IEEE TAC-25, 399.

[30] El-Sakkary, A.K., 1985, IEEE TAC-30, 240.

[31] Gohberg, I.C., and Kreĭn, M.G., 1960, Amer. Math. Soc. Transl., Vol. 2, 217.

[32] Kailath, T., 1980, Linear Systems (Englewood Cliffs : Prentice Hall).

[33] Kučera, V., 1981, Automatica, Vol. 17, 745.

[34] Vidyasagar, M., and Kimura, H., 1986, Automatica, Vol. 22, 85.

[35] Youla, D.C., and Kazanjian, N.N., 1978, IEEE CAS-25, 57.

[36] Bellman, R.E., and Roth, R.S., 1984, The Laplace Transform, Vol. 3 (Singapore : World Scientific).

[37] Banks, S.P., 1983, State-Space and Frequency-Domain Methods in the Control of Distributed Parameter Systems, Vol. 3 (London : Peter Peregrinus).

[38] Stakgold, I., 1968, Boundary Value Problems of Mathematical Physics, Vol. 2 (London : The McMillan Company).

[39] Callier, F.M., and Winkin, J., 1985, IEEE CAS-32, 1178, (correction : 1986, Ibid., Vol. 33, 356).

[40] Antsaklis, P.J., 1986, IEEE TAC-31, 634.

[41] Callier, F.M., and Desoer, C.A., 1972, IEEE TAC-17, 773.

[42] Callier, F.M., and Desoer, C.A., 1976, IEEE TAC-21, 128.

LIST OF PARTICIPANTS

The lecturers are indicated by an asterisk.

R. Baker	United Kingdom
* J.A. Ball	United States of America
* L. Baratchart	France
H. Bart	Netherlands
* S. Bittanti	Italy
D. Biss	United Kingdom
P. Boekhoudt	Netherlands
J. Bontsema	Netherlands
G.D. Brown	United Kingdom
* F.M. Callier	Belgium
D. Clements	Australia
R.F. Curtain	Netherlands
* P. Dewilde	Netherlands
* J.C. Doyle	United States of America
R. Eising	Netherlands
* B.A. Francis	Canada
M. Green	Australia
K. Gregsen	United Kingdom
* K. Glover	United Kingdom
* M. J. Grimble	United Kingdom
* G. Grübel	West Germany
G. Halikias	United Kingdom
* B. Hanzon	Netherlands
* C. Heij	Netherlands
* W. Helton	United States of America
P.S.C. Heuberger	Netherlands
A.J.E.M. Janssen	Netherlands
* E. Jonckheere	United States of America
M.A. Kaashoek	Netherlands
J. Komornik	Czechoslovakia
* H. Kwakernaak	Netherlands

	J. Lam	United Kingdom
*	D. Limebeer	United Kingdom
*	L. Ljung	Sweden
	H. Logemann	West Germany
	W. Manthey	West Germany
	H.H. Martens	Norway
*	S.K. Mitter	United States of America
	J.W. Nieuwenhuis	Netherlands
*	D.H. Owens	United Kingdom
*	A.B. Özgüler	Turkey
*	J. Partington	United Kingdom
*	J.B. Pearson	United States of America
	W. Philippsen	West Germany
*	I. Postlethwaite	United Kingdom
	D. Prätzel-Wolters	West Germany
*	A. Ran	Netherlands
	M.P.M. Rocha	Portugal
*	M.G. Safonov	United States of America
	J.M. Schumacher	Netherlands
	J.H. van Schuppen	Netherlands
	M. Steinbuch	Netherlands
	M. Stöhr	West Germany
	S. Szumko	United Kingdom
*	A. Tannenbaum	Canada
	S. Townley	United Kingdom
	H.L. Trentelman	Netherlands
*	A.I.G. Vardulakis	Greece
	I. Westdorp	Netherlands
*	J.C. Willems	Netherlands
	J.L. Willems	Belgium
*	J. Winkin	Belgium
*	N. Young	United Kingdom
*	G. Zames	Canada
	Siguan Zhu	People's Republic of China
	Y.C. Zhu	People's Republic of China
	H.J. Zwart	Netherlands

NATO ASI Series F

Vol. 1: Issues in Acoustic Signal – Image Processing and Recognition. Edited by C. H. Chen. VIII, 333 pages. 1983.

Vol. 2: Image Sequence Processing and Dynamic Scene Analysis. Edited by T. S. Huang. IX, 749 pages. 1983.

Vol. 3: Electronic Systems Effectiveness and Life Cycle Costing. Edited by J. K. Skwirzynski. XVII, 732 pages. 1983.

Vol. 4: Pictorial Data Analysis. Edited by R. M. Haralick. VIII, 468 pages. 1983.

Vol. 5: International Calibration Study of Traffic Conflict Techniques. Edited by E. Asmussen. VII, 229 pages. 1984.

Vol. 6: Information Technology and the Computer Network. Edited by K. G. Beauchamp. VIII, 271 pages. 1984.

Vol. 7: High-Speed Computation. Edited by J. S. Kowalik. IX, 441 pages. 1984.

Vol. 8: Program Transformation and Programming Environments. Report on an Workshop directed by F. L. Bauer and H. Remus. Edited by P. Pepper. XIV, 378 pages. 1984.

Vol. 9: Computer Aided Analysis and Optimization of Mechanical System Dynamics. Edited by E. J. Haug. XXII, 700 pages. 1984.

Vol. 10: Simulation and Model-Based Methodologies: An Integrative View. Edited by T. I. Ören, B. P. Zeigler, M. S. Elzas. XIII, 651 pages. 1984.

Vol. 11: Robotics and Artificial Intelligence. Edited by M. Brady, L. A. Gerhardt, H. F. Davidson. XVII, 693 pages. 1984.

Vol. 12: Combinatorial Algorithms on Words. Edited by A. Apostolico, Z. Galil. VIII, 361 pages. 1985.

Vol. 13: Logics and Models of Concurrent Systems. Edited by K. R. Apt. VIII, 498 pages. 1985.

Vol. 14: Control Flow and Data Flow: Concepts of Distributed Programming. Edited by M. Broy. VIII, 525 pages. 1985.

Vol. 15: Computational Mathematical Programming. Edited by K. Schittkowski. VIII, 451 pages. 1985.

Vol. 16: New Systems and Architectures for Automatic Speech Recognition and Synthesis. Edited by R. De Mori, C.Y. Suen. XIII, 630 pages. 1985.

Vol. 17: Fundamental Algorithms for Computer Graphics. Edited by R. A. Earnshaw. XVI, 1042 pages. 1985.

Vol. 18: Computer Architectures for Spatially Distributed Data. Edited by H. Freeman and G. G. Pieroni. VIII, 391 pages. 1985.

Vol. 19: Pictorial Information Systems in Medicine. Edited by K. H. Höhne. XII, 525 pages. 1986.

Vol. 20: Disordered Systems and Biological Organization. Edited by E. Bienenstock, F. Fogelman Soulié, G. Weisbuch. XXI, 405 pages. 1986.

Vol. 21: Intelligent Decision Support in Process Environments. Edited by E. Hollnagel, G. Mancini, D.D. Woods. XV, 524 pages. 1986.

Vol. 22: Software System Design Methods. The Challenge of Advanced Computing Technology. Edited by J.K. Skwirzynski. XIII, 747 pages. 1986.

NATO ASI Series F

Vol. 23: Designing Computer-Based Learning Materials. Edited by H. Weinstock and A. Bork. IX, 285 pages. 1986.

Vol. 24: Database Machines. Modern Trends and Applications. Edited by A.K. Sood and A.H. Qureshi. VIII, 570 pages. 1986.

Vol. 25: Pyramidal Systems for Computer Vision. Edited by V. Cantoni and S. Levialdi. VIII, 392 pages. 1986.

Vol. 26: Modelling and Analysis in Arms Control. Edited by R. Avenhaus, R.K. Huber and J.D. Kettelle. VIII, 488 pages. 1986.

Vol. 27: Computer Aided Optimal Design: Structural and Mechanical Systems. Edited by C.A. Mota Soares. XIII, 1029 pages. 1987.

Vol. 28: Distributed Operating Systems. Theory und Practice. Edited by Y. Paker, J.-P. Banatre and M. Bozyiğit. X, 379 pages. 1987.

Vol. 29: Languages for Sensor-Based Control in Robotics. Edited by U. Rembold and K. Hörmann. IX, 625 pages. 1987.

Vol. 30: Pattern Recognition Theory and Applications. Edited by P.A. Devijver and J. Kittler. XI, 543 pages. 1987.

Vol. 31: Decision Support Systems: Theory and Application. Edited by C.W. Holsapple and A.B. Whinston. X, 500 pages. 1987.

Vol. 32: Information Systems: Failure Analysis. Edited by J.A. Wise and A. Debons. XV, 338 pages. 1987.

Vol. 33: Machine Intelligence and Knowledge Engineering for Robotic Applications. Edited by A.K.C. Wong and A. Pugh. XIV, 486 pages. 1987.

Vol. 34: Modelling, Robustness and Sensitivity Reduction in Control Systems. Edited by R.F. Curtain. IX, 492 pages. 1987.